MODERN

ANALYTICAL

CHEMISTRY

MODERN

ANALYTICAL

CHEMISTRY

W. F. PICKERING

THE UNIVERSITY OF NEWCASTLE
NEW SOUTH WALES, AUSTRALIA

MARCEL DEKKER, INC., NEW YORK

MARCEL DEKKER, INC. 95 Madison Avenue, New York, New York 10016

LIBRARY OF CONGRESS CATALOG CARD NUMBER 77-138500

ISBN 0-8247-1547-0

PRINTED IN THE UNITED STATES OF AMERICA

PREFACE

The advent of the nuclear age has created a need for reappraisal of all past ideas and preconceived notions. Technological advances have revolutionized all aspects of life; and education has been hailed as the prime requisite in preparing the citizens of the world for the future. In most fields of endeavor, the frontiers of knowledge are being enlarged at an alarming rate and radical changes have been made in the teaching methods and aims of the courses given at university level.

However, within the discipline of chemistry, there appears to be one branch in which the necessary dichotomy, i.e., simultaneous advancement of knowledge and revised educational approach, has not been maintained. Despite a great deal of discussion, and the increased use of semiautomatic equipment in laboratories, most new textbooks on quantitative chemistry or analytical chemistry cling to a traditional pattern set in the past. These books are invariably of high standard, generally comprehensive, and contain a great deal of experimental detail, descriptions of apparatus, and laboratory exercises.

In the days when analytical chemistry was accepted as an essential, occasionally major, part of a chemistry course, these books provided an excellent educational approach to this branch of the discipline. In the new era, however, many specialities compete for the time available within an undergraduate curriculum, and opponents of traditional quantitative analysis regularly campaign for elimination of this subject. Some institutions have responded to this pressure by introducing "elective courses," others have reduced the time available to an ineffective minimum. In many, the retention of traditional patterns by analytical chemists (despite enlargement to include instrumentation) is the major weapon used by antagonists to discredit this segment.

A much abused, but quite appropriate phrase is "attack is the best means of defence." Analytical chemists must take the initiative and produce a new approach to their subject. Chemistry is an experimental discipline but why does analytical chemistry choose to be the one branch of this discipline which is technique and experiment oriented? In most textbooks describing inorganic, organic, or physical chemistry, descriptions of how to do things are usually subsidiary.

Time has proved that undergraduates can follow "cookbook instructions" and satisfactorily perform laboratory experiments in all branches of chemistry without necessarily understanding what they are doing. Maximum benefit from practical classes is only obtained if students are forced to simultaneously study the background theory of the cookbook method, e.g., by prelaboratory tutorials or by probing questions to be answered in the final report.

During an undergraduate course, it is to be hoped that the student will encounter most techniques used in quantitative and instrumental analysis at least once. If this student is given sufficient theory to enable understanding of the manual manipulations involved every time a new technique is met, analytical laboratory sessions can be confined to procedures not adequately covered elsewhere. It is often more convenient and more interesting to the student to have a large proportion of the required quantitative determinations associated with projects which form part of the practical courses of other branches of chemistry.

This is tantamount to suggesting that the laboratory sessions within a chemistry department should be carefully coordinated, with carefully prepared laboratory manuals which incorporate abstracts of theory, operating instructions, and cookbook methods. For some courses such manuals exist, e.g., Reilley and Sawyer's "Experiments in Instrumental Analysis" (McGraw-Hill) and many coordinated courses have been proposed at the freshman level.

If one accepts the possibility of an undergraduate gaining a satisfactory (even if minimal) introduction to instruments and quantitative experiments during laboratory sessions, there is little to be gained by loading a teaching text with comparisons of different models of instruments (many superseded before publication) or detailed methods of analysis. If an inorganic or organic chemist wishes to synthesize a particular compound, he consults the appropriate reference work, e.g., "Inorganic Syntheses," "Beilstein," etc. For detailed information on a technique, reference is usually made to an appropriate monograph.

For chemical analysis, an ample number of reference works is available and there is an abundance of monographs. Yet, by tradition, analytical texts try to be a combination of theoretical text, reference book, and instrument catalog.

If analytical chemistry is to come of age, it is the author's view that this pattern must be superseded. A lecture course in this branch of chemistry must provide the undergraduates with the background necessary to achieve maximum benefit from the large amount of information currently available. The major problem in many scientific investigations is selection of the appropriate analytical procedure. Technique selection and critical evaluation of the results requires the graduate to have an understanding of the basic principles of the methods available. For the professional analytical chemist (and to a lesser extent, other chemistry majors), a secondary task is the development of new recipes, adaption of known methods to new problems, etc.

This text represents an attempt by the author to put his views into practice. Its aim is to create a new image of analytical chemistry, to provoke thought and discussion, and to challenge tradition.

In this project I have been aided and abetted by many, and for their help and encouragement, I particularly wish to mention Marcel Dekker and his advisers such as Dr. Joseph Jordan, etc.; Professor J. A. Allen and other colleagues at the University of Newcastle; Mrs. A. Rowley who patiently transcribed my scribble into presentable typescript; and my wife and family who ungrudgingly assisted in all the proofreading.

It is also a pleasure to acknowledge the cooperation of the following copyright holders who agreed to reproduction of material in this text, namely: Johann Barth, Elsevier Publishing Co., Journal of the American Chemical Society, McGraw-Hill Book Company, Orion Research Incorporated, Dr. A. R. Pinder, and Technicon Instruments Corporation.

It may be, that an apt quotation at this stage is:

> "we see but through a glass darkly"

but should this book succeed in giving graduates an appreciation of modern analytical chemistry; or should it inspire some chemists to follow the discipline of analytical chemistry; or merely challenge other authors to produce a better modern approach

> "then my effort shall not have been in vain."

Newcastle   W. F. PICKERING

INTRODUCTION

The teaching of chemistry in American colleges and universities has been effectively "radicalized" in recent years. The Westheimer Report has advocated drastic changes abandoning entirely fragmentation into "vertical" subject matter areas, such as inorganic and organic chemistry, and substituting "horizontal" subdivisions, e.g., "chemical dynamics." Fifteen years ago a sophomore "first course" in quantitative analysis was omnipresent. It served chemistry majors (who followed up by more advanced offerings at a later stage) as well as majors in other areas including biology (who hated it!). As of 1970, the sophomore "quant course" of yesteryear has virtually disappeared. To meet the contemporary educational requirement a new type of offering must be devised, viz., an advanced terminal course for a beginning student. This paradox has created a genuine "textbook gap." Professor Pickering's "Modern Analytical Chemistry" is an interesting attempt to bridge this gap. Pickering's book is a concise lecture text suitable in junior, four-year colleges, or universities. However, the treatment of the subject matter is sophisticated and has a lucid emphasis on fundamental principles. The student is well protected from the insidious germs of cookbook chemistry. Pickering focuses—in tune with the times—on analytical methodologies based on matter-energy interactions ("instrumental" analysis). Of the analytical methods based on matter-matter interactions ("chemical" analysis), Pickering has retained those which remain of contemporary interest (e.g., titrimetry). Altogether, Pickering has written a remarkable little book which manages to cover—without becoming shallow —topics such as: thermal methods; radiochemical analysis; emission, atomic and molecular absorption, Raman, microwave, NMR, and ESR spectroscopy; mass spectrometry; optical rotatory dispersion; X-ray diffraction; organic structure euclidation by cooperative use of electronic, IR and magnetic spectroscopy; kinetic methods of analysis; polarography; gas and liquid chromatography; and gel filtration. A well-written chapter on statistical treatment of experimental data and an excellent synoptic "guide" to the literature of analytical chemistry are added features of Pickering's book. It is hard to believe that all this was competently accomplished in a 600 page volume. Perhaps Pickering's Australian background has provided him with just the right perspective to write a book which appears to fit rather well current American educational requirements. A sprinkling of typically British style peculiarities is a virtue rather than a vice: It makes the book eminently readable.

University Park, Pennsylvania   Joseph Jordan

CONTENTS

Preface   iii
Introduction   vii

1. Modern Methods of Chemical Analysis   **1**

   I. Modern Trends in Chemical Analysis   1
  II. Summary of Current Techniques   3

2. Selection of Chemical Methods   **23**

   I. Problem Definition   23
  II. The Selection of Methods   25
 III. Literature Guide   30
 IV. Library Exercises   36

3. Factors Influencing the Accuracy of Results   **37**

   I. Accuracy and Precision   37
  II. Statistical Treatment of Experimental Results   39
 III. Standard Samples   43
 IV. Calibration of Equipment   44
  V. Variable Effects Introduced by Experimental Conditions   45
 VI. Sampling   48
VII. Survey Questions and Revision Problems   51

4. Thermal Transformations   **55**

   I. Thermodynamic Relationships   55
  II. Differential Thermal Analysis and Thermogravimetry   59
 III. Techniques Based on Temperature Measurement   62
 IV. Miscellaneous Thermal Methods   68
  V. Exercises in Information Retrieval   70
     References   71

5. Radiation Emission and Radioactivity   **73**

   I. Emission Spectra   73
  II. Flame Excitation   84
 III. Emission Spectrometry   90

  IV. Excitation by Accelerated Particles or High Energy Radiation   100
   V. Radiochemical Procedures   110
  VI. Revision Questions and Assignments   116
       References   119

6. The Absorption of Radiation   121

    I. General Principles   121
   II. Absorption by Atoms   131
  III. Molecular Electronic Transitions   138
  IV. Infrared Absorption Spectrometry   153
   V. Raman and Microwave Spectroscopy   167
  VI. Tutorial and Revision Questions   171
       References   174

7. Interactions with Magnetic Fields   177

    I. Basic Concepts   177
   II. Magnetic Susceptibility   183
  III. Nuclear Magnetic Resonance Spectroscopy   189
  IV. Electron Spin Resonance Spectroscopy   201
   V. Mössbauer Spectroscopy   208
  VI. Mass Spectroscopy   212
 VII. Tutorial and Revision Problems   222
       References   223

8. Reflection, Refraction, and Rotation of Light   225

    I. Physical Optics   226
   II. Refractometry   234
  III. Polarimetry and Optical Rotatory Dispersion   238
  IV. Light Scattering Photometry   244
   V. Microscopy   249
  VI. Revision and Tutorial Exercises   255
       References   256

9. Structure   259

    I. Introduction   259
   II. Elucidation of Structure by Means of Spectroscopic and Optical Methods   264
  III. X-Ray Diffraction   288
       References   295

10. Basic Solution Chemistry   **299**

    I. Ions in Solution   299
   II. Ion Association, Complex Formation, and Precipitation   303
  III. Solvent Extraction and Adsorption   306
  IV. Competing Equilibria   308
   V. Effects of Solvent   328
  VI. Tutorial Problems   330
     References   333

11. Ionic Reactions—Titrimetric Analysis   **335**

    I. Introduction   335
   II. Acid-Base Titrations   338
  III. Nonaqueous Titrations   350
  IV. Complexometric Titrations   358
   V. Precipitation Titrations   367
  VI. Oxidation-Reduction Titrations   374
 VII. Tutorial Questions and Problems   385
     References   389

12. Ionic Reactions—Selective Procedures and Kinetics   **391**

    I. Selectivity through Preoxidation or Reduction   391
   II. Organic Reagents   394
  III. Precipitate Formation   399
  IV. Analytical Methods Based on Solution Kinetics   407
   V. Literature Assignments   424
     References   426

13. Electrical Transformations   **429**

    I. Equilibrium and Kinetic Considerations   430
   II. Conductivity of Electrolyte Solutions   435
  III. Potentiometric Methods   440
  IV. Electrodeposition   453
   V. Coulometric Analysis   462
  VI. Polarography   467
 VII. Amperometric Titrations   482
VIII. Revision and Review Questions   482
     References   486

14. Adsorption, Diffusion, and Ion Exchange   **489**

    I. Separation Techniques   489
   II. Adsorption   491

   III. Separation Based on Electromigration   502
   IV. Diffusion   508
    V. Ion Exchange   517
   VI. Literature Exercises   531
       References   533

15. Heterogeneous Equilibria   **553**

    I. Two Phase Systems   535
   II. Solvent Extraction   542
  III. Factors Influencing the Separation of Mixtures by Multiple
       Distribution Processes   556
   IV. Partition Chromatography   563
    V. Gas-Liquid Chromatography   569
   VI. Assignments and Questions   580
       References   584

16. The Challenge of Automation   **587**

    I. Modern Instrumentation   587
   II. Automated Chemical Analysis   589
  III. Analytical Research   596
   IV. Conclusions   599

    Subject Index   601

MODERN

ANALYTICAL

CHEMISTRY

1

MODERN METHODS OF CHEMICAL ANALYSIS

I. Modern Trends in Chemical Analysis   1
II. Summary of Current Techniques   3

I. MODERN TRENDS IN CHEMICAL ANALYSIS

Less than three decades ago, the analysis of chemical compounds was achieved mainly by methods which are today given the euphemistic name of "classical." The components of complex samples were determined by a series of independent analyses, and in order to overcome interference effects, preliminary steps in the chemical methods were often long and time consuming. The determination of individual elements was generally based on the formation of chemical compounds having limited solubility in water or on ionic reactions in solution. The ionic reactions were utilized in titrimetric procedures; the sparingly soluble salts were isolated and weighed, i.e., gravimetric analysis.

Since many of the recognized methods consisted of a series of steps, and since errors could occur at each step, accurate results were only achieved if the analyst carefully adhered to established procedures. Chemical analysis required great manual dexterity, extreme patience, and systematic application of the principles of selective precipitation. Analysis was regarded as an "art" and training in "analytical chemistry" was based on "passing on the art"; i.e., emphasis was placed on HOW to analyze particular types of material and on HOW to separate the components of different types of materials.

However, advances in technology placed greater emphasis on quality control and manufacturers started seeking faster and simpler methods of analysis. Simultaneously, there was an increased demand for trace analyses (i.e., determination of components present in amounts less than 0.01%). Instruments which compared the physical properties of samples with the same physical property of standard samples were found to be the answer to both these problems.

The net result has been an "instrumental revolution," and today, probably more analyses are based on physical measurements than on chemical reactions.

1

This trend has prompted some cynics to remark that analysis is now merely a matter of pressing buttons and involves more electronics than chemistry. For routine analysis, this comment is becoming increasingly valid, and accordingly any course in analytical chemistry should commence with a consideration of future trends.

It seems highly probable that within the next decade, most routine chemical analyses will be carried out by means of automatic or semi-automatic instruments operated by technicians or semiskilled labor. The majority of instruments will undoubtedly be based on the measurement of *physical* phenomena, e.g., emission of X rays and other electromagnetic radiation, but *chemical* processes will also be more widely automated. Autoanalyzers for wet chemical procedures are currently very popular in clinical laboratories, and are regularly being adapted for other types of operations.

The introduction of these instruments will eliminate a major proportion of the dull, uninteresting, repetitive analyses on which many chemists are currently engaged. For large organizations, the cost of the equipment will be more than offset by the greater efficiency with which the results can be used. For example, the analytical results may be fed from the laboratory instrument directly into a computer, the task of which is to monitor and control a complete manufacturing process.

This delegation of routine analysis to semiskilled workers and expensive "direct-readers" does not relieve chemists of all their responsibilities in this direction. The chemist of the future must be capable of advising his management in regard to the best instrument for a particular operation; he must be capable of providing standard samples for the calibration of the machine; and he must accept responsibility for calibrating and checking the validity of the results produced by a range of instruments.

Since validity tests require standards and other materials to be analyzed by alternative procedures where accuracy and not speed is the prime requisite, the chemist of the future will need to be familiar with a wide range of experimental techniques.

While routine work accounts for a large proportion of the output from a laboratory, the greatest challenge comes from the need to service research and development sections.

In any vigorous organization there is a continuing demand for nonroutine analysis. It may involve analysis of a competitor's product; the introduction of a new or modified product may necessitate the development of a completely new method of analysis; or a team of pure scientists may be planning a large scale research project.

In each case, the chemist must be able to select the most appropriate technique for the problem; he must develop new methods rapidly, using theoretical calculations to reduce the amount of trial and error necessary; and he must be capable of critically evaluating the results obtained in all the analyses.

In future, fewer chemists will be employed on routine work, but the quality required for the other tasks will be much greater than the current average.

Chemistry is essentially an experimental discipline and few professional chemists escape the need to select analytical procedures related to their chosen tasks. Thus, while a course in modern analytical chemistry is a necessity for those electing to major in this field, it can also be classified as highly desirable for all students undertaking courses in which chemistry is an integral part. It is the view of the author that all such people should know the basic principles of a wide range of techniques and appreciate the limitations, possible interferants, and factors contributing to the overall accuracy.

The aim of this book is to provide an outline of the aspects of analytical chemistry which may assist the graduate of today prepare for his role of tomorrow.

II. SUMMARY OF CURRENT TECHNIQUES

During an undergraduate course most students encounter a variety of experimental techniques including a fair selection of procedures for quantitative chemical analysis. This experimental introduction is rarely comprehensive, and accordingly the aim of this section is to inform the reader of the tremendous variety of techniques currently in use.

Analytical procedures have been developed which utilize almost all the known chemical and physical properties of atomic and molecular species.

For discussion purposes, it is desirable that the various techniques be considered in groups; and it has become traditional for methods of chemical analysis to be subdivided in accordance with the procedure used for the final evaluation of the amount of the selected element or compound in the sample.

However, as indicated in Table 1.1, evaluation is only one of the components

TABLE 1.1
THE COMPONENTS OF AN ANALYTICAL METHOD

SAMPLING PROCEDURE and PREPARATION OF SAMPLE for chemical analysis.

METHOD SELECTION. The technique proposed must be justifiable in terms of the chemical and/or physical properties of the sample.

INTERFERENCE ELIMINATION. In a complex sample, several components may behave in a similar manner to the species of interest, so causing errors in the final analysis. Procedures are required to overcome such interference.

LINK REACTIONS. These are represented by separation procedures or selective reactions which contribute to an increase in the reliability of the final result.

EVALUATION. Determination, by a suitable measuring procedure, of the amount of some component present in the sample submitted for analysis.

---

of an analytical procedure, and subdivision on any other principle is equally valid.

For the purpose of this survey, classification is arranged on the basis of any property or general phenomenon that facilitates correlation of a series of different approaches.

The properties of a chemical compound can be described in terms of its color, response to heat, solubility in various liquids, electrical and/or magnetic behavior, ability to react with other chemical species, and so on; and many of the succeeding tables list techniques which utilize differences in these properties.

It should be apparent from Table 1.1, however, that several different properties may be involved in the various steps of a typical analytical procedure.

For example, chemical interactions proceed most rapidly in solution, hence in techniques based on selective reactions solubility is an important initial consideration since it strongly influences the nature of the sample preparation procedure. The first selective reaction in the method may yield a precipitate which is subsequently discarded; in this case solubility is also the key procedure in overcoming an interference effect. Finally, the evaluation of the element of interest may be based on the color of its compounds, this color being developed by another selective solution reaction. Alternatively the element may be titrated, precipitated, oxidized, or reduced.

In considering a particular technique, one has therefore to consider many aspects concurrently, e.g., the most suitable sample preparation procedure, appropriate methods of overcoming interference problems, and potential sources of error in the evaluation stage.

These facets are discussed in the more detailed treatments of individual techniques provided in later chapters of this text.

On the other hand similarities and differences between techniques are readily demonstrated as in Table 1.2. The physical processes involved in solution reactions may be divided into three categories, namely, proton transfer, electron transfer, and association of ions with other ions or molecules. Table 1.2 summarizes the common analytical techniques based on these various processes. It may be noted that this list includes all of the techniques commonly grouped together as "classical" procedures.

In all cases, the original sample is brought into solution prior to the addition of the reactant species needed for the determination of the sample composition.

Gravimetric procedures (in which some reaction product is accurately weighed) are regarded as absolute methods and are commonly used to standardize alternative techniques. Their popularity for routine determinations has waned because they are invariably tedious and time consuming.

Titrimetric procedures, on the other hand, are usually fast and simple. The scope of these methods is determined mainly by the availability of means of accurately determining the equivalence point in the chemical reaction of interest. The accuracy of the procedures is normally high, being limited by the

TABLE 1.2

ANALYTICAL TECHNIQUES BASED ON SOLUTION REACTIONS

Process	Name of technique	Procedure	Nature of reaction product	Mode of measurement	Comments
Proton transfer	Acid–base titration	Controlled addition of reactant solution from a burette	Salt and water	Determination of volume (or weight) of titrant required for complete reaction	Essential that some means of detecting equivalence point be available
	Nonaqueous titration	Controlled addition of reactant solution from a burette	Salt and solvent		
Electron transfer	Redox titrations	Controlled addition of reactant solution from a burette	Ions of different valency state to reactants	Determination of volume (or weight) of titrant required for complete reaction	Essential that some means of detecting equivalence point be available
	Electrodeposition	Excess quantity of electricity is passed between two electrodes	Metal film on cathode	Determination of weight of deposit	Limited number of metals readily reduced
	Colorimetry	Excess oxidizing or reducing agent added to assay solution	Highly colored ion	Evaluation of color intensity	Limited number of applications, e.g., transition elements only

(continued overleaf)

TABLE 1.2 *(continued)*

Process	Name of technique	Procedure	Nature of reaction product	Mode of measurement	Comments
Association	Complexometric titrations	Controlled addition of reactant solution from a burette	Stable complex ion	Determination of volume of titrant required for complete reaction	A means of detecting the equivalence point is an essential requirement
	Precipitation titrations	Controlled addition of reactant solution from a burette	Sparingly soluble compound		
	Gravimetry	Excess precipitating reagent added to the solution	Sparingly soluble compound or metal chelate	Determination of weight of treated precipitate	Composition of species weighed must be known
	Colorimetry	Excess color forming reagent added	Highly colored derivatives or complexes	Evaluation of color intensity	pH control may be required, together with extraction into non-aqueous solvent
	Turbidimetry and nephelometry	Excess precipitating reagent added to solution in presence of a dispersing agent	Suspension of sparingly soluble compound	Evaluation of amount of light stopped or scattered by the suspension	Trace analysis only—limited to number of suitable systems

quality of the equivalence point detection system and the presence of competing reactants.

Colorimetric methods are also generally fast and relatively simple, but for accurate results the color intensity meter has to be of high quality. The procedures are normally restricted to studies of trace (i.e., $<1\%$) components of samples and the range of applications is restricted by the availability of selective color forming reagents and/or methods of overcoming interference effects.

Colorimetry is the simplest of the techniques based on the absorption of electromagnetic radiation.

In these absorption techniques a sample is subjected to radiation of an appropriate wavelength and the fraction absorbed is determined in a suitable instrument. Under ideal conditions, the negative logarithm of this fraction is directly related to the amount of absorbing species in the path of the incident beam. If the intensity of radiation reaching the sample is represented by the symbol P_o, and the intensity finally transmitted is represented by P, then

$$\log P_o/P = a \cdot b \cdot c$$

where a is the proportionality constant, b is the length of the absorption path, and c is the concentration of absorbing species.

Radiations of different wavelengths induce different types of changes in the absorbing species, and these effects are summarized in Table 1.3.

Atomic and molecular species are selective in regard to the wavelengths of radiation absorbed. This property makes the absorption techniques selective, and they are widely used for the identification and characterization of molecular species.

With some samples, the absorption of radiation is followed by the emission of secondary radiation of lower energy content. This phenomenon is known as fluorescence, and the intensity of the secondary emission can be used as a means of quantitative analysis.

This technique is only one of many procedures based on the measurement of the intensity of emitted radiations (cf. Table 1.4).

Exposing a sample to electromagnetic radiation increases the energy content of the material. The total energy content of atomic and molecular species can also be increased by the application of heat or by bombardment with high energy electrons or charged particles.

In the high energy (excited) state, all species are unstable and tend to revert rapidly to a lower energy state. In this process, some of the excess energy may be released in the form of radiation. The radiation so emitted is characteristic of the species being excited and can be used for identification purposes or for quantitative analysis.

The intensity of the emitted radiation is very sensitive to experimental variables; hence, for quantitative measurements, standard procedures must be adopted and careful calibration is essential.

TABLE 1.3

ANALYTICAL TECHNIQUES BASED ON THE ABSORPTION OF RADIATION

Name of technique	Type of radiation absorbed	Species absorbing radiation	Change induced by absorption	Applications
X-ray absorption spectroscopy	X rays	The constituent atoms of solid or liquid samples	Ejection of electron from inner orbital of atom	Determination of heavy atoms in a light matrix—thickness of metal films
Atomic absorption spectroscopy	Ultraviolet or visible resonance radiation	Gaseous atoms formed by introducing an aerosol into a flame	Electrons in outer orbital of atom promoted to higher energy orbitals	Highly selective and sensitive procedure for determination of traces of metallic elements
Ultraviolet spectrophotometry	Ultraviolet light	Molecular species present as pure liquids or in solution	Electrons in molecular orbitals promoted to antibonding orbitals	Identification of chromophores, and organic compounds
Colorimetry	Visible radiations	Colored solutions	Electrons promoted to anti-bonding orbitals or to a higher energy d orbital	Trace metal analysis
Infrared spectrometry	Infrared radiation	Solution, thin film or pressed disk containing molecular species	The amplitude of the vibrations associated with atomic groups is increased	Identification of functional groups and organic compounds—used in structure determinations

TABLE 1.4

ANALYTICAL TECHNIQUES BASED ON THE EMISSION OF RADIATION

Name of technique	Mode of excitation	Type of emission	Dispersion and measuring circuit	Characteristics	Applications
Atomic fluorescence	Characteristic UV-visible emissions derived from elements of interest	UV-visible resonance lines of atoms excited	Intensity measured by photoelectric detector, mounted at right angles to excitation beam Stray light stopped by filter	Highly selective and sensitive. Requires intense source and gaseous atoms (e.g., in flame)	Determination of extremely low concentrations of metal ions (<1 ppm in aerosol solution)
Fluorimetry	Ultraviolet radiation	Visible light from excited molecule	Intensity measured by photoelectric detector, mounted at right angles to excitation beam. Stray light stopped by filter	Sensitive procedure. Requires careful blanking and special reagents for metal ion studies	Quantitative determination of a select group of organic compounds; trace metal analysis
X-ray emission spectroscopy	High energy X-radiations	X rays characteristic of atoms in sample	Secondary X-ray beam is dispersed by a crystal, the intensity of the individual rays being determined by a radiation detector (e.g., proportional counter)	X-ray emission spectrum is simple; used for qualitative and quantitative analysis. Multiple determinations on single sample	Determination of major and minor constituents of, e.g., minerals and alloys

(continued overleaf)

TABLE 1.4 (*continued*)

Name of technique	Mode of excitation	Type of emission	Dispersion and measuring circuit	Characteristics	Applications
Electron–probe microscopy	High energy electron beam	X rays characteristic of atoms in sample		Excitation beam concentrated on small area (e.g., square micron)	Analysis of small zones of samples, e.g., inclusions in minerals and grain boundaries
Flame photometry	High temperature fuel gas–oxygen flame	Ultraviolet or visible radiations characteristic of sample components	Characteristic emissions of a particular element isolated by a monochromator. Intensity determined by a photoelectric detector	Solutions containing samples introduced into the flame as an aerosol. Emissions from most elements very weak	Used mainly for the determination of the alkali and alkaline earth metals
Emission spectroscopy	Electrical arc or spark discharge	Ultraviolet or visible radiations characteristic of sample components	Emissions dispersed by high quality prism or grating. Either whole spectrum is photographed or the intensity of isolated characteristic radiations measured by photoelectric detector	Solid samples used as electrodes. The spectra are complex but they contain lines attributable to most elements in the sample. Multiple analysis on single sample	Identification of the major and minor constituents of samples Determination of minor constituents

As indicated in Table 1.4, the various emission techniques differ in respect to the mode of excitation and fields of application.

Several analytical techniques are based on the measurement of the intensity of the radiations emitted by radioactive materials. In one procedure, a known amount of a radioactive isotope of an element is mixed with the sample and the isotope is assumed to react in an identical manner with the excess of stable isotope present in the test sample. The course of a reaction is then monitored by measuring the activity of the labeled sample.

A technique of wider applicability is that known as neutron activation. Samples are exposed to a high density neutron flux which converts some or all of the constituent elements into unstable species. The radiations emitted by these treated samples are separated in terms of their energy content, and the intensities of the radiations are studied as a function of time to yield decay curves. These are used for both qualitative and quantitative analysis. This technique is extremely sensitive, but one needs access to an appropriate activating source (e.g., a reactor pile) and access to the specialized equipment associated with all radioactive studies.

In some of the techniques listed in Table 1.4 the application of heat results in the emission of visible light.

Heating a sample can also induce chemical and physical changes, and these changes can be used to characterize materials. Some of the procedures based on thermal transformations are summarized in Table 1.5.

The physical form or chemical structure of materials is occasionally best elucidated through the reflection or rotation of beams of light, and the characteristics of these optical techniques are indicated in Table 1.6.

Other modes of characterization include techniques in which the samples or derived particles are subjected to the influence of a strong magnetic field. The better known of these techniques are listed in Table 1.7. While these procedures are used primarily in structure elucidation studies, they are also suitable for quantitative analysis.

In order to be influenced by the imposition of a strong magnetic field, a sample must possess a magnetic dipole. A magnetic dipole is a macroscopic or microscopic magnetic system in which the north and south poles, equal and opposite in character, are separated by a short but definite distance.

When placed in a uniform magnetic field, unmagnetized material can become magnetized through alignment of the dipoles associated with the orbital motion of electrons and nuclei within the system. The intensity of magnetization at any point within a body is directly proportional to the field strength, and the proportionality constant for a given material is known as its magnetic susceptibility.

The contributions of nuclei to the measured magnetic susceptibility of chemical compounds can usually be neglected; hence, susceptibility measurements provide information on the spin pattern of orbital electrons. For example,

TABLE 1.5

SAMPLE CHARACTERIZATION BY THERMAL PROCEDURES

Name of technique	Property observed	General procedure	Characteristics of technique	Comments
Thermogravimetric analysis	Chemical decomposition induced by heating	Sample is weighed continually as temperature is increased linearly	Indicates the thermal stability of chemical compounds or complex materials	Weight changes can be used for characterization and quantitative analysis
Differential thermal analysis	Physical and chemical changes induced by heating	Temperature of the sample compared with that of an inert material during a prolonged heating cycle	Indicates the thermal stability of chemical compounds or complex materials	Useful for characterizing solid materials such as clays, minerals, ceramics
Thermal analysis	Melting points	Melting point curve of samples compared with that of pure compound	Melting point depression is proportional to mole fraction of impurities present	Nonselective, but useful for confirming sample purity
Thermometric titration	Enthalpy change associated with ionic reactions	The temperature changes occurring during a titration are carefully recorded	The observed temperature change is proportional to the heat of reaction; gradient changes at equivalence point	Useful for determining endpoint of titration and ΔH values

TABLE 1.6
SAMPLE CHARACTERIZATION BY OPTICAL PROCEDURES

Name of technique	Property observed	General procedure	Characteristics of technique	Comments
Chemical microscopy	Physical shape or form of solids	Sample illuminated by light to facilitate viewing of magnified image	The solid studied may be produced by chemical reaction	Useful for microanalysis and for identifying particle size, nature of compounds, fibers, etc.
Electron microscopy	Physical shape or form of solids	Sample subjected to a beam of high velocity electrons. The image formed by scattered electrons is viewed on a fluorescent screen	Identification based on characteristic diffraction patterns	Gives enlarged image of extremely small objects. Useful for elucidating molecular structure
X-ray crystallography	Spatial distribution of atoms in pure solid	Sample exposed to beam of monochromatic X rays. The image formed by scattered X rays is recorded on a photographic film	Identification based on characteristic diffraction patterns	Ultimate method of structure determination

(continued overleaf)

TABLE 1.6 *(continued)*

Name of technique	Property observed	General procedure	Characteristics of technique	Comments
Polarimetry	Rotation of beam of polarized light by optically active substances	A change in the direction of light propagation is induced by placing a sample solution in a beam of plane polarized light	The angle of rotation is proportional to the concentration of optically active species present	Useful for the determination of the purity of optically active preparations, or for estimation of small amounts of active impurities
Optical rotatory dispersion	Rotation of beam of polarized light by optically active substances	The angle of rotation caused by a sample is determined as a function of the wavelength of the incident polarized light	The manner in which the angle of rotation varies with wavelength is determined by the structure of the sample	Useful means of determining structures of certain organic compounds, e.g., steroids
Refractometry	Refractive index	The refraction of a light beam is accurately measured	Index of refraction is characteristic of sample and may be used for identification purposes	May be applied to quantitative analysis of binary liquid mixtures

TABLE 1.7
MAGNETIC FIELD EFFECTS

Name of technique	Species responding	Procedure	General principles	Comments
Magnetic susceptibiliy	Unpaired orbital electrons	A finely ground sample is weighed in and out of a strong magnetic field	Unpaired electrons are attracted by a magnetic field and cause an apparent increase in weight	Used for the identification of the electron spin arrangement of metal atoms in molecular species
Electron spin resonance spectroscopy	Unpaired orbital electrons	Sample is exposed to a micro-wave signal of fixed frequency and a variable magnetic field. The intensity of the signal is plotted against field strength	Some of the microwave signal is absorbed when the combined effect of frequency and field strength provides the energy required to change the orien-tation of free spinning electrons	Magnetic effects associated with nuclei can cause splitting of the absorption signal, and this helps in elucidating the nature of the free radicals detected by this technique
Nuclear magnetic resonance spectrometry	Nuclei such as ^{1}H, ^{19}F, ^{31}P, and ^{11}B	Sample exposed to oscillating field (radiofrequency signal) at right angles to strong magnetic field. The intensity of the signal is plotted against field strength	Some signal is absorbed when the frequency and magnetic field combine to provide the energy required to cause the nuclei to change orientation in respect to the field	Used predominantly in proton studies. Adjacent nuclei cause splitting of the signal. This allows determination of the spatial relationships of the nuclei and indicates the nature of near neighbours
Mass spectrometry	Ions and ion frag–ments derived from gaseous sample	Very low pressure of vapor is subjected to a high energy electron beam. The resulting ions are accelerated by a potential through a strong magnetic field	By varying the magnitude of the accelerating potential or applied field, ions of different m/e values can be brought to focus on the detector slit. This provides a spectrum indicating the mass of the fragments and their abundance	The mass spectrum can be used for identification purposes or for structure elucidation. Ideal method for determining molecular weights. Can be adapted to quantitative analysis

Name of technique	Mode of measurement	General procedure
Electrogravimetry	The amount of material deposited on the electrode(s) is weighed accurately	A potential which ensures continuous electrolysis is applied across the tared electrodes. The current is maintained at a high value till some species is quantitatively deposited
Controlled potential electrogravimetry	The amount of material deposited on the electrode(s) is weighed accurately	The potential applied to the circuit electrodes is controlled at a limiting value to ensure deposition of one species only. The current flowing decays with time to zero
Coulometry	The quantity of electricity required for quantitative reduction is ascertained	(a) In *constant potential* technique, the applied potential is fixed and electrolysis is continued till current reaches zero. (b) In *constant current* technique, the current is fixed and completion of desired electrode reaction is detected by an auxiliary circuit
Chronopotentiometry	Potential of one electrode is plotted as a function of time	A constant current is applied to an unstirred solution containing inert electrolyte and a small amount of electroactive species. Time required for potential values to change from one value to another is noted
Polarography	The current flowing in the electrolysis circuit is recorded as a function of the applied potential	The potential applied to a microelectrode immersed in a solution of inert electrolyte (containing small amount of electroactive species) is gradually increased and the increase in diffusion current noted
Amperometric titrations	The diffusion current observed in a polarographic circuit is plotted as a function of the volume of titrant added	The amount of electroactive species in solution is varied by the addition of a titrant solution. The applied potential is fixed at some appropriate value
Potentiometric titrations	The potential of an indicating electrode is plotted as a function of the volume of titrant added	Species in solution are titrated with a suitable reagent, the equivalence point being located from a potential vs. volume plot
Conductometric titrations	The electrical conductivity of a solution is plotted as a function of the volume titrant added	Standard titration procedure. Equivalence point is detected from the conductivity plot

Principle of technique	Comments	Applications
Metal ions are reduced at cathode to give metal film $M^{n+} + ne \rightleftarrows M$ A few elements (Ag, Pb, Mn) give oxide deposit on anode	Deposit may be impure if several reducible species present in solution and applied potential is too high	Quantitative determination of small range of elements. Removal of interfering metals ions from solution (e.g., on Hg cathode)
Metal ions are reduced at cathode to give metal film $M^{n+} + ne = M$ A few elements (Ag, Pb, Mn) give oxide deposit on anode	Deposit free of impurity but longer time required for quantitative deposition	Quantitative metal determinations. By varying the controlled potential several elements may be successively determined
One Faraday (96,487 coulombs) of electricity is required to reduce a gram equivalent of any species in solution. Therefore quantity of electricity consumed proportional to amount present if only one species involved	Results more accurate and precise than titrimetric and gravimetric procedures, especially with low concentration levels	Particularly useful for reactions in which reactants are difficult to handle, e.g., Cl_2, hazardous materials, molten salts
The transition time is related to the amount of electro- active species present	For some systems, e.g., molten salts, this approach is preferred to the other electroanalytical techniques	To date, not widely used for analytical purposes
With no stirring, and excess inert electrolyte, the current flow is controlled by the rate of diffusion of electro- active ions to the micro- electrode. With controlled conditions, the diffusion rate is proportional to concentration	Through control of the chemical composition of the solution, the pro- cedure can be made reasonably selective and is capable of multiple analyses on single sample	Applicable to any reaction in which electrons are transferred. Normal concentration range studied 10^{-3} to 10^{-4} M
As the concentration of re- ducible species changes, the diffusion current alters. A distinct change in gradient occurs at the equivalence point of the titration	Selectivity achieved by means of potential used and nature of titrant. Very precise	Used for the titration of 10^{-3} M solutions in presence of foreign salt
The potential of specific electrode materials can vary with the concentration of some species in solution	Eliminates need for visual indicators, and gives greater accuracy	Applicable to all forms of titration procedures
The conductivity of a solution depends on the number and type of ions present	Precision and accuracy increases as concentration decreases	Useful for titration of very dilute solutions

if all the electrons in an inorganic compound are spin paired, an external field causes a feeble repulsion effect. On the other hand, in the presence of unpaired electrons there is an attractive effect, because by virtue of its charge and spin, the electron behaves as a bar magnet.

Since the free electron can have a spin quantum number of either $+1/2$ or $-1/2$, it has two possible orientations in the field, differing in energy by $g\beta H$ where g is the spectroscopic splitting factor (equals 2 for a free electron), β is the magnetic moment of the free electron, and H is the magnetic field. The imposition of a radiation of frequency ν (where $\nu = g\beta H/h$) can cause transitions between these two orientations.

This effect is utilized in electron spin resonance spectrometry.

Any nucleus which possesses angular momentum, a magnetic moment, together with a charge may also act as a small bar magnet and orient itself in a magnetic field. The number of possible orientations depends upon the spin. For protons where $I = 1/2$, there are two orientations, parallel and antiparallel to the field, these orientations differing in energy by an amount μHo where μ is the component of the magnetic moment in the direction of the magnetic field (Ho). Application of an oscillating field H_1 of frequency ν at right angles to the magnetic field can result in absorption of the energy required to cause the nucleus to transfer from the lower to the higher energy state. The frequency ν required for absorption equals $\mu Ho/h$ where h is Planck's constant.

In most nuclear magnetic resonance spectrometers, a sample is subjected to a constant frequency and the magnetic field is varied until resonance conditions (i.e., absorption of radiation) are realized. Protons which are differently situated in a molecule may be magnetically nonequivalent and come successively into resonance as the applied field strength is changed. A plot of the amount of radiation absorbed against the change in applied field thus yields a series of peaks. The position of the peaks can be associated with the chemical environment of the protons, and the height of the peaks is a measure of the number of protons in this environment. Fine structure, present in many peaks, tells much about the spatial relationships of the nuclei and the nature of near neighbors.

In mass spectrometry, a small amount of sample vapor is converted into ions and ion fragments (usually by bombardment with high energy electrons in a vacuum system) and the positively charged species are accelerated towards a magnetic field (H) by means of electrically charged grids (potential V). In the magnetic field, the ions describe a circular path, the radius of which is determined by the mass to charge ratio of the ion

$$r = \frac{1}{H}\frac{\sqrt{2VM}}{e}$$

By varying the magnitude of the magnetic field in a precise manner, fragments of different mass are sequentially brought to focus on a detector electrode.

TABLE 1.9
SEPARATION PROCEDURES

Name of technique	Basic principle	General procedure	Comments
Selective precipitation	Compounds differ in respect to their solubility in aqueous solution	By controlling the solution pH and/or adding a chemical which forms precipitates with a limited number of species, selected groups of elements are isolated by filtration	Sparingly soluble inorganic compounds tend to adsorb appreciable amounts of impurity. Greater selectivity and less adsorption is obtained using organic precipitating agents
Selective oxidation or reduction	Electroactive species differ in respect of the potential required to cause electron transfer	The test sample is treated with reagents capable of reacting with only a few of the electroactive species present. Alternatively, the potential applied in an electrolysis circuit is carefully controlled	Selectivity based on choice of chemical reagents is applicable only to a few systems, and the electrolysis approach has greater scope
Solvent extraction	"Uncharged" species tend to be more soluble in organic solvents than water	A mixture of solutes in aqueous solution is shaken with some water-immiscible organic solvent. Covalent or "uncharged" compounds extract into the organic layer which is subsequently isolated	Solvent extraction is simpler and faster than precipitation but care is needed to ensure that extraction is 100% efficient. The extractable species are formed through chelate formation or ion pairing

(continued overleaf)

TABLE 1.9 *(continued)*

Name of technique	Basic principle	General procedure	Comments
Masking	The chemical reactivity of a particular ion can be "neutralised" by the formation of a stable complex	Complexing agents which react selectively with a few ions are added to solutions containing a mixture of solutes. The species forming stable complexes then do not react with the reagents added subsequently as part of the final quantitative evaluation step	Masking eliminates the need to physically separate interfering species. It is often used in conjunction with selective precipitation and solvent extraction
Selective volatilisation	Liquids differ in respect to vapor pressure; in mixtures or solutions the vapor pressure is related to the mole fraction of solvent present	The solvent mixture is heated to give vapors richer in the more volatile compound. With the aid of a distillation column mixtures of vapors can be separated completely	Some separations are simple, e.g., water is removed from solids by heating; while others require a cyclic heating and condensation process, e.g., fractionation of petroleum crudes
Chromatography	The separation achieved in a single distribution between two phases is greatly enhanced if the distribution is repeated many times in a counter-current fashion	Microgram amounts of sample are added to a column packed with one phase. Over this is maintained a flow of mobile phase (gas or liquid). The components separate and leave the column at different times	The separations are classified in terms of the phases involved, e.g., gas-liquid or liquid-solid, and on the type of column used. The procedure is very efficient but the minute amounts of separated components require sensitive methods of detection

The output of the detector is recorded, and this provides a mass spectrum which can be used to calculate both the mass of the constituent fragments and their relative abundance. The spectrum can be used for identification purposes, for determination of molecular weights, and for elucidation of molecular structure. Quantitative analytical determinations of mixtures can be performed by measuring the intensity of an ion fragment unique to a specific structure.

Not included in any of the preceding tables are the large number of techniques which utilize the electrochemical properties of species in solution.

In these techniques, the sample is taken into solution, and this solution is placed in an electrolysis cell unit. The unit normally consists of two electrodes together with the electrical circuitry required to measure and control the potential difference applied to the electrodes and the current flowing in the circuit.

If the applied potential is great enough, some electroactive species are reduced at the cathode, another is oxidized at the anode. As a result of this electrolysis, current flows in the circuit.

As indicated in Table 1.8, the electroanalytical techniques differ in respect to the electrical variable measured, and each has a range of applications to which it is best suited.

TABLE 1.10

COMPARATIVE COST OF BASIC EQUIPMENT

Cheap	Low	Medium	High
Titrimetry	Electrogravimetry	Flame photometry	Direct-reading emission spectrographs
Gravimetry	Fluorimetry	Emission spectroscopy	X-ray emission spectroscopy
Solvent extraction	Potentiometry	UV–visible spectrometry	Neutron activation
	Coulometry	Atomic absorption spectroscopy	Electron microscopy
	Conductivity		
	Thermometric titrations	Radiation analysis	Nuclear magnetic resonance spectrometry
	Microscopy	Infrared spectroscopy	Mass spectrometry
	Polarimetry	Chronopotentiometry	Electron spin resonance spectroscopy
	Refractometry	Polarography	
	Magnetic susceptibility	Thermal analysis	
		Gas chromatography	

Because electrical potentials are involved, these techniques are readily amenable to sophisticated instrumentation and semiautomatic operation.

A large number of the analytical procedures described in the preceding tables are nonselective, and as a result the analysis of complex mixtures often requires the sample to be broken up into groups of components prior to individual quantitative determinations. The separation procedures adopted depend on the type of sample being examined, and six major types are summarized in Table 1.9. All these separation procedures are subject to errors, and the overall accuracy of a chemical analysis includes the deviations introduced by these steps. Accordingly, selection of a suitable separation procedure can be as important as selection of the final evaluation technique.

Many factors influence the selection of a particular method of analysis for a particular application, and in some cases financial aspects sway the final decision.

An appropriate conclusion to this introductory section may, therefore, be a list of comparative costs as shown in Table 1.10. In this table, the capital cost of different approaches has been organized into four broad categories which correspond to approximate divisions such as <$200, <$2000, <$6000, and >$6000, but it must be emphasized that for most techniques one has a choice of instruments which vary in cost and quality, and accordingly price divisions are far from rigid.

SELECTION OF CHEMICAL METHODS

I. Problem Definition 23
II. The Selection of Methods 25
III. Literature Guide 30
    A. Handbooks and Compilations 31
    B. Multiple Volume Reference Books 32
    C. Monographs, Analysis of Specific Types of Material 32
    D. General Teaching Texts 33
    E. Journals, Advances, and Reviews 35
IV. Library Exercises 36

I. PROBLEM DEFINITION

The summary provided in Chapter 1, which is far from exhaustive, lists nearly fifty different techniques which a chemist might elect to use in various scientific studies.

The techniques differ in selectivity, sensitivity, simplicity, and equipment requirements; and an important aspect of many chemical studies is the selection of the most appropriate procedure for the given problem.

By adjustment of experimental conditions, almost every technique can be used for the analysis of a wide range of products. In most cases, however, there is a restricted area in which one particular technique is preferable to all others. When this technique is extended to include a wider range of applications, there is usually an increase in complexity of the method, an increase in the time required, and/or a decrease in the overall accuracy.

The prime requisite in selecting an analytical technique for a new problem is DEFINITION OF THE PROBLEM. The factors to be considered include:

1. Nature of the analysis—elemental or molecular, repetitive or intermittent.
2. Restrictions due to the physical and chemical properties of the species of interest, e.g., corrosive gases, radioactive solids.
3. Potential interference effects arising from similar physical or chemical properties of other species in the sample.
4. Concentration range to be studied.
5. Accuracy desired and time available for determinations.
6. Analytical facilities available, e.g., access to specialized equipment.

Since it is highly probable that no single technique will meet all of the specifications, a secondary step is the allocation of priorities. For example, in a kinetic study or in industrial process control, speed in a series of repetitive determinations may be regarded as the prime requisite. In order to achieve this aim, accuracy may have to be sacrificed or expensive equipment may have to be purchased. Errors (discussed in Chapter 3) can occur in every operation, and the best method is often the simplest method.

The need for problem definition applies equally to routine determinations and exotic research studies.

On some occasions the selection of the analytical method is influenced by the location or environment in which the analyses are to be performed.

In this regard, consideration has to be given to:

1. Type of power and heating supplies available
   These items are important when the testing units are to be located away from public supplies. For example, at a field station power may be based on dc generators and batteries while heating may be supplied by bottled gas units. In such situations it is desirable to select methods which reduce the amount of heating required to a minimum, and most procedures should be essentially noninstrumental.

2. Type of water supply on hand
   Chemical analysis by "wet" methods (and these constitute a majority) requires an ample supply of distilled and/or demineralized water. If heating facilities or purification facilities are limited, preference must be given to those procedures which require a minimum of distilled water.

3. Restricted spaces
   The smaller the space available, the simpler the equipment and techniques used. For inaccessible areas, or areas containing toxic fumes, radioactivity, etc., the techniques adopted must be amenable to remote control and automatic recording.

4. Environment
   Dust, excessive changes in temperature, high humidity, presence of corrosive gases, transmitted vibrations, etc., all detract from the effectiveness of many instrumental methods. In permanent central laboratories all these factors can be overcome by locating instruments in rooms separated from dusty and corrosive atmospheres and by air-conditioning. This is not feasible in most temporary or mobile units, and the methods selected in such situations therefore need to show little sensitivity to the prevailing atmospheric conditions.

As a generalization, one can, therefore, suggest that the methods selected for use in a mobile laboratory van or a marine survey ship could be much different from the methods used for the same analyses at a large, base laboratory.

At a central laboratory one would not anticipate any physical restriction on the type of technique which might be attempted, and in many cases the volume of work would probably justify the installation of direct reading spectrometers or AutoAnalyzer units.

In a mobile van, it would be difficult to use any technique which requires air-conditioning, or a stable power supply, or delicate instruments. The climatic conditions could change dramatically overnight and bumpy roads could create major vibrational problems. Under these conditions, the methods used would be mainly those which require simple and/or rugged equipment.

On board a ship, the problems of power supply and temperature control could be overcome, but the motion of the vessel would still restrict the list of suitable techniques.

II.   THE SELECTION OF METHODS

Once the analytical problem has been defined and environmental restrictions noted, the selection of provisional methods can begin.

This process is greatly facilitated by judicious use of the tabulated data now available. For example, in the "Handbook of Analytical Chemistry," the material is organized both in terms of specific techniques and in terms of the analysis of different types of materials.

Thus, if one has to select a method for the determination of element X in a particular type of material, one can refer to the appropriate page of the Handbook and find a summary of suitable procedures.

A typical page of the Handbook is reproduced in Fig. 2.1. It can be seen that for each element, a number of alternative procedures involving different techniques are listed, together with a reference number which permits the reader to consult the original literature for full experimental details.

The Handbook is a valuable source of information. It is extremely useful for making a quick survey of possible approaches, and it provides much of the data required for calculations. However, for detailed experimental procedures, one must refer to other publications.

If methods of ultimate accuracy and reliability are desired, one is well advised to consult the publications of national organizations, e.g., the American Society for Testing Materials, the British Standards Association, etc. The methods quoted in these works have been tested by a large number of chemists and finally selected for inclusion in these publications after comparison with a number of alternative procedures.

For routine operations, these standard methods are sometimes considered to be too lengthy and time consuming, and alternative procedures (shorter and less accurate) are preferred. Thus, in books with titles such as "The Analysis of Iron and Steel" one can expect to find experimental procedures which have been found satisfactory for process control in a particular industry.

Table 12-1. Methods for the Determination of the Elements (*Continued*)

Element determined	Technique	General approach	References (p. 12–34)
Ga	Gravimetric	Ppt. with aq. NH_3, cupferron, or NH_4HSO_3, and ignite to Ga_2O_3. Interferences: Cl^- (because of volatility of $GaCl_3$) and subts. that ppt. or coppt.	106
	Volumetric	Titr. with EDTA, using gallocyanine indicator. Range, 0.25–50 mg. Ga. Can be used after sepn. of Ga by Et_2O extn.	172
		Titr. with EDTA at pH 3.0–3.5 with Cu-PAN as indicator (end pt. = violet → clear yellow). Cd, Pb, and Zn, which may be similarly detd., interfere unless removed.	75
	Colorimetric	Sep. Ga by ether extn., add rhodamine B to Ga soln. contg. 6 M HCl, and ext. with C_6H_6. Meas. A_{565}.	214
	Fluorometric	At pH 2.6–3.0 add 8-hydroxyquinoline and ext. with $CHCl_3$. Interferences: Cu(II), Fe(III), Mo(VI), and V(V), also Ti(IV) unless first sepd. by ether extn.	214
Gd		See Table 12-2.	
Ge	Gravimetric	Sep. Ge by distn. of $GeCl_4$ in presence of $KMnO_4$, ppt. twice with H_2S from 3 M H_2SO_4 (to remove Cl^- completely), and ignite to GeO_2.	214
	Colorimetric	Usually after sepn. by distn. of $GeCl_4$. Rxn. with phenylfluorone and extn. into CCl_4. Meas. A_{510}.	53, 214, 270
H	Gravimetric	Vacuum fusion in C, oxdn. of H. Weigh H_2O.	8
	Volumetric	Combustion in O_2; collect resulting H_2O on Ascarite and weigh.	55, 106
		Vacuum hot extn.: only H_2 is evolved on heating to an appropriate temp. Calcn. is based on measurement of temp. and press. of evolved gas at known vol.	272
He	Spectrographic	In air. High-voltage condensed-spark discharge. Line pairs, He 5876, C 5880. Sensitivity 0.1%, probable error 5–8%.	22
	Mass spectrometric	For trace amts. See ref. for details of instrumentation, including modifications to increase sensitivity.	60
Hf		Zr + Hf is generally concd. by suitable sepns. (*e.g.*, by pptn. with cupferron, PO_4^{-3}, NH_3, tannin, etc., depending on compn. of samp.), then ign. A suitable carrier element may be added first if Zr + Hf concn. in samp. is very low.	
	Spectrographic	Burn finely powdd. samp. in d.-c. arc (15 amps., 220 v.) and meas. ratio of intensities of lines at 2861.696 Å. (Hf) and 2856.065 Å. (Zr).	63
	X-ray fluorescence	Meas. intensity of Hf $L\beta_2$ radiation with Hf-Zr oxide mixt. and compare with stds. similarly prepared.	62
Hg	Gravimetric	In minerals, by volatilization and condensation on Ag or Au or in H_2O. Weigh met. Hg.	50, 71, 106, 112
	Volumetric	Titr. with SCN^- in dil. HNO_3 soln., using	106

12-16

FIGURE 2.1   Typical page from the "Handbook of Analytical Chemistry" showing summaries of some of the alternative procedures available for the determination of elemental species. Used by permission of McGraw-Hill Book Company, Copyright (©) 1963, McGraw-Hill Book Company.

Table 5-23. Polarographic Characteristics of the Elements in Various
Supporting Electrolytes (*Continued*)

Element and oxidation state	Polarographic characteristics in supporting electrolyte number				
	1 (OAc⁻, pH 5)	2 (OAc⁻, pH 12)	3 (NH₃—NH₄Cl)	4 (Br⁻—HCl)	5 (CO₃⁻, pH 11)
Cu(I).......	....	....	(−0.22),—, −1,+56/ −0.50,—,1, −56,w	....	....
Cu(II)......	−0.07,3.1,2, —,w	pptn.	−0.24,—,1, −56,fw/ −0.51,3.75T, 1,−56,w,MS	>0	−0.24,—,2,—, w; 1 F K₂CO₃, pH 9.5–11: −0.201,—,2, −28,w
Eu(III)......	....	....	....	....	....
Fe(II)......	NR	....	(−0.34),—, −1/−1.49,—, 2,irr.	....	(−0.53),—, [−1]
Fe(III).....	>0,—,1,—,i, pptn.	pptn.	pptn.	>0,—,[1]	−0.86,—,[1]
Ga(III).....	NR	....	−1.6	....	NR
Gd(III).....	....	....	....	....	NR
Ge(II)......	....	....	....	....	....
Ge(IV).....	....	....	0.5 F NH₃,1 F NH₄Cl: −1.45,—, (2)/−1.70,—, cat.?	....	−1.44
H(I)........	....	....	....	....	....
Hg(I)......	....	....	....	....	....
Hg(II)......	>0,—,2; 0.1 F OAc⁻,pH 5: >0,3.9,2,—, w	....	....	....	>0,—,[2]
In(III)......	−0.708,3.7,3, [−19],w	....	....	....	−1.24,—,[3]
Ir(IV)......	Ir(III) in 0.1 F NaOAc, pH 5: >0,0.8D,3, —,w	....	NR; Ir(III): NR	....	....
Mn(II).....	NR	pptn.	−1.66,—,2, —,w	....	(−0.1),—, [−1]
Mo(VI).....	−0.6,—,?,—, i/−1.1,—,?, —,i/−1.2,—, ?,—,i	....	−1.71,—,(1), −57,fw	....	NR
Nb(V)......	....	....	....	....	....
Nd(III).....	....	....	....	....	NR
Ni(II)......	−1.1,—.2,—,i	pptn.	−1.10,3.56,2, irr.,w; 0.1 F NH₃,0.1 F NH₄Cl: −0.92	....	−1.2,—.[2], —,i
Np(IV).....	....	....	....	....	....
Os(IV)......	....	....	....	....	....
Os(VI)......	....	....	....	....	....
Os(VIII)....	>0,—,2,—,i/ ±0.0,—,2,—, i/−0.45,8.6T, 1,—,w/−1.3, —,→H₂	....	>0,—,2,—, fw/−0.241, 6.7T,2–3,—, fi	....	....
Pb(II)......	−0.50,2.7,2, [−28],w	−0.72,—,[2], −76	pptn.	−0.57,3.35,2, —,w; 1 F KBr, MS: −0.457, —,2,[−28],w	−0.66,—,[2]
Pd(II)......	−0.6,—,2,—,i	....	−0.75,3.8,2, irr.,w	....	....

5-56

FIGURE 2.2 Typical page from the "Handbook of Analytical Chemistry" showing polarographic characteristics of a few elements in various supporting electrolytes. Used by permission of McGraw-Hill Book Company, Copyright © 1963, McGraw-Hill Book Company.

For teaching purposes, analyses are generally restricted to fairly pure, simple compounds, and the methods required are correspondingly less complicated. Procedures which have proved to be satisfactory in teaching laboratories are described in most quantitative chemistry textbooks.

Thus, in selecting methods from books, one must consider the aim of the author at the time of writing. A procedure which works well in a teaching laboratory may be almost useless when applied to a complex industrial material.

In multivolume reference works such as the "Comprehensive Treatise on Analytical Chemistry," details are to be found of both simple and complex procedures. These treatises or reference works allow one to choose between the alternatives and should be consulted if the experimental problem does not fall into one of the categories described above.

These works provide good coverage of the information available at the time of writing (about twelve months before publication). However, new procedures appear every week, and for an important new project a complete literature survey may be necessary.

To make a complete survey, use is made of "Chemical Abstracts." By looking at the appropriate sections of the decennial indexes, references to abstracts of papers which have appeared on a given topic over a ten-year period can be obtained. The abstracts referred to have then to be consulted in the bound volumes of "Chemical Abstracts." Many of the abstracts will be found to be inappropriate to the problem being considered; others will be highly relevant, and in these cases, one attempts to locate and read the original papers, the details of which are given in the abstract.

To cover the years for which decennial indexes have yet to appear, annual indexes are available and can be consulted in the same way. For the current volume of abstracts (unless it is an issue indexed version) it may be necessary to read through all the abstracts in the Analytical Section. An alternative source of recent abstracts is the publication known as "Analytical Abstracts."

Since the preparation of abstract listings lags behind the appearance of original papers, a literature survey is not complete until recent issues of all the important analytical journals have been scanned for possible new developments. References given in recent papers can be backtracked to check on the efficiency of the survey.

A further check can be made by using the cumulative indexes put out at regular intervals by different journals. For example, the journal, "Analytical Chemistry," produces an index covering each five-year period while the "Analyst" produces a decennial index.

Making a complete survey can be a time-consuming and a boring process, but many chemists have learned that a day spent in the library can often save a wasted month in the laboratory.

The need to make personal literature surveys has decreased with increased publication of review articles and books of reviews. For example, every year the journal, "Analytical Chemistry," prepares a special review issue. In one year, the reviews are technique oriented; in the alternate year the reviews are application oriented. A large number of the analytical journals publish reviews on different topics at irregular intervals, and these are supplemented by a growing number of annual "Advances in ..." publications.

Whatever the means used to obtain coverage of possible methods of analysis for a given project, the final step is evaluation of the merits of the many alternative procedures. Many possibilities are readily excluded by comparing the literature methods with the initial definition of the analytical problem.

The remaining alternatives may then be considered in terms of questions such as:

1. What other species could interfere with this particular method and what steps are recommended to overcome interference problems?
2. What is the maximum accuracy one might expect from the published procedure?
3. Are all the alternative procedures subject to the same degree of interference?
4. What special equipment is required and how long should the procedure take?
5. How skilled would the analyst need to be to produce reliable results?

To provide answers to some of these questions, the basic principles of the various procedures have to be understood reasonably well.

In most situations, a suitable experimental procedure is usually suggested by the literature survey.

There are occasions, however, when no method appears to meet the desired specifications and under these circumstances a new procedure must be developed.

In this event, the second step, after defining the problem, is the selection of a probable technique. Reference to the technique oriented sections of the "Handbook of Analytical Chemistry" then provides some clues to the experimental conditions which may be required. For example, Table 5.23 of the Handbook (Fig. 2.2 shows a typical page) lists the polarographic characteristics of eighty elemental species in sixty-four different supporting electrolytes. Reference to these particular pages allows one to predict how each element in a sample will behave under similar conditions. This information indicates which species are probable interferants and allows selection of suitable supporting media.

A tentative method can now be developed in outline. It should consist of the following stages:

1. Sampling procedure (if applicable).
2. Physical preparation, or mode of solution, of the sample.
3. Separation procedures needed to eliminate interferants.

4. Adjustment of conditions to suit the requirements of the evaluation procedure.
5. Evaluation procedure for element X using the selected technique.

If, perchance, it appears that elimination of interference effects could be difficult, it is advisable to consider alternative evaluation techniques, with repetition of the above process.

Before testing the selected tentative method experimentally, some time may be saved by using equilibrium constant data or other physical data (available in the Handbook and other reference books) for some approximate calculations. These calculations help define suitable concentrations of reactants, suitable operating conditions for instruments, etc. Examples of the type of calculations involved are given in later chapters.

The final stage is to test the tentative method in the laboratory using standard samples and assay materials. The experimental results have then to be evaluated, and the procedure modified if necessary.

The whole process of method selection can, therefore, be described in terms of four main steps:

1. Definition of the analytical problem.
2. Survey of the relevant literature.
3. Theoretical evaluation of the most suitable method.
4. Experimental testing of the selected method.

While many excellent methods have been developed from individuals' hunches, the pattern to be encouraged for solving set problems is that based on systematic use of available data and information. However, the men and women of "inspired ideas" still have a very important role to play as free researchers (i.e., not restricted to fixed problems to be solved in a minimum time) since their work can lead to completely new approaches and new techniques of analysis.

III. LITERATURE GUIDE

The advent of data retrieval systems, new science indexes, etc. will tend to reduce the time needed to make a library survey, but these new aids will not eliminate the desirability of all chemists regarding the library as part of his basic equipment.

Every competent chemist should be familiar with the catalogs in the libraries available to him, and he should be capable of readily locating any data or procedures required. To provide some experience in this aspect, a series of exercises are given in Section IV, and the reader is recommended to prepare brief answers to these problems.

The range of available books varies from library to library, and the publications listed in subsequent pages of this text are merely a representative selection of the very large number of analytical books currently on the market.

The publications can be divided into about five categories.

A small, but important, group are the compilations and data books, such as those listed in Section A.

Another basic group is composed of multivolume reference books. Some of these attempt to provide a comprehensive coverage of each of the available analytical techniques. This leads to treatises, several thousand pages in length. Other reference works are devoted to a detailed cover of the various procedures used in studying particular types of materials. A few of the multivolume issues deal specifically with a particular technique or some particular type of material. Examples of this group are quoted in Section B.

The largest group of books in a library, however, are the monographs which deal in some detail with one particular technique or one type of material.

Examples of technique oriented monographs are given in the bibliography sections of the succeeding chapters of this book while Section C of the book list which follows indicates some of the books which describe the analysis of particular types of materials. In an industrial library, this type of book usually predominates.

On the other hand, in an educational institution, teaching texts occupy much of the shelf space and Section D lists many of the texts currently available.

Last, but certainly not least, are the journals, reviews, and "Advances in..." series which provide the detailed coverage from which the various books are prepared.

A.   HANDBOOKS AND COMPILATIONS

A. Albert and E. P. Serjeant, *Ionisation Constants of Acids and Bases*, Methuen, London, 1962.

K. Kodoma, *Methods of Quantitative Inorganic Analysis: An Encyclopaedia of Gravimetric, Titrimetric and Colorimetric Methods*, Wiley (Interscience), New York, 1963.

G. Kortum, W. Vogel, and K. Andrussow, *Dissociation Constants of Organic Acids in Aqueous Solution*, Butterworth, London, 1961.

N. A. Lange, *Handbook of Chemistry*, 10th Ed., McGraw-Hill, New York, 1966.

W. M. Latimer, *The Oxidation States of the Elements and Their Potentials in Aqueous Solutions*, 2nd Ed., Prentice-Hall, Englewood Cliffs, New Jersey, 1952.

L. Meites (ed.), *Handbook of Analytical Chemistry*, McGraw-Hill, New York, 1963.

N. L. Parr (ed.), *Laboratory Handbook*, Newnes, London, 1963.

A. Seidell (ed.), *Solubilities of Organic Compounds*, Vols. 1 and 2, 3rd Ed., Van Nostrand, Princeton, New Jersey, 1941, Supplement 1952.

A. Seidell and W. F. Linke (eds.), *Solubilities of Inorganic and Metal-Organic Compounds*, Vols. 1 and 2, 4th Ed., Van Nostrand, Princeton, New Jersey, 1958.

L. G. Sillen and A. E. Martell, *Stability Constants of Metal Ion Complexes*, Chem. Soc. London, 1964.

R. C. Weast (ed.), *Handbook of Chemistry and Physics*, 49th Ed., Chem. Rubber Pub., Cleveland, 1968.

B.   MULTIPLE VOLUME REFERENCE BOOKS

ASTM Standards, American Society for Testing Materials, Philadelphia, 1964.
W. G. Berl (ed.), *Physical Methods in Chemical Analysis*, 2nd Ed., Academic Press, New
    York, 1960.
J. A. Dean and T. C. Rains, *Flame Emission and Atomic Absorption Methods*, Vols. 1 to 3,
    Marcel Dekker, New York, 1969.
L. Erdey, *Gravimetric Analysis*, Vols. I, II and III, Macmillan (Pergamon), New York, 1963.
H. A. Flaschka and A. J. Barnard, *Chelates in Analytical Chemistry*, Vols. 1 and 2, Marcel
    Dekker, New York, 1969.
N. H. Furman and F. J. Welcher (eds.), *Standard Methods of Chemical Analysis*, 6th Ed.,
    Van Nostrand, Princeton, New Jersey, 1962.
D. Glick (ed.), *Methods of Biochemical Analysis*, Wiley, New York, 1954.
I. M. Kolthoff and P. J. Elving (eds.), *Treatise on Analytical Chemistry*, Wiley (Interscience),
    New York, 1968.
J. Mitchell, I. M. Kolthoff, E. S. Proskauer, and A. Weissberger (eds.), *Organic Analysis*,
    Vols. 1–4, Wiley (Interscience), New York, 1960.
F. D. Snell and C. L. Hilton (eds.), *Encyclopaedia of Industrial Chemical Analysis*, Vols. 1–8,
    Wiley (Interscience), New York, 1969.
F. D. Snell and C. T. Snell, *Colorimetric Methods of Analysis*, 4 Vols., 3rd Ed., Van Nostrand,
    Princeton, New Jersey, 1954.
C. R. N. Strouts, H. N. Wilson, and R. T. Parry-Jones (eds.), *Chemical Analysis: The
    Working Tools*, Oxford Univ. Press (Clarendon), Oxford, 1962.
A. Weissberger (ed.), *Techniques of Organic Chemistry*, 2nd Ed., Wiley (Interscience), New
    York, 1949.
F. J. Welcher (ed.), *Organic Analytical Reagents*, Vols. I–IV, Van Nostrand, Princeton,
    New Jersey, 1947.
C. L. Wilson and D. W. Wilson (eds.), *Comprehensive Analytical Chemistry*, Elsevier,
    Amsterdam, 1959.

C.   MONOGRAPHS CONCERNED WITH THE ANALYSIS
     OF SPECIFIC TYPES OF MATERIALS

American Public Health Association, *Standard Methods for the Examination of Water, Sewerage
    and Industrial Wastes*, 10th Ed., Am. Pub. Health Assoc., New York, 1955.
Association of Official Agricultural Chemists, *Official Methods of Analysis*, 9th Ed., Washing-
    ton, 1960.
F. E. Beamish, *The Analytic   stry of the Noble Metals*, Macmillan (Pergamon), New
    York, 1966.
N. D. Cheronis and T. S. Ma, *Organic Functional Group Analysis*, Wiley (Interscience),
    New York, 1964.
K. A. Connors, *A Textbook of Pharmaceutical Analysis*, Wiley, New York, 1967.
C. M. Dozinel, *Modern Methods of Analysis of Copper and Its Alloys*, Elsevier, Amsterdam,
    1963.
W. T. Elwell and I. R. Scholes, *Analysis of Copper and Its Alloys*, Macmillan (Pergamon),
    New York, 1967.
W. T. Elwell and D. F. Wood, *The Analysis of Titanium, Zirconium and Their Alloys*,
    Wiley, New York, 1961.
W. T. Elwell and D. F. Wood, *Analysis of the New Metals*, Macmillan (Pergamon), New
    York, 1966.
F. Feigl, *Spot Tests in Inorganic Analysis*, 5th Ed., Elsevier, Amsterdam, 1958.

F. Feigl, *Spot Tests in Organic Analysis*, 7th Ed., Elsevier, Amsterdam, 1969.

D. C. Garratt, *The Quantitative Analysis of Drugs*, 3rd Ed., Chapman & Hall, London, 1964.

W. F. Hillebrand, G. E. F. Lundell, H. A. Bright, and J. I. Hoffman, *Applied Inorganic Analysis*, 2nd Ed., Wiley, New York, 1953.

M. B. Jacobs, *The Chemical Analysis of Air Pollutants*, Wiley (Interscience), New York, 1960.

C. L. Lewis, W. L. Ott, and N. N. Sine, *The Analysis of Nickel*, Macmillan (Pergamon), New York, 1966.

J. J. Lingane, *Analytical Chemistry of Selected Metal Elements*, Reinhold, New York, 1966.

J. A. Maxwell, *Rock and Mineral Analysis*, Wiley (Interscience), New York, 1968.

A. J. Moses, *Analytical Chemistry of the Actinide Elements*, Macmillan, New York, 1963.

P. W. Mullen, *Modern Gas Analysis*, Wiley (Interscience), New York, 1955.

E. C. Pigott, *Ferrous Analysis*, Chapman & Hall, London, 1953.

F. D. Snell and F. M. Biffen, *Commerical Methods of Analysis*, McGraw-Hill, New York, 1944.

W. R. Schoeller and A. R. Powell, *The Analysis of Minerals and Ores of the Rarer Elements*, 3rd. Ed., Griffin, London, 1955.

F. Versagi, *The Routine Analysis of Copper Base Alloys*, Chemical Publishing, New York, 1960.

R. C. Vickery, *Analytical Chemistry of the Rare Earths*, Macmillan (Pergamon), New York, 1961.

A. Volborth, *Elemental Analysis Geochemistry*, Elsevier, Amsterdam, 1968.

R. S. Young, *Industrial Inorganic Analysis*, Chapman & Hall, London, 1953.

R. S. Young, *The Analytical Chemistry of Cobalt*, Macmillan (Pergamon), New York, 1966.

G. Zweig, Ed., *Analytical Methods for Pesticides, Plant Growth Regulators, and Food Additives*, Vols. 1–5, Academic Press, New York, 1967.

Microanalysis

R. Belcher, *Submicro Methods of Organic Analysis*, Elsevier, Amsterdam, 1966.

R. F. Milton and W. A. Walters, *Methods of Quantitative Micro-Analysis*, Arnold, London, 1955.

C. J. van Nieuwenburg and J. W. L. van Ligten, *Quantitative Chemical Micro-analysis*, Elsevier, Amsterdam, 1963.

H. Weiss, *Microanalysis by the Ringoven Technique*, Macmillan (Pergamon), New York, 1961.

D.   GENERAL TEACHING TEXTS

1.   *Quantitative Chemical Analysis*

G. H. Ayres, *Quantitative Chemical Analysis*, 2nd Ed., Harper, New York, 1968.

J. A. Barnard and R. Chayen, *Modern Methods of Chemical Analysis*, McGraw-Hill, London, 1965.

R. Belcher and C. L. Wilson, *New Methods of Analytical Chemistry*, 2nd Ed., Chapman & Hall, London, 1964.

W. J. Blaedel and V. W. Meloche, *Elementary Quantitative Analysis*, 2nd Ed., Harper, New York, 1963.

G. H. Brown and E. M. Sallee, *Quantitative Chemistry*, Prentice-Hall, Englewood Cliffs, New Jersey, 1963.

R. A. Day and A. L. Underwood, *Quantitative Analysis*, 2nd Ed., Prentice-Hall, Englewood Cliffs, New Jersey, 1967.

R. B. Fischer and D. G. Peters, *Quantitative Chemical Analysis*, 3rd Ed., Saunders, Philadelphia, Pennsylvania, 1968.

J. S. Fritz and G. H. Schenk, *Quantitative Analytical Chemistry*, Allyn and Bacon, Boston, 1966.

J. J. Klingenberg and K. P. Reed, *Introduction to Quantitative Chemistry*, Reinhold, New York, 1965.

I. M. Kolthoff, E. B. Sandell, and E. J. Meehan, *Quantitative Chemical Analysis*, 4th Ed., Macmillan, New York, 1969.

H. A. Laitinen, *Chemical Analysis*, McGraw-Hill, New York, 1960.

L. Meites, H. C. Thomas, and R. P. Bauman, *Advanced Analytical Chemistry*, McGraw-Hill, New York, 1958.

W. C. Pierce, E. L. Haenisch, and D. T. Sawyer, *Quantitative Analysis*, 4th Ed., Wiley, New York, 1958.

W. F. Pickering, *Fundamental Principles of Chemical Analysis*, Elsevier, Amsterdam, 1966.

S. Siggia, *Survey of Analytical Chemistry*, McGraw-Hill, New York, 1968.

D. A. Skoog and D. M. West, *Fundamentals of Analytical Chemistry*, 2nd Ed., Holt, New York, 1969.

A. I. Vogel, *Textbook of Quantitative Inorganic Analysis*, 3rd Ed., Longmans, Green, New York, 1961.

H. F. Walton, *Principles and Methods of Chemical Analysis*, 2nd Ed., Prentice-Hall, Englewood Cliffs, New Jersey, 1964.

H. H. Willard, N. H. Furman, and C. E. Bricker, *Elements of Quantitative Analysis*, 4th Ed., Van Nostrand, Princeton, New Jersey, 1956.

2. *Instrumental Analysis*

E. J. Bair, *Introduction to Chemical Instrumentation*, McGraw-Hill, New York, 1962.

G. W. Ewing, *Instrumental Methods of Chemical Analysis*, 3rd Ed., McGraw-Hill, New York, 1968.

G. Ewing, *Analytical Instrumentation*, Plenum, New York, 1966.

L. Fowler, *Analysis Instrumentation*, Vol. IV., Plenum, New York, 1967.

M. St. C. Flett, *Physical Aids to the Organic Chemist*, Elsevier, Amsterdam, 1962.

J. Krugers and A. I. M. Keulemans (eds.), *Practical Instrumental Analysis*, Elsevier, Amsterdam, 1965.

H. A. Strobel, *Chemical Instrumentation*, Addison-Wesley, Reading, Massachusetts, 1960.

H. H. Willard, L. L. Merritt and J. A. Dean, *Instrumental Methods of Analysis*, 3rd Ed., Van Nostrand, Princeton, New Jersey, 1958.

3. *Laboratory Manuals*

R. Belcher, *Submicro Methods of Organic Analysis*, Elsevier, Amsterdam, 1966.

R. Belcher and A. J. Nutten, *Quantitative Inorganic Analysis*, 2nd Ed., Butterworth, London, 1960.

D. A. Lambie, *Techniques for the Use of Radioisotopes in Analysis*, Spon, London, 1963.

C. E. Meloan and R. W. Kiser, *Problems and Experiments in Instrumental Analysis*, Merrill, Columbus, Ohio, 1963.

O. Mikes, *Laboratory Handbook of Chromatographic Methods*, Van Nostrand, Princeton, New Jersey, 1967.

R. L. Pecsok, *Experiments in Modern Methods of Chemical Analysis*, Wiley, New York, 1968.

C. N. Reilley and D. T. Sawyer, *Experiments for Instrumental Methods*, McGraw-Hill, New York, 1961.

J. Waser, *Quantitative Chemistry*, Benjamin, New York, 1964.

4.   Problems in Structure Determination

A. Ault, *Problems in Organic Structure Determination*, McGraw-Hill, New York, 1967.

R. H. Shapiro, *Spectral Exercises in Structural Determination of Organic Compounds*, Holt, New York, 1969.

E.   JOURNALS, ADVANCES, AND REVIEWS

1.   Analytical Journals

Analytica Chimica Acta
Analytical Chemistry
Analytical Letters
The Analyst
Association of Official Analytical Chemists Journal
Chemist-Analyst
Chemia Analityczna
Chimie Analytique
Clinica Chimica Acta
Collection of Czechoslovak Chemical Communications
Helvetica Chimica Acta
Journal of Analytical Chemistry of the U.S.S.R.
Journal of Chromatography
Journal of Electroanalytical Chemistry and Interfacial Electrochemistry
Journal of Gas Chromatography
Journal of Radioanalytical Chemistry
Microchimica et Ichnoanalytica Acta
Radiochemical and Radioanalytical Letters
Talanta
Zavodskaya Laboratoriya
Zeitschrift für analytische Chemie
Zhurnal Analiticheskoi Khimii

2.   Abstracts, Advances and Reviews

Analytical Abstracts
Chemical Abstracts
Chemical Titles
Current Chemical Papers
Electroanalytical Abstracts
Gas Chromatography Abstracts
Advances in Analytical Chemistry and Instrumentation
Advances in Chromatography
Advances in Electroanalysis
Advances in Mass Spectrometry
Advances in X-ray Analysis
Electroanalytical Chemistry: A Series of Advances
Ion Exchange: A Series of Advances
Applied Spectroscopy Reviews
Analytical Chemistry—April Review Issue
Annual Reports of the Chemical Society, London
Chromatographic Reviews
Progress in Nuclear Energy—Series IX—Analytical Chemistry
Thermal Analysis Reviews

IV. LIBRARY EXERCISES

1. Describe a titrimetric procedure for the routine determination of the chromium content of a stainless steel containing between 18 and 20% of chromium.

2. Outline the methods commonly used for the routine determination of the C, H, and N content of an organic compound.

3. The nickel content of basic nickel carbonate [$Ni_2CO_3 \cdot 2Ni(OH)_2 \cdot 6H_2O$] is to be determined by a group of undergraduate students. The nickel is to be determined both gravimetrically by precipitation with dimethylglyoxime and titrimetrically by titration with potassium cyanide solution. Write out the experimental procedures which should be issued to the students.

4. Outline some of the alternative procedures which can be used for the determination of boron and suggest the most appropriate method for determining the boron content of a special steel containing about 0.001% of boron.

5. The trace impurities in zinc (e.g., Pb, Cd, Cu, Fe, Bi) are to be determined polarographically. Indicate suitable base solutions for these determinations.

6. Write a brief summary of the applications of thermometric titrations in chemical analysis.

7. Prepare a list of the books in your library which may be used as reference books for:
   a. Analysis of metals by emission spectroscopy.
   b. Analysis of hydrocarbon mixtures by gas chromatography.
   c. The analysis of fertilizers.

8. A new preparation is to contain significant amounts of thiomalic acid (mercaptosalicylic acid). Make a complete literature survey of the analytical methods which may be used for this purpose.

9. Prepare a list of sources of review articles dealing with analytical topics which are available in your library.

10. About 1% of magnesium is to be added to cast iron (3% C and remainder Fe). Using the Handbook and other relevant reference books, propose a method for the determination of the magnesium content based on solvent extraction followed by colorimetric comparison. What methods of analysis are recommended in the literature?

3

FACTORS INFLUENCING
THE ACCURACY OF RESULTS

  I. Accuracy and Precision.   37
 II. Statistical Treatment of Experimental Results   39
III. Standard Samples   43
 IV. Calibration of Equipment   44
  V. Variable Effects Introduced by Experimental Conditions   45
      A. Contamination   45
      B. Instrument Variables   46
      C. Secondary Effects   47
      D. Solvent Effects   47
      E. Temperature Effects   48
 VI. Sampling.   48
      A. Sampling Procedure   49
      B. Preparation Errors   50
VII. Survey Questions and Revision Problems   51

I. ACCURACY AND PRECISION

The aim in any quantitative determination is to obtain experimental results which closely resemble the most probable composition of the sample.

While it is impossible to state the composition of a complex sample with one hundred percent certainty, a good approximation can be obtained by repeated analysis of the sample, preferably using different techniques. The care and time required for such an approach is rarely justified for routine work, and it is restricted mainly to samples which are subsequently sold (at some cost) as standards.

These standards are used to test the reliability of the techniques used for the routine analysis of similar materials. Multiple analyses of the standards are performed using the selected technique and the difference between the answer obtained and the "true" answer for the standard is a measure of the accuracy of the technique being studied. Since the apparent accuracy varies from determination to determination (i.e., there is usually a range of experimental results), it is normal practice to treat the results as an exercise in simple statistics. In this way, it is possible to express the accuracy as the probability of the answer being within certain limits.

For general comparison, as in the previous chapter, accuracies are expressed as a percentage of the total content of the component.

That is,

$$\% \text{ Accuracy} = \pm 100 \text{ (True result} - \text{experimental mean)/True result}$$

For the determination of the accuracy of a procedure, a standard sample of known composition must be available, and it is fortunate that in a large number of cases, very pure chemical compounds or metals can be substituted for the analyzed samples discussed above.

An alternative means of describing the reliability of an analytical procedure is in terms of the reproducibility of results.

The scatter observed when a sample is analyzed a large number of times is treated statistically and the result is described as the precision of the method.

This term, precision, represents the probability that successive results will fall within a narrow range of results centered around the median value.

The median value may be grossly different from the "true" result, i.e., there can be a constant positive or negative error in the method being studied.

A satisfactory method of analysis should possess both reasonable precision and accuracy. On the other hand, a high degree of precision does not necessarily indicate that the method is equally accurate.

It is essential that chemists distinguish carefully between these two terms. All the high quality instruments currently in use are designed to operate with a high degree of precision. Thus, if the same sample is submitted to a particular machine, the dials tend to give exactly the same reading every time. It is, therefore, possible for manufacturers to claim a reproducibility of $\pm 0.1\%$, but chemists should never interpret this claim as automatically meaning that this is the accuracy they will achieve in routine analyses.

Most instruments are superlative comparators, and even though dials on the instrument may quote figures involving three or four decimal places, the chemical significance of these figures is directly related to the accuracy of the calibration procedure used to set up the machine.

In calibration, the reading on an instrument is compared with the known composition of standard samples. By using a series of standards, the relationship between the instrument reading and chemical composition can be established. If the analyses of the standard samples are in error, then the calibration relationship will be in error and all subsequent determinations based on this relationship will be in error.

It is therefore essential that all measuring equipment be calibrated, and the standards used should be of the highest quality available.

Apart from errors arising from poor calibration, instrument variables can introduce further inaccuracies. More significant than the frailties of machines, however, are the errors introduced by faulty human manipulation.

The effect of variables on accuracy and precision are summarized in Table 3.1.

TABLE 3.1

FACTORS INFLUENCING ACCURACY AND PRECISION

Contributing Factor				
Manipulation	Calibration	Instrument	Accuracy	Precision
Careless	Faulty	Faulty	Poor	Poor
Careless	Faulty	Efficient	Poor	Poor
Careless	Exact	Efficient	Reasonable	Poor
Careful	Faulty	Faulty	Poor	Poor
Careful	Faulty	Efficient	Poor	Good
Careful	Exact	Efficient	Good	Good

II. STATISTICAL TREATMENT OF EXPERIMENTAL RESULTS

In any series of repetitive analyses the experimental results differ in value by large or small amounts, and the task facing the chemist is to decide if the scatter is excessive. The tendency for individual values to vary also makes it difficult to compare the reliability of different techniques.

More reliable decisions can be made if the experimental data are subjected to some form of standard statistical treatment.

If the individual results obtained in a series of determinations are represented by $x_1, x_2, x_3 \ldots x_n$, the arithmetical mean (m) represents an estimate of the most probable value of x (true value represented by μ).

$$m = \sum x/n$$

The precision of the analytical procedure may be estimated by calculating the standard deviation s, where

$$s = [\sum (x - m)^2/(n - 1)]^{\frac{1}{2}} \quad \text{or} \quad s = \{[\sum x^2 - (\sum x)^2/n]/(n - 1)\}^{\frac{1}{2}}$$

The divisor $(n - 1)$, termed the degrees of freedom, is a reminder that there are only $(n - 1)$ independent deviations from the mean.

Reporting the precision in terms of a calculated standard deviation is equivalent to stating that about two thirds of the experimental results have values lying between the limits $m \pm s$.

In situations where n is very large, the experimental values can be sub-divided into smaller groups having similar values, e.g., $1 \pm 0.1\%$, $1.2 \pm 0.1\%$, $1.4 \pm 0.1\%, \ldots 5.0 \pm 0.1\%$. The number of experimental results (N) falling into each subgroup can then be plotted against the central value for the group to yield a distribution curve.

If the scatter is purely random, the distribution curve should resemble a Gaussian curve as shown in Fig. 3.1, having a peak at a concentration value equal to *m* and the limits ±*s* enclosing about 65% of the total number of determinations.

If the limits are broadened to ±*ts* (where *t* > 1), the percentage of the total number (*n*) enclosed is increased, e.g., to 90%. Conversely, it may be stated that the PROBABILITY of a single determination falling within the limits of *m* ± *ts* is 90%, that is, ninety out of every hundred results are expected to fall within these limits.

The actual numerical value of *t* varies with the number of degrees of freedom involved in the study and the probability limits sought. For example, if the

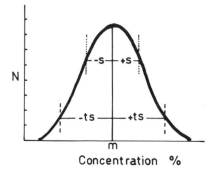

FIGURE 3.1   Random distribution curve.

probability (*P*) is assigned a value of 0.01, for 30 degrees of freedom, *t* equals 2.75. This means that if *s* is calculated from the results of 31 tests, one out of every hundred subsequent answers would be expected on the average to be outside the limits of *m* ± 2.75*s*.

Some chemists discard as "erratic," results which lie outside the 95% probability limits; others discard results lying outside the ±3*s* limits. Whatever the arbitrary standard chosen, standard deviation calculations do provide some mathematical justification for discarding results. In addition the numerical value of *s* provides a term which can be used to compare the scatter obtained using different procedures or techniques.

As stated earlier, *m* (and accordingly *m* ± *s*) is an estimate of the most probable value of *x*. Should several series of analyses be performed on the same sample, slightly different values of *m* may be obtained in each series.

When considering the standard deviation associated with a series of mean values, a slightly different equation is used, since the limits or range of distributions is much narrower.

$$s_m = \text{Calc. } s/\sqrt{y}$$

where *y* is the number of mean values being considered.

The estimate obtained in this way is still not necessarily similar to the "true value" (μ), since the method used to obtain the results may introduce a distinct positive or negative bias.

The accuracy of different procedures has to be evaluated by examining standard samples in which μ can be assumed to be known. μ is substituted for m in the standard deviation equation and the numerical term so obtained can then be used to compare accuracies.

EXAMPLE

A skilled analyst is asked by a group preparing new metallochromic indicators to comment on the value of five compounds (Code numbers 1, 2, 3, 4, and 5).

The titration figures obtained by the analyst using the different indicators with a standard titrant and a standard assay solution are recorded in Table 3.2. Knowing the "true" value from calibration studies, the analyst calculated standard deviations to describe both the precision and accuracy of the five series of results.

TABLE 3.2
TITRATION FIGURES USING DIFFERENT INDICATORS[a]

	1	2	3	4	5
	22.25	22.50	20.60	22.30	22.05
	22.30	22.70	22.00	22.35	22.15
	22.35	22.80	22.35	22.45	22.10
	22.35	22.60	21.75	22.40	22.20
	22.40	22.80	21.95	22.35	22.15
	22.35	22.60	22.20	22.45	22.15
	22.40	22.80	24.25	22.25	22.20
	22.45	22.50	21.80	22.35	22.25
	22.40	22.30	22.35	22.30	22.20
	22.25	22.40	21.65	22.30	22.05
Mean m =	22.35	22.60	22.09	22.35	22.15
Precision s =	0.07	0.18	0.91	0.07	0.07
Accuracy s' =	0.07	0.32	0.95	0.07	0.22

[a] Theoretical titration value = 22.35 ml.

The statistical calculations show that indicators 1, 4, and 5 are equally precise and yield more reproducible results than indicators 2 and 3. The value for compound 3 is large and several results in this series look erratic, but only one has a deviation from the mean which is $>2s$. In regard to accuracy, indicators 1 and 4 are shown to be far superior to the other three species investigated.

On the basis of these studies, the analyst commented that indicators 1 and 4 could prove to be eminently satisfactory while compounds 2 and 3 should be eliminated from subsequent testing. Although the observed negative bias with indicator 5 could be compensated by the use of an "indicator correction," this practice was not recommended.

A constant error in a series of results (cf. Indicator 5 in the above example) can also be detected by first calculating a value of t using the equation

$$t = \sqrt{n} \ (\text{True result} - \text{Experimental mean})/s.$$

Published tables (e.g., Table 14.5, "Handbook of Analytical Chemistry," McGraw-Hill) show the values of t exceeded once in ten (10% probability), twenty (5%), fifty (2%) or hundred times (1%) by chance for different degrees of freedom. A calculated value for t in excess of the appropriate value in the table is interpreted as evidence of a constant error at the indicated level of significance. In most cases, it is sufficient to know that the probability is less than 5%, and this level of significance is generally used in comparing calculated and tabulated results.

In comprehensive evaluation studies, the calculation of the precision (and accuracy) of a procedure should be based upon the combined evidence of a number of small sets of data. For example, n_1 determinations may be made on sample 1, from which a value S_1 can be calculated $[S_1 = \sum (x - m)^2]$. Similarly, values of S_2 may be obtained from n_2 determinations on sample 2, etc. If k samples are examined, the individual values of S and degrees of freedom should be summed, i.e., $\sum S = S_1 + S_2 + \ldots S_k;$ $\sum n - k = (n_1 - 1) + (n_2 - 1)$ $\ldots (n_k - 1)$.

The variance V (or squared standard deviation) for the procedure is then given by

$$V = s^2 = \sum S/(\sum n - k).$$

The variance provides a convenient means of comparing the precision of different sets of results. For example, if it is suspected that the precision of a procedure varies with the percentage of the substance being determined, several statistical studies should be made using materials which as a class differ in percent substance present.

The calculated variances should then be taken in pairs, and an F value calculated by dividing the larger variance value by the smaller value, i.e.,

$$F = V_n/V_d$$

where the subscripts, n and d, indicate the degrees of freedom in the two studies being considered.

The calculated F values are then compared with literature values. These tabulations are arranged in order of probabilities (e.g., 5%) and indicate appropriate F values for different values of n and d. If the calculated F value is larger than the corresponding tabulated F value, this may be taken as evidence (at the selected probability level) that the difference between the variances is significant.

The same approach can be used to compare the precision of different procedures, the precision of different chemists, the precision of different laboratories, etc.

To compensate for any uncertainty in the estimate of variance, confidence limits may be calculated using tables of chi-square (χ^2) values. These tables list numerical values for the term χ^2 for different degrees of freedom and different probabilities. If the system being studied has N' degrees of freedom, the 90% confidence limits for the estimate of variation are obtained by taking the entries for 5 and 95% with N' degrees of freedom. If S equals $\sum (x - m)^2$, then S/χ^2_5 gives the upper, and S/χ^2_{95} gives the lower, of the 90% confidence limits for $s^2 = S/N' = V$.

The χ^2 table is also useful for detecting excessive variation in the counts recorded by radiation detectors, e.g., a Geiger counter. Such counts generally follow a Poisson distribution, and in this type of distribution the variance is equal in magnitude to the mean.

If repeat counts $x_1, x_2 \ldots x_k$ have a mean m, χ^2 can be calculated from the equation

$$\chi^2 = \sum (x_i - m)^2/m$$

The value of χ^2 so obtained is then compared with the values shown in the table for $(k - 1)$ degrees of freedom. The χ^2 value for 95% confidence limits will be exceeded 5% of the time even when the data follows the Poisson distribution, but larger values of χ^2 are usually taken to indicate excessive variation among the repeat counts.

III. STANDARD SAMPLES

From the foregoing section, it is obvious that while the precision of a procedure can be estimated by using any sample available, estimates of the accuracy require the availability of a standard.

Standard materials must closely resemble the composition of the unknown samples to be subsequently analyzed by the selected procedure. In addition, the probable composition of the standard must be known with a fairly high degree of certainty, and the material must be stable, i.e., the known composition must not change with time.

For many analytical procedures suitable standards are pure (>99.9%) stable chemical compounds which can be purchased quite cheaply from reputable chemical manufacturers. Chemicals which can be used directly are often termed PRIMARY STANDARDS. Examples of primary standards are dry sodium chloride, dry potassium dichromate, potassium hydrogen phthalate, anhydrous sodium carbonate, potassium iodate, metallic zinc, etc. Chemical standards are widely used in titrimetric analysis.

For more complex analyses, such as alloy determinations in metal samples or determination of the mineral content of rocks, pure chemicals do not provide a true comparative test of the procedures, and in such cases the equivalent of the primary standard is the ANALYZED STANDARD issued by various "National Standards" laboratories. The standards issued by these organizations are subjected to repeated analysis by a range of techniques, the accompanying certificate giving some indication of the probable error in the analysis quoted.

Analyzed standards are expensive and most organizations tend to build up a collection of SECONDARY STANDARDS. These are materials whose composition is carefully determined within the individual organization after standardization of the procedures with primary or analyzed standards.

Another approach is to "manufacture" secondary standards by adding known amounts of a pure chemical species to a suitable matrix. Thus a known amount of pure metal A and a known excess of metal B may be melted together to give a series of alloys of A in B. Alternatively, two powders may be ground to a common particle size and intimately mixed. With both approaches there is a need to ensure that no segregation of the components occurs during the solidification of the melt or during storage of powdered materials.

This "doping" technique is closely related to a procedure known as "internal standardization." In its simplest form, internal standardization consists of adding a known amount of pure chemical to an aliquot of the assay solution. This is followed by simultaneous analysis of aliquots which are untreated and those which contain known additional amounts of the selected elements. Polarography is one analytical technique in which internal standardization is widely practiced. A specially prepared assay solution (sample solution plus bulk electrolyte) is examined in a polarograph to yield a current versus applied voltage plot. To this analyzed solution is added a small volume of solution containing a known amount of the same reducible species, and this mixture is examined in the polarograph. After allowing for dilution in the treated solution, the change in observed current can be related to the amount of standard added.

This example also serves to introduce the next section, the calibration of equipment, since the response or readings given by most instruments must be regularly checked and related to chemical composition by means of standards.

IV. CALIBRATION OF EQUIPMENT

Many instrument manufacturers remind the user of the need to regularly calibrate the machine by placing a CALIBRATION knob in a prominent position. Thus, few chemists attempt to use a pH meter without setting the machine with a standard buffer solution.

However, there is a tendency to forget the general maxim that all equipment should be standardized regularly. For example, a pipette is designed to deliver a certain volume of liquid within a given time. When new, and clean, the volume delivered is usually very close to the figure quoted on the glassware, but with continual use the tip can become chipped, the outlet restricted with grease or dirt, and the internal surface contaminated with past solutions. Unfortunately, many chemists fail to consider the errors which may be introduced by the use of dirty or damaged glassware. For the majority, calibration of glassware is a tedious chore sometimes undertaken, unwillingly, as an undergraduate. Some solve the problem by purchasing A grade calibrated glassware, complete with a certificate of errors; and then forget that a certificate does not eliminate the need for continued cleanliness and occasional checking.

Similarly, the most basic piece of equipment for chemical analysis is the balance and the accompanying set of masses. Although the advent of modern one-pan balances has reduced much of the need for regular checking of the interrelationship between the individual mass units, it has not eliminated the desirability of occasionally checking dial readings against a standard set of masses.

For the majority of instruments, calibration is a simple process—the response of the instrument to standard materials is observed and compared against meter or knob settings.

Thus the wavelength settings on a spectrophotometer (UV, visible, or IR) are checked by comparing the spectra produced by the instrument for selected pure materials against the published spectra for the same materials; potential readings on electroanalytical equipment are checked against the output of standard (e.g., Weston) cells; refractive index calibrations are checked against published values for selected solvents; etc.

The calibration of apparatus ensures that the errors introduced by the instruments are minimal, but it does not necessarily compensate for other variables encountered in the routine analysis of unknown materials. These latter effects are minimized by observing the behavior of standards whose composition closely resembles that of the unknown. Some examples of experimental variables are given in the next section.

V. VARIABLE EFFECTS INTRODUCED
BY EXPERIMENTAL CONDITIONS

At least five additional types of variables can influence the accuracy and precision of analytical results.

A. CONTAMINATION

The degree of contamination present at the evaluation stage can be significant. The contamination may arise from inefficiency in preliminary separation procedures, e.g., incomplete precipitation of an interferant species, or through "memory effects" in the equipment. By memory effects is meant the accidental retention of some of the previous sample in the equipment being used. For example, it is sometimes difficult to completely free a mass spectrometer of traces of a compound of low volatility; in gas chromatography a component of a mixture may migrate so slowly through the column that it suddenly appears as a peak several samples later; or solvent spilt in the cell compartment can lead to erroneous infrared spectra in subsequent studies. Most contamination effects can be eliminated by due care, but many a sudden erratic result can be attributed to this cause.

B. INSTRUMENTAL VARIABLES

While many of the errors in instrumental procedures are attributable to faulty calibration, variable behavior can often be ascribed to the use of equipment beyond its designed range. This can be exemplified by considering a very simple example. For routine titrations, a B grade 50 ml burette in which the volume delivered is within ± 0.10 ml of the volume indicated by the markings is quite satisfactory. On the other hand, for a research project one may be attempting to distinguish between titrations differing by no more than 0.05 ml. In this case, the B grade equipment must be replaced by calibrated A class material if the results are to have any real validity.

To minimize costs, the instruments provided in laboratories are usually as simple as the average usage will permit, and caution is advised whenever the equipment is subsequently used to handle more complex systems.

For example, in the determination of the sodium ion content of a simple salt solution by flame photometry, the weak emissions of other components are usually screened out by the use of a color filter, the amount of extraneous radiation transmitted being insignificant in comparison with the intensity of the yellow sodium light. On the other hand, if the same instrument is used with a hotter flame in studies of elements less readily excited, the broad range of frequencies transmitted by a filter may permit transmission of significant amounts of radiations emanating from several different types of atoms. This leads to reduced sensitivity, nonlinear calibration graphs, and variable results if different types of material are studied.

A similar sort of effect occurs in absorption spectroscopy. The frequency range or band of radiation which is characteristically absorbed by a given sample is usually quite narrow. On the other hand the general arrangement of the slit and dispersion unit of an instrument normally leads to samples being subjected to a wider range of frequencies than is actually required for selective absorption, so that only a fraction of the total radiation reaching the detector is liable to sample absorption. This again leads to loss of sensitivity and nonlinear calibration curves.

In operating such instruments some degree of compromise has to be reached. For example, reducing the width of the slit in the spectrophotometer can be used to increase the selectivity (and sensitivity to change) of the equipment, but it also reduces the transmitted intensity and a point is reached where this transmitted intensity is too weak to cause adequate response by the detection unit.

As another example one can consider the effect of carrier gas flow and temperature on the efficiency of separations based on gas chromatography. Increasing these two variables speeds up the process, but because separation is based on multiple equilibrium steps, the additional speed can lead to incomplete resolution, and some components may not be detected.

Even the scale of an instrument is a potential source of variable results, because the significance of parallax errors depends on the portion of the scale being read. For example, an error of 0.05 ml in reading a burette constitutes a 2.5% error in a total titration of 2.0 ml or a mere 0.1% error in a 50.0 ml titration. For a similar reason, in absorption spectroscopy it is preferable to control conditions so that the sample absorbance ($\log P_0/P$) lies between 0.3 and 0.8 (where $P_0 =$ intensity of radiant power entering the sample and $P =$ intensity of radiant power leaving the sample).

C. Secondary Effects

Typical examples of variations attributable to secondary effects can be drawn from analytical techniques based on the emission of radiation.

In one technique, samples are heated in a hot gas flame. Under these conditions, only a small fraction of the total number of atoms present are normally excited and emit radiation, and during propagation of these radiations towards the detection unit, some fraction can be absorbed by cooler atoms present in the propagation path. The fraction absorbed in this self-absorption process tends to increase nonlinearly with the total concentration of atoms present in the source region.

In flames, compound formation (e.g., metal oxide) can also occur. This leads to the excitation of a different type of species, and a reduction in the intensity of the atomic emissions being studied.

When the initial excitation of species is based on the absorption of X rays, a cascade or multiple enhancement effect may be observed. On irradiation with high energy X rays, several types of atoms in the sample may begin emitting characteristic X rays of lower energy. These secondary X rays may themselves then be absorbed by the sample, leading to a decrease in intensity of the higher energy components and an increase in intensity of the lower energy radiations through excitation of more atoms of the latter elements by absorbed higher energy secondary rays.

D. Solvent Effects

In electroanalytical techniques many of the observed variations can be attributed to some property of the solution phase being studied. For example, potentiometric measurements reflect the activity of species in the solution phase. The activity in turn is related not only to the concentration of the species present, but also to the total concentration of electrolytes in the solution. A change in the dielectric constant of the solvent (introduced by the addition of another miscible liquid, e.g., alcohol, acetone), also causes the response to vary.

Solutes can dissociate or polymerize on solution, the degree of change varying with the concentration of material present and with the solvent being used. A simple example of this is a solution of potassium dichromate. On dilution an increasing proportion of the original dichromate is converted to the chromate form, and if the absorbance of the solution (at a fixed wavelength) is plotted against the total concentration of chromium present, a nonlinear relationship is observed.

In any technique using liquid samples, one must consider the possible influence of solute–solvent interaction.

E.   TEMPERATURE EFFECTS

When one of the basic processes in an analytical technique involves the absorption of energy or diffusion of species, temperature changes can alter instrument response. Thus, in order to compensate for the fluctuations in temperature which can occur during excitation of samples by electrical discharges, it has been found necessary to compare the emitted intensity of a characteristic line of the element of interest with the intensity of some selected line of the base material or internal standard.

In gas chromatography, where the species to be separated diffuse in and out of the vapor phase, instrument designers try to ensure close temperature control. In polarography, diffusion is equally important, and during electrolysis the magnitude of the diffusion current is related to the concentration of reducible species and to the temperature, yet the latter aspect is sometimes overlooked by newcomers to the technique.

Temperature changes can also cause instruments to move out of alignment and so necessitate recalibration.

VI. SAMPLING

Careful standardization of the technique, accurate calibration of the apparatus, and due consideration of the effect of all possible experimental variables minimizes the magnitude of the probable error in the analysis of the sample being examined. In other words, the experimental result finally obtained is a good estimate of the composition of the material subjected to analysis.

The amount of material studied may be measured in grams or milligrams— the decision to be made on the basis of the experimental results may refer to ton quantities of material. For example, the carbon content of a one-gram sample of stainless steel may be used to decide if 20 tons of this product is of saleable quality or should be recycled as scrap metal.

This poses a new problem for the chemist.   How can one ensure that the few grams studied is truly representative of the whole mass of matter on which some final decision has to be made?

This introduces the concept of sampling and sample preparation, two very important aspects of any chemical study.   Unfortunately, many scientists ignore the importance of efficient sampling even though they should be aware that the overall variance of an analytical procedure (V) is the sum of the variances attributable to individual steps, i.e.,

$$\text{Total Variance } V = V_s + V_p + V_a$$

where

$$V_s = \text{variance of sampling}$$

$$V_p = \text{variance of preparation}$$

$$V_a = \text{variance of analysis}$$

A.   SAMPLING PROCEDURE

The method used to collect a sample for analysis is determined, to some extent, by the state of the material being sampled, i.e., gas, liquid, or solid.

Gaseous and liquid products are generally essentially homogeneous and the collection of a single sample suffices.   On the other hand, where there is any possibility of stratification (e.g., within a smokestack or eddies in a moving stream), a modified form of the procedures recommended for solids should be attempted.

The procedure for sampling solid materials is determined, in part, by the physical form (e.g., lumps, solid mass, fine grain) and the degree of heterogeneity (i.e., purity or contamination with extraneous matter).

The initial step involves the collection of a gross sample, this sample being obtained by combining a large number of separate increments of the material. The increments should all be of the same size and should be collected in some manner which eliminates any chance of personal selection.   One way of eliminating bias is to use a random sampling procedure, although for physical and chemical tests the increments are usually collected systematically from areas spaced evenly over the material being studied.

The minimum satisfactory weight for the gross sample depends on the quality of the material and its physical properties.   The more heterogeneous and impure the material, the larger the sample must be.   For crushed solids, the smaller the particle size, the smaller the gross sample required.

The individual increments must themselves be large enough to avoid bias in respect to particles of different size and nature.   The number of these increments (N) required to give the gross sample can be calculated from a relationship such as $N = 4V/A^2$ where V is the total variance of the sampling plus analytical procedures and A is the arbitrary accuracy required in subsequent determinations. V has to be determined experimentally by analyzing a large number of samples taken from the particular type of material under consideration.

When all the increments have been collected together, it is generally desirable to reduce the actual magnitude of the sample prior to despatch to the laboratory. With gases and liquids, subsamples are taken after an efficient mixing stage. With particulate matter, it is necessary to reduce the particle size of the material before mixing and resampling. This is a corollary of the recommendation given above, viz., the smaller the particles, the smaller the gross sample required. For example, for a particular mineral it may be suggested that the minimum gross sample needed for 1-inch lumps is 400 lbs; for 0.1-inch lumps, 4 lbs. On this basis, if only 4 lbs of sample are to be delivered to the laboratory, the 400 lbs of 1-inch lumps must be reduced to 0.1-inch particles and this fine material randomly sampled to give a new gross sample of 4 lbs. This 4 lbs of 0.1-inch material may then (if desired) be reduced in particle size, and a smaller sample collected.

B. PREPARATION ERRORS

Errors introduced in the preparation stage are as important as errors introduced in the initial sampling or final analysis.

Thus if a number of liquid samples are to be combined, the vessel used to hold the combined bulk of liquid must be free of any form of contamination and it must be possible to adequately stir the contents. A final sample should not be drawn until all the increments have been thoroughly mixed together.

With solid particles efficient mixing is difficult to achieve, and alternative procedures have to be adopted in many cases to ensure random selection of material. For example, the solids may be fed into a chute which divides the initial flow into two streams, only one of which is continually recycled, the process continuing until the new reduced gross sample weight has been isolated. Alternatively, a heap of the material may be shovelled away, every tenth shovelful being set aside to form part of a new heap.

Whatever the mixing and subsampling procedure adopted, results are not exactly reproducible, i.e., there is a definite "variance of preparation" which contributes to the overall variation of the analytical procedure.

Apart from variations introduced in the mixing stages, errors can arise from preliminary steps, e.g., particle size reduction and storage. The equipment used for particle size reduction must not introduce contamination nor facilitate chemical changes in the material. A hot grinder can cause marked loss of water and facilitate aerial oxidation of species such as metal sulfides.

With careless storage between reduction stages some fine powder (possibly of different composition to the bulk) may be lost as a dust cloud and chemical changes may be introduced if the material is exposed to water or other vapors.

Sampling and preparation procedures are usually entrusted to unskilled labor, but the analytical chemist must accept responsibility for ensuring that the conditions selected are followed adequately. An artisan cannot be expected to realize that the preparation of metal drillings using an oily coolant or the transfer of samples by means of greasy hands nullifies the results of subsequent carbon analyses, but the chemist must be aware that such a probability exists and take steps to avoid such accidental contamination.

Every sampling problem has to be considered on its merits, and some indication of the procedures which have been adopted in different cases will be derived from preparing answers to the questions listed in the next section.

VII. SURVEY QUESTIONS AND REVISION PROBLEMS

1. Using reference sources such as the following:

   C. L. Wilson and D. W. Wilson (eds.), *Comprehensive Analytical Chemistry*, Vol. 1A, Elsevier, Amsterdam, 1959.

   I. M. Kolthoff and P. J. Elving (eds.), *Treatise on Analytical Chemistry*, Vols. 1 and 2, Wiley (Interscience), New York, 1959 and 1961.

   W. F. Hillebrand, G. E. F. Lundell, H. A. Bright, and J. I. Hoffman, *Applied Inorganic Analysis*, 2nd Ed., Wiley, New York, 1953.

   N.'H. Furman and F. J. Welcher (eds.), *Standard Methods of Chemical Analysis*, 6th Ed., Van Nostrand, Princeton, New Jersey, 1962.

   prepare a short (500–1000 word) essay on each of the following topics:
   a. Sampling of gases.
   b. Sampling of liquids.
   c. Equipment used to reduce the size of solids.
   d. Methods of sampling metal objects.
   e. Role of the chemist in eliminating sampling errors.
2. Using reference sources such as the following:

   ASTM Standards, American Society for Testing Materials, Philadelphia, 1964.

   National Standards Association Publications, e.g., B.S.A. standards and other specialized books dealing with particular products,

   prepare an outline of the sampling and preparation procedures required in the following situations:
   a. The average quality of a train load of coal is to be assessed.
   b. A flour mill receives some of its wheat in bulk and some in bags. Indicate the most suitable approach for each type of delivery.
   c. Laboratory samples are to be provided from each of several heaps of pig iron.

d. The average analysis of a batch of one hundred scrap white metal bearings is requested by the plant manager.

e. The oil flowing past a given point in a pipeline has to be sampled every fifteen minutes and a composite sample submitted for analysis every four hours.

f. The output of a cement plant flows along a conveyer belt on its way to the packing department. Regular samples are required for process control purposes.

g. A mechanical coal mining machine produces lumps ranging from >6-inch diameter to fine dust. After reducing the large pieces to <2-inch, the screening plant divides the product according to size into six fractions. If the average ash content of about 10 % is to be reported with a maximum deviation of $\pm0.3\%$, use some of the published tables to indicate the magnitude of the gross sample which should be collected from the +1-inch, +1/2-inch and 1/8-inch fractions. What procedures should be followed in preparing the laboratory sample from each fraction?

h. An extensive ore body is suspected to be below the subsoil of a remote region. How could the extent and quality of the ore body be explored?

3. Ten samples of limestone, collected by a laborer from a stockpile, gave analytical results of 50.6, 50.9, 48.5, 50.0, 46.5, 50.5, 49.2, 49.7, 43.2, 50.5% CaO. Ten further samples were collected after the stock pile had passed through a crushing plant. The results obtained for the smaller crushed material were 50.5, 50.3, 50.9, 50.7, 50.3, 51.4, 49.2, 49.9, 50.4, 50.6% CaO. Using pure calcium carbonate, the results obtained using the same analytical procedure were 56.0, 55.8, 55.8, 56.1, 55.9, 55.6, 55.9, 55.8, 56.0, 55.7% CaO.

a. Calculate the precision and accuracy of the analytical procedure.

b. Calculate the total variance of the samples taken from the stockpile. Should any results be discarded? Why is there such a scatter? How might the precision be improved?

c. Calculate the combined variance of sampling plus preparation in the increments taken from the crushed material.

4. The quality of an industrial raw material, produced in a hot and dusty region, is monitored by analyses performed in a temporary, "on-site" laboratory with check analyses being performed at the industry's air-conditioned central laboratory. Despite the use of similar types of equipment and the same techniques, e.g., titrimetry and colorimetry, the results obtained at the two centers tend to differ by significant amounts. Suggest reasons for the observed variations.

5. Preliminary tests indicate that the concentration of a reactive species being studied in a kinetic experiment will not vary from the previous

reading by more than 10 % of content if the time interval between tests is fixed at ten minutes. The kinetic project is part of a research program and maximum accuracy is required. Suggest a suitable value for the order of precision and accuracy to be sought in the selected analytical technique. What techniques might be suitable if the reaction involves (a) a liquid phase, or (b) a gas phase.

4

THERMAL TRANSFORMATIONS

  I. Thermodynamic Relationships 55
 II. Differential Thermal Analysis and Thermogravimetry 59
III. Techniques Based on Temperature Measurement 62
 IV. Miscellaneous Thermal Methods 68
  V. Exercises in Information Retrieval 70
     References 71

I. THERMODYNAMIC RELATIONSHIPS

The majority of chemical reactions of interest to analytical chemists are accompanied by changes in the heat content of the system. Usually, the chemist is not concerned with the thermal changes, but is primarily interested in the concentrations of the chemical species present at equilibrium. However, there are a small number of analytical techniques which are based on the measurement of the heat changes associated with chemical reactions and phase transformations.

The theoretical treatment of heat changes, and chemical equilibrium, constitutes one aspect of the field of study known as thermodynamics. Hence an appropriate introduction to this segment of modern analytical chemistry is a brief revision of relevant thermodynamic relationships.

The heat content or enthalpy (H) of a compound in its equilibrium state at 298° K (25° C) is defined as being equal to its heat of formation at this temperature.

Exposure of the compound to an external source of heat results in an increase in temperature, and the amount of heat required to increase the temperature of one mole by one degree centigrade is termed the heat capacity of the material (Cp if measurements made at constant pressure). The magnitude of Cp varies with the temperature (T) and the relationship can be described by an equation of the form:

$$Cp = a + bT + (cT^2 \text{ or } cT^{-2}).$$

a, b, and c are constants which have different values for different substances. Over fairly short temperature ranges the power terms in T can often be neglected.

The enthalpy of a system at temperature T_2 (H_{T_2}) is therefore equal to the heat content at temperature T_1 (H_{T_1}) plus the heat required to increase the temperature. That is,

$$H_{T_2} = H_{T_1} + \int_{T_1}^{T_2} Cp \, dT$$

or

$$H_{T_2} = H_{T_1} + a(T_2 - T_1) + \frac{b}{2}(T_2^2 - T_1^2)$$

If the specimen being heated changes to some other physical form, additional heat is absorbed by the system. For example, if the compound melts at a temperature between T_1 and T_2, energy equivalent to the heat of fusion (ΔH) is required, and the total enthalpy at temperature T_2 (H_{T_2}) is given by:

$$H_{T_2} = H_{T_1} + \int_{T_1}^{T_2} Cp \, dT + \Delta H$$

The occurrence of a change of state is readily detected, because during the period of change, heat transfer causes no significant change in temperature.

The total energy (H) of a system at some selected temperature is composed of two major units, the free energy (G) and entropy (S)

$$H = G + TS$$

The free energy is the component which determines the capacity of a given substance to enter into a chemical reaction, and at a given temperature, the thermodynamically stable state is that with the least free energy. Any other state tends to change to the stable form.

The entropy is essentially a measure of atomic disorder. It is defined as zero in the state of perfect order existing at $0°$ K and increases with increasing temperature. The entropy of a substance at temperature T is defined by the equation

$$S_T = \int_0^T (Cp/T) \, dT$$

If in going from 0 to $T°$ K, a physical transformation occurs, then the entropy at $T°$ is the sum of the entropy change associated with heating the original material from $0°$ to the transition temperature, the entropy change associated with the transformation, and the entropy change arising from heating the new physical form from the transition temperature to T.

Heating a substance to an elevated temperature can also induce chemical transformations, for example, decomposition resulting in the liberation of gases or vapors, oxidation, or some other interaction with the surrounding atmosphere. In the chemical transformation, heat may be evolved (exothermic reactions) or absorbed (endothermic reactions).

The extent to which the chemical reaction proceeds at a given temperature can be described in terms of the magnitude of the equilibrium constant (K), e.g.,

$$CaCO_3 \rightleftarrows CaO + CO_2; \quad K_p = p_{CO_2}$$

Alternatively, the position at equilibrium can be defined in terms of the standard free energy change (ΔG_T°) since

$$\Delta G_T^\circ = -RT \ln K$$

ΔG° represents the change in free energy which occurs in the reaction, that is, ΔG° = [Free energy content of the product(s)] − [Free energy content of reactant(s)].

During the course of the reaction there are corresponding changes in enthalpy and entropy, since it can be shown that

$$\Delta G = \Delta H - T\Delta S$$

The change in enthalpy, ΔH, for a reaction depends on the physical states of the reactants and products. For example, the heat of combustion of graphite is different from that of diamond.

To facilitate the tabulation of thermodynamic data, certain standard states have been adopted and thermodynamic properties are tabulated for these standard states. When the reactants and products are in the standard states, a superscript degree sign is placed on the ΔH symbol (e.g., ΔH°). Similarly ΔG° and ΔS° represent standard state values.

The standard state of a solid substance is a specified crystalline state at 1 atm pressure at the temperature concerned; for a liquid it is the pure liquid at 1 atm at the temperature concerned; and for a gas the standard state is the ideal gas at 1 atm pressure at the temperature concerned. Unless otherwise specified, the standard state refers to a temperature of 25°.

ΔH° and ΔS° (and hence ΔG°) values vary with temperature, but over a range of a few hundred degrees the change is small and

$$\Delta G_T^\circ \simeq \Delta H_{298}^\circ - T\Delta S_{298}^\circ$$

Thus, using thermodynamic data tabulated for 25°, it is possible to obtain an approximate value for ΔG° at elevated temperatures. This ΔG° value can then be used to estimate the relative concentrations of the species at equilibrium.

The enthalpy change associated with a reaction can also be used to calculate the effect of temperature on the magnitude of the equilibrium constant.

$$d \ln K/dT = \Delta H^\circ/RT^2$$

When reactions occur in solution, there are several additional processes which contribute to the total enthalpy change. For example, when a solute is dissolved in a solvent, heat may be evolved or absorbed; the magnitude of the thermal change depends, among other things, on the concentration of the final solution.

The observed enthalpy change per mole of solute is called the "integral heat of solution."

During this solution process, the solute may be wholly or partially dissociated into solvated ions. Where the species are fully dissociated, the enthalpy change associated with subsequent reactions in solution can be attributed to ionic reactions, e.g.,

$$H^+_{(aq)} + OH^-_{(aq)} \rightleftarrows H_2O(l); \quad \Delta H^\circ_{298} = -13.7 \text{ kcal}$$

The enthalpies of formation of individual ions cannot be separately determined, but by the adoption of an arbitrary scale the standard enthalpies of formation of other ions in aqueous solutions may be tabulated relative to this standard. The arbitrary scale adopted defines the heat of formation of the hydrated hydrogen ion, H^+_{aq}, as zero.

$$\tfrac{1}{2} H_2(g) + aq \rightleftarrows H^+_{(aq)} + e; \quad \Delta H^\circ_{298} = 0$$

If a molecular species is not fully ionized in solution, additional energy is absorbed in the dissociation process which accompanies the overall chemical reaction.

There are several extensive compilations of thermochemical data (1-4), and substitution into one or more of the relationships summarized in this section permits calculation of enthalpy changes and equilibrium distributions for an extremely wide range of reactions.

Analytical chemists thus have available a means of confirming the feasibility of any proposed chemical reactions. The mathematical treatment need not be rigorous and the less formal approach of Allen (5) is probably sufficient for most purposes.

The simple thermodynamic calculations can be used to predict the effect of temperature on a reaction, to estimate the amount of heat evolved or absorbed in the process, and to indicate the relative concentrations of different species at equilibrium.

On the other hand, thermodynamic calculations take no account of rate processes, and it cannot be assumed that all reactions proceed extremely fast. On the contrary, many of the chemical reactions which are thermodynamically feasible are never observed in the laboratory because of kinetic limitations.

Before considering the analytical procedures which are based on thermal transformations, one or two other relationships should be mentioned.

For a liquid phase in equilibrium with its vapor, e.g.,

$$S(l) \rightleftarrows S(g); \quad K_p = p_s$$

the effect of temperature on the vapor pressure, p_s, is described by the equation

$$d \ln p_s/dT = \Delta H^\circ/RT^2$$

where ΔH° is the latent heat of vaporization per mole.

For a mixture of ideal liquid phases, the vapor pressure of one particular component at a given temperature is related to its molar fraction in the liquid phase

$$p_A = \frac{n_A \cdot p_A^\circ}{n_A + n_B + n_C + \dots}$$

where p_A° is the vapor pressure of pure A at the fixed temperature, n_A is the number of moles of A in the liquid phase and $n_A + n_B + n_C + \dots$ is the total number of moles of liquid.

Due to intermolecular attraction effects, deviations from this relationship are often observed, but this simple form is useful for predicting the effect of adding components to a solvent system.

For example, if n_B moles of nonvolatile solute are added to pure solvent A, the vapor pressure above the solvent is reduced to $(n_A/n_A + n_B)p_A^\circ$. Since the boiling point of the liquid phase represents the temperature at which the total vapor pressure equals the applied pressure, the presence of the solute elevates the boiling point.

The boiling point elevation (ΔT) for low concentrations of solute is given to a first approximation by the expression

$$\Delta T = \frac{R T_0^2}{\Delta H_{vap}} \cdot X_2$$

where T_0 is the boiling point of the pure solvent, ΔH_{vap} is the molal enthalpy of vaporization and X_2 is the mole fraction of solute.

The addition of a solute can also reduce the freezing point of a liquid system and the depression in freezing point (ΔT_f) can be described by an identical type of equation except that in this case T_0 represents the freezing point of the pure liquid phase and ΔH_f, the molal heat of fusion.

II. DIFFERENTIAL THERMAL
ANALYSIS AND THERMOGRAVIMETRY

If the temperature surrounding a solid is increased in a uniform manner, heat energy is transferred to the solid.

At certain points in the temperature cycle the transferred heat may provide the energy required for a physical transformation or a chemical reaction. The absorption of energy in these transformations causes the sample temperature to differ from that of the surroundings. In differential thermal analysis (DTA) this temperature difference is recorded as a function of the temperature in the heating furnace.

Chemical transformations during the heating cycle cause changes in mass, and in thermogravimetric analysis (TGA) the mass of a solid sample is measured as a function of the temperature surrounding it.

With pure compounds both techniques provide experimental records which are reasonably characteristic of the material being studied, the behavior of compounds often being sufficiently specific to permit their identification in simple mixtures.

The general principles of these two techniques are shown in Fig. 4.1.

In both cases the samples are heated in a suitable furnace at some uniform rate (e.g., 200°/hr), the temperature of the surroundings being recorded on a chart. In DTA the temperature of the sample is compared with the temperature

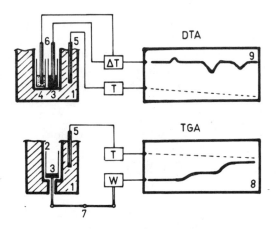

FIGURE 4.1   Basic principles of differential thermal analysis (a) and thermogravimetry (b). 1. Heating furnace, 2. Sample holder, 3. Sample, 4. Inert reference solid, 5. Thermocouple for measuring the temperature (T) surrounding the sample, 6. Thermocouples for measuring the temperature difference (ΔT) between 3 and 4, 7. Balance for recording weight changes (W) in heated sample, 8. Thermogravimetric curve showing two distinct chemical changes, 9. DTA curve showing two endothermic and one exothermic change.

of some inert material (e.g., alumina), and the temperature difference (ΔT) is continually recorded during the heating cycle. In TGA the weight of the sample is monitored during the heating cycle.

The actual form of the thermal spectrum is controlled to a certain extent by experimental factors. Heat has to be transferred through the solid and this introduces heat transfer problems which can result in the shape and position of the transformation peaks being functions of the rate of heating, the shape of the furnace, the nature of the sample container, the nature and size of the sample, etc.

Other important instrumental factors are the speed and response of the recording unit, the furnace atmosphere, and thermocouple location. The sample characteristics which are significant include the density of the packed sample, its thermal conductivity, the heat of reaction, the nature of the sample, the solubility of evolved gases in the sample, and the degree of swelling during the heating cycle.

The influence of each of these factors is discussed in some detail in the monographs by Wendlandt (6) and Garn (7).

Because of the large number of variables, DTA and TGA curves are not as strictly reproducible as the infrared spectrum or X-ray diffraction pattern of a substance. Different instruments located in different laboratories do not, in general, yield identical curves.

The thermal effects observed in the DTA record can be caused by such physical phenomena as fusion, crystalline structure inversions, destruction of the crystal lattice, vaporization, boiling, and sublimation. Other thermal effects observed in both the DTA and TGA records are caused by chemical reactions such as dissociation or decomposition, dehydration, oxidation and reduction, direct combination, and displacement. Most of the transformations produce endothermic heat effects except those of oxidation and certain crystalline structure inversions which give exothermic heat effects.

The basic assumption for quantitative DTA studies is that the area enclosed by the curve peak is proportional to the heat of reaction. This assumption is not usually strictly valid but if the ΔH of the peak forming reaction is known, DTA can be used to determine semiquantitatively the composition of mixtures containing this component.

The major application of DTA to problems in analytical chemistry involves sample characterization. For example, raw materials which are similar but not identical can be compared with ease and any variations liable to cause trouble in the manufacturing process may be readily spotted.

Some of the earliest uses of this technique included the identification of metals, minerals, clays, and ceramic materials. Current applications include the identification of amino acids and proteins, carbohydrates, coordination compounds, coal and lignite, fats and oils, greases, metal salts and oxides, polymers, and wood products.

In contrast to DTA, thermogravimetry has proved an invaluable aid to quantitative analysis. The thermal stabilities of the original sample, the intermediate compounds, and the final product can be ascertained from the various regions of the weight-loss–temperature curve.

The plateaus, which represent a species which is stable over a range of temperatures, are reasonably characteristic of the material undergoing thermal decomposition. Starting with a known weight of a pure sample (A), the weight remaining at each plateau can be used to calculate the composition of the species present at this stage (preferably this composition should be confirmed by some other technique).

When the composition corresponding to a temperature plateau is known, the weight loss observed in this plateau temperature range on heating an impure sample can be related to the amount of A in the initial material.

The technique has proved particularly useful for defining the temperature ranges required for the ignition of analytical precipitates. The TGA curve provides information on the composition of the species being subsequently weighed and specifies the temperature limits necessary to produce a particular chemical composition. A comprehensive listing and discussion of the TGA curves of a large number of compounds of analytical interest has been published (8).

Other applications of TGA include automatic gravimetric analysis, the determination of the composition of a complex mixture, the determination of the purity and thermal stability of analytical reagents, and studies of the behavior of materials in a vacuum or in atmospheres of various gases. TGA and DTA have also been used to elucidate the reaction kinetics of solid or liquid state decomposition reactions.

Because the temperature associated with a thermal transformation is more clearly defined if one plots the first derivative of the reaction variable (e.g., dw/dt), a number of workers advocate the adoption of the derivative-temperature curve for recording experimental data. The two techniques then become known as differential thermogravimetric analysis and derivative differential thermal analysis.

III.   TECHNIQUES BASED ON TEMPERATURE MEASUREMENT

Combustion is an exothermic process; hence if a small amount of combustible material is introduced into a hot flame, the temperature of the flame increases. The magnitude of the temperature increase is related to the amount of additional combustible material added and the heat of combustion associated with this material. Since small temperature changes can be readily measured, it is possible to detect small amounts of foreign material in a flame.

This effect has been used to study variations in the composition of gas streams.

For example, gas chromatographs (cf. Chapter 14,II,C and Chapter 15,V) separate mixtures of volatile compounds into individual zones, the separated components leaving the apparatus in a stream of some inert carrier gas. The microgram quantities of each component present in this stream are usually sufficient to raise the temperature of a hydrogen-air flame by an appreciable amount. For this reason a flame-temperature detector has been used on some models to monitor the gas stream leaving the chromatograph. The effluent gas is fed into the flame and the change in temperature as each separated component emerges is plotted against the volume of gas which has passed through the detector after the introduction of the sample mixture into the column. This type of detector can be used to study concentrations of combustible vapor ranging from 100 to 10^{-2} μg/ml of carrier gas, and through standardization with known amounts of pure sample the detector response can be related to the total amount of vapor reacting. The sensitivity increases with increasing heat of combustion.

The presence of different vapors in the gas stream leaving a chromatograph produces pockets of gas in which the thermal conductivity of the mixture differs from the thermal conductivity of the pure carrier gas. If the stream is passed over a heated filament, the cooling effect of the gas varies as different mixtures pass by. Any change in the cooling rate causes the temperature of the filament to vary, and this temperature can be determined indirectly through its effect on the electrical resistance of the filament. In thermal conductivity detectors, two similar filaments, heated by the application of an electrical potential, are formed into two arms of a conventional Wheatstone bridge electrical circuit. The other two arms of the bridge are composed of

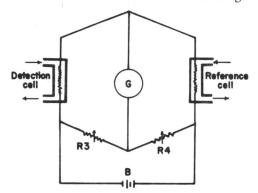

FIGURE 4.2   Schematic diagram of a thermal conductivity cell. With the same gas in the reference and detection cells, the variable resistors (R_3, R_4) are adjusted to yield zero current through the galvanometer (G) on imposition of a potential from the battery (B).

variable resistors. A stream of pure carrier or reference gas is allowed to flow over one filament while the second filament is cooled by the effluent gas stream from the chromatographic column. The electrical circuit is balanced by passing the same pure carrier gas over both filaments. Any subsequent inclusion of impurity in the gas passing through the detector cell causes an electrical inbalance which can be amplified and recorded.

The general principles of the detector are shown in Fig. 4.2.

The thermal conductivity detector (or katharometer) is no more sensitive than the flame-temperature method but it is far more convenient to use and it has been fitted to a wide range of commercial gas chromatographs. Its sensitivity is controlled by the cell design and by the difference in thermal conductivity of the major and minor components of the mixtures studied.

Apart from its widespread application in association with gas chromatography, thermal conductivity measurements provide a convenient means of monitoring the composition of industrial gas flows (e.g., flue gas analysis).

Any heat evolved in a chemical reaction in solution causes the temperature of the liquid phase to increase.

If the change in enthalpy associated with the reaction of one mole of species A, to form various products, is ΔH cal mole^{-1}, then reaction of N_m moles of A will liberate $N_m \cdot \Delta H$ calories. This quantity of heat is capable of raising the temperature of the solution and the containing vessel by an amount ΔT which is inversely related to the heat capacity (Q) of the system.

$$-\Delta T = N_m \cdot \Delta H / Q$$

If there are no heat losses to the surroundings during the course of the reaction and if the heat capacity of the system is known, then the rise in temperature can be used to calculate ΔH (N_m being known) or N_m (ΔH known).

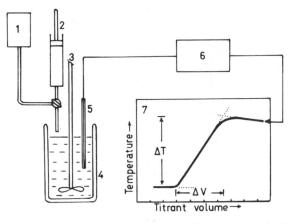

FIGURE 4.3 Thermometric titration apparatus. 1. Titrant reservoir, 2. Motor driven constant flow burette, 3. Stirring device, 4. Adiabatic titration cell containing titrate solution, 5. Temperature detector, 6. Temperature measuring circuit, 7. Recorder showing typical titration curve.

Since the majority of chemical reactions result in enthalpy changes, the procedure theoretically possesses few limitations in respect to type of reaction or few interference effects attributable to color, turbidity, etc., of reactants.

For the study of reactions in solution, a titrimetric procedure has been developed which permits the determination of N_m and ΔH.

The basic components of the apparatus required are shown in Fig. 4.3.

A known volume of a solution of a chemical compound is placed in an insulated vessel (e.g., a Dewar flask) and the temperature of the solution is recorded over a period of minutes to observe the rate of heat loss to the surroundings. A titrant solution (usually 10–100 times more concentrated than the test solution) is then added at a constant rate. Heat is evolved in the ensuing chemical reaction and the temperature rises (or falls for an endothermic reaction) until equivalence point is reached. The addition of excess titrant may cause small subsequent changes in the temperature if the titrant temperature differs from that of the reaction mixture.

The equivalence point in the titration is readily estimated from the plot of temperature against volume of titrant added.

The technique is known as thermometric titrimetry, and the published applications include studies of acid-base, oxidation-reduction, complexation and precipitation reactions. A number of thermometric titrations have also been carried out in non-aqueous solvents. As the only limiting factor is the need for an enthalpy change during the reaction, this method of end-point detection is applicable when other methods fail.

Since the titration step permits calculation of the amount of reactant species (i.e., N_m), the net rise in temperature can be used to calculate ΔH, provided that the heat capacity of the system is known through some standardization procedure. The determination of Q is complicated by the fact that the capacity of the apparatus varies during the reaction due to the volume of titrant added. This effect is minimized by using concentrated titrant solutions and by extrapolating the first part of the curve where volume changes are minimal.

The "Handbook of Analytical Chemistry" contains an excellent review of thermometric titrimetry (9a) and further information is available in monographs (6) and recent literature.

In the more general technique known as calorimetry, a known amount of reactant is placed in a special heat conductive unit which is later immersed in a known volume of water retained in an insulated vessel.

The second component required for the desired chemical reaction is also placed in the reaction cell, but initially interaction is prevented by physical separation or by unfavorable temperature conditions.

After the cell and surrounding water have reached thermal equilibrium, reaction is allowed to occur, and the rise in temperature of the water is measured accurately. Great care is required to correct for small heat losses and the measured temperature change has to be corrected for the heat evolved in secondary processes. For example, if gaseous products dissolve in a liquid phase, allowance must be made for the heat of solution; if concentrated reactants are diluted, the heat of dilution must be considered.

Analytical chemists often become involved in calorimetric studies. For example, the determination of the calorific value of a fuel is a standard procedure undertaken by most industrial chemists at some stage of their career. The heat capacity of the laboratory apparatus is usually determined by igniting a known amount of a standard material (e.g., benzoic acid) for which ΔH is accurately known. The heating capacity of the fuel under test is then measured in the same apparatus using similar experimental conditions. To ensure rapid combustion of the standards and fuels the reaction is performed in a metal vessel under a high pressure of oxygen (e.g., 20 atm). The metal vessel is immersed in a known volume of water held in an insulated tank, and ignition of the combustible materials is achieved by passing an electric current through a wire dipping into the sample. The heat evolved is transferred to the surrounding water, and the increase in temperature, after correction for heat losses, is used to calculate the unknown calorific value.

Thermal analysis is the name commonly applied to the cryoscopic evaluation of sample purity. In this technique the amount of impurity in a substance is determined from an analysis of the temperature-time curve which covers the melting range of the material. In a typical melting curve there is an initial linear segment which slopes steeply and represents the increase in temperature of the solid as heat is applied. In the vicinity of the melting point, heat is absorbed in the fusion process and the sample maintains a near constant temperature for a period of time. After complete fusion, there is a second steep-rising linear segment which records the rise in temperature of the liquid on further heating.

In the presence of small amounts of impurity, the constant-temperature plateau shifts to lower temperature values. With suitable apparatus, this depression of the freezing point $(T_0 - T)$ is directly proportional to the mole fraction of impurity present (N_0).

$$N_0 = K(T_0 - T)$$

T_0 is the freezing point of pure sample and T is the freezing point of the impure material. The proportionality constant K equals $100 \, \Delta H_f / R T_0^2$, ΔH_f being the heat of fusion.

Two basic methods are used to determine the temperature-time or temperature-heat content curves of a substance. In the static or calorimetric method a precision adiabatic calorimeter is required. In the dynamic or thermometric method, heat evolution or absorption occurs continuously and preferably at a constant rate. The constant rate of heat transfer is attained by maintaining the surroundings at some constant temperature difference (e.g., $2°$) from the sample.

Thermal analysis is nonspecific, that is, it does not distinguish between the types of impurity present, but it is applicable to all substances which are stable at their melting point. It is used to its greatest advantage in the case of high purity compounds, that is, $>99\%$ of the main component. Melting points of pure compounds and other items of interest are tabulated in the "Handbook of Analytical Chemistry."

The determination of the molecular weight of compounds soluble in stable solvents can be based on the fact that the presence of a nonvolatile solute in a pure solvent elevates the observed boiling point. The elevation (ΔT_b) is directly related to the mole fraction of solute present.

If the molal concentration of solute, m (moles of solute per 1000 gm of solvent), is negligible in comparison with the number of moles of solvent, the elevation of boiling point is given by the expression

$$\Delta T_b = R T_0^2 M_1 m / 1000\, \Delta H_{vap} = K_b m$$

where R is the gas constant; T_0 is the boiling point of pure solvent; M_1 is the molecular weight of the solvent; ΔH_{vap} is the molal enthalpy of vaporization; and K_b is called the molal boiling point constant.

If w_2 g of solute having a molecular weight M_2 is dissolved in w_1 g of solvent, the solute molality, m, is $1000\, w_2 / M_2 w_1$ and

$$\Delta T_b = K_b m = K_b \cdot w_2 \cdot 1000 / w_1 M_2$$

This equation can be used to calculate M_2.

Molal boiling point constants have been tabulated for many materials but they have to be used with caution since a change in pressure of 0.3 or 0.4 mm changes the boiling point of most solvents by as much as 0.01°.

Hence the preferable procedure is to determine the boiling points of the pure solvent and the solution simultaneously in two pieces of apparatus. It is difficult to measure the boiling point of a solution accurately, but one can obtain reasonable values by placing an inverted funnel below the surface of the liquid. The mixture of liquid and vapor pumped up the stem can then be directed on to the bulb of the thermometer being used as the measuring device.

The determination of the critical solution temperature (CST) for two partially miscible liquids provides a means of characterizing organic solvents, and it can be used for the rapid analysis of binary mixtures.

When two liquids which are not completely miscible are brought in contact they form two phases, each being the saturated solution of one component in the other. In the majority of cases a rise in temperature increases the mutual solubilities of the two components and a temperature may be reached at which the two liquids become completely soluble in each other. This temperature is known as the upper critical solution temperature. The difference in the CST values of different compounds with the same solvent can be taken as a measure of the relative affinity or selectivity of the solvent. This information can then be used to screen solvents in terms of their selectivity for desired and undesired components. The CST method has also been used to determine the amount of water present in organic solvents.

For characterizing hydrocarbons, the aniline point is frequently used instead of CST. The aniline point is the temperature at which equal volumes of pure aniline and another liquid, usually a petroleum product, become miscible.

The microdetermination of critical solution temperatures permits very small amounts of sample to be characterized. Minute amounts of sample and test liquid are sealed into a capillary glass tube. The tube is then heated on a hot stage and the interface between the two liquids is observed under a microscope.

Fuller details of the technique and range of applications are given in the review by Fischer and Schmid (10).

IV.   MISCELLANEOUS THERMAL METHODS

Fractionation is primarily a process for the separation of the components of a mixture of liquids, but it is included in this chapter because the process is based on the transformation of mixtures of liquids into mixtures of vapors through the application of heat.

In this separation procedure, a liquid mixture is boiled and the resulting vapors are passed through a long packed column before being condensed and collected in a separate vessel. The vapors moving up the column become progressively richer in the more volatile component, since the liquid phases which condense in the cooler upper regions contain higher molar fractions of this component. This experimental fact can be explained by considering a simple binary solution which is partially vaporized.

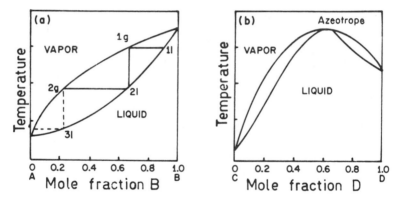

FIGURE 4.4   Boiling point curves showing the liquid and vapor compositions of a binary mixture as a function of temperature, (a) an ideal solution (b) a nonideal system which forms an azeotrope.

In the boiling flask the composition of the two phases may be represented by the points 1 *l* and 1 *g* in Fig. 4.4. In passing up the column the vapor reaches a zone of lower temperature and some of the material condenses to give a liquid phase of composition 2 *l* in equilibrium with a vapor phase of composition 2 *g*. This vapor is much richer in the more volatile component than the mixture leaving the boiling flask. Migration of the vapor 2 *g* to cooler regions yields further condensation and a residual vapor which is again richer in the more volatile component. In fractional distillation this process of successive vaporization and condensation is repeated a large number of times as the vapors move up the column. The vapors leaving the top of the column are condensed and the bulk of the liquid is recycled down the column (reflux ratio) while some pure isolated component is continually removed. When most of the more volatile component has been collected, the temperature of the vapor discharge increases and this signifies the presence of the second species.

The same type of argument applies to mixtures containing more than two components.

When the boiling point curve for a binary mixture has a maximum or a minimum, the solutions having the maximum or minimum boiling points are called azeotropes. These solutions distil without a change in composition because the liquid and the vapor have the same composition. Fractionation of a solution that yields an azeotrope can, in principle, yield the azeotrope and one of the pure liquids depending on the composition of the initial solution. Lists of azeotropic mixtures and other data of interest are tabulated in the "Handbook of Analytical Chemistry."

A process somewhat akin to fractionation is involved in the extremely important analytical technique known as gas chromatography (discussed in Chapter 15).

Very special cases of thermal transformations associated with analytical procedures are the "weight loss" method and "separation by vaporization."

The determination of the moisture content of a sample by drying the material at 110° to constant weight is a good example of the weight loss technique. The heat supplied transforms water from the liquid to the vapor state, i.e., the separation is based on a physical change. On heating limestone at 1000° to constant weight a weight loss is observed which can be attributed to the liberation of carbon dioxide. The chemical transformation induced by heat thus provides a simple, if inaccurate, means of analyzing carbonate minerals.

In separation by vaporization, one component in a mixture of inorganic species is converted into a chemical form which has a lower boiling point than any other species present. The volatile component is then isolated by carefully heating the mixture. Examples of this approach include the conversion of nitrogen compounds into ammonia, the volatilization of silica as silicon tetra-fluoride, the distillation of arsenic as a trichloride or hydride, and the elimination of chromium from a mixture through the formation of chromyl chloride.

The term "pyrolysis" has been used to describe the technique in which the sample is thermally decomposed and the decomposition products are analyzed by some physical or chemical means. Samples may be identified by the temperatures at which products appear, by the number of products detected, and by observing the temperatures at which the production of certain products is a maximum.

The pyrolysis chamber is usually connected directly to the analytical instrument which may be a mass spectrometer, a vapor phase gas chromatograph, or an infrared spectrometer.

Finally, it seems desirable to mention "dilatometry" which is the name applied to the determination of the volume change of a substance, usually as a function of temperature. The technique involves the continuous measurement of the sample length as it is heated or cooled at a constant known rate. Pronounced changes in sample volume occur during most solid → solid phase transitions and the thermal expansion of a metallic or ceramic substance is a physical parameter of utmost importance in the development of new alloys and ceramic materials.

V.   EXERCISES IN INFORMATION RETRIEVAL

1. Using recent review articles prepare short summaries which illustrate current trends in
   a. differential thermal analysis
   b. thermometric titrations
   c. pyrolytic techniques.
2. Briefly describe the principles of the major types of thermobalances described in monographs. What advantages would arise from control of the gaseous atmosphere surrounding the samples?
3. Use monographs and reviews to prepare an essay on the characterization of minerals by differential thermal analysis.
4. Discuss the statement "A single thermal property is not sufficient to characterize a chemical reaction or system; indeed, as many thermal methods as possible should be employed."
5. a. Compare and contrast the analysis of gas mixtures by flame detectors and katharometers.
   b. Prepare a list of applications of thermal conductivity cells, excluding gas chromatography.
   c. What are the advantages of having a pyrolysis unit preceding a gas chromatograph?
6. In a natural product survey an organic chemist isolated small amounts of several new compounds. Describe microtechniques which he could use to determine the melting point, affinity for solvents and molecular weight of each of these compounds.
7. Briefly describe the apparatus and procedure involved in the determination of the calorific value of
   a. kerosene,
   b. a breakfast cereal,
   c. a powdered solid fuel.
8. a. Using published thermogravimetric curves, define the best conditions for drying and/or igniting the following analytical precipitates:

   calcium oxalate; ammonium 12-molybdophosphate; potassium tetraphenylborate; thorium benzoate; aluminum 8-hydroxyquinolate.

   b. Indicate how the calcium and magnesium content of a dolomite (Ca, Mg carbonate) may be estimated from a single thermogravimetric curve.

c. Outline how the purity and thermal stability of primary and secondary standards may be checked by TGA.

d. What are the advantages of using derivative plots to record DTA and TGA curves?

9. a. In a thermometric titration, 50 ml of acid solution (about 0.01 M) is to be titrated with 1.0 M NaOH. Using 0.01 M HCl for which the heat of neutralization (ΔH°_{298}) is -13.5 kcal mole^{-1} a maximum temperature rise of $0.0614°$ was observed. What rises could be predicted if $\Delta H°$ for the acid being titrated was

a. -10.2 kcal mole^{-1} b. -12.8 kcal mole^{-1}?

b. In this particular apparatus, to gain a possible accuracy of $\pm 2\%$ the temperature rise observed must be $>0.01°$. What would be the minimum concentration of the three acids which could be titrated under these circumstances?

c. The heat of neutralization of an acid (ΔH°_{n}) with a strong base can be additively equated to its heat of ionization, ΔH°_{i}, plus the heat of neutralization of strong acids $(-13.5$ kcal mole$^{-1})$. That is, $-\Delta H^{\circ}_{n} = -(\Delta H^{\circ}_{i} + 13.5)$ kcal mole^{-1}. The heat of neutralization of acetic acid $(K_{a} = 10^{-5})$ is reported as -13.1 kcal mole^{-1} and for boric acid $(K_{a} = 10^{-10})$ $-\Delta H^{\circ}_{n}$ equals -10.2 kcal mole^{-1}. Calculate the free energy change and entropy change associated with the neutralization of these two acids at $25°$ C.

10. a. At $25°$ C the entropies of $AgCl_{(s)}$, $Ag^{+}_{(aq)}$ and $Cl^{-}_{(aq)}$ are quoted as 22.97, 17.67, and 13.17 cal deg^{-1} mole^{-1} respectively. The formation constant for the reaction $Ag^{+}_{(aq)} + Cl^{-}_{(aq)} \rightleftarrows AgCl_{(s)}$ at this temperature is quoted as log $K = 9.7$. If one milliequivalent of chloride is to be titrated with silver nitrate in a cell having a heat capacity of 45 cal deg^{-1}, what rise in temperature can be expected during the precipitation process?

b. When 0.904 g of an organic compound was dissolved in 100 g of benzene, the boiling point of the solution was $0.12°$ higher than that of pure solvent. The molal boiling point constant (K_{b}) for benzene is 2.53 deg molal^{-1}. What was the molecular weight of the organic compound?

References

1. F. D. Rossini, D. D. Wagman, W. H. Evans, S. Levine, and I. Jaffe, *Selected Values of Chemical Thermodynamic Properties*, Nat. Bur. Standards Circular 500, U.S. Govt. Printing Office, Washington, D.C., 1952.

2. F. D. Rossini, K. S. Pitzer, R. L. Arnott, R. M. Braun and G. C. Pimentel, *Selected Values of Physical and Thermodynamic Properties of Hydrocarbons and Related Compounds*, Am. Petrol. Inst. Res. Project 44, Carnegie Press, Pittsburgh, Penn., 1953.

3. D. R. Stull and G. C. Sinke, *Thermodynamic Properties of the Elements*, Am. Chem. Soc., Washington, 1956.
4. O. Kubaschewski and E. L. Evans, *Metallurgical Thermochemistry*, 3rd Ed., Pergamon, London, 1958.
5. J. A. Allen, *Energy Changes in Chemistry*, Blackie, Glasgow and London, 1965.
6. W. W. Wendlandt, *Thermal Methods of Analysis*, Wiley (Interscience), New York, 1964.
7. P. D. Garn, *Thermoanalytical Methods of Investigation*, Academic Press, New York, 1965.
8. C. Duval, *Inorganic Thermogravimetric Analysis*, 2nd Ed., Elsevier, Amsterdam, 1963.
9. L. Meites, ed., *Handbook of Analytical Chemistry*, McGraw-Hill, New York, 1963.
10. R. W. Fischer and H. H. O. Schmid, *Standard Methods of Chemical Analysis*, Volume IIIA, F. J. Welcher (ed.), Van Nostrand, Princeton, New Jersey, 1966.

ADDITIONAL READING

P. E. Slade and L. T. Jenkins, *Thermal Analysis*, Marcel Dekker, New York, 1966.
W. J. Smothers and Yao Chiang, *Differential Thermal Analysis*, Chemical Publishing Co., New York, 1958.
H. J. V. Tyrrell and A. E. Beezer, *Thermometric Titrimetry*, Chapman and Hall, London, 1968.

5

RADIATION EMISSION AND RADIOACTIVITY

I. Emission Spectra	73
A. Radiation Emission	73
B. Atomic Spectra	76
C. Molecular Spectra	81
D. Modes of Excitation	83
II. Flame Excitation	84
A. Flame Characteristics	84
B. Flame Photometry	87
III. Emission Spectrometry	90
A. Excitation by Electrical Arcs and Sparks	90
B. Gaseous Discharges and Plasmas	93
C. Emission Spectroscopy	95
IV. Excitation by Accelerated Particles or High Energy Radiation	100
A. X-Ray Emission	100
B. X-Ray Emission Spectroscopy	101
C. Atomic Fluorescence Spectrophotometry	106
D. Fluorometric Analysis	106
E. Chemiluminescence	109
V. Radiochemical Procedures	110
A. Radioactivity	110
B. Activation Analysis	112
C. Radioactive Tracer Techniques	115
VI. Revision Questions and Assignments	116
References	119

I. EMISSION SPECTRA

A. RADIATION EMISSION

In the preceding chapter a number of effects arising from the absorption of heat have been considered, but one important phenomenon, photoemission, was omitted.

The bulk emission of light by heated solids is a common observation. For example, when a piece of steel is heated, it first glows with the emission of red light and finally, at higher temperatures, it emits white light. It can be shown that this white light consists of all the colors of the visible spectrum and the emission is said to be continuous. Similarly, if some sodium chloride crystals are heated (bunsen flame), yellow light is emitted. This radiation, however, can be shown to consist of a few characteristic radiations only.

Detailed studies of this phenomenon have shown that the nature of the radiations emanating from a heated sample varies from species to species. It has also been observed that different amounts of energy are required to cause emissions.

When reasonably isolated, every elementary system appears to possess a number of discrete energy states, and to change from one state to another of higher energy, the system has to be exposed to heat, or radiation, or bombardment with high velocity particles. The energy supplied must equal or exceed the energy difference between the two states, and because of the discrete nature of the states, only definite fixed amounts of energy (defined as quanta) are absorbed.

After absorbing extra energy a system is said to be "excited," and such a system tends to revert quickly to a lower energy state. The quantum of energy lost in this process may be liberated as heat, radiation of a particular frequency, or as a mixture of both.

The multiple energy transitions which a system can undergo are characteristic of the species being excited; hence the radiation emitted is also characteristic of the system.

By means of a monochromator, sample emissions can be dispersed into components which differ in frequency, and by studying the frequency and intensity of the isolated components, the chemical composition of the sample can be ascertained. The optical dispersion systems used in such studies produce sharp images of the initial slit at the focal plane of the equipment, and on a photographic plate these images appear as a series of parallel lines, each line corresponding to a particular energy transition in an excited atom. Accordingly, the characteristic frequencies (or wavelengths) of emitted radiations are often referred to as "lines."

The emission line most readily obtained is that which corresponds to the transition involving the excited state of lowest energy (E_1), and such a line is commonly referred to as the resonance line (cf. Fig. 5.5).

The most characteristic property of any electromagnetic radiation is its frequency (that is, the number of wave crests which pass any given point per second), and the frequency (ν) of the radiation emitted on the deactivation of an excited species is related to the energy change in the system, ΔE, by the equation:

$$\nu = \Delta E/h \quad \text{or} \quad \Delta E = h\nu,$$

where h is Planck's constant $(6.624 \times 10^{-27}$ erg sec).

For emissions originating from excited atoms it is more traditional to describe the radiations in terms of the wavelength λ. For a particular medium, m, the two are related by the equation

$$v = v_m/\lambda_m$$

where v_m is the velocity of wave propagation. The highest, or limiting, value of the velocity is the speed of light in vacuo, c, viz., 2.998×10^{10} cm sec^{-1}.

Thus in vacuo (and approximately in air)

$$v = c/\lambda$$

Frequency is reported in terms of cycles per second, cps, while the units used to describe spectroscopic wavelengths are the Angstrom (10^{-10} m); millimicron, mμ, or nanometer, nm (10^{-9} m); and the micron, μ, or micrometer, μm (10^{-6} m).

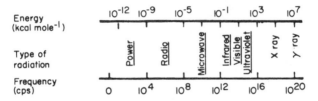

FIGURE 5.1   The frequency and energy scale of electromagnetic radiation.

Some spectroscopists prefer to describe radiations in terms of their wave numbers. The wave number is the reciprocal of the wavelength in centimeters, and is directly related to the energy involved in the transition producing the emission. The unit for wave numbers is the "reciprocal centimeter" (cm^{-1}).

Special names such as γ rays, X rays, ultraviolet light, visible light, infrared radiation, etc., have been given to radiations which can be grouped within particular ranges of frequencies, and Fig. 5.1 indicates the general relationship between frequency, type of radiation, and energy equivalence (expressed as kcal mole^{-1}).

When the system being excited is an atom, energy is absorbed through the promotion of electrons to orbital states of higher energy. The difference in energy between the outer (i.e., higher principal quantum number) orbitals of an atom are generally of the order of 40 to 140 kcal mole^{-1}. This corresponds to the energy associated with visible and ultraviolet radiations, and electronic transitions between these orbitals yields emissions of this frequency type. On the other hand, the energy difference between inner orbitals (low principal quantum number) is much higher (about 10^5 kcal mole^{-1}) and is equivalent to the energy content of X rays. Thus when a high energy electron collides with an atom with sufficient force to cause ejection of an inner electron and a neighboring electron falls into the vacant site, the excess energy is emitted as characteristic X rays.

In the Bohr model of the atom, electrons were considered to be arranged around the nucleus in discrete orbits designated by the letters K, L, M, N, etc. (corresponding to principal quantum numbers 1, 2, 3, 4 etc.). Thus when an electron in a K orbit is ejected, the radiation emitted on replacement of this electron is known as K-radiation. Similarly, if the electron is initially ejected from the L shell, the resultant X-radiation is known as L-radiation. The source of the replacement electron is designated by a Greek letter suffix. Thus when the electron falls from the next highest orbit the letter α is used; β represents a change of two in principal quantum number. Some of the transitions commonly observed are shown diagrammatically in Fig. 5.2.

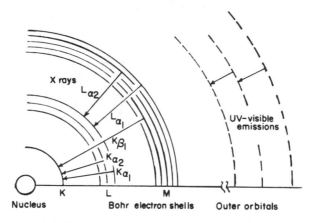

FIGURE 5.2   Origin of spectral lines.

The frequency of the X-radiation emitted when an electron changes from one energy level designated N_2 to another level N_1 is given by the equation

$$\text{Frequency } \nu = Z^2 \cdot \frac{2\pi^2 m e^4}{h^3} \cdot \left(\frac{1}{N_1{}^2} - \frac{1}{N_2{}^2} \right)$$

where Z is the atomic number of the atom involved, and m and e are the mass and charge of the electron. For a K_α emission (cf. Fig. 5.2) $N_2 = 2$, $N_1 = 1$; for a K_β emission $N_2 = 3$, $N_1 = 1$; for a L_α emission $N_2 = 3$, $N_1 = 2$; etc.

B.   ATOMIC SPECTRA

The number of characteristic frequencies emitted by an excited atom is controlled partially by the magnitude of the initial energy input and partially by the number of energy transitions which are statistically feasible.

The energy of electrons in atomic orbitals can be described in terms of the principal quantum number, n (1, 2, 3, . . .), angular momentum quantum number, l (0, 1, 2, 3, . . . or $s, p, d, f, . . .$), and spin quantum number, $s (\pm 1/2)$. The magnetic quantum number is also significant if the excited atom is exposed to an electrical or magnetic field.

On excitation of an atom, the number of energy changes which can occur appears to be restricted, and the electron transitions which are permitted can be defined by a series of selection rules. For example, in atoms with one valence electron, the predominant or permitted transitions are those in which the angular momentum quantum numbers of the two states differ by unity, that is, $\Delta l = \pm 1$. The change in principal quantum number, Δn, can be zero or any integral value.

The total angular momentum of an electron involves l and s, and this is described by a new term, j. For atoms with one valence electron, $j = l + s = l \pm 1/2$. Where $l > 0$, j has two values and these correspond to two different energy states. For example, with an electron in the p state, $l = 1$ and $j = 3/2$ or $1/2$. The well-known yellow doublet of sodium arises from a $3\,p - 3\,s$ transition.

In atoms of low atomic number possessing more than one valence electron, the quantum numbers for the valence electrons can be summed to give new terms defined by the relationships:

$$L = \sum l; \qquad S = \sum s; \qquad J = L + S$$

(With heavier atoms, $l_i + s_i = j_i$; $\sum j_i = J$.)

For these systems, permitted transitions are those in which $\Delta S = 0$, $\Delta L = \pm 1$ or 0, and $\Delta J = \pm 1$ or 0 (but transition $J = 0$ to $J' = 0$ is forbidden). Δn is unrestricted.

The number of ways the spin can interact with the angular momentum is equal to $2\,S + 1$ and this describes the *multiplicity* of the transition (e.g., with the sodium atom, $s = 1/2$; hence multiplicity equals two and a doublet is observed).

The spectroscopic notation for energy levels is based on writing the capital letters S, P, D, F etc. to represent electrons having l values of 0, 1, 2, 3 etc. The value of the multiplicity ($2\,S + 1$) is then written as a prefix superscript. The value of j is written as a suffix subscript. For a complete description this notation is preceded by the value of n.

Thus the valence electron of the sodium atom, in its ground (i.e., 3 s, or $n = 3$, $l = 0$, $j = \frac{1}{2}$) state is described by the term $3^2\mathrm{S}_{1/2}$.

The notation for this valence electron in the first excited state (3 p) is $3^2\mathrm{P}_{1/2}$ or $3^2\mathrm{P}_{3/2}$ (Here $n = 3$, $l = 1$, $j = \frac{1}{2}$ or $\frac{3}{2}$).

In closed shells like s^2, p^6, d^{10} the total of l and s equals zero and all closed shells are termed $^1\mathrm{S}_0$ states.

Where the total angular momentum (j or J) does not equal zero, there are $(2\,J + 1)$ possible orientations of J in respect to an applied magnetic field, each orientation involving a slightly different energy change. The term $(2J + 1)$ is known as the degeneracy or statistical weight (symbol P) for a particular state. For example, in the excited sodium atom where $j = 3/2$ or $1/2$, the statistical weights are 4 and 2. Since the total difference in energy above the ground state is practically the same, it may be concluded that at any elevated temperature there are twice as many atoms with electrons in the $j = 3/2$ level as in the $j = 1/2$ level. Thus one line of the doublet should be twice as intense as the other.

The relative intensities of lines in a given multiplet are quite frequently given by the sum rule which states that the sum of the intensities of all lines of a multiplet which belong to the same initial, or final, state is proportional to the degeneracy of the state not common to each line.

Consider the case of a single electron which is excited into an f level. Here the multiplicity is 2 and j equals $7/2$ or $5/2$. The transition to the allowed ground state $(\Delta l = -1)$ is then represented by $^2F \rightarrow {}^2D$. The selection rules for ΔJ permit three transitions, namely, $^2F_{7/2} \rightarrow {}^2D_{5/2}$, $^2F_{5/2} \rightarrow {}^2D_{5/2}$, $^2F_{5/2} \rightarrow {}^2D_{3/2}$. From the sum rule, the relative intensities of the emissions arising from the transitions in which $^2D_{5/2}$ is common is 8/6. The relative intensity of the emissions in which the $^2F_{5/2}$ state is common is 6/4. If the relative intensities of the three permitted transitions are $a:b:c$, then $a/(b + c) = 8/6$ and $(a + b)/c = 6/4$ which simplifies to $a:b:c = 20:1:14$.

When a sample is exposed to heat energy, only a small proportion of the atoms present become excited. The relation between the number of atoms, N_j, in a particular excited state and the number, N_o, in the ground or final state is given by

$$N_j/N_o = P_j/P_o \exp\left(-E_j/kT\right)$$

where P_j and P_o are the statistical weights of the two states, E_j is their energy difference, k is Boltzmann's constant and T is the temperature.

The fraction (N_j/N_o) in the first (and most probable) excited state is very small for most systems. For example, sodium atoms are easily excited yet the fraction in the excited state at any given moment is circa 10^{-5} at $2000°$ K and 10^{-2} at $5000°$ K. For zinc the corresponding fractions are 10^{-14} and 10^{-6}.

The number of atoms in the excited state determines the intensity of the emissions. Since P_j, P_o, and E_j are fixed for a particular system, the intensity becomes proportional to the total number of atoms (i.e., concentration of element in the sample) and the temperature. Conversely, to obtain the same intensity of emission (e.g., the minimum detectable by the apparatus) from different elements at the same temperature, different concentrations are required. Thus, in a flame photometer the excitation of 0.001 ppm of sodium may yield the same meter deflection as 10 ppm of lead.

If, during excitation, an atom is ionized, the resultant ion may also be excited and emit a spectrum. The spectrum of the ion differs from that of the atom because the atom and the corresponding ion have different electronic structures.

In fact, the spectrum of a singly ionized element resembles that emitted by the atom of the element of the preceding atomic number; a doubly ionized element yields a spectrum similar to that of the element whose atomic number is two less.

The degree of ionization of any element increases with the temperature of the excitation source. At atmospheric pressure the relationship is given by Saha's equation

$$\log K^2/1 - K^2 = -5050 \, V_i/T + 2.5 \log T - 6.5$$

where T is the absolute temperature, V_i is the ionization potential, and K is the fraction of the atoms ionized. A high degree of ionization depletes the number of available neutral atoms with a resulting weakening of the intensity of the atom spectrum and a brightening of the ion spectrum.

Another factor which influences the observed intensity of radiations is the phenomenon known as self-absorption.

The natural shape of a line as emitted from a source is of the form shown in curve 1, Fig. 5.3. The line has a definite width and the width of the line at half the peak intensity is usually used for comparison purposes. Two main factors control this width, Doppler broadening and collisional broadening.

The Doppler half-width (Δ) is given by the equation

$$\Delta = 0.72 \times 10^{-6} \lambda \sqrt{T/M}$$

where T is the absolute temperature of the emitting gas, λ is the wavelength of the line, and M is the atomic weight.

Sometimes lines are considerably broader than can be explained by the Doppler effect. Much of this broadening is attributable to collision of the emitting atoms with other atomic species (collisional broadening). Emitted lines are also broadened if the excited atoms are subject to a strong electric field (the Stark effect).

The line emitted at the source usually has to pass through a cooler outer fringe before traveling through space to the dispersion and recording instrument. In the low temperature outer zone many atoms are present which are still in the ground state or very low energy states. These low energy atoms are capable of absorbing the characteristic emission, leading to self-absorption. Self-absorption is greatest at the peak of a line and its effect on the shape and intensity is indicated by curve 2, Fig. 5.3.

The amount of self-absorption increases rapidly with the concentration of absorbing vapor, and in the extreme case the intensity at the center of a line is almost wholly absorbed and looks like curve 3, Fig. 5.3. This is "self-reversal."

A final factor influencing the intensity of an emission is selective volatilization. When a specimen which contains several elements is heated very strongly, each element may volatilize selectively. In extreme cases, one or more elements may distil almost completely before others begin to volatilize. Since it is the vapors which are excited, the intensity of emission becomes a function of the excitation time under these circumstances.

From this very brief discussion, it can be appreciated that exciting an atomic species can result in a large number of permitted energy transitions, each transition giving rise to a characteristic emission. The intensity of the different emissions varies greatly and depends on the temperature, the energy change involved, the total concentration of atoms present, and the type of electron energy levels involved.

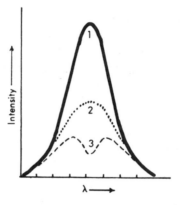

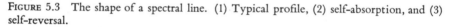

FIGURE 5.3   The shape of a spectral line. (1) Typical profile, (2) self-absorption, and (3) self-reversal.

For each element, some emissions persist as the total concentration of atoms is reduced to extremely low levels. These radiations are very suitable for identifying the presence of trace amounts of different elements in a sample. Other radiations show a near linear relationship between the intensity of emission and the concentration of element in the sample. These are most suitable for quantitative analytical studies.

Thus, while complete tabulations of all the emissions attributable to different atoms exist, the analytical chemist can operate with short lists which record the most persistent and characteristic wavelengths. One source of this information is the "Handbook of Analytical Chemistry." Other reference books are included in the bibliography.

The X-ray emission spectra of elements is much simpler than the corresponding visible–ultraviolet emission spectra. This arises in part from the fact that only inner electrons are involved and the number of permissible transitions is thus far more limited.

In practice, the sharp, characteristic X-radiations are superimposed on a continuous background emission spectrum, this continuous spectrum resulting from the general interaction of high energy electrons with matter.

The intensity of the characteristic emissions can be increased by increasing the energy of the exciting source. For example, if the X-radiations are produced by bombarding a solid target with electrons accelerated by a high potential, the intensity (I) is related to the applied potential by the expression

$$I = K_i(V - V_o)^n$$

where i is the X-ray tube current, V is the applied voltage, V_o is the minimum voltage for excitation of the particular radiation, and n is an integer (ranging between 1 and 2) whose magnitude depends on the atomic number of the element and the ratio V/V_o.

Part of the initial characteristic radiation can be absorbed by other parts of the sample before entering the measuring instrument. The extent of this absorption varies with the sample matrix, since different elements have different tendencies towards absorption. This absorption process can be followed by the emission of secondary X rays, that is, X rays of longer wavelength which are characteristic of the absorbing element.

C. Molecular Spectra

In a molecule, the constituent atoms vibrate to and from each other in a characteristic manner and the molecule as a whole can rotate in space. The whole system is thus quantized in respect to electronic energy levels, vibrational frequencies and rotational energy states.

The total energy of the molecule is given by

$$E_{total} = E_{electronic} + E_{vibrational} + E_{rotational}$$

The three components of the total energy vary in magnitude by factors of about ten. For example, energy equivalent to about 100 kcal mole^{-1} is required to cause an electronic transition, about 10 kcal mole^{-1} is needed to cause vibrational changes, and changes in rotational states absorb energy measured in kcal mole^{-1}. The difference in energy requirements for a diatomic molecule is shown schematically in Fig. 5.4.

A molecule is normally in the ground electronic state. This is designated by E_o in the diagram, but the electronic quantum number of molecular species is usually represented by the symbols Σ, π, Δ, ϕ (corresponding to values of 0, 1, 2, 3 respectively).

At room temperatures, a molecule is also normally in the ground vibrational state (V_o). The energy of the various vibrational levels is given by the expression

$$E_{vibrational} = (V + 1/2)h\nu_0$$

where V is the vibrational quantum number and can take values 0, 1, 2, 3 etc; and ν_0 is the characteristic vibrational frequency of the bonded atoms under consideration.

Vibrational changes are generally accompanied by changes in the rotational state. To a first approximation, the rotational energy of a diatomic molecule is given by

$$E_{\text{rotation}} = J(J+1)h^2/8\pi^2 I$$

where J is the rotational quantum number (0, 1, 2, 3, etc.) and I is the moment of inertia of the molecule. The absorption of small amounts of energy, e.g., room temperatures or radiations from the microwave region, can cause changes in the rotational levels without inducing vibrational or electronic transitions.

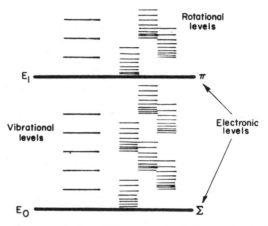

FIGURE 5.4   Schematic representation of the energy levels in a molecule. E_o represents the ground electronic state and E_1 represents an excited electronic state. Vibrational levels are indicated by V_o, V_1, etc., and rotational levels are shown as finely spaced lines.

The energy required to cause emission is often greater than the energy required to cause molecular dissociation. In these circumstances either no emission is observed, or the radiations detected are characteristic of molecular fragments.

As in atomic studies, only a limited fraction of the total number of possible energy transitions appear to occur and this has led to the pronouncement of various "selection rules" for "allowed" transitions: A rather oversimplified way of describing these rules is to state that allowed transitions include those in which the electronic, vibrational, and/or rotational states change by ±1.

The lifetimes of electronically excited species average about 10^{-8} sec, a time determined by the probability of transitions to lower energy levels. In the majority of molecular systems the absorbed energy is degraded to heat and is never released as electromagnetic radiations. Where radiation is emitted, the emissions have a banded structure, because each electronic transition is accompanied by vibrational and rotational changes. Transitions which differ only in respect to the rotational levels involved yield a series of lines which differ only slightly in wavelength. These lines converge towards particular points in the spectrum where the "piled-up" lines form what are called band heads. Consequently, a molecular band, viewed as a whole, shows an edge at the band head and a shading in the direction of either longer or shorter wavelengths. Each permissible vibrational level transition yields a distinct band.

There are molecular systems in which excitation is followed by an initial loss of energy as heat, with subsequent emission of radiation. These systems generally possess more than the usual degree of stability in the excited state,

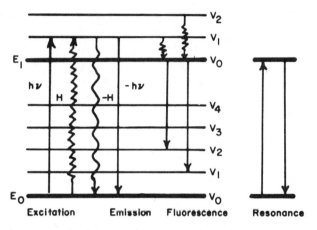

FIGURE 5.5 Absorption of radiation and subsequent loss of energy. The system can be excited by the absorption of heat (*H*) or radiation (*hv*). The energy can be subsequently lost as heat, or radiation, or both (fluorescence).

which allows excess vibrational energy to be transferred to neighboring molecules by collision. The observed radiation then results from a transition involving the lowest vibration state of the electronically excited species and an excited vibration level in the ground electronic state. This phenomenon is known as fluorescence and the energy changes involved are shown diagrammatically in Fig. 5.5.

D. MODES OF EXCITATION

An ideal mode of excitation should:

1. Excite the sample to such an extent that characteristic lines of all the elements of interest are emitted with sufficient intensity to be detected in the instrument being used.
2. Provide uniform excitation conditions from sample to sample, ensuring uniform and reproducible vaporization, and constancy over a period of time.
3. Provide a low spectral background.

In most instances no one procedure is capable of fulfilling all these requirements, but with care it is possible to optimize the characteristics which are most important in specific applications.

The most common sources of radiation used in analytical emission spectrometry are hot flames, electrical arc or spark discharges, and high energy X rays. In addition to these, there are special sources such as plasma jets, microwave discharges, hollow cathode discharges, high energy electrons and chemical reactions (chemiluminescence).

Since each source possesses some individual characteristics, it is convenient to discuss simultaneously the source and the analytical technique in which it is used.

II. FLAME EXCITATION

A. FLAME CHARACTERISTICS

A hot flame provides sufficient energy to excite many elements, in fact, over sixty elements emit sharp lines or bands in conventional flames. With modification of the flame conditions, this number can be increased to nearly eighty. However, unless a very hot flame is used, most elements emit few intense lines and the sensitivity is very poor.

The flame is particularly useful for the excitation of metals having low ionization potentials (e.g., the alkali metals) since the fraction of the total number of atoms ionized is minimal. In addition, the temperature of the flame can be maintained reasonably constant over a period of time by keeping the gas pressures and gas flows to the burner constant.

Sample introduction is comparatively simple, since solutions can be added to the flame in the form of an aerosol, a steady flow of solution being achieved by incorporating an atomizer in the burner unit. A typical arrangement is shown in Fig. 5.6. This type of assembly feeds only the finer droplets, i.e., a homogeneous fog, to the burner. Another type of unit, known as an integral atomizer-burner, injects the entire spray directly into the flame.

When the aerosol enters the flame the following events occur in rapid succession.

1. The solvent is vaporized leaving minute particles of dry salt.
2. The heat of the flame melts and/or vaporizes the salt and part or all of the gaseous molecules dissociate to yield neutral atoms.
3. Some of the atoms unite with other atoms or radicals present in the flame.
4. Other atoms or molecules are excited by the thermal energy of the flame.
5. Radiation is emitted by excited atoms or molecules.

In order to excite atoms to the state where radiation is emitted, a medium of high temperature, with sufficient heat content to avoid being cooled too greatly by the preceding steps, is required.

Any fuel gas with a high combustion temperature can be used, the most common being acetylene, propane, butane, hydrogen, natural gas, and coal gas. When burned with air or oxygen these gases yield flames with temperatures between 1700 and 3200° C. In recent years, higher temperatures have been realized using cyanogen as the fuel gas. The higher the temperature, the greater is the number of elements excited. Increasing the temperature also increases the sensitivity attainable. The nature of a flame is controlled in part by the burner design.

In atomizer-burners the fuel gas and oxygen are fed through concentric ports and meet at the tip where they burn with a "turbulent" flame. The sample is atomized directly into the flame from a third orifice at the center. This type of burner allows highly explosive gas mixtures (oxy–hydrogen, oxy–acetylene etc.) to be handled safely, and combustible solvents, such as gasoline, can be fed in directly as a spray.

In the type of burner shown in Fig. 5.6, the fuel gas, oxygen, and aerosol are thoroughly mixed before combustion. This yields a premixed or laminar type flame in which three distinct zones can be identified.

The primary combustion zone or inner cone is only about 0.1 mm thick at atmospheric pressure, and if hydrocarbons are being burned, this zone is visible by virtue of the strong blue-green light which emanates from the radicals C_2 and CH. Although light emission, ionization, and the concentration of radicals are all extraordinarily high in this zone, it is rarely used for flame photometric analysis because thermodynamic equilibrium does not prevail.

Immediately adjacent to the primary combustion zone is the interconal zone. This zone is reaction free, it can be characterized in terms of its temperature, and many of the species present are in a state of thermodynamic equilibrium. This is the part of the flame used in analytical studies.

When the flame is surrounded by air, a secondary combustion zone forms around the flame. At the border, carbon monoxide is oxidized to carbon dioxide and there is a weak emission of blue-violet light. If the primary combustion is incomplete owing to an insufficiency of premixed air, the temperature of this outer cone can become higher than the interior of the flame. Radiation from this zone creates a background which must be heeded in practical analysis.

The spectrum associated with a flame depends on the fuel, the fuel oxygen ratio and the temperature. Prominent OH band structures are observed between 280 and 285 mμ and 306 and 320 mμ; dissociating CO molecules yield a continuum extending between 350 and 550 mμ.

The observed intensity of emitted spectral lines is different in different parts of the flame and the behavior of elements can vary with the nature of the flame or with the solvent used. In addition, the majority of errors in flame photometry arise from phenomena that develop in the flame.

Two causes of nonlinear calibration curves are self-absorption and ionization, both of which have been discussed in Section I. Ionization occurs to a measurable extent with all elements having ionization potentials of less than 6 eV but the ionization of a test element can be repressed by adding a second, more easily ionized, element to the test solution. This is known as the radiation-buffer technique.

Chemical interference arises from the formation in the flame of condensed phases which are difficult to volatilize and dissociate into free metal atoms. For example, the presence of phosphate, sulfate, and other oxyanions interferes markedly with the emission of magnesium and the alkaline earth elements. The use of releasing agents or protective chelating agents has been recommended to overcome these effects. The first approach involves adding a competing ion which preferentially combines with the interferant. Thus the addition of large amounts of strontium or lanthanum ions tends to free calcium from the depressant action of phosphate. Alternatively, the metal ion is converted into a stable complex (e.g., by addition of EDTA) which is readily decomposed by the flame. The use of a fuel-rich flame provides an environment which favors the continued existence of the free atoms of those elements which have a strong tendency to form stable oxide molecules in ordinary flames.

Added salts and acids can cause the droplets of solution entering the flame to increase in size and subsequently hinder the evaporation of the solvent. The time (t) required for complete evaporation is determined by the initial drop diameter (d_o) according to the equation, $d_o^2 = Ct$, where C is a constant which depends on the flame temperature, the boiling point of the solvent, and the thermal conductivity of the solvent vapor. The position in the flame of the dry solute and subsequent emission is thus a function of these factors and the speed of ascent of the hot gases. When elements are aspirated into the flame from an organic solution instead of an aqueous solution there is an increase in emission intensity. This can be attributed partly to the influence of the solvent on the atomizer efficiency, partly to the ease of solvent volatilization, and partly to chemical effects in the flame.

In the emission process, an important intermediate step is the vaporization of the solute particles. The equilibrium for this stage may be described by a standard type of equation, e.g.,

$$K_p = p_A \cdot p_B / p_{AB}$$

where p_A is the partial pressure of element A in equilibrium with its parent compound AB.

The degree of dissociation thus increases with temperature and with a decrease in the total concentration of material present. The presence of some B from another source (e.g., a second salt) tends to reduce p_A and accordingly the number of atoms present. Reaction of B with another species in the flame (e.g., forming HB) can increase p_A and subsequently the intensity of the A emission.

Combination of A with another flame species (e.g., forming AO or AOH) reduces the signal intensity by reducing the number of free A atoms capable of being excited.

In some circumstances, the molecular species so formed in the flame are excited and the band emissions are used in quantitative studies. In most cases, however, molecular emissions are undesirable since they tend to mask the sharp, characteristic lines of free atoms.

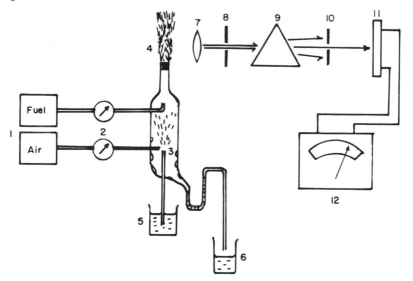

FIGURE 5.6   Emission system for flame spectrometry. 1. Cylinders of fuel and air or oxygen, 2. pressure regulating valves and flow meters, 3. spray chamber, 4. burner, 5. test solution, 6. spray chamber drainage system, 7. focusing lens, 8. entrance slit, 9. wavelength selector, 10. exit slit, 11. photoelectric detector, and 12. recording meter.

B. FLAME PHOTOMETRY

The technique in which the intensity of a characteristic radiation emanating from a flame is correlated with the concentration of that element in a solution is known as flame emission spectrometry or flame photometry.

The basic equipment required is shown in Fig. 5.6 and consists of the following components:

1. Pressure regulators and flow meters for the fuel gases.
2. An atomizing device associated with the flame source.
3. An optical system preceding appropriate photosensitive detectors.
4. An electrical circuit for measuring or recording the intensity of the radiation.

For the determination of easily excited elements, such as sodium and potassium, the optical system can be relatively simple (e.g., an interference filter preceding a photodetector) and these instruments are known as "flame photometers." In the more sophisticated flame spectrometers the optical system includes a prism or grating monochromator, and the electrical circuit includes a signal amplifier. The monochromator allows selected characteristic radiations of different elements to be successively focused on the exit slit. This permits multiple analyses and minimizes spectral interference effects. The detectors employed are either vacuum phototubes or photomultiplier tubes. The photomultiplier tube provides the maximum signal, and allows flame emission methods to be applied to systems in which the intensity of emissions is very weak, either because of a small concentration of the test element or because of difficulty in exciting any appreciable fraction of the test atoms.

Monographs dealing with flame photometry, and the "Handbook of Analytical Chemistry," list the wavelength and the emission sensitivity of representative groups of lines for different elements.

Since the sensitivity varies with the type of flame used, this information is included in the data tables. The sensitivity is usually recorded in terms of the units μg ml^{-1} (%T)$^{-1}$ and indicates the concentration of test solution (μg ml^{-1}) required to cause a 1% deflection of the recording meter. Using a specified line and flame type, the limit of detection for an element is about twice the sensitivity, while the minimum concentration required to give a full scale deflection is approximately one hundred times the sensitivity.

The tabulated data thus facilitate the selection of the experimental conditions required for a given study, since it nominates suitable flame types and wavelengths, and indicates the approximate concentration range required in the test solution. However, many operational variables affect the effective light flux; hence the final conditions for any new application have to be determined by experiment.

For example, measurements should be conducted with varying oxygen and fuel flows to ascertain the optimum ratio of oxygen to fuel. The optimum conditions are those which yield a maximum emission intensity and a steady background. The region of maximum emission in the flame should then be determined by moving the flame relative to the entrance slit.

To minimize interference from the emissions of other elements in the sample, the slit widths used should be kept as small as possible while still retaining reasonable response from the detector.

For quantitative determinations, the aim is to compare the intensity of a characteristic line or band with the observed intensities of the radiations emitted by standard solutions of the test element.

The first correction which is required is for the flame background. In one approach the instrument is adjusted to zero while aspirating pure solvent only. Alternatively, the reading obtained while aspirating pure solvent may be subtracted from the reading observed with the test sample. In a third method, the correction for the background is ascertained by measuring the recorder output when the wavelength dial is shifted slightly to one side or other of the characteristic peak.

The net emission intensities of the standard solutions are then used to prepare a calibration graph in which indicator response is plotted against the concentration of the element in the standards.

Because of the many interfering or competing effects which can arise in the flame, the composition of the standards must closely resemble the composition of the analyte.

As indicated in Section II,A, some interfering effects can be minimized by the addition of competing cations or anions or by the presence of a radiation buffer. The same additions must be added to any standards being prepared.

Since the exact composition of the test solution may vary from sample to sample, one recommended approach is the "standard-addition method" of calibration. In this method, a known amount of a salt of the element of interest is added to an aliquot of the test solution. The difference in reading caused by adding this salt is noted and by adding different amounts of standard to separate aliquots, a calibration curve can be prepared. If this curve is linear and passes through the origin, it yields reasonable values for the unknown sample, the unknown concentration being determined by comparing its emission intensity with the corresponding value on the graph.

Errors in quantitative determinations arising from variations in spray rates, changes in droplet size, fluctuations in gas pressures and variations in viscosity and surface tension can be minimized by using the "internal-standard method." In this technique a fixed quantity of some element which has spectral properties similar to the species of interest is added to samples and standards alike. The radiant energy emitted by the standard (S) and test element (T) upon excitation is then measured simultaneously by means of dual detectors (double beam instrument) or by scanning successively the two emission lines.

The ratio of the intensities observed for the two elements corresponds to the equation:

$$I_T/I_S = (\alpha_T C_x/\alpha_s C_s)(\exp -[\Delta E_T - \Delta E_S]/RT)$$

where C is the concentration of atoms in the vapor state and α is a proportionality constant.

If ΔE_T and ΔE_S (the energy changes associated with the emission transitions) are nearly equal, the intensity ratio is independent of temperature.

The ideal internal standard is, therefore, an element whose ΔE_S factor is similar to the unknown, and a plot of log (I_T/I_S) versus log (concentration of test element) should give a straight line.

The choice of a suitable elemental spectral line for use as an internal standard in flame photometry is rather limited because of the small numbers of lines excited. Lithium is frequently used, particularly when analyzing for sodium and potassium.

The determination of sodium and potassium constitutes the major application of flame emission spectroscopy, although methods for some fifty different elements have been published.

The types of materials which have been analyzed by this method include agronomic materials, biological fluids and tissues, cement, glasses, metallurgical products, petroleum products, and waters.

In some of these materials the concentration of the element is so small that a prior concentration step (e.g., sorption on an exchange resin) is necessary. For other materials the interference effects of the matrix are eliminated by a preliminary separation procedure. Thus solvent extraction into an organic phase can be used to isolate and/or concentrate the test element. The presence of the organic phase also tends to enhance the intensity of the emission in the flame.

Most metals, however, can be determined in a single solution without preliminary chemical separations, and a sequence of flame analyses takes much less time than the corresponding chemical procedures. The amount of material required is small, and with efficient calibration an accuracy of $\pm 2\%$ of content can be achieved.

Where interference effects are observed the initial establishment of a new method may take some time, although the problems encountered are usually fewer than in other spectrochemical emission methods.

Some workers have sprayed powdered solid samples directly into an atomizer-burner flame, but most of the standard procedures use liquid samples. Thus unless the sample is a liquid, preliminary dissolution procedures are a necessary adjunct to the technique.

While newer procedures such as atomic absorption spectroscopy may tend to reduce the number of applications of flame spectrometry, it is doubtful if a more convenient method of determining the alkali metals will be readily devised.

III. EMISSION SPECTROMETRY

A. Excitation by Electrical Arcs and Sparks

Electrical discharges have proved very effective for volatilizing and exciting solid samples. They are usually subdivided into two major types—arcs or sparks.

An "arc" is an electrical discharge which has to be initiated by momentary mechanical connection across the electrode gap. (The discharge can also be started by means of an auxiliary spark.) It is a more energetic source than the flame, being capable of exciting over seventy elements. Arc discharges give emission spectra which contain relatively few lines due to high energy transitions or ionic species, but because of the temperature developed at the electrodes, very minute amounts of material are volatilized and excited. The arc is, therefore, highly suitable for detecting minute traces of materials and for qualitative analysis.

The highest energy of excitation is achieved in a "spark" source, by means of which elements difficult to excite, such as the noble gases and the halogens, can be so energized that they emit their spectra. Spark discharges jump electrode gaps unassisted and yield spectra rich in high energy lines. Spark sources have been widely used for the quantitative analysis of metals, alloys, minerals, rocks, and ores, a reproducibility of $\pm 1\%$ being claimed in many systems.

The mechanism of the "arc discharge" is complex and as yet is not fully understood. The bulk of the current maintaining the arc is carried by electrons from the cathode. These electrons are accelerated by the applied potential and reach the anode with sufficiently high velocities to make it incandescent. The accelerated electrons ionize vapor from the heated anode, and the resultant cations, plus those which may be produced in the arc column, migrate towards the cathode. Bombardment of the cathode by these particles raises the temperature of this electrode sufficiently to produce thermionic electrons.

When any new substance is introduced into the arc (e.g., by placing it in a cavity in the lower electrode) the high temperature causes the constituent elements to volatilize into the arc column (arcs operate at temperatures between 3000 and 8000° K). The atoms of these elements can become excited through collision with one of the several types of missiles present in the discharge. The missiles can be atoms, ions, molecules, or electrons, the velocity (hence energy) of these particles being inversely proportional to the square root of the particle mass.

Emission of characteristic elemental lines occurs when the energy of the colliding missiles exceeds the excitation energy required to cause electronic transitions. The amount of energy available in the dc arc depends largely on the temperature of the gas in the arc column, this temperature being determined mainly by the ionization potential of the element which most readily ejects an electron. The lower the minimum ionization potential, the lower is the arc temperature.

The current–voltage relationship for an arc is described by the equation,

$$V = A + B/I^x$$

where A and B are constants, I is current, V is voltage and x depends on the composition of the anode. The applied voltage is usually of the order of 100 V dc which yields a current of 1 to 50 amp between electrodes spaced 1 to 15 mm apart. Once initiated the dc arc is self-sustaining and provides a continuous light output. However, since the resistance decreases with increasing current, the current tends to increase without limit unless a ballast resistor is placed in series with the gap.

The arc usually discharges between the anode and a small spot on the cathode surface, and during operation it tends to wander from spot to spot. The current fluctuations resulting from these changes in arc length and temperature can be opposed, but not eliminated, by having an inductance in the electrical circuit. The emitted radiations thus tend to vary in intensity from moment to moment since any sample introduced into the arc is not vaporized or excited uniformly.

While the dc arc is capable of detecting very low concentrations of most metals, its application to quantitative analysis has been limited by its erratic behavior and poor reproducibility. These undesirable effects are less marked in some of the newer techniques such as gas sheathing, inert atmosphere excitation and low pressure stabilization. For example, in the inert gas–jet method, a

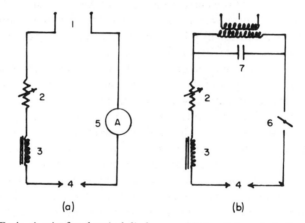

(a)   (b)

FIGURE 5.7   Basic circuits for electrical discharges. (a) Arc. (b) spark. 1. Power source—dc generator for arc and transformer for the spark circuit, 2. variable resistor, 3. inductance, 4. analytical gap, 5. ammeter, 6. spark control gap, and 7. condenser.

vertical current of gas is caused to flow upward along the sample electrode. This removes the outer flamy fringe of the discharge and greatly reduces self-absorption and flame background.

For quantitative studies, the measurement of the radiation intensity is integrated over a period of time, the timing being carefully controlled to ensure that selective volatilization does not cause spurious results.

The basic circuit for a dc arc is shown in Fig. 5.7(a).

In theory the reproducibility of excitation can be improved by alternating the electrode polarity. This can be achieved by using a high ac voltage (2–5 kV) as the energizing source in a more or less conventional arc circuit. In practice, the ac arc has not found wide acceptance.

To initiate a "spark discharge" a voltage of 10–30 kV is applied across the two electrodes which form the analytical gap, and to ensure a controllable high discharge current, a condenser is included in the circuit. The condenser is charged from the secondary winding of a high voltage transformer and charging continues until the voltage at which the analytical gap begins to conduct is reached. At this point the discharge commences and the energy stored in the condenser flows through the gap producing instantaneous currents of several hundred amperes. This current pulse takes material from the electrodes into the spark gap where the atoms are excited by collision with the high energy electrons.

By damping the discharge with a series resistor, the gap can be made non-conducting before the next half-cycle of the line voltage recharges the condenser. This simple spark circuit does not provide the uniform and reproducible results desired in quantitative spectrochemical analysis. Improved stability has been achieved by incorporating in the circuit an auxiliary spark gap (having uniform characteristics) whose role is to control the discharges. In the Feussner type circuit the auxiliary gap is closed by a rotating electrode driven by a synchronous motor. By this means the discharge across the analytical gap occurs only during the brief interval in which the auxiliary gap is shorted. Instead of a rotary gap, a thyratron circuit can be substituted.

In another type of source unit, a stream of air is blown across a fixed auxiliary gap. When the applied voltage exceeds the breakdown of this gap discharge can also occur across the analytical gap. The auxiliary spark is rapidly quenched by the stream of air which carries away the ionized particles, and several individual discharges may occur in each half-cycle of the input voltage.

Figure 5.7b indicates the basic components of a spark source. Most modern commercial spectroscopic source units usually combine a spark and a dc arc unit.

Discharge characteristics and vaporization processes depend somewhat on the atmosphere within the discharge gap. For this reason, many workers have examined the effect of replacing air with some inert atmosphere. Apart from reducing the spectral background by eliminating many of the molecular species which contribute greatly to this background (e.g., cyanogen bands), the accuracy and reproducibility of results is improved greatly.

B. Gaseous Discharges and Plasmas

Ordinary gases (at low pressures) are also conveniently excited by means of an electrical discharge. In one type of discharge tube, a high frequency alternating field is applied between a central insulated electrode and the outer wall of the gas container. Under the influence of the HF alternating field, free electrons in the gas acquire sufficient energy to excite and ionize gas molecules and eventually the ionization process becomes cumulative. This leads to a breakdown in the insulating properties of the gas and the production of a luminous discharge. At low frequencies the breakdown process is substantially the same as that found in dc discharges except that the alternating field leads to somewhat simpler spectra. Since there are no actual metal conductors in contact with the gas, the spectra obtained contain no lines arising from the metal atoms of the electrodes.

One of the earliest high frequency sources was the Tesla coil which produces a HF output of about 50 kV by means of a mechanical vibrator. In the ring discharge tube an inductance is wound on the glass or quartz tube which contains the gas sample. This inductance is then made part of a conventional condensed spark circuit. The strength of the discharge can be controlled by varying the size of the spark gap, and by progressively increasing the strength, the emitted spectra can be made to change from the molecular band type to the line spectra of neutral atoms and ionized atoms. In valve oscillator discharge units, continuous radiofrequency oscillations are maintained by having a triode valve in a tuned circuit.

In recent years, hollow cathode discharge tubes have become highly developed, interest in this type of excitation arising mainly from the popularity of the technique known as atomic absorption spectroscopy.

A hollow cathode tube consists of a small sealed chamber which contains a cathode in the form of a small metal tube. The tube is normally evacuated and then filled with a low pressure of rare gas (e.g., argon). By careful adjustment of the pressure the electrical discharge can be confined within the cathode. The material to be excited is either placed inside the cathode or the cathode is made from the material itself. When a voltage is applied across the electrodes, positive ions collide with the cathode and cause spluttering of the cathode material. This yields metal atoms, some of which are excited by the electrical discharge to yield resonance radiation. The intensity of this emission can be increased a hundredfold by introducing isolated auxiliary electrodes across which a potential of the order of 500 V can be applied for excitation purposes. The hollow cathode produces very narrow spectral lines and is very useful for high resolution work such as isotope analysis.

Another form of spectrometric excitation which is currently receiving much attention is the "plasma torch."

In the dc-arc plasma jet, a closed chamber contains at one end an anode and at the other end a cathode which has a small opening in it (the polarity of these electrodes is sometimes in the reverse order). Gaseous argon is introduced tangentially through the chamber wall, and swirls around the chamber before leaving by means of the hole in the electrode. When an arc is struck in the chamber the outer layers are cooled by the argon stream giving a "thermal pinch" effect which causes the arc column to contract. The resultant increase in current density gives a higher arc temperature. The pressure generated leads to the ejection of hot plasma through the electrode opening where it appears as a flamelike jet. At higher operating currents, the arc suffers a "magnetic pinch" effect due to the self-induced magnetic field.

In the high frequency plasma torch a stream of ionizable gas (argon) is passed through a circular quartz tube which is surrounded by a coil carrying high frequency alternating electric current. The gas is heated by induction and the kinetic energy imparted to the electrons by the electric field is shared with plasma atoms through collision. The temperature finally achieved in the plasma is much higher than that observed with any other source. With pure argon a temperature of about 16,000° K may be obtained; with samples in the discharge the temperature is lower, probably about 8000°.

The plasma torch is ignited by placing a carbon rod in the high frequency field. The field heats the carbon which in turn heats and ionizes the argon. Once the main discharge has started, the carbon rod is removed and the gas stream carries the ionized gas plasma down the tube away from the rod.

The plasma has the general appearance of a bright flame with three regions or zones. The small central core is centered just above the radiofrequency coil and is brilliant in color, but nontransparent. The larger second region surrounds the core, and extends upwards. It is bright but only slightly transparent. The third region or tail-flame extends up above this second region and when pure argon is being used is barely visible. The radiation from the core includes an intense continuum extending from 3000–5000Å. The intensity of this continuum decreases in the second zone and is negligible in the tail-flame. However, when the tail-flame extends outside the surrounding coolant tube emissions attributable to the band systems of O_2, N_2, NH, N_2^+, and OH are observed.

Systems to be excited are usually introduced into the plasma in the form of an aerosol although direct injection into the plasma of liquids (aqueous and organic), powdered solids, and slurries has also been carried out. The higher temperatures and the longer residence time is considered to lead to a greater degree of conversion of the aerosol into atoms and to greater dissociation of "stable" molecular compounds.

By studying the emissions emanating in the tail-flame, spectral background is minimal and the sensitivity of detection is increased. It has been suggested that the plasma torch provides the high degree of stability of the ac spark with the sensitivity of the dc arc.

One advantage of the plasma excitation is the virtual elimination of molecular bands attributable to metal oxides. This is particularly important when considering elements which form refractory oxides in air–fuel flames. However, it has been shown that the amount of oxide formation can be greatly reduced in many cases merely by exciting the aerosol in a luminous (fuel rich) air–acetylene flame. Another procedure, which is proving particularly useful in atomic absorption studies, uses a nonluminous nitrous oxide–acetylene flame. This flame is oxygen-free and is sufficiently hot to promote dissociation of most salts.

C. Emission Spectroscopy

The general terms, emission spectrometry or emission spectroscopy, are widely used to describe the technique in which the radiations emanating from an electrical discharge are dispersed and recorded.

The basic components of the equipment required for this technique are shown in Fig. 5.8.

Some of the emissions from the excitation source are isolated by a slit before being dispersed into distinct spectral lines by transmission through a prism or reflection from a ruled grating. The separated radiations are then detected by placing a photographic plate or film at the focal plane of the optical system. Each characteristic radiation produces an image of the slit at some particular point on the focal plane.

The quality of the dispersion unit is usually described in terms of the resolution and dispersion of the system. The resolution of an instrument indicates the wavelength separation ($\Delta\lambda$) of two lines which are just distinguishable as two.

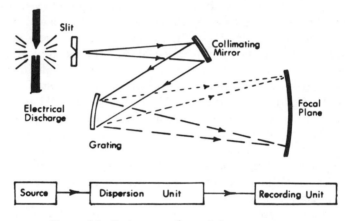

FIGURE 5.8   Basic system for emission spectrometry.

It is defined by $\lambda/\Delta\lambda$ where λ is the average wavelength of the two lines. The degree of spreading of the spectral lines over the focal plane is a function of the dispersive power of the instrument. Dispersion is usually expressed as A/mm and is obtained by measuring the wavelength separation of two lines in A ($\Delta\lambda$) and dividing by the linear separation in mm (Δd), i.e., dispersion equals $\Delta\lambda/\Delta d$.

Instruments are classified in terms of the type of dispersive element used, the optical mounting of the spectrometer components, and the manner in which the spectra are recorded.

The principal methods of recording the spectra are photographic and photoelectric. In photographic recording a photographic emulsion is placed at the focal plane and images of the slit corresponding to all the dispersed radiations are recorded simultaneously. In photoelectric recording, narrow exit slits are placed at selected positions on the focal plane. The radiations transmitted by these slits (usually characteristic of some particular element) are allowed to fall on to sensitive photodetection tubes. The output of these tubes is amplified and fed to some form of integrator before final recording on a meter or chart.

The electrodes used in emission spectroscopy are generally made of graphite although metal samples may be cast into discs or rods and used directly. It is quite common practice to use some form of pointed upper electrode, but each laboratory appears to develop its own preferred shape and size for the lower electrode. The simplest form has a small cavity drilled into the top of a one-fourth inch graphite rod.

Since this technique is very sensitive and uses very small samples (mg) it is often difficult to ensure that the sample used is truly representative and great care is required to prevent contamination.

Powders are a common sample form and are extensively used for both qualitative and quantitative analysis. Finely ground material ($\simeq$100 mesh) can be placed directly in the cavity of a graphite electrode. Premixing with powdered graphite improves the stability of the discharge, and with spark excitation preliminary conversion of the graphite-sample mixture into a pellet is recommended. For some materials, the sample is first fused (e.g., with a mixture of boric oxide and lithium carbonate) before being ground and mixed with graphite powder. Another approach applies a thin film of powder to a moving band of adhesive tape which passes slowly between two metal electrodes.

Metal samples are commonly cast into discs of about $1\frac{1}{4}$ inch in diameter and the surface is machined to a flat, smooth finish. Samples of this type are ordinarily analyzed by a spark method. Small metal samples may be placed in the cavity of a graphite electrode, or the sample may be dissolved and treated as a solution.

There are several methods of handling solutions, the most widely used being the porous cup, vacuum cup, and rotating disc. In the vacuum cup a small hole in the lower electrode leads down to a liquid reservoir. The low pressure created by a spark discharge draws the solution up this hole and sprays it into the discharge zone. With a plasma torch the argon flow can be used to introduce samples with the aid of a standard type of atomizer. The rotating disc method makes use of a wheel-shaped graphite electrode which dips into a reservoir of the solution. The electrode is turned slowly in the vertical plane, thus bringing fresh solution up into the spark discharge which is struck to the upper rim of the wheel.

The greatest advantages of emission spectroscopy are the ease, speed and accuracy with which qualitative analyses of a wide range of materials can be performed.

It is not necessary for the analyst to identify each line in the spectrum of an unknown sample; the identification of only a few of the most persistent or intense lines of each element is all that is necessary to establish the presence of an element beyond all reasonable doubt.

Wavelength tables which record the wavelength and relative intensity of the characteristic emissions of the various elements are included in most of the monographs devoted to this technique. The "Handbook of Analytical

Chemistry" contains a table which lists the most persistent lines of the elements and it provides an indication of the minimum concentration of element which has to be present in order to observe a particular line. Since many entries are shown to have a sensitivity of 0.0001%, the technique can be seen to be an extremely valuable means of locating small traces of impurity in materials.

The major problem in qualitative analysis is defining accurately the wavelength of the various lines observed. This can be resolved by photographing on the same emulsion the emission spectra of pure samples of the elements thought to be present in the unknown. Alternatively, the spectra of pure metals such as iron or copper may be recorded.

The wavelengths of characteristic groupings in these spectra (e.g., intense lines, doublets, and triplets) are then used to define the position of lines in the unknown spectra.

A widely adopted method for qualitative analysis utilizes the juxtaposition of two standard spectra with the spectrum of the sample. The two standard spectra used are the iron arc and a mixture known as "principal line powder." This is a mixture of about seventy elements present in such proportions that the powder emits the five most persistent lines of each element on excitation.

For quantitative analysis the intensity of characteristic lines is used as a measure of the amount of particular elements present. As indicated in Section I,B, the intensity of emission is a function of many factors besides the actual concentration; hence, it is essential that an internal standard be used to compensate for many of the experimental variables.

The internal standard must be present in a constant concentration and the standard line used should meet the energy requirements outlined in Section II,B. For metal alloys, a suitable line of the base metal is frequently selected as the internal standard.

Thus almost all quantitative determinations are based on a comparison of the intensities of specially selected pairs of lines known as homologous pairs. One line of the homologous pair is selected from the spectrum of the element being determined, the other line is taken from the spectrum of the internal standard.

Where photographic recording is being used the exposure time has to be carefully adjusted to suit the emulsion type and the nature of the emitted spectrum. Because of differences in component volatility, the period of excitation before exposure has to be closely controlled, and excitation conditions should be maintained as constant as possible. To obtain reproducible results developing procedures for the photographic films have also to be standardized.

The intensity of selected emissions can be compared in several ways. In one approach, a rotating step sector is placed in front of the entrance slit to the spectrometer. The number of steps observed on the record for each line provides an indication of the relative intensity of the emission. Alternatively, a logarithmic sector can be substituted for the step sector.

More quantitative data is obtained by measuring the density of the photographic image of selected lines with a microphotometer. The ratio of the intensities of the components of a homologous pair can then be recorded. A calibration graph which relates this intensity ratio to concentration can be readily prepared by exciting standard samples. The errors observed in quantitative studies using photographic recording are often of the order of ±5–10% of content, and considerable time can be required for each analysis.

With photoelectric recording instruments multiple determinations on a given sample may be determined in a matter of minutes with a precision of about ±2%. The output of the photomultiplier tube detectors is integrated by means of suitable capacitor-resistor circuits over a period of 25 to 40 sec. The intensity of the line of interest is measured by comparing the voltage developed across the capacitor with the voltage developed by the selected reference line. By means of calibration graphs the observed emission intensity ratio can be related to concentration. Most commercial instruments simultaneously isolate the characteristic lines of a number of elements so that multiple analysis on a sample devolves into a systematic recording of the intensity ratios for a series of homologous pairs. Accordingly, direct reading emission spectrographs using photoelectric recording are widely used in industry for routine analysis.

The accuracy of the quantitative determinations depends greatly on the quality of the calibration curve. Varying the chemical composition and the physical properties of the matrix material can have a significant effect on line intensity.

This matrix effect can be attributed to many causes. For example, the rate of volatilization of an element is usually altered by a change in chemical composition of the sample. A change in composition of the specimen also alters the composition of the arc gas and hence its effective ionization potential. This in turn influences the temperature of the arc and accordingly the intensity of emissions. The probability of excited atoms becoming involved in radiationless transformations through collision with other particles varies with the types of species in the arc gas. The greater the number of this type of collision, the weaker the emitted line becomes.

For these reasons, for accurate quantitative work it is essential that the general chemical composition and the physical properties of standards and unknowns should be the same or closely similar.

The effect of matrix variations can sometimes be offset by adding a substance termed a spectroscopic buffer to analysis specimens. The added buffer may act primarily by keeping the temperature of the source fairly uniform. Salts of the alkali metals, K, Rb, and Cs, are most effective for this purpose because their first ionization potentials are lower than those of any other element. Calcium has also proved to be an effective spectroscopic buffer for many materials.

The range of applications of emission spectrometry is extremely broad, the technique having been used successfully for the analysis of rocks, minerals, metal alloys, and commercial products of all types.

For qualitative analysis and identification of trace impurities, emission spectrometry is undoubtedly superior to most other procedures. For the quantitative determination of trace amounts of material the accuracy obtainable at least equals that of most other techniques. In addition, multiple analyses can be performed on milligram amounts of sample and many of the elements determined readily by this technique are difficult to analyze by conventional procedures.

IV. EXCITATION BY ACCELERATED
PARTICLES OR HIGH ENERGY RADIATION

A. X-Ray Emission

As indicated in Section I, collision of high speed electrons with a solid target can result in the emission of X-radiations.

In chemical analysis, this phenomenon is used in two ways. In standard X-ray emission spectroscopy (see Section III,B) the target is made of a metal, such as tungsten, which emits high energy radiations. The radiations from this primary tube are then used to excite emissions from an analytical sample mounted in a spectrometer unit. In the alternative procedure the sample is used as the target in an X-ray tube.

The basic components of an X-ray generator are shown in Fig. 5.9.

Electrons from an incandescent filament, heated by a low voltage source, are accelerated in a vacuum towards a metal target by an applied potential of many kilovolts. The minimum voltage is fixed by the energy required to excite X-radiations. This increases with increasing atomic number and for the excitation of the K series varies from 1.3 kV for magnesium to 39 kV for lanthanum. The intensity of the characteristic emissions is related to the applied voltage; hence spectrographic tubes are usually operated at voltages up to 60 kV, and units operating at even higher voltages are available.

The heat generated in the target has to be removed continually and this is achieved by means of a water cooled support. Targets of high atomic number (tungsten, platinum, or gold) are preferred for most operations because the intensity of the continuum varies with the atomic number. When the X-ray tube is being used as a primary radiation source (cf. Fig. 5.9) it is also important to ensure that the characteristic emissions of the target do not overlap any of the secondary radiations emanating from the assay sample.

Windows (transparent to X rays) are provided in the evacuated envelope to permit transmission of the primary radiations.

A good example of the use of primary X rays for analysis is the ELECTRON-PROBE X-RAY MICROANALYZER. In this instrument the electrons emitted on heating a fine tungsten filament (0.004 inch diameter) are accelerated by potentials of 5 to 40 kV towards the analytical specimen which forms the target. The system is maintained under a high vacuum and en route to the sample the beam is directed through one or two magnetic lenses. These lenses consist of a small borehole in a spool which is surrounded by a strong magnetic field. By varying the applied magnetic field the electron beam can be focused on to a very small area (e.g., about one square micron) of the target. The emissions from this small target area are then dispersed and studied quantitatively to elucidate the composition of the sample in this area.

The intensity of the emitted X rays is a function of many variables, among which may be listed the magnitude of the accelerating potential, the fraction of the electrons absorbed, the depth of penetration, the atomic number of the elements being excited, the nature of the matrix, etc. In some analyses the intensity of characteristic emissions varies linearly with composition, and results good to about 3% of the amount present can be obtained. In other studies, several series of standards are required to prepare suitable calibration curves.

The probe provides a unique method of analysis, because it can pinpoint a very small area of a sample for study and simultaneously determine several elements within this area.

The technique has been applied in many fields. For example, in the metallurgical field it has been used to study the nature of the species present in grain boundaries and the composition of corrosion pits; geologists have used it to examine minute inclusions in minerals; and its application to the analysis of biological tissue has been reported. While the equipment is expensive to purchase, the determination of an elemental composition distribution pattern is sufficiently essential in many research fields to warrant the investment.

On exposure to high energy X rays, the component elements of solid analytical samples all tend to emit their own characteristic radiations. Each element emits the same frequencies as it would if it were made the target in a separate X-ray tube, although energy considerations dictate that the secondary X rays must have less energy (that is, longer wavelength) than the exciting source.

This characteristic behavior of solids exposed to high energy radiation has been exploited in the technique known as X-ray emission spectroscopy or X-ray fluorescence.

B. X-Ray Emission Spectroscopy

A study of the frequency and intensity of the secondary X-ray emissions provides both qualitative and quantitative data on the sample without being destructive. In most commercial units, however, detection is limited to elements having an atomic number greater than twelve.

The basic components of an X-ray spectrometer are shown in Fig. 5.9.

The instrument consists of a high intensity X-ray tube, a sample chamber, a goniometer with collimator and analyzer crystal and a radiation detector with its associated electronics. There are various designs. The most common type uses a flat crystal analyzer and the radiation beam is collimated by means of a series of parallel plates. Other models use a curved or focusing crystal.

Crystals are capable of dispersing X-radiations when the spacings between the planes of the atoms in the crystal are of the correct order of magnitude for diffraction of this type of radiation.

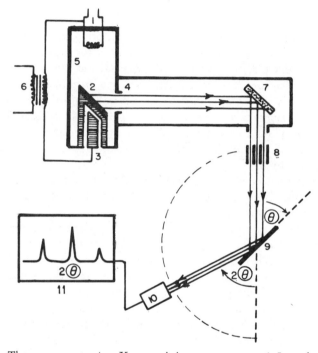

FIGURE 5.9   The components of an X-ray emission spectrometer. 1. Incandescent filament heated by a source of low voltage, 2. Metal target, 3. Water cooled support for the target, 4. X-ray tube window, 5. Evacuated envelope, 6. High tension transformer supplying 20–100 kV, 7. Sample for analysis, 8. Collimator slits, 9. Flat analyzing crystal, 10. Radiation detector unit, and 11. Count meter or chart recorder.

Analyzer crystals must have $2d$ spacing values which satisfy the Bragg equation

$$n\lambda = 2d \sin \theta$$

without exceeding the maximum 2θ angle of the goniometer. In practice the dispersed radiations can usually be detected over a range of angles for θ varying from 4° to 75°.

In the Bragg equation n is an integer, λ is the wavelength of the incident radiation, d is the distance between the reflecting planes in the crystal, and θ is the angle made by the incident beam with the direction of the reflecting plane. For instruments in which θ ranges from 5° to 75°, the minimum wavelength which may be conveniently detected is equal to $0.15d$ while the maximum wavelength detectable is $1.90d$. Thus in order to disperse radiations varying in wavelength from 0.2 to 10 Å, several different crystals are needed. For $\lambda < 3$ Å, important analyzer crystal materials are topaz ($d = 1.356$ Å) and lithium fluoride ($d = 2.015$ Å). Compounds suitable for dispersing longer wavelengths include quartz ($d = 3.135$ Å), ammonium dihydrogen phosphate ($d = 5.324$ Å) and gypsum ($d = 7.66$ Å).

The flat analyzing crystals are rotated about a central axis at one half the speed of rotation of the detector unit.

The detector unit may consist of a Geiger counter, a proportional counter, or a scintillation counter. Detector tubes have different relative efficiencies for detecting radiations of different wavelengths, and the choice can be influenced by the type of samples to be studied. Geiger tubes have a nonlinear response at moderate X-ray intensities and the pulse output is independent of the energy of the incident radiation. On the other hand they are simple to operate and hence tend to be used in routine analysis where low counting rates and physical discrimination of radiation are adequate.

The spectral sensitivities of proportion detectors are similar to Geiger tubes but they possess the advantages of high counting rates, and output pulse voltages which are proportional to the energy of the incident X rays. This permits electronic discrimination of any mixed signals. For wavelengths longer than 2 Å flow-proportional counters are used. The windows of these detectors, which have to be transparent to long wavelength radiations, tend to be somewhat porous; hence the counting gas has to be continually replenished by being passed in a continuous stream through the detector. These counters are suitable for the determination of all elements of atomic number 24 or below.

For the determination of elements of atomic number 25 or above scintillation counters are preferred. They have a high spectral response for X rays of wavelengths less than 2 Å and the output pulse voltage is proportional to the energy of the incident beam.

In order to realize maximum intensity in the recorded spectrum, it is desirable to irradiate as large a sample area as possible, the sample being as close to the primary source as convenient (usually about 1 inch distant). Metal samples have to be flat and smooth and since the intensity depends on surface smoothness, both standard and unknown must be subjected to the same surface treatment. Rotation of the sample during analysis tends to minimize variations in intensity due to preferential polishing and localized inhomogeneities. Metal drillings and scraps have to be compressed into flat wafers.

Powders may be compressed into briquets, formed into borax discs or packed into a cell, leaving a flat surface for the X rays to strike. Alternatively the powder can be mixed with a binder and spread over a flat surface.

Liquids may be evaporated on to a surface or placed in a cell with thin windows made of material transparent to X rays. The most widely used material for windows is Mylar, although other organic compounds such as Formvar or nitrocellulose have been used. Samples in the gaseous or vapor state require sample cells that have X-ray transparent windows capable of withstanding high pressure differentials.

General qualitative analysis of an unknown sample can be performed by automatically scanning the emission spectrum and recording the counter output on a strip chart calibrated in degrees 2θ. If the d spacing of the crystal is known, the Bragg equation can be solved for each observed peak. The elements present may then be identified by reference to a standard table of the principal K and L lines of the elements, arranged in order of increasing atomic number.

For quantitative analysis the major problem is correlation of the measured peak intensity with the amount of the element present in the sample. The sample matrix absorbs both primary and secondary X rays and for an infinitely thick sample the observed intensity (I_A) is related to concentration (Weight fraction W_A) by an expression such as

$$I_A = KW_A/\mu_1 + \mu_2$$

where K is a constant, μ_1 is the linear absorption coefficient of the sample for the incident primary radiation and μ_2 is the linear absorption coefficient of the sample for the secondary emissions.

The magnitude of μ varies with wavelength, and at any given wavelength μ equals the summation of the linear absorption coefficients of each element times its weight fraction.

$$\mu_{\text{sample}} = \sum^{i} \mu_i W_i$$

These two equations show that the intensity of a spectral line depends on both the concentration of the element being determined and the overall sample composition.

The complications arising from matrix absorption can be enhanced by another effect known as multiple excitation. If perchance the radiation from one element in the sample is of shorter wavelength than an absorption edge of a second element in the sample, it can be absorbed and cause the second element to radiate. This increases the intensity of the radiation attributable to the second element and decreases the intensity of the emission from the first element.

With powdered samples matrix effects have also been observed to vary with particle size. This effect can be minimized by using finely ground material (<300–400 mesh) of uniform size.

Five general procedures are used to correct for matrix effects.

In the "comparison standard" method the measured intensity (I_u) of a given line of the unknown is compared with the intensity (I_s) of that same line emitted from a standard sample of known composition. When the unknowns and standard are very close in composition, calculations can be based on the simple ratio $I_u/I_s = C_u/C_s$. However where the unknowns vary over a range of concentrations, it is necessary to prepare calibration graphs which relate I_u/I_s to concentration. To prepare these graphs a series of samples of known concentration must be available.

In the "internal standard" method a known amount of an element which has an atomic number close to that of the unknown is added to an aliquot of the sample. The characteristic radiation of this reference element must be excited and absorbed in the same manner as the characteristic radiation of the element being determined.

If a linear relationship between line intensity and concentration can be assumed to hold over a limited concentration range, matrix effects can be minimized by using the "addition technique." In this procedure, analysis is accomplished by measuring the intensity of a characteristic line before and after the addition of a known amount of the element to the sample.

All dilute solutions have approximately the same absorption coefficient and in the "dilution technique" the sample is intimately mixed with an excess of some base. Dilution may be achieved by fusion with a flux (borax, carbonate or pyrosulfate) or by dissolution with a suitable solvent. A heavy absorber is sometimes added as well. In the diluted samples emission varies almost linearly with concentration.

In thin film type samples neither the primary nor the secondary X rays are strongly absorbed and matrix effects are minimal. Generally this method involves preliminary separation of the desired elements from the host compound. For example, metal ions may be collected on ion exchange paper which is subsequently excited in the spectrometer.

Chapter 6 of the "Handbook of Analytical Chemistry" contains tables which list the important analytical lines of the elements, the wavelengths of absorption edges, and the total mass absorption coefficients for different wavelengths of incident radiation.

Low energy (i.e., longer wavelength) X rays are readily absorbed by air. Thus in studying the emissions from elements having atomic numbers less than twenty, the air in the optical path has to be removed or replaced by a gas (e.g., helium) which is much less absorbing than air. Another problem associated with these longer wavelengths is the difficulty of finding crystals with large enough "d" spacings to disperse the radiations in the spectrometer.

X-ray methods are not destructive and the emission spectra are reasonably simple; hence, despite the limitations associated with the technique it is becoming increasingly popular for routine analysis. Among the list of applications may be included the analysis of heat resistant alloys, glass, high alloy steels and minerals. Detection limits vary from parts per million to parts per thousand, the determination of low percentages of constituents being limited by the relationship between the background radiation and the radiant power of the desired element.

For routine control purposes, X-ray emission spectrometry is being used for the analysis of major and minor constituents as well as trace elements. In favorable circumstances an accuracy of $\pm 0.5\%$ of the amount present can be achieved while in other systems the accuracy may be only $\pm 5\%$.

C. ATOMIC-FLUORESCENCE SPECTROPHOTOMETRY

Isolated metal atoms, such as are formed in the burner–atomizer unit of a flame photometer, are capable of absorbing radiation to reach the first excited state of the atom. On reverting to the ground state, light of the same wavelength as that absorbed is emitted.

The basic equipment for atomic fluorescence spectrometry is similar to that shown in Fig. 5.10 except that the sample cell is replaced by an atomizer–burner unit.

A fairly intense source of radiation of the requisite energy is placed opposite the flame, and the fluorescence emitted by atoms of the element, introduced into the flame as a fine mist, is measured at right angles to the incident beam of light. To eliminate interference due to thermal emission and flame background it is desirable to modulate the source and amplification system.

The intensity (F) of the observed signal can be increased by increasing the intensity of the exciting source, but with standard conditions F is proportional to the concentration of metal ions in the solution being sprayed into the flame.

The method is quite specific and extremely sensitive, being applicable to solutions containing less than 1 ppm of metal ion. The technique is comparatively new and its full range of applicability has yet to be determined.

D. FLUOROMETRIC ANALYSIS

X-ray emission spectrometry and atomic fluorescence can be regarded as special cases of the more general technique known as fluorimetry.

In fluorimetry, molecular species are excited through irradiation with visible or ultraviolet light. With particular types of compounds absorption is followed by the emission of characteristic radiations of longer wavelength than the exciting source.

Allowed electronic levels in a molecule are classified into types such as sigma (σ) and pi (π), sigma electrons being those present in the single bonds formed by the on-axis overlap of electron clouds while pi electrons are characteristic of aromatic systems and the multiple bonds formed by off-axis overlap of electron clouds. If the two electrons in each energy level have opposite spin directions, they are said to be paired and the system is in a "singlet" state. However, if the molecule contains two unpaired electrons, both having the same spin, it is said to be in a "triplet" state.

On excitation of a molecule a variety of electron-energy transitions are possible. For example, if one of a pair of pi electrons is excited to a higher pi level, designated as pi-star (π^*), the resulting state is a π,π^* singlet if no change of spin occurs or is a π,π^* triplet if the spin changes direction on excitation.

When a molecule in a π,π^* state reverts to the ground state, some of the energy is lost as characteristic radiation. Fluorescence is the light emission which arises from deactivation of a π,π^* singlet system. The lower probability of a change in spin causes a delay in light emission when the excited state is of the π,π^* triplet type, and the delayed emission is known as phosphorescence.

Molecules containing oxygen or nitrogen atoms also contain n type electrons, these being the nonbonded electrons associated with these atoms. When one of these n electrons is excited to a π^* level (giving an n,π^* state) other energy-dissipating processes compete successfully against the emission of light and no fluorescence occurs. The probability of these electrons being excited decreases markedly if the n electrons are utilized to form a bond, as for example, when the molecule combines with a metal ion to form a complex. On excitation the π,π^* state then becomes the more probable and the result can be strong fluorescence.

The types of substances which fluoresce are therefore mainly aromatic organic compounds (benzene, naphthalene, anthracene and their derivatives) or a metal-fluorogenic reagent complex. The fluorogenic reagents usually contain two or more aromatic rings joined by an unsaturated linkage containing either oxygen or nitrogen, and to facilitate complex formation the molecule may also contain sulfonate, hydroxyl, or amino groups.

The most popular sources for exciting fluorescence are special fluorescent lighting tubes, mercury vapor lamps, and the xenon arc lamp. By means of filters or a monochromator specific bands of radiation may be isolated. These are fed into the sample which is usually present in solution form.

The fluorescent emissions from the sample are then isolated by means of a filter or second monochromator, the intensity being measured by a sensitive phototube. The design of the instrument generally ensures that exciting radiation does not reach the detector, and shutter systems are included to eliminate undue excitation and fatigue of the detector. A schematic representation of a fluorometric unit is shown in Fig. 5.10.

Selection of the best wavelength for excitation is governed by several considerations. It should be strongly emitted by the source and strongly absorbed by the sample. It should also be capable of easy separation (e.g., by filters) from the fluorescent emission. Most fluorescing solutions contain colloids which cause scattering of the incident light; hence the wavelength of the exciting radiation should vary from that of the secondary emission by at least 50 mμ.

Selection of the best wavelength for intensity measurement may be achieved by fixing the nature of the exciting source and subsequently observing the intensity of emission as a function of wavelength.

Fluorescence analysis is especially applicable to trace quantities of substances;

hence care must be taken to eliminate contamination of the samples. Common solvents often contain fluorescent substances; stoppers can contain extractable fluorescent materials; and grease or paper fibers can introduce interfering species. While most studies involve species in solution, some instruments are designed to study solid samples.

In every set of fluorometric determinations, the analyst is well advised to run a blank and at least two standards of known composition covering the range of

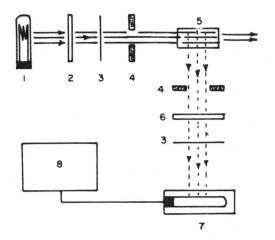

FIGURE 5.10   Schematic representation of a fluorometer. 1. Radiation source, 2. Excitation filter or monochromator, 3. Shutters, 4. Variable apertures, 5. Sample cell, 6. Emission filter or monochromator, 7. Photo tube detection unit, and 8. Recorder or meter.

concentrations expected. For very dilute solutions, subject to a low degree of light absorption, it can be shown that the relationship between the intensity of fluorescence (*F*) and concentration (*c*) should be nearly linear, i.e.,

$$F = \phi I_o abc = Kc$$

where ϕ is the quantum yield for fluorescence (ratio of light emitted to light absorbed), I_o is the intensity of the incident light, a is 2.303 times the absorptivity, and b is the length of the optical path. K is the proportionality constant observed using standard experimental conditions.

A number of assumptions are involved in the derivation of this relationship; hence it is advisable to check its validity for each new system through preparation of a calibration graph.

The main advantage of fluorescence methods is their high sensitivity, about 0.01 ppm being detectable in many systems. With good instruments a precision of $\pm 1\%$ can be achieved but in order to obtain an accuracy of similar magnitude very careful attention to detail is needed and all possible sources of error have to be eliminated. Sources of error are numerous. For example, if the fluorescent intensity is strongly pH dependent, careful buffering is necessary. The intensity of other systems has been found to vary with time due to photochemical destruction of the fluorescent molecule.

Only a limited number of elements and compounds exhibit definite fluorescence so the scope of the analytical technique is limited. Among the procedures which have been published are determinations involving trace elements in metal alloys and silicates. Of greater interest, however, are its applications in clinical medicine and food analysis.

Relevant information concerning a wide range of applications is listed in Chapter 6 of the "Handbook of Analytical Chemistry."

E. CHEMILUMINESCENCE

While not very widely used in chemical analysis, the phenomenon of chemiluminescence is worthy of brief mention. In this case, the source of excitation energy is the energy change associated with a chemical reaction.

A number of hydrazides of aromatic acids emit visible light when oxidized (e.g., with hydrogen peroxide) in the presence of free base. The intensity of these emissions can be markedly enhanced by the presence of a catalyst. For example, one of the most brilliant chemiluminescent materials is luminol (3-aminophthalhydrazide), for which hemin is a very efficient catalyst. In fact, the emission of light by this material has been used to identify blood stains.

The emission process is sensitive to pH, and the appearance (or disappearance) of light has been used to detect the equivalence point in acid-base titrations. To the test solution is added an excess of oxidant, a small amount of catalyst, and a chemiluminescent material (e.g., luminol, lophine, lucigenin). The solution is then titrated in the dark with standard acid or base until a change in the emission state occurs. In other applications of these chemiluminescent indicators, the pH is fixed at a suitable value and an oxidant solution is used as the titrant.

The detection of the equivalence point through observation of a dark–light transition is particularly useful when dealing with colored or turbid solutions, and it provides an alternative to instrumental methods of end-point detection for these systems.

Chemiluminescence may be regarded as a special type of phosphorescence. For example, with lucigenin and luminol, the primary step in the mechanism is envisaged as the formation of a transannular peroxide. The triplet state or magnetic field associated with the oxygen is then considered to facilitate singlet–triplet conversion within the molecule. When the intermediate decomposes the remaining molecule is left in an excited unstable triplet state and reversion to the ground singlet state is accompanied by the emission of light.

V. RADIOCHEMICAL PROCEDURES

A. RADIOACTIVITY

The phenomenon of radioactivity results from the transformation of the nucleus of an atom into another nucleus.

With the exception of naturally occurring radioactive nuclei, atomic nuclei are generally quite stable. However, on bombardment with high energy particles, some of the particles can be captured to yield unstable nuclides. The excited nuclei regain stability through the emission of one or more of several types of radiation. These emissions can involve the release of alpha or beta particles, positrons, neutrons, X rays and/or γ rays.

An alpha particle is the nucleus of the helium atom, having a mass of four atomic units and a positive charge of two units. Owing to their large mass these particles have a range of only a few centimeters in air and a fraction of a millimeter in solid samples.

Beta particles are electrons possessing a continuous spectrum of energies, the peak energy being sufficient to penetrate about 1 cm of aluminum. Due to their negative charge and small mass they are easily scattered or rejected.

Positrons are beta particles of unit positive charge which decay to yield two photons of definite energy content.

Sometimes the excess energy associated with a disintegration process is released directly as photons in the form of gamma rays. In other systems, a nuclear proton captures a K orbital electron to form a neutron. This electron capture is accompanied by the emission of X rays. Gamma rays and X rays have no charge or mass, and they are very penetrating.

Neutrons also possess very high powers of penetration, the particles having zero charge but unit mass.

The particles and radiations emitted from different radioactive nuclei vary widely in their energy content and in the frequency with which they are produced. Both of these properties are characteristic of the particular isotope which is disintegrating.

The energy of nuclear radiation is measured in electron volts, eV; an electron volt is the energy acquired by an electron accelerated through a potential difference of one volt. The energy of alpha particles varies from 3 to 8 million electron volts (MeV) while the energy of the photons emitted by positrons equals 0.51 MeV.

The rate of decay of radioactive material has proved to be proportional to the number (N) of excited atoms present, i.e.,

$$-dN/dt = \lambda N$$

λ is the proportionality or decay constant which can be shown to equal $0.693/T$, T being the half-life of the material.

The half-life is a characteristic property of each active isotope. It is the time required for any given sample of the isotope to be reduced to one half of its initial quantity, and it varies in magnitude from millions of years to millionths of a second.

A more useful form of the decay law is

$$A = A_0 \exp\left(-0.693 t / T\right)$$

where A is the activity after time t and A_0 is the initial activity.

The relevant properties of important radio nuclides are tabulated in the "Handbook of Analytical Chemistry." These tables list the half-life of each isotope together with an indication of the mode of decay and the energies of the emissions.

The emissions can be detected in several ways, the simplest being the darkening of a photographic emulsion. For quantitative studies the principal methods are based on the measurement of either the degree of ionization created in a gas tube (Geiger tubes or proportional counters) or the amount of visible light produced from a phosphor by the radiation (scintillation counters).

Radioactive decay is a random process; hence the counters deliver their information as irregular pulses. These pulses have to be counted by auxiliary electronic equipment. Being a random process, the error associated with the measurement decreases as the magnitude of the count (e.g., counts per second, n) increases and results may generally be expressed as $n \pm \sqrt{n}$.

As the emitted particles can differ in energy and have different effects on a detector tube, it is necessary to determine the efficiency of the tube for various types of applications. The difference in energy content can be used to advantage in separating mixed pulses. For example, because α particles possess a high ionizing power and β and γ emissions low ionizing power, alpha counters are generally insensitive to moderate amounts of β and γ radiation. This permits the study of alpha emitters in the presence of other species decaying by alternative modes. Alpha radiation can be prevented from entering a beta counter by interposing a thin metal foil. Gamma counters can be protected from beta radiation by interposing a plate of aluminum, beryllium or plastic.

The intensity of the signal from proportional and scintillation counters is related to the energy of the incident radiation; hence a more elegant way of studying the emissions from a multiple source is to use an electronic pulse height analyzer. This apparatus sorts out the radiations according to their energy content.

In a mixture of independently decaying radionuclides, the total activity A is given by

$$A = A' + A'' + \cdots$$
$$= A_o{'} \exp\left(-0.693\, t / T'\right) + A_o{''} \exp\left(-0.693\, t / T''\right) + \cdots,$$

where A', A'' represent the activity of the separate species and A_o', A_o'' are the respective initial activities. T', T'' are the respective half-lives.

Provided the half-lives are sufficiently different the decay of the mixture can be resolved graphically or analytically by simultaneous equations.

The number of disintegrations that a radionuclide emits in a given time is a measure of its activity. The basic unit is the curie or the quantity of radioactivity equivalent to 3.70×10^{10} disintegrations per second (d/s). In tracer applications it is more common to use quantities of activity that are decimal fractions of a curie, e.g., 1 millicurie (mC) or 1 microcurie (μC).

If a sample contains a single radioactive nuclide only, a plot of log (dA/dt) versus t yields a straight line. The half-life determined from this plot provides a useful means of identifying the radioactive species. (The half-life is the time required for log (dA/dt) to change by log 2 units.) When more than one radioactive material is present there is a departure from a straight line plot. In many cases the composite line can be resolved into a number of straight lines, each having a slope characteristic of the particular radioactive nuclide present. Figure 5.11 is an idealized representation of the decay curve for a mixture.

B.   ACTIVATION ANALYSIS

A great many elements become radioactive when bombarded by energetic particles such as protons, deuterons, alpha particles or neutrons, and measurement of the activity of a sample after bombardment provides a means of quantitatively determining the composition of the material.

The majority of analytical studies have used the method of neutron activation analysis but the same basic principles apply if other particles are used. On bombardment by neutrons possessing thermal velocities, atomic nuclei can capture neutrons to give larger nuclei. In many cases these new nuclei are unstable and spontaneously decompose with the emission of a particle or a gamma ray, e.g.,

$$_{11}Na^{23} + _{0}n^{1} \rightarrow _{11}Na^{24} + \gamma$$

At least 70 of the elements can be activated by thermal neutron capture reactions. The active isotopes formed in this way differ widely in half-life values and in many instances can be identified by determination of this constant, along with other pertinent information such as the mode of decay, energy of the emissions, etc.

The activity produced by a given element can be calculated from the equation

$$A = N\sigma f[1 - \exp(-\lambda t)]$$

where A is the observed radioactivity (d/s), N is the number of target atoms in the sample, σ is the atomic activation cross section for the nuclear particle reaction (10^{-24} cm^2), f is the nuclear particle flux (particles cm^{-2} sec^{-1}), and λ is the decay constant of the radioisotope produced.

For quantitative determinations, it is often more convenient to replace N by WN_0/M where W is the weight of the element present (g), N_0 is Avogadro's number (6.02×10^{23}), and M is the atomic weight. Since detectors are not fully efficient, for comparative results a further term ψ should be added to the equation, ψ indicating the efficiency of the detector system.

The term $[1 - \exp(-\lambda t)]$ is sometimes called the growth factor. It has a value of 0.5 when the time of irradiation equals the half-life of the nuclide

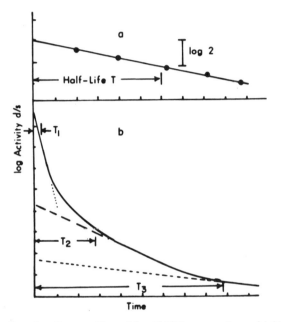

FIGURE 5.11   Radioactive Decay. Plot of log (dA/dt) versus time. (a) Single radioactive species and (b) mixture of three radiation emitters.

formed. When the irradiation time is six half-lives or greater this factor approaches its maximum value of unity.

It can be seen from the above equation that for a given mass of substance (W fixed) the intensity will be greatest when the flux and activation cross section are both high. The sensitivity of analysis by means of neutron activation thus depends upon the intensity of the source, the ability of the element to capture neutrons (σ), the atomic weight of the element, the half-life of the nuclide formed, and the character of the radiation emitted since this influences detector efficiency.

A nuclear reactor can produce a flux as great as 10^{13} n cm^{-2} sec^{-1} and this is sufficient to activate nearly all elements heavier than neon, but with greatly differing sensitivities. An ^{124}SbBe source which produces a flux of 10^3 to 10^4

n cm^{-2} sec^{-1} activates about twenty elements while a flux of about 10^2 n cm^{-2} sec^{-1} activates only Rh, Ag, In, Ir, and Dy.

With a nuclear pile as the neutron source, as little as 10^{-10} gm of some elements can be detected. The "Handbook of Analytical Chemistry" contains a list of neutron activation analysis sensitivity limits covering most elements.

Where more than one element is activated by irradiation, the components have to be identified from the decay curve as indicated in Section III,A, or have to be isolated by chemical separation procedures. Dependence on chemical separations is rapidly being replaced by new instrumental procedures, such as γ-spectrometry.

As indicated in the preceding section, emissions differ in energy content, and instruments (known as spectrometers) have been designed which separate groups of pulses on the basis of the energy content.

With mixtures of alpha emitting nuclides, selectivity is achieved by using a gridded ionization chamber or a semiconductor detector. For other types of radiation some form of scintillation spectrometer is usually used. In a scintillation unit, the impact of radiation on an organic phosphor, special liquid, or a selected crystal (usually sodium iodide containing a trace of thallium) results in the emission of flashes of light corresponding to the impact radiation. The light pulses are converted to electrical pulses by means of a photomultiplier tube, the intensity of the signal varying with the energy of the incident radiation. A pulse height analyzer sorts out the signals of different intensity, to yield an energy spectrum. The photopeaks in this spectrum can then be used for qualitative and quantitative analysis. Since many radionuclides decay with the emission of gamma radiation, a large number of determinations are based on gamma-ray scintillation spectrometry. Detectors which are sensitive only to the γ radiation appropriate to the element of interest are now available.

Since it is difficult to determine accurately all the factors which influence the intensity of radiation emission, the weight of a given element present is usually determined by a comparison technique.

In this technique a known amount of the element to be determined is bombarded simultaneously with the test sample and is processed after its activation in the same manner as the test sample. Since all the variables except W and A are the same for both the comparator and the test sample, a simplified equation can be used in subsequent calculations, namely,

W in test sample/W in comparator = A in test sample/A in comparator.

It is common practice to use samples ranging from 1 to 100 g in magnitude but much smaller samples can be employed when necessary.

On being irradiated with neutrons, a nucleus can emit some radiation during the deexciting period in which the new radioactive nuclide is formed. This radiation, which may be a particle or photon, is referred to as prompt radiation. Monitoring the prompt radiation emitted by a sample provides another means of analysis.

Neutron activation analysis has been used to determine a very large number of elements in a wide variety of sample materials. Many of the methods are currently being applied routinely to assay samples from all branches of science, and the scope is being enhanced by the availability of neutron generators designed specifically for activation analysis.

The major applications can be divided into three groups:

1. Trace analysis, that is, procedures concerned with measuring elements present at <0.001%. The technique is more sensitive than conventional methods in this range.
2. Minor constituent analysis (0.001–1%). With species having half-lives of the order of minutes to hours, the methods for some elements can compete in cost and simplicity with conventional techniques.
3. Fast automatic analysis where chemical separation is avoided through the use of sophisticated instrumentation. The automated procedure has been adapted to special routine analyses (e.g., oxygen in steel determinations) and the analysis of process streams.

Apart from the usual precautions required in handling radioactive material, activation analysis is subject to a number of limitations including self-absorption effects and sample destruction.

A few elements contain naturally radioactive isotopes and since some have long half-lives the radioactivity is relative to the weight of the element present in the sample. Direct measurement of the alpha or gamma emission can thus be used as a quantitative method for these elements. Elements which have been determined in this way include francium, lutecium, potassium, rhenium, rubidium, samarium, thorium, and uranium.

C. Radioactive Tracer Techniques

Much simpler measuring equipment is required when radioisotopes are used as "tracers" in chemical reactions. A radioactive isotope of an element behaves identically with the stable isotopes of the given element in all chemical processes. Thus if a small amount of radioactive isotope is mixed with an excess of the more stable elemental forms, it can be used to trace the behavior of the atoms throughout a whole series of simple or complex processes.

Some elements have one or more radioactive species of varying half-lives which can be used in tracer experiments. The selection of the most suitable isotope depends primarily on the duration of the experiment, since it is highly desirable that the activity observed at the end of the experiment should be at least ten times greater than any background signal.

In radiometric methods of analysis, an element is labeled with a radioisotope and the radioactivity is determined per unit weight. If this specific radioactivity remains virtually constant throughout the experiment, then the radioactivity observed after some chemical process is directly proportional to the quantity of the element involved. This procedure has been used to determine solubilities, to measure the extent of coprecipitation, to determine distribution coefficients and to observe adsorption processes. It has also been used in the determination of elemental species in a range of products.

In some chemical reactions, products are readily isolated in a pure state but with a poor yield. If a known amount of the same substance containing a small amount of an active isotope is added prior to the chemical treatment, the same fraction of the radioactive species is isolated as for the original sample. This is the basic principle of the technique known as isotope dilution.

By measuring the specific activity of the initial tracer material (counts min^{-1} mg^{-1}) and the activity of the isolated product, it is possible to calculate the amount of material (W) in the original sample.

$$W = g(A_0/A_r) - w$$

where g is the weight of the isolated product, w is the weight of tagged material added, A_0 is the amount of radioactivity added, and A_r is the amount of radioactivity recovered.

VI. REVISION QUESTIONS AND ASSIGNMENTS

1. After photographing with a uniform exposure the spectra emitted when solutions of (a) calcium chloride, (b) calcium chloride contaminated with phosphoric acid, and (c) sodium chloride were aspirated into a hot gas flame, a group of students mixed up the plates. They found that they had one photograph which showed a series of distinct lines with a very intense doublet near the 590 mμ mark on the wavelength guide scale; a second photograph showed some distinct lines and some broad bands of lesser intensity; and the third photograph had fainter lines and bands.

   a. How should the students label the three photographs? Give reasons.
   b. Why do two of the photographs show distinct bands?
   c. What electronic transition is responsible for the observed doublet?
   d. In what region of the spectrum would one expect to find the most intense lines and/or bands in the calcium studies?
   e. If some lanthanum nitrate was added to each of the three solutions, what effect, if any could be predicted?
   f. Explain why the addition of some potassium chloride to the calcium solutions was found to reduce the intensity of the calcium spectrum.
   g. If the students had repeated the experiment using (i) a much hotter flame, and (ii) an acetylene-nitrous oxide flame, what effect might this have had on the nature and intensity of the photographs?

    h. Which of the observed emissions are recommended for use in
       quantitative studies?
    i. Reference books list the sensitivity attainable in the determination of
       different elements by flame photometry. If it can be assumed that
       this-sensitivity limit corresponds to a 1% deflection on the instru-
       ment meter, what range of concentrations should be covered in the
       preparation of the calibration graphs for the quantitative determina-
       tion of sodium and calcium?
    j. Using the calculated concentration ranges, the students found that
       the calibration curve was near linear in the segment covering the
       most dilute solutions but developed distinct curvature as the concen-
       tration was increased towards the higher values. How might this
       be explained?
2. a. In emission spectrometry and X-ray emission spectrometry the
       intensity of characteristic radiations is measured by integrating over
       a period of time. In flame photometry and fluorometric analysis
       instantaneous readings are taken. Why is integration so desirable
       in the first two techniques?
    b. It is stated that the standards used in all the above techniques must
       be as similar as possible to the unknown sample. Why is this so
       important?
    c. Arc discharges are recommended for qualitative analysis, spark
       discharges for quantitative studies. Why?
3. Write brief notes to describe the following terms.
    a. Radiation buffer
    b. The multiplicity of a transition
    c. Doppler broadening
    d. Self-absorption
    e. Hollow-cathode tube
    f. Dispersion of a spectrograph
    g. Bragg's equation
    h. Internal standards
    i. Multiple excitation in X-ray spectrometry
    j. Phosphorescence.
4. To accelerate the routine analysis of their alloy steel production
   section, a large company invested in an X-ray emission spectrometer
   and a direct reading vacuum emission spectrograph. The steels
   produced contained C ($<$1%), Mn (0.5–2%), P ($<$0.1%), S ($<$0.1%),
   Si ($<$1%), Cu ($<$1%), Ni (2–10%), Cr (1–20%), Mo ($<$1%),
   Ti ($<$1%), and V ($<$1%).
     Before delivery of the instruments the process control chemist was
   asked to indicate the wavelengths of the emissions he considered should
   be measured quantitatively. The X-ray spectrometer was to be fitted

with two analyzing crystals (e.g., topaz and NaCl) but restricted to studying elements of atomic number greater than 20. The vacuum emission spectrograph was to be used for the determination of the elements present in amounts of $<1\%$, iron lines being used as the internal standards.

Using the data available in the "Handbook of Analytical Chemistry" and the wavelength tables available in emission spectroscopy monographs, prepare a list of the conditions you would choose if faced with the same task. For the X-ray instrument tabulate the element, nature of the line, wavelength, crystal and 2θ value. The 2θ values for the different elements should be as widely separated as possible. If overlapping occurs, suggest a suitable replacement for one of the crystals.

For the vacuum direct-reader, list two possible emission lines for each element and indicate iron lines which might be suitable for pairing with several elemental emissions.

Write brief notes on some of the problems the process chemist might encounter after installation of the instruments, e.g., curved calibration graphs, different types of response with different types of steel, and overlapping background emissions.

5. Write an essay on the analytical uses of radioisotopes.

6. Atomic fluorescence spectroscopy is a comparatively recent technique, and its scope is still being assessed. Using chemical abstracts and any relevant review articles, prepare a list of the types of applications currently being advocated and comment on possible extensions in the future.

7. a. Describe the components of an X-ray generator and an emission spectrometer.

   b. Prepare a brief description of each of the major types of radiation detectors viz., photomultiplier tubes, Geiger counters, proportional counters, scintillation counters, neutron detectors, etc.

   c. Describe the precautions and procedures associated with the use of photographic emulsions for recording emission spectra. How can these records be used in quantitative studies?

   d. Outline the principles of neutron activation analysis.

8. A technical salesman employed by the manufacturer of a neutron source capable of providing a flux of 10^8 neutrons cm^{-2} sec^{-1} has the task of convincing a laboratory manager that he should introduce neutron activation as an additional technique. The large and well-equipped laboratory services an organization which produces a wide range of metals and chemical compounds of ultra-high purity. What arguments could the salesman make and how could these be countered by the laboratory manager?

9. a. A public analyst, specializing in food and drug analyses, is persuaded to purchase a high quality fluorometer. By referring to the

"Handbook of Analytical Chemistry," relevant reviews, and/or other source books, prepare a report which indicates the type of problems which could be solved by the public analyst using this instrument.

b. Critically evaluate the advantages and limitations of fluorometric methods of analysis.

10. "The emission of electromagnetic radiation can be induced by exposing chemical species to a source of energy capable of causing electronic transitions in some or all of the component atoms. The energy can be in the form of heat, high energy particles or high frequency radiation." Discuss this statement.

References

American Society for Testing and Materials, *Methods for Emission Spectrochemical Analyses*, 4th Ed., Am. Soc. Testing Mater., Philadelphia, 1964.

F. J. Welcher (ed.), *Standard Methods of Chemical Analysis*, 6th Ed., Vol. IIIA, Van Nostrand, Princeton, New Jersey, 1966.

FLAME PHOTOMETRY

F. Burriel-Marti and J. Ramirez-Munoz, *Flame Photometry*, Elsevier, Amsterdam, 1964.

J. A. Dean, *Flame Photometry*, McGraw-Hill, New York, 1960.

R. Herrmann and C. Th. J. Alkemade, *Chemical Analysis by Flame Photometry*, English translation, Wiley (Interscience), New York, 1963.

R. Mavrodineanu and H. Boiteux, *Flame Emission Spectroscopy*, Wiley, New York, 1964.

E. Pungor, *Flame Photometry Theory*, Van Nostrand, Princeton, New Jersey, 1967.

EMISSION SPECTROSCOPY

L. H. Ahrens and S. R. Taylor, *Spectrochemical Analysis*, 2nd Ed., Addison-Wesley, Reading, Massachusetts, 1961.

W. R. Brode, *Chemical Spectroscopy*, 2nd Ed., Wiley, New York, 1952.

B. F. Scribner and M. Margoshes, *Treatise on Analytical Chemistry*, Part I, Vol. V. (I. M. Kolthoff and P. J. Elving, eds.),Wiley (Interscience), New York, 1964.

F. Twyman, *Metal Spectroscopy*, Griffin, London, 1951.

X-RAY EMISSION AND FLUORESCENCE

H. A. Liebhafsky, H. G. Pfeiffer, E. H. Winslow, and P. D. Zemany, *X-ray Absorption and Emission in Analytical Chemistry*, Wiley, New York, 1960.

L. S. Birks, *Electron Probe Microanalysis*, Wiley (Interscience), New York, 1963.

L. S. Birks, *X-ray Spectrochemical Analysis*, Wiley (Interscience), New York, 1959.

E. J. Bowen and F. Wokes, *Fluorescence of Solutions*, Longmans Green, New York, 1953.

J. A. Radley and J. Grant, *Fluorescence Analysis in Ultraviolet Light*, 4th Ed., Van Nostrand, Princeton, New Jersey, 1954.

R. Theisen, *Quantitative Electron Micro-probe Analysis*, Springer, Berlin, 1966.

S. Undenfriend, *Fluorescence Assay in Biology and Medicine*, Academic Press, New York, 1962.

RADIATION TECHNIQUES

H. J. M. Bowen and D. Gibbons, *Radioactivation Analysis*, Oxford Univ. Press, (Clarendon), 1963.
W. S. Lyon (ed.), *Guide to Activation Analysis*, Van Nostrand, Princeton, New Jersey, 1964.
R. T. Overman and H. M. Clark, *Radioisotope Techniques*, McGraw-Hill, London, 1960.
D. Taylor, *Neutron Irradiation and Activation Analysis*, Newnes, London, 1964.

6

THE ABSORPTION OF RADIATION

I. General Principles 121
    A. Absorption Spectra 121
    B. Quantitative Laws 126
II. Absorption by Atoms 131
    A. Absorption of X-Rays 131
    B. Atomic Absorption Spectroscopy 135
III. Molecular Electronic Transitions 138
    A. Origin of Spectra 138
    B. Ultraviolet Spectrometry 145
    C. Visible Spectrometry 147
    D. Special Techniques and Modified Procedures 151
IV. Infrared Absorption Spectrometry 153
    A. Vibrational Energy Changes 153
    B. Infrared Spectrometry 157
    C. Applications 163
V. Raman and Microwave Spectroscopy 167
    A. Raman Spectroscopy 167
    B. Microwave Spectroscopy 169
VI. Tutorial and Revision Questions 171
    References 174

I. GENERAL PRINCIPLES

A. ABSORPTION SPECTRA

As indicated in Chapter 5, Section I, atomic and molecular species can absorb quanta of energy, the resultant excited states reverting to the ground state in 10^{-10} to 10^{-14} sec.

The energy required for excitation can be provided by electromagnetic radiation since these wave motions have an energy content defined by the relationship

$$E = h\nu$$

where h is Planck's constant and ν is the frequency of the radiation.

With atomic species, the absorption of energy involves the transfer of orbital electrons to higher energy levels. The quantity of energy required to cause such an electronic transition is comparatively large and is only possessed by radiations having frequencies greater than 10^{14} cycles per second (e.g., visible and ultraviolet light). Radiation of similar energy content is required for electronic transitions within molecular species.

With molecular species, however, there are many more permitted excited states, since energy can also be consumed in changing the rotational and/or vibrational levels of the compound of interest. Changes in rotational levels alone can be induced by microwave or far infrared radiations. Infrared radiations generally possess sufficient energy to change both the rotational and vibrational states of a molecule. Absorption of visible or ultraviolet light causes changes in the electronic, vibrational, and rotational levels of the molecular species.

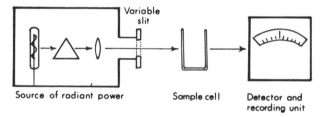

FIGURE 6.1   The basic components of an absorptiometer unit.

Identification of the frequency of the radiations selectively absorbed by chemical species provides a form of qualitative analysis which is reasonably unique. With molecular species the presence of particular types of atomic groupings can be confirmed and in many cases the shape and structure of the molecule can be predicted. Quantitative determinations can be made by measuring the fraction of some characteristic incident radiation absorbed by the sample.

The type of equipment required is shown schematically in Fig. 6.1. It consists of a source of radiant energy, an adjustable slit to control the radiant power passing through the sample, a cell to retain the sample, and a detector to measure the intensity of the radiation. Instruments containing these components are known as absorptiometers, or as spectrophotometers if the source includes a monochromator unit.

The range of wavelengths being transmitted through the sample at the time of measurement should be very small. To achieve this narrow "band width" the radiations from a multienergy or broad spectrum source are usually passed into a monochromator. In the monochromator, the radiations are dispersed by means of either a prism or a diffraction grating, and by rotating these components successive portions of the dispersed radiation can be focused on an exit slit. The range of wavelengths simultaneously transmitted through the slit depends on the dispersive power of the prism or grating and the slit width.

With prism instruments the angle of deviation for any wavelength is determined by the refractive index of the prism material for that particular wavelength. The greatest values, hence the highest degree of dispersion, are observed using wavelengths near to those absorbed by the prism material. For example, the refractive indexes of glass and quartz increase markedly in magnitude at wavelengths shorter than 400 mμ. The dispersive power of prisms made of these materials is thus greatest in the ultraviolet region; the range of usefulness, however, is limited by the onset of strong absorption (350 mμ for glass and 200 mμ for quartz). With wavelengths longer than 400 mμ the degree of dispersion decreases and glass or quartz prisms are rarely used to disperse radiations other than those falling in the visible and ultraviolet region. Prisms made from ionic crystalline solids such as NaCl, LiF, CaF$_2$, and KBr are used to disperse infrared radiations.

Diffraction gratings, on the other hand, exhibit linear angular dispersion over the entire region of wavelengths being studied and the band width of the radiation supplied by a grating monochromator is virtually independent of wavelength. Diffraction gratings consist of a highly reflective aluminized surface on which are drawn or etched a large number of equally spaced parallel grooves (600 to 2000 lines per millimeter), the degree of dispersion achieved being determined by the number of lines per unit length.

The specifications in regard to an acceptable band width vary with the type of applications involved. For example, in atomic absorption studies the half-width of the radiation being absorbed should be less than 0.01 Å. This specific requirement cannot be met by monochromator dispersion units alone and is only achieved by using a hollow cathode tube as the source. In this case, the source emits radiations having the desired wavelength and dimensions, and the dispersion unit merely serves to isolate the required line from extraneous background emissions.

In molecular studies greater half-widths can be tolerated, the desirable limit being fixed mainly by the breadth of the absorption peak under study. In the ultraviolet region band widths are usually measured in Ångstroms while in colored solutions the bands can be 20 to 30 mμ wide. These requirements are ably met by an efficient monochromator fed by a broad spectrum source.

The nature of the source unit varies with the type of radiation being sought. Thus a tungsten lamp is commonly used to provide visible radiations while a hydrogen or deuterium lamp is a rich source of ultraviolet light. Heated rods of silicon carbide or the oxides of cerium, zirconium, thorium, and yttrium are adequate sources of infrared radiation.

Microwave radiations are normally obtained from special electronic tubes (usually of a type known as a klystron oscillator) and in this particular case the wavelength of the emitted radiation is selected by varying the voltage applied to the tube.

At the other end of the electromagnetic scale are X rays. With this type of radiation, selectivity is achieved by varying the nature of the metal in the primary X-ray tube, followed in some cases by isolation of required wavelengths by metal filters.

The sample containers and the optics used in the absorption units have to be transparent to the incident radiation. Thus while glass components are satisfactory for visible colorimetric studies, they have to be replaced by quartz if readings are to be extended into the ultraviolet region. Similarly, for infrared radiations having wavelengths between 1 and 16 μ, sodium chloride components are quite satisfactory but once this range is exceeded other halide salts (such as AgCl, KBr, CsBr, CsI, TlBr, TlI) have to be substituted for the rock salt.

The type of detector required is also defined by the nature and the intensity of the radiation entering the sample cells. For example, with ionizing radiations such as X rays, suitable detectors are Geiger, scintillation or proportional counters. The presence of low frequency radiations (e.g., far infrared) is observed principally through a heating effect; hence the detector unit for these generally consists of a sensitive thermocouple or resistance thermometer. The sensing element in the cells widely used for the detection of near infrared radiation (wavelength 0.8 μ to 3 μ) is a semiconductor (e.g., lead sulfide, lead telluride, or germanium). On illumination with this range of radiation the electrons of the semiconductor are raised to conduction bands and the electrical resistance drops. The current which flows on imposition of a small potential is thus related to the intensity of the incident radiation.

The most widely used detector for ultraviolet and visible radiations is the phototube which consists essentially of a semicylindrical cathode coated on the inside with an alkali or alkaline earth oxide and a central wire anode, both electrodes being mounted in an evacuated glass envelope fitted with a quartz window. When ultraviolet or visible radiations strike the coated cathode surface there is photoejection of electrons from the surface. By applying a potential of about 90 V between the electrodes, these electrons are collected at the anode and the resultant current is a measure of the intensity of the incident radiation.

With very weak initial signals, i.e., low radiant power levels, a photomultiplier tube is used instead of the simple phototube. A photomultiplier tube contains a number of photoemissive surfaces, each successive unit being at a higher potential. Electrons released at the first surface by incident radiation are accelerated by the potential difference towards the second surface where much of the additional kinetic energy is transferred through the liberation of further electrons. These electrons in turn are accelerated towards the next unit. After nine stages of amplification the original signal can have increased by a factor of a million.

The means used to record the detector signals varies with the nature of the commercial unit. For studies in which the detector response is to be measured as a function of the incident wavelength, recording usually involves some type of strip chart recorder. This is particularly necessary in infrared studies since the absorption pattern is generally complex.

Typical absorption spectra involving different types of radiation are shown in Fig. 6.2. These spectra (intensity versus wavelength plots) are used for

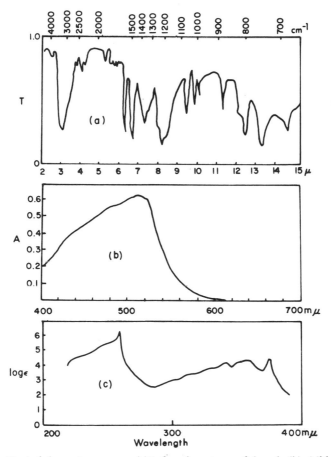

FIGURE 6.2 Typical absorption spectra. (a) Infrared spectrum of phenol; (b) visible spectrum of iron (II)—1:10 phenanthroline solution; (c) ultraviolet spectrum of naphthalene.

identification purposes and for identifying the wavelength most strongly absorbed. The latter is known as λ_{max} for the particular absorbing species involved.

Other instruments are designed primarily for the measurement of the amount of absorption at some selected wavelength. In this case the detector output is fed into a direct reading meter. Alternatively, the signal is opposed by a current derived from a precision potentiometer circuit; the position of balance is indicated by a sensitive galvanometer. The readings obtained at a single wavelength are widely used in quantitative determinations since there is a definite relationship between the amount of radiation absorbed and the amount of absorbing species in the sample cell. Alternatively, the readings made at a series of wavelengths can be combined to give an absorption spectrum.

The determination of an absorption spectrum involves the same principles whether it be performed automatically by an instrument or by the plotting of individual points. At each wavelength in the range being studied the intensity of the radiation leaving the source unit is measured with the sample in, and out, of the beam. The difference in the two intensity levels represents the amount of radiation absorbed by the sample. The absorption level is then plotted against wavelength. With monochromator units, the peak wavelength being transmitted through the exit slit is indicated by an instrument dial, and it is a wise precaution to regularly check the validity of the dial readings by determining the spectrum of standard materials whose absorption peaks are well tabulated.

As indicated in Fig. 6.2 the shape of the absorption spectrum varies with the type of material being studied and with the type of radiation being absorbed. For this reason the analytical techniques based on radiation absorption are generally subdivided on the basis of the type of electromagnetic vibration involved and the nature of the various spectra are discussed in more detail under the appropriate technique subheadings.

B. QUANTITATIVE LAWS

With monochromatic radiation (i.e., very narrow bandwidths) the quantitative relationship between the absorption of radiation and other experimental factors is reasonably well described by an equation derived from what is known as the Beer-Lambert or Bouguer-Beer law.

For the system shown in Fig. 6.3 the appropriate mathematical expression is

$$\log P_0/P = A = abc$$

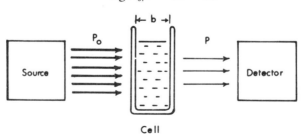

Cell

FIGURE 6.3   Symbols associated with the absorption of radiation. P_0 is the radiant power entering the sample cell; P is the radiant power leaving the cell. The length of the light path through the sample is defined by the cell thickness, b cm. The concentration of absorbing species, c, may be expressed in grams per liter or moles per liter.

P_o is the intensity level of the radiant power entering the cell and P is the intensity level of the radiant power which reaches the detector after absorption. The term P/P_o is defined as the transmittance (symbol T) and represents the fraction of the incident radiation which is transmitted by the sample. The percent transmittance is thus equal to $100 \, T$. The absorbance (symbol A) or optical density (symbol OD) of the sample is mathematically equivalent to $-\log T$.

The length of the cell, or the absorption path, b, is usually measured in centimeter units. The concentration, c, of the absorbing species is sometimes expressed in terms of grams per liter and sometimes as moles per liter. Where the concentration units are grams/liter, the proportionality constant, a, is known as the absorptivity for the species. When concentrations are expressed in moles/liter, the proportionality constant is represented by the symbol ε and is known as the molar absorptivity or molar extinction coefficient for the species of interest.

The absorptivity or molar extinction coefficient is a characteristic property of the solute–solvent system. The magnitude of this term varies with wavelength and many chemists record absorption spectra by plotting ε (or $\log \varepsilon$), against wavelength (cf. Fig. 6.2c). Commercial instruments, however, are usually designed to record absorbance or transmittance values; hence, most spectral curves are displayed as plots of A or T (or $\%T$) against wavelength (cf. Fig. 6.2,a,b) or wavenumber. Because of the direct relationship between wavenumbers (cm^{-1}) and energy, a large proportion of chemists now favor substitution of wavenumber for wavelength in the spectral plots.

The walls of the sample cell cause some scattering of the incident radiation and together with the solvent are responsible for a certain amount of absorption. To compensate for this effect it is standard practice to measure P_o using an identical cell filled with pure solvent.

If more than one species capable of absorbing radiation is present, the total absorbance measured is the sum of the absorbances due to the individual components.

$$\text{Total absorbance, } A_T = \sum A_i = b \sum_i a_i c_i$$

The additivity of absorbances is a useful property since it provides a means of studying samples which contain more than one absorbing species.

Consider a solution containing two absorbents (I and II) each possessing a characteristic absorption spectrum. The spectrum of the mixture is a composite of the two individual curves, and at any selected wavelength the observed reading can be described by an equation, i.e.,

$$A_T^{\lambda_1} = A_I^{\lambda_1} + A_{II}^{\lambda_1} = a_I^{\lambda_1} b c_I + a_{II}^{\lambda_1} b c_{II}$$

where $A_T^{\lambda_1}$ is the observed total absorbance; $a_I^{\lambda_1}$ and $a_{II}^{\lambda_1}$ are the absorptivities of the species I and II at wavelength λ_1; b is the radiation path length; and c_I and c_{II} are the concentrations (grams/liter) of the two species present.

A similar equation can be set up for any other wavelength. For example; at wavelength λ_2

$$A_T^{\lambda_2} = A_I^{\lambda_2} + A_{II}^{\lambda_2} = a_I^{\lambda_2} b c_I + a_{II}^{\lambda_2} b c_{II}$$

Values for $a_I^{\lambda_1}$, $a_I^{\lambda_2}$, $a_{II}^{\lambda_1}$, $a_{II}^{\lambda_2}$ can be ascertained from the spectrum of the pure components of the mixture. When this information is available values for c_I and c_{II} can be calculated by solving the pair of equations simultaneously.

If possible, the wavelengths selected should yield absorptivity values for the component species which are significantly different in magnitude.

For a solution containing only two absorbing species it is sufficient to measure the total absorbance at two wavelengths and subsequently solve two simultaneous equations. With three components, measurements need to be made at three wavelengths followed by solution of three simultaneous equations. In principle, n absorbance measurements at n different wavelengths would permit the determination of n components in a mixture. Where n is >3 the mathematical treatment becomes complex but can often be simplified by selecting some wavelengths where one or more of the components do not absorb.

This calculation of a concentration value from experimental absorbance readings is of primary interest to the analytical chemist since it provides a very convenient method for quantitative analysis. In fact, spectrophotometric methods are extremely popular; they are sensitive, selective, and yet reasonably simple.

Quantitative analysis is based on the measurement of the amount of a selected wavelength absorbed by the sample solution, and calculations are based on the Beer–Lambert law ($A = abc$). This law is usually applied through the medium of a calibration graph, the absorbance of a series of standard solutions being measured and plotted against concentration. Although the equation suggests that a plot of A versus c should be linear, the plots obtained in experimental studies are often distinctly curved.

These "apparent" deviations from Beer's law can be attributed to a number of causes which fall into two broad groups—chemical and instrumental.

In the Beer–Lambert equation c represents the concentration of the absorbing species, and because of chemical equilibrium effects this concentration is not necessarily equal, or even proportional, to the analytical concentration.

Consider the case where a known amount of some absorbing species (X) is added to the solvent phase (S) and undergoes some form of association or dissociation reaction to form a related species Z, i.e.,

$$X + S \rightleftarrows Z + Y$$

For example,

$$Cr_2O_7{}^{2-} + H_2O \rightleftarrows 2\,HCrO_4{}^-$$

$$HA + H_2O \rightleftarrows A^- + H_3O^+$$

$$ML_x + \gamma H_2O \rightleftarrows M(H_2O)_y L_{(x-y)} + \gamma L$$

The analytical concentration is given by $(C_x + C_z)$ but the absorbance reading is related to C_x.

From the general equilibrium equation, $K = C_z C_y / C_x C_s$,

$$C_x / C_z = C_y / KC_s \quad \text{or} \quad C_x = (C_y / C_y + KC_s)(C_x + C_z)$$

where K is the equilibrium constant.

The concentration of absorbing species C_x is, therefore, related to the amount initially added by the expression $(C_y / C_y + KC_s)$. The smaller the value of K, the closer this expression approaches to unity. Where K has a significant value, the conditions at equilibrium are controlled by C_y and C_s. In this case the absorbance (given by $\varepsilon_x b C_x$) may be found to vary with the concentration of the solvent (i.e., dilution of the solution), pH of the solvent phase, concentration of ligands present, etc.

The equilibrium effects are enhanced by the fact that the other species in solution may also absorb some of the incident radiation, the total absorbance observed being equal to

$$A_{\text{(total)}} = b(\varepsilon_x C_x + \varepsilon_s C_s + \varepsilon_z C_z + \varepsilon_y C_y)$$

In order to obtain a linear calibration curve the ratio C_x / C_z must remain virtually constant. Thus where proton transfer is involved in the equilibrium, the effect of dilution on the C_x / C_z ratio can be minimized by having present an excess of acid or base or by maintaining a constant pH through the addition of a buffer. For example, in studying chromium (VI) solutions, straight line graphs are obtained using strongly acid solutions ($Cr_2O_7^{2-}$ predominates) or strongly basic solutions (CrO_4^{2-} predominates). An aqueous solution with no pH control yields a curved calibration plot. Where metal complexes are being studied it is often necessary to control the pH and the concentration of excess ligand present.

A state of equilibrium is not necessarily achieved instantaneously and in some systems the magnitude of the observed absorbance is time dependent. The effect may be positive or negative. For example, if the rate of formation of Z is measurably slow, then absorbance readings based on C_z increase with time. On the other hand, readings based on C_x decrease with time.

Among the other factors which can contribute to the curvature of calibration graphs are indeterminate instrumental deviations such as stray radiations reaching the detector (reflected within the instrument), power fluctuations in the radiation source or detector amplifying system, changes in detector sensitivity, etc. Double beam operation tends to cancel out most of these random causes of deviation.

Reflection losses constitute another source of error. Whenever a beam of radiation passes from one medium to another of different refractive index, some radiation is reflected and the reflection losses at the solid–liquid interface during the passage of a light beam through a cell containing a liquid can be quite significant. Compensation for this effect is made in spectrophotometric measurements by using a comparison cell containing a solution of similar refractive index. The difference between the refractive index of a solvent and a solution is negligible except at high concentrations of solute (e.g., $>10^{-2} M$); hence for dilute solutions pure solvent is usually used in the comparison cell. With more concentrated solutions the comparison cell should contain an absorbing solution of similar composition if deviations due to variable reflectance losses are to be avoided.

Beer's law is only strictly valid for monochromatic radiation and a common cause of "apparent" deviations from this law is the polychromatic beam of radiation produced by many commercial instruments. The band width of the beam emanating from a high quality monochromator unit may be of the order of one Ångstrom in the ultraviolet increasing to a few millimicrons in the visible region. With lower quality monochromators the band widths are ten to twenty times broader. Unless the molar absorptivity is invariant within the wavelength band used, the absorbance measured is an "average" absorbance over the band and due to the logarithmic nature of absorbance this is not a true average. The greater the slope of the absorption curve through the wavelength band, the greater the deviation. Accordingly, it is recommended that absorbance measurements be made using radiation corresponding to λ_{max} for the test species, since it is in this region that molar absorptivity is relatively constant over a wavelength region corresponding to the band width. A second advantage of choosing λ_{max} for measurement purposes lies in the fact that ε has a maximum value at this point which increases the sensitivity and accuracy of the readings.

While narrow band widths are obviously desirable for best accuracy, decreasing the band width reduces the amount of energy reaching the detector. Consequently there is always a compromise between accuracy, sensitivity, and detector requirements.

The best precision and accuracy is obtained when the sample absorbs about 40% of the incident radiation, as shown by the following argument.

From Beer's law $c = A/\varepsilon b = -\log T/\varepsilon b$ and for a given species ε and b are constant. All the uncertainty in a measured concentration is, therefore, attributable to the error, dA, in a measured absorbance A or due to the error, dT, in a measured transmittance T.

Transmittance readings are generally displayed on a linear scale and if the uncertainty in reading the scale is considered to be the limiting factor, the error (dT) can be assumed to remain constant over the whole scale (e.g., $\pm 1\%$).

The relative error in the concentration, c, is dc/c where dc represents the absolute error.

Differentiating the transmission form of Beer's law yields

$$dc/dT = -\log e/\varepsilon b T$$

Dividing both sides by c, and rearranging gives

$$dc/c = -\log e \cdot dT/T(\varepsilon bc)$$
$$= (\log e)\, dT/T \log T$$

If dT remains virtually constant, dc/c is inversely proportional to $T \log T$. A plot of this function against T yields a flattened parabola with a center at $T = 0.37$, the values of the function being within 20% of the minimum value for all transmission readings lying between 0.17 and 0.61.

In order to keep the relative error to a minimum, one should, therefore, attempt to keep the scale readings between 17% and 61% transmission, corresponding to absorbance readings of 0.77 and 0.21.

In making this calculation it has been assumed that the response of the detector is directly proportional to the radiant power striking the detector. In most cases this is a reasonable assumption and if possible the concentration of the absorbing species or the cell thickness, or both, should be adjusted to keep the transmittance in the range 0.2 to 0.6.

For the absorption of a monochromatic beam of X-radiation by a solid sample, the basic equation is expressed in different terminology, namely:

$$\log_e P_o/P_x = \mu x = \mu_m dx$$

P_o is the initial power of the radiation and P_x is the power after passage through an absorbing sample x cm in thickness. The linear absorption coefficient, μ, represents the fraction of the energy absorbed per cubic centimeter of the sample. A more convenient unit is the mass absorption coefficient μ_m which equals μ/d where d is the density of the material. Values for μ_m are listed in standard reference tables.

The magnitude of the mass absorption coefficient is related to the atomic properties of the absorbing substance and varies with wavelength

$$\mu_m = B \cdot \lambda^{\frac{3}{2}} \cdot (Z^4/A)$$

B is a proportionality constant which includes Avogadro's number. The same value is obtained for all elements heavier than aluminium when measured under comparable conditions. Lambda, λ, represents the wavelength being absorbed while Z and A are the atomic number and atomic weight, respectively, of the absorbing element.

II.   ABSORPTION BY ATOMS

A.   ABSORPTION OF X RAYS

The absorption of X rays can be attributed to two main processes. One of these is scattering of the incident radiation. This decreases the incident X-ray intensity but usually it only becomes significant using systems in which the values of Z are low and/or the values of λ are high. The second type of absorption is photoelectric, whereby all the energy of the incident X-radiation quantum is transferred into kinetic energy through ejection of a photoelectron. One result of this is the emission of characteristic X rays (cf. Chapter 5, Section IV,A).

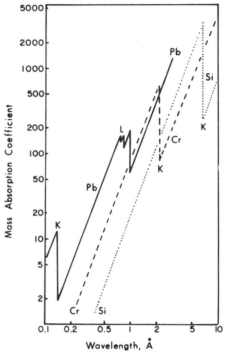

FIGURE 6.4   The absorption of X rays. A logarithmic plot of mass absorption versus wavelength. The discontinuities are known as absorption edges. Based on data from the "Handbook of Analytical Chemistry" edited by L. Meites. Used with permission of McGraw-Hill Book Co., copyright © 1963, McGraw-Hill Book Co.

On the basis of the equation quoted above, a plot of log μ_m against λ for any element should give a straight line of slope 5/2. The plots obtained by experiment, however, possess discontinuities as shown in Fig. 6.4.

The first discontinuity is known as the K critical absorption wavelength and it can be attributed to the fact that radiation of greater wavelength possesses insufficient energy to eject the K electrons of the element. Accordingly, radiation having a wavelength slightly greater than the K absorption edge is not absorbed as strongly as X-radiation possessing a shorter wavelength. The discontinuities which appear at longer wavelengths correspond to the photoelectric ejection of L and M electrons.

Tables listing the wavelengths of the absorption edges of all the elements are to be found in reference books including the "Handbook of Analytical Chemistry" and the detection of these edges has been used as a means of identification.

Another application of absorption edge data is in the selection of filter media for X-radiations.

As explained in the previous chapter, X-ray tubes yield an emission spectrum which is characteristic of the target, the two most intense emissions being usually the K_α and K_β lines. A beam of near monochromatic radiation can be obtained by filtering out the higher energy K_β species. This is achieved by passing the mixed radiation through a thin strip of a metal possessing an absorption edge which lies between the wavelengths of the α and β radiation. An indication of suitable metals and the thickness required to give 96–99% absorption of several different K_β radiations is provided by the list of "filters" quoted in the "Handbook of Analytical Chemistry."

For analytical purposes the secondary effects of X-ray absorption are possibly more important, e.g., the majority of quantitative studies are based on the measurement of the wavelength and intensity of the secondary X rays emitted after absorption. However, some analytical procedures based on absorption measurement have been developed and these can be divided into three classes: polychromatic, monochromatic, and differential absorption across an absorption edge.

The absorption of polychromatic radiation is not specific for any element; hence, cannot be used to identify elements or to obtain quantitative values for samples of unknown composition. However, if the same material is continually being monitored, μ_m remains essentially constant and any observed variations in absorbance can be attributed to changes in thickness. This provides a very convenient method for measuring film thicknesses. Alternatively, with constant thickness, an abrupt change in absorbance indicates a change in material composition. For control analysis by this method, the absorptiometer unit may consist of little more than a high intensity X-ray tube whose output is fed directly through the sample to the detector. A modern trend is to use radioactive isotopes as the X-ray source, since these tend to lower the cost and increase the reliability of the monitoring units.

For monochromatic radiation studies the equipment required can be more complex. In some cases suitable monochromatic radiation can be isolated by means of a metal filter, but for greater flexibility it is necessary to duplicate the equipment required for emission analysis (cf. Chapter 5, Section IV,A). In this case the primary X rays are used to excite emissions from a secondary emitter which are subsequently dispersed by an analyzing crystal. By varying the nature of the secondary emitter and using different Bragg angles a wide range of X-radiations are available for the absorption studies.

Quantitative determinations are made by measuring the intensity of a selected wavelength after transmission through samples and unknowns that are similar in composition. The X rays are absorbed by the constituent atoms of samples and the technique is of particular value when the element to be determined is the sole heavy component in a material of low atomic weight, for example, the determination of sulfur or lead in petroleum products.

The mass absorption coefficient of a sample containing three components (A, B, and C) is given by

$$(\mu_m)_{\text{sample}} = (\mu_m)_{\text{A}}W_{\text{A}} + (\mu_m)_{\text{B}}W_{\text{B}} + (\mu_m)_{\text{C}}W_{\text{C}}$$

where $(\mu_m)_{\text{A}}$ is the mass absorption coefficient of component A, W_{A} is the weight of component A; components B and C are similarly defined.

If A and B are the major constituents and vary little from sample to sample and C is the minor variable to be determined, then this equation simplifies to

$$(\mu_m)_{\text{sample}} = (\mu_m)_{\text{matrix}}(1 - X) + (\mu_m)_{\text{C}}X$$

where X is the weight fraction of C. In such cases a plot of $\log_e P_o/P$ against the concentration of C is linear, the slope being related to the difference in the mass absorption coefficients of C and the matrix.

Absorption edge analysis is based on the large change in the mass absorption coefficient across the absorption edge. The method uses two X-ray spectral lines, one below (λ_1) and the other above (λ_2), the absorption edge of the element to be determined. The sample can be considered to consist of two components, the matrix (M) and element of interest (A), and for each wavelength one can write a simplified equation similar to that shown in the preceding paragraph. From the fundamental absorption law $(\mu_m)_{\text{sample}}$ can be expressed in terms of $\log_e P_o/P$ and through mathematical manipulation one obtains the relationship

$$\log_e (P_{\lambda_1}/P_{\lambda_2}) = \log_e (P^{\circ}_{\lambda_1}/P^{\circ}_{\lambda_2}) - [AW_{\text{A}} + BW_{\text{M}}]\, dx$$

where

$$A = {}_{\lambda_1}(\mu_m)_{\text{A}} - {}_{\lambda_2}(\mu_m)_{\text{A}}$$

$$B = {}_{\lambda_1}(\mu_m)_{\text{M}} - {}_{\lambda_2}(\mu_m)_{\text{M}}$$

W_{A} = Weight fraction of the element being determined

W_{M} = Weight fraction of the matrix = $1 - W_{\text{A}}$

d = density of the sample

x = absorption path length

Where the composition of the matrix and the density of the samples remain reasonably constant from sample to sample, $\log_e (P_{\lambda_1}/P_{\lambda_2})$ varies linearly with the concentration (i.e., W_{A}), the slope being equal to Adx.

X-ray absorption is not as versatile as emission for general elemental analysis and the number of published applications is somewhat limited. Like emission, however, it possesses two distinct advantages—the method is nondestructive and the response of the elements is independent of their chemical form. On the other hand, in order to achieve selectivity and sensitivity it is necessary to have a matrix which does not absorb the incident radiation to any appreciable extent.

B. ATOMIC ABSORPTION SPECTROSCOPY

Chemical analysis by atomic absorption spectrometry involves converting part of the sample into an atomic vapor and measuring the absorption, by this vapor, of a radiation which is characteristic of some particular element.

At present, in almost all the analytical applications of this technique, the atomic vapor is formed by spraying a solution of the sample into a flame. As explained in the preceding chapter (Section 5,II,A,1) the flame volatilizes the solvent and ultimately causes dissociation of the minute solid particles into atoms. A small fraction of these atoms become excited by the flame and emit radiation while the majority remain in the ground state. The ground state atoms can also be excited by the absorption of radiation, but for the process to be efficient the energy content (i.e., $h\nu$) of the radiation must correspond exactly to the energy jump associated with the electronic transition. It is also desirable that the half-width of the line being absorbed should be narrower than the half-width of the absorption band. With the types of flame commonly used the width of the absorption band is generally less than 0.05 Å; hence the radiation passing through the flame should ideally have a half-width of about 0.01 Å.

The stringent requirements in regard to energy content and half-width are most readily met by using the characteristic atomic resonance radiation of the particular element of interest as the light source in the absorption unit. The use of resonance radiation makes the process very selective, and since the majority of atoms are initially in the ground state the method is quite sensitive (ppm range).

The requirements of the light source are therefore manifold. It has to emit the spectrum of the element to be determined, the lines must be sharp and of constant intensity, and the background emission must be minimal. A hollow cathode electrical discharge lamp fed by a stabilized power source meets these specifications. The components of a hollow cathode tube are shown schematically in Fig. 6.5 and have been described in Section 5,II,B,1. The exciting current applied to the tube is alternated so that a tuned amplifier in the detection unit responds only to resonance radiation from the lamp. In this way the small amount of emission from excited atoms in the flame is not recorded. Of the large number of lines emitted by any source only a small number show any appreciable absorption by atomic vapor. Accordingly, a monochromator is usually used to isolate the line to be measured from all other lines in the spectrum of the light source.

The basic components of an atomic absorption spectrophotometer are therefore a hollow cathode lamp, a flame-atomizer unit, a monochromator, a photoelectric detector, an ac amplifier and rectifier, and an output meter. The units are shown in diagrammatic form in Fig. 6.5.

The flame can be considered to play the same role as the sample cell in other forms of absorption spectrophotometry. The absorbance by the flame of the characteristic radiation can, therefore, be expected to vary with the length of the absorption path and with the concentration of absorbing species present. The latter is invariably a function of the atomizer unit, the flame, and the concentration of the material being sprayed into the flame.

The flame conditions required to produce atoms of different elements vary; hence, it is important that the various gas supplies can be controlled and

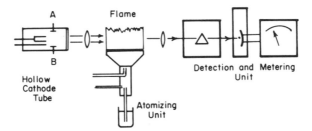

FIGURE 6.5   Schematic representation of the basic components of an atomic absorption spectrophotometer.

metered.   Air–coal gas, air–propane, air–hydrogen, air–acetylene, oxy-acetylene, and nitrous oxide–acetylene flames have all been used.   For each combination there is a desirable burner design and for each element a preferred combustion system and an optimum fuel to oxidant ratio.   The maximum concentration of atoms appears in different zones in different flames; hence, the burner height is usually adjusted until a maximum absorption value is obtained. Since the sensitivity of the technique is also a function of the length of flame in the optical path, many units use a 10 cm burner while other instruments increase sensitivity through multiple reflection of the radiation through the flame.   An alternative procedure extends the interconal zone of the flame in a horizontal direction by means of a long Pyrex tube.

The absorption lines studied generally fall in the ultraviolet or visible region of the spectrum and the intensity of the source is usually weak.   Accordingly, the most suitable detectors are photomultiplier tubes, connected via appropriate amplifiers to recording meters.

In single beam instruments the recording meter is adjusted to show full scale deflection when only the solvent is being aspirated into the flame.   A sample solution is then introduced into the flame through the atomizing unit and the change in meter reading is observed.   By calibration with standard solutions the meter reading can subsequently be correlated with the concentration of the element present.   For maximum accuracy the power supply to the lamp and detection systems must be highly stabilized.

The effects of power supply variations are minimized in double-beam operation where the initial source is split and the detector compares the intensity of the two beams, only one of which passes through the flame.

For the determination of one or two elements only, a simple instrument has been designed which does not contain a monochromator. The radiation from a high intensity hollow cathode lamp passes through the flame into the window of a low intensity lamp containing the same elemental vapors. The intensity of the characteristic radiation emitted in this lamp by atomic fluorescence is then measured at right angles to the original beam path by a photomultiplier tube. Absorption in the flame reduces the intensity of the fluorescent radiation.

Atomic absorption spectroscopy is applied mainly to the determination of the minor constituents of samples. The only limitation in regard to the type of sample handled is that it must be capable of giving a solution of the metal concerned in either an aqueous or nonaqueous solvent. Although investigations of the suitability of alternative methods of vaporizing solids directly continue, almost all analytical applications still use a flame as the source of atoms. The sensitivity obtainable with various elements is, therefore, usually compared in terms of parts per million of the element in solution.

As in flame photometry, monographs on atomic absorption spectroscopy list sensitivities in terms of $\mu g/ml \ (\%T)^{-1}$. The values refer to a particular type of flame, path length and type of instrument, but they do provide a guide as to the concentration ranges which should be studied (e.g., maximum concentration approximately equals sensitivity $\times$ 100) and the tables specify suitable experimental conditions (e.g., preferable flame type and resonance line).

Where it is desired to measure higher concentrations, the absorption may be brought within the range of the instrument by using a less strongly absorbed line, a shorter path length (narrower flame) or by dilution of the original sample solution.

The number of elements which can be determined by this technique is restricted by two main factors:

1. The availability of suitable radiation sources. Hollow cathode or discharge tubes for the determination of over fifty elements are available commercially; hence this limitation is fast disappearing.
2. The ability to produce atomic vapors. For example, in the normal type of air–fuel flame, metals which form highly refractory oxides (e.g., Be, Al, V, Ti, Ta etc.), cannot be determined since the concentration of free atoms in the flame is almost nonexistent.

This particular problem has been partially solved by varying the nature of the flame. For example, the use of an acetylene–nitrous oxide flame has extended the scope of the technique by over thirty elements bringing the total number of determinable elements to nearly seventy.

Atomic absorption spectroscopy measures the amount of a narrow line source absorbed by ground state atoms exhibiting a narrow absorption peak. These factors make the method highly specific and the technique is basically free of all the spectral interference effects observed in emission studies. Since most atoms are present in the ground state the procedure is sensitive and comparatively free of effects arising from small changes in flame temperature. A small amount of one element can be determined in the presence of large amounts of other substances.

Because the technique is sensitive and specific, small amounts of material can be handled, pretreatment of samples is minimal, and the final methods are usually rapid and simple. To obtain the best performance from the instrumental unit the concentrations studied should cause 40–80% of the incident radiation to be absorbed. Over this optimum range, the accuracy obtainable is of the order of $\pm 2\%$ of the concentration of the element being determined.

The method is subject to some interference effects, which are associated primarily with processes which occur in the flame. As in emission flame photometry, the number of atoms present in the flame can be reduced significantly through the recombination of species to yield compounds which are incompletely dissociated or atomized. Typical examples occur in the determination of magnesium in the presence of aluminum and calcium in the presence of phosphorus. As in flame photometry these effects can be minimized by the addition of competing reagents (e.g., strontium or lanthanum ions to associate with phosphates) or complexing agents.

In some circumstances it is considered desirable to separate the element of interest in a preliminary step. Solvent extraction is very suitable for this purpose since the use of a solution in an organic solvent can increase the sensitivity by a factor of three or more.

Because of its many favorable aspects, interest in atomic absorption spectroscopy is widespread; and the technique has been used in the analysis of water samples, soil extracts, plant materials, fertilizers, blood sera, urine, human tissues, oil samples, coal ash, mineral ores, ferrous and nonferrous alloys, and sundry other materials found in the fields of agriculture, biochemistry, fuel science, geochemistry, and metallurgy.

III.   MOLECULAR ELECTRONIC TRANSITIONS

A.   Origin of Spectra

In the absorption of visible or ultraviolet light by atoms (as described in the preceding section) the outer orbital electrons of the atoms are promoted to a limited number of quantized atomic orbital states. The frequency of the radiation required for absorption is then related to the difference in the energy content of the ground state and the various permitted excited states.

When atoms are joined by a chemical bond, as in a molecule, electrons from both atoms participate in the bond. The bonding electrons are associated with the molecule as a whole and not with any particular nucleus, and they are said to occupy a molecular orbital.

It can be shown that the combination of two atomic orbitals from two bonding atoms leads to the formation of one "bonding" molecular orbital of low energy and one "antibonding" molecular orbital of very high energy.

Covalent bonds, consisting of electron pairs, are characterized as σ (sigma) bonds when there is "head-on" atomic orbital overlap and π (pi) bonds when

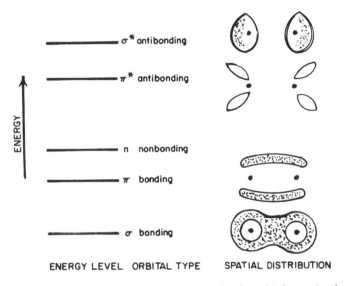

ENERGY LEVEL   ORBITAL TYPE   SPATIAL DISTRIBUTION

FIGURE 6.6   Schematic representation of electronic molecular orbital energies showing the electron–cloud probability distributions for some of the orbitals.

there is "parallel" atomic orbital overlap. By molecular orbital theory, each bonding σ orbital must have a corresponding σ^* antibonding orbital, and each π bonding orbital must have a corresponding π^* antibonding orbital. Electron-cloud probability distributions for these four types of molecular orbitals are illustrated in Fig. 6.6. Valence electrons which do not participate in chemical bonding are referred to as nonbonding or "n" electrons.

Bonding electrons do not normally occupy the antibonding orbitals but they can be promoted to this state through the absorption of energy.

For organic molecules, absorption of ultraviolet or visible radiation can be sufficient to promote electrons in n, σ, or π orbitals to some higher energy antibonding orbital (excited state). The order of the energy levels of the various molecular orbitals is indicated in Fig. 6.6, and this diagram indicates that the ΔE values for transitions are in the order:

$$n \to \pi^* < \pi \to \pi^* < n \to \sigma^* \ll \sigma - \sigma^*$$

Since $\Delta E = hv = hc/\lambda$, the larger the value of ΔE, the shorter the wavelength of the radiation required to cause the appropriate electronic transition.

For example, the energy required for a $\sigma \rightarrow \sigma^*$ transition is very large and molecules possessing only σ bonds (e.g., saturated hydrocarbons) have absorption maxima in the far ultraviolet region (e.g., λ about 120–130 mμ). At the

TABLE 6.1

CLASSIFICATION OF ELECTRONIC ENERGY TRANSITIONS FOR ORGANIC MOLECULES[a]

Transition and symbols	Notation	Example	λ_{max} mμ
1. Between bonding and antibonding orbitals of same type	$N \rightarrow V$		
$\sigma \rightarrow \sigma^*$		Methane	125
		Ethane	135
$\pi \rightarrow \pi^*$		Ethylene	165
		Acetone	188
		Benzene	203
		Toluene	208
		1,2-Butadiene	220
		Acetophenone	240
2. Between nonbonding atomic orbitals and antibonding molecular orbitals	$N \rightarrow Q$		
$n \rightarrow \pi^*$		Acetone	277
		Nitroso–butane	665
$n \rightarrow \sigma^*$		Methanol	183
		Acetone	190
		Methylamine	213
		n-Butyliodide	257
3. Between ground state orbital and a high energy orbital far removed from the nuclear framework	$N \rightarrow R$		Vac. UV

[a] Based on table in "Physical Methods of Organic Chemistry" by A. R. Pinder. Copyright © 1964, The English Universities Press. Used by permission of the copyright owner.

other end of the scale, a $n \rightarrow \pi^*$ transition can result from the absorption of ultraviolet radiation having wavelengths between 250 and 300 mμ.

Some representative molecular electronic transitions are shown in Table 6.1.

Since electronic transitions in molecules are accompanied by a series of vibrational and rotational changes, bands of radiation are absorbed and the frequency tabulated in Table 6.1 is the peak of the absorption curve, i.e., λ_{max} (cf. broad absorption peaks in Fig. 6.2c).

A molecule may possess all three types of bonding electrons. A simple example of this is formaldehyde, HCHO.

Type of electrons	*Structural formula*

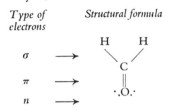

All the various types of electrons can be promoted by energy absorption and interchanges are possible between the groups. For example, it can be observed from Table 6.1 that acetone has three absorption peaks corresponding to $\pi \rightarrow \pi^*$, $n \rightarrow \sigma^*$, and $n \rightarrow \pi^*$ transitions.

Only organic molecules which contain unsaturated or polar groups have been found to absorb radiations from the 180 to 800 mμ region of the electromagnetic spectrum, and it has become customary to ascribe the name chromophore to the unsaturated groups responsible for much of the observed absorption. Typical chromophoric groups are listed in Table 6.2.

When two or more chromophores are present in the same molecule and are separated by more than one single bond, the absorption is usually additive, that is, the absorption characteristics of each chromophore are observed. For example, diallyl, H_2C=CHCH$_2$CH$_2$CH=CH$_2$, absorbs radiation of about the same frequency as ethylene, H_2C=CH$_2$, but the amount absorbed for similar concentrations is approximately doubled.

When the chromophores are separated by only one single bond, conjugation occurs. In conjugated systems, the π electron distribution is delocalized over a minimum of four atoms. This causes a decrease in the $\pi \rightarrow \pi^*$ transition energy and an increase in the probability of the transition. The net result is an increase in the wavelength of the radiations absorbed and a higher value for the molar extinction coefficient (ε).

The effect of conjugation can be gauged by noting that λ_{max} for divinyl ethylene, H_2C=CHHC=CHHC=CH$_2$, is 265 mμ with an ε_{max} value of 50,000. The nonconjugated counterpart of this compound is diallyl which has a λ_{max} of 180 mμ and an ε_{max} value of 20,000. The wavelength shift is sometimes termed a bathochromic effect.

When the conjugated system is extensive, the shift in λ_{max} can be sufficiently great to cause the compound to appear colored and absorb radiations from the visible part of the spectrum. Thus tetraterpene α-carotene, which has ten conjugated ethylenic linkages, is red in color and absorbs at the violet end of the visible spectrum (λ_{max} 445 mμ).

Aromatic systems have characteristic ultraviolet absorption patterns of a more complex nature. The effect of combining two or more aromatic rings

TABLE 6.2
CHROMOPHORIC GROUPS

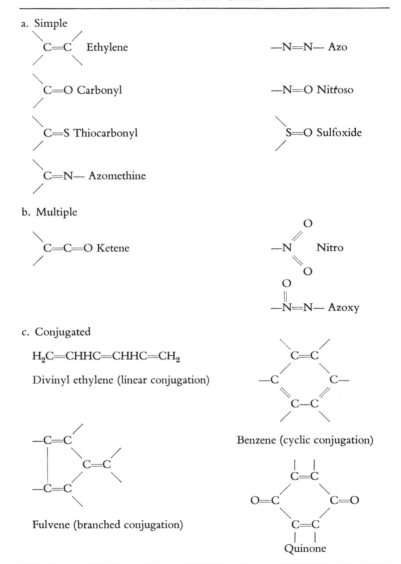

a. Simple

C=C Ethylene   —N=N— Azo

C=O Carbonyl   —N=O Nitroso

C=S Thiocarbonyl   S=O Sulfoxide

C=N— Azomethine

b. Multiple

C=C=O Ketene   —N Nitro

                                                            —N=N— Azoxy

c. Conjugated

H_2C=CHHC=CHHC=CH_2

Divinyl ethylene (linear conjugation)

Benzene (cyclic conjugation)

Fulvene (branched conjugation)

Quinone

to form a polycyclic compound is similar to that observed in aliphatic conjugation. For example, the corresponding values of λ_{max} and ε_{max} for benzene, naphthalene, and anthracene are 268 mμ (39), 311 mμ (280), and 476 mμ (9700) respectively. However, when chromophores are substituted into the benzene nucleus there is mutual electrostatic interaction. Consequently the absorption spectrum of the resulting compound is somewhat different from that of isolated benzene and substituent chromophores.

In organic molecules, nonbonding (n type) electrons are located principally in the atomic orbitals of nitrogen, oxygen, sulfur, and the halogens. Such elements are often present in the molecule in the form of polar groups, e.g., —OH, —SH, —NH$_2$, —OCH$_3$, —Cl, —Br, and —I.

Although some polar groups absorb at longer wavelengths, the above groups usually absorb with varying intensity in the far ultraviolet ($n \rightarrow \sigma^*$ transitions) and their individual spectral behavior is not very important. However, when polar groups are suitably positioned in respect to chromophoric groups they tend to extensively modify the absorption characteristics of the chromophore. In this capacity they are termed "auxochromes."

The introduction of a positive charge into a molecule or a change from a nonpolar to a polar solvent can cause the major absorption band to shift to a shorter wavelength. This is known as a hypochromic effect. Polar solvents interact electrostatically with polar chromophores (e.g., carbonyl) and tend to stabilize both nonbonding electronic states and π^* states. The $n \rightarrow \pi^*$ and $\pi \rightarrow \pi^*$ absorptions of polar chromophores, therefore, move closer to each other as the polarity of the solvent is increased.

Steric effects can also influence the character of electronic spectra and spectral studies have been used to provide information on stereochemical aspects of organic structures. In many instances, the UV-visible spectra of compounds are sufficiently characteristic to permit identification of species isolated during chemical synthesis.

Most organic compounds do not absorb radiations from the visible segment of the spectrum (exceptions are the azo dyes) but inorganic compounds, and particularly transition metal complexes, do commonly absorb radiations of this type.

Among inorganic substances, selective absorption may be expected whenever an unfilled electronic energy level is covered or protected by a stable, filled energy level, usually formed by means of coordination with other atoms.

This observation can be explained in terms of the Ligand Field Theory.

In a typical octahedral complex $(\mathrm{ML_6})^{n+}$, the σ lone pairs of the six ligands surrounding the ion exert an electrostatic field along the x, y, and z axes. The $e_g d$ orbitals (d_{z^2} and $d_{x^2-y^2}$) of the central atom are concentrated along these axes, and electrons in these orbitals can be strongly repelled by the ligand field. Electrons in the $t_{2g} d$ orbitals (d_{xy}, d_{xz}, d_{yz}) are concentrated between the axes and do not feel the same effect. The five degenerate d levels, therefore, split into an upper doublet and a lower triplet, the energy difference between them being termed ΔE (cf. Fig. 6.7). The magnitude of the energy difference, ΔE, between the two sets depends on the nature of the central metal atom and the nature of the ligands. For a given metal ion there is large variation in ΔE with changes in ligand type, since the field produced by the ligands is related to the ease with which the ligand electron clouds are distorted by the ion.

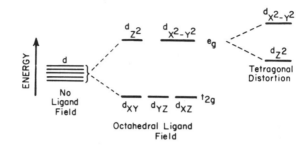

FIGURE 6.7   Ligand-field splitting of *d* orbitals.

Ligands can be arranged in order of their crystal field effect and this order, sometimes referred to as the "spectrochemical series," is independent of the nature of the central metal ion. $I^- < Br^- < -SCN^- < Cl^- < NO_3^- < F^- < $ urea $\simeq OH^- \simeq O\dot{N}O^- < HCOO^- < C_2O_4^{-2} < H_2O < NCS^- < $ glycine $\simeq EDTA^{-4} < $ pyridine $\simeq NH_3 < $ en $\simeq$ dien $\simeq$ trien $<$ dipyridyl $< o$-phenanthroline $< -NO_2^- \ll CN^-$; where $EDTA^{-4} = $ anion of ethylenediaminetetraacetic acid, en = ethylenediamine, dien = diethylenetriamine and trien = triethylenetetramine.

Other transition metal complexes have tetrahedral, tetragonal, or square planar symmetries. In such cases there are different splitting patterns for the five *d* orbitals of the metal atom.

The absorption spectra of transition metal compounds show peaks in the visible and near infrared region (4000–10,000 Å). The peaks are of low intensity (ε_{max} between 1 and 50) and arise from the promotion of electrons from one *d* level to another. These are referred to as *d–d* or $t_{2g} \rightarrow e_g$ transitions. The wavelength of the absorption peak is a measure of the magnitude of the splitting energy, ΔE. Where the complex is distorted from the regular octahedral arrangement, the e_g doublet can split into two levels. For example, if the four ligands in the *xy* plane move toward the metal ion, and simultaneously the two ligands on the *z* axis move away, the d_{z^2} orbital is stabilized and the $d_{x^2-y^2}$ orbital is destabilized. The movement tends to cease when the energy gained (by putting the odd e_g electron in the lower energy d_{z^2} orbital) balances the energy required to compress and stretch the bonds. Square planar compounds can be thought of as the limiting case where the two *z* axis ligands are so far removed that they exert no significant ligand field on the metal ion.

In addition to the low intensity *d–d* peaks, transition metal complexes generally show intense "charge transfer" peaks in the ultraviolet region (2000–4000 Å) with ε_{max} values of 1000–10,000. The charge transfer peaks are probably associated with the transition of an electron from a nonbonding metal *d* orbital to an antibonding orbital on the ligand.

B.   ULTRAVIOLET SPECTROMETRY

The ultraviolet region of the spectrum is frequently subdivided into the far or vacuum ultraviolet region (10–200 mμ) and the near ultraviolet region which extends from 200 to 400 mμ.   Since the number of analytical applications based on the absorption of far ultraviolet radiations is few, this section is concerned solely with studies using the longer wavelengths.

Hydrogen or deuterium discharge lamps provide a continuous spectrum of radiant energy in the 185–375 mμ region which can be dispersed by either prisms of quartz (or fused silica) or by echlette gratings.

In commercial spectrophotometers selected frequencies are isolated by means of a narrow slit and transmitted to a photoelectric detection and recording unit, passing en route through an absorption cell fitted with silica windows.

The general procedure for preparing a solution for ultraviolet measurements is similar to that followed in visible spectrometry, but there are a few unique aspects.

For minimum error in absorbance readings, the maximum absorbance of the test solutions should lie between 0.2 and 0.7.   Since most of the samples produce colorless solutions, suitable dilutions often have to be found by trial and error.

The solvent used to prepare the solutions must be chosen with great care since it has been noted earlier that the solvent can influence the nature of the absorption spectrum of some types of compounds.   It is also necessary to ensure that the absorption bands of the sample occur at longer wavelengths than the absorption bands of the solvent.   Monographs and handbooks have tabulations which note the recommended lower wavelength limits for various solvents.

High molecular weight materials have been examined by dispersing them in melted polythene or by pressing the material as a KBr disc.

Ultraviolet spectrometry is widely used to study organic compounds.   It follows from the principles outlined in Section III,A that any organic compound containing a chromophoric group should yield a characteristic absorption spectrum.   Thus by studying a spectrum, one should be able to identify structural components in the molecule.   The correlation of ultraviolet absorption bands with specific structures is made chiefly by analogy, and Table 6.3 lists several common chromophoric groups and the approximate wavelengths of maximum absorbance (λ_{max}).   The molar extinction coefficient (ε_{max}) is also an important diagnostic aid, since it is sometimes possible to differentiate by intensity measurements between two chromophoric groupings which absorb at the same wavelength.

However, it has been noted earlier that substitution in the molecule can modify the position of λ_{max} and the magnitude of ε_{max}.   The spectral curve can

TABLE 6.3

SELECTED LIST OF CHROMOPHORIC GROUPS

Group	Wavelength of absorbance maximum (mμ)[a]
C≡C	180
C=C	190
C=C—C=C	220
C≡N	170
C=N	190
C=O	280
C=S	330
—COOH	210
$C_6H_5^-$ (phenyl)	270
$C_{10}H_7^-$ (naphthyl)	310
C_9H_7N (quinoline)	310
N=N	370
—NO_2	270
—ON=O	370

[a] Approximate values.

also be altered by changes in solvent, pH, or temperature. In addition some metal chelates and organic substances fluoresce when irradiated with ultraviolet radiation.

Quite dissimilar substances can give almost identical spectra and it is highly desirable to use additional diagnostic techniques and methods in the identification of specific absorbers. The presence of a small amount of a substance with a high extinction coefficient as an impurity may lead to erroneous conclusions.

Great care is, therefore, required in interpreting ultraviolet absorption spectra. Despite these limitations the technique has been widely used. Listed applications include simple qualitative analysis, the determination of molecular structure, and the investigation of isomerism and steric hindrance effects.

For qualitative analysis the test spectra are compared with standard spectra. Several extensive collections of spectra and indexes of ultraviolet spectra are now available for this purpose.

Among the species regularly identified by this means are aromatic hydrocarbons, conjugated aliphatics, heterocyclics, steroids, and vitamins.

Ultraviolet spectrometry has also been applied extensively to the quantitative determination of organic substances. A summary of many of these quantitative procedures is given in Table 6.13 of the "Handbook of Analytical Chemistry." The type of products studied include foods, fats, lacquers, paints, pesticides, petroleum products, plastics, rubber products, soaps, and detergents.

Many inorganic substances can also be determined by ultraviolet spectrometry, but the absorption of visible radiations is more extensively used for this type of analysis.

C. VISIBLE SPECTROMETRY

Visible spectrometry involves the determination of the capacity of a chemical system to absorb radiations falling in the visible segment of the electromagnetic spectrum, viz. 380 to 750 mμ.

Chemical systems which absorb this type of radiation are themselves colored and hence the terms colorimetry and colorimetric analysis are often applied to this analytical technique.

The analytical procedure can be divided into three steps:

1. Preparation of a suitable colored solution.
2. Measurement of the light absorptive capacity (absorbance) of the solution.
3. Calculation of the concentration of colored species present.

1. *Preparation of the Colored Solution*

As explained in a preceding section, for optimum results the absorbance of the test sample should lie between 0.2 and 0.7.

By the Beer-Lambert law $A = \varepsilon \cdot b \cdot c$; hence, $0.7 > \varepsilon b c > 0.2$. The range of cell thicknesses (b) used commonly vary from 0.5 to 4 cm, the 1 cm cell predominating.

Self-colored constituents (e.g., permanganate ions) usually have ε_{max} values of the order of 10 to 100 liter mole^{-1} cm^{-1}; hence in order to obtain absorbance readings within the desired range, the concentration (c) needs to be of the order of 10^{-2} to 10^{-3} molar.

The molar extinction coefficients of the intensely colored species formed by adding selected complexing agents can be over 100 times larger than those of the simple hydrated species, and in these circumstances the concentration limits are accordingly lower (e.g., 10^{-4} to 10^{-5} molar which is equivalent to units of mg liter^{-1} or parts per million of solute in the solvent).

For the determination of the minor constituents of samples, these are the concentration conditions liable to be encountered; hence, the preliminary step in colorimetric analysis is generally the formation of some intensely colored complex species.

This color forming stage, in which a suitable reagent is added to the test solution, involves a chemical process and as such is subject to the influence of competing equilibria.

The primary problem is to transform all of the desired constituent (A), or something chemically equivalent to it, into the colored state. If the chromogenic reagent is represented by R, the equilibrium constant for the reaction

$$aA + rR \rightleftarrows A_aR_r \text{ (colored)}$$

should have a large numerical value to ensure that $[A_aR_r]/[A] > 100$.

With the extremely small amount of the species A initially present, the reaction can be slow or nonstoichiometric, and usually a large excess of reagent is added to both accelerate the reaction and ensure maximum conversion to the colored species. In some color forming processes the solutions are heated to increase the reaction rates. If true equilibrium has not been established at the time of color intensity measurement, variable results are obtained. Hence some systems are reported to increase in intensity with time of standing while others fade (due to sudsequent destruction of A_aR_r). It is highly desirable that the solution be sufficiently stable to allow reliable and reproducible measurements to be made.

Since the color reaction is an equilibrium process, the magnitude of the term $[A_aR_r]/[A]$ must vary with any factor that influences the magnitude of $[A]$ and $[R]$ in the original solution.

For example, in the presence of another ligand capable of reacting with A there is competition for A and this can result in either partial or minimal conversion to A_aR_r. The final equilibrium position is determined by the respective concentrations of the competing ligands and the relative stabilities of the complex species formed. When the solution contains a number of components capable of reacting with the chromogenic reagent, it is often possible to restrict reaction to one component by adding a competing ligand system which preferentially reacts with the interfering species. This "masking" procedure is explained in more detail in Chapter 10.

The concentration of the reagent R available for reaction can also be influenced by conditions in the solution phase. Of major importance is the pH of the solution, since complex formation quite regularly involves displacement of a proton from the reagent and introduces a $[H^+]^x$ term in the equilibrium equation. On other occasions, other constituents of the sample form stable complexes (not necessarily colored) which consume most of the reagent and leave little for interaction with the desired species (A).

Control of pH, the addition of masking agents, and extraction of the species A_aR_r into a nonaqueous solvent are some of the means used, either singularly or in conjunction, to ensure that the final colored product represents a derivative of constituent A free of interference effects.

An indication of the extremely large number of combinations of these factors currently in use may be obtained by perusing Tables 6.10 to 6.12 in the "Handbook of Analytical Chemistry." These tables summarize the principles of the methods recommended for the colorimetric determination of metal ions, nonmetallic species and organic substances.

TABLE 6.4
SPECTROPHOTOMETRIC METHODS[a]

Component determined	Method or reagent conditions	Wavelength (mμ)	Range (ppm)	Interferences	References
Co	Nitroso-R salt (I); 0.5% I/H_2O, pH 5.5 ± 0.5, destroy excess I with 3% $KBrO_3$, Co:R = 1:3	425	0.1–1.0	Fe(III); Cl⁻, F⁻	AC, *28*, 1151 (1956)
	SCN⁻; 44% aq NH_4SCN, pH 3–5, ext. with i-$C_5H_{11}OH$	312	0.2–10	Bi, Cu, Fe(III), Ni, Ti, U(VI)	AC, 27, 1731 (1955)
	2-Nitroso-1-naphthol (I); 1% I/glacial HOAc, Cit. buffer, H_2O, pH 3–4 ext. with $CHCl_3$, wash with NaOH soln. to remove excess I	530	0.2–4	Fe(II), Pd, Sn(II)	ACA, *12*, 547 (1955)
	2,2′,2″-Terpyridine (I); 0.1% I/H_2O, pH 2–10, Co:R = 1:2	505	0.5–50	Cr, Cu, Fe(III), Ni, V(V)	IEC, AE, *15*, 74 (1943)
Ascorbic acid	Phosphomolybdate; 0.25% NH_4 phosphomolybdate/0.85 M H_2SO_4, 90°—10 min	825–837	0.1–3	Reducing agents, urea	AA, *4*, 3139 (1957)
	2,6-Dichlorophenolindophenol(I); 0.34% I, buffer (pH 3.6), measured at 15 and 30 sec	520	0–15	Reducing agents	IEC, AE, *13*, 793 (1941); *17*, 754 (1945)

[a] Extracted from the "Handbook of Analytical Chemistry," edited by L. Meites. Used with permission of McGraw-Hill Book Company; copyright © 1963, McGraw-Hill Book Company.

In each case the summary indicates a suitable chromogenic reagent, mentions masking agents which are recommended to minimize interference, suggests pH limits and solvents for extraction. In addition the recommended wavelength for colorimetric comparison and concentration range is listed.

For example, for the determination of cobalt, nine colorimetric methods are summarized and the data for four of these is shown in Table 6.4. Also shown in this table is an example covering the analysis of an organic compound, viz., ascorbic acid. For each method, a reference is given which allows the original literature to be consulted if more details are desired. By means of such extensive tabulations it is possible to rapidly assess the feasibility of using colorimetric analysis for any new problem encountered. Admittedly, these tables do not include the very latest methods available in the current literature, but for a large proportion of analytical requirements, the tabulated material is sufficiently comprehensive.

Other major sources of information on colorimetric methods are the extensive monographs now on the market. These reference books usually provide full experimental details for all the important alternative methods known at the time of publication.

2.   *Absorbance Measurements*

The absorbance of the prepared colored solution is measured with a filter photometer or spectrophotometer.

By means of colored filters (photometer) or a monochromator unit (spectrophotometer) selected bands of visible radiations are isolated from the continuum emitted by a tungsten lamp and are fed through a sample cell to a photoelectric detector. The response of the detector with pure solvent in the cell provides a measure of P_o, the response when the colored solution is placed in the light deam is a measure of P, the transmitted radiant power. Instead of using these values directly the circuits for most instruments are arranged so that values of P/P_o (transmittance) or $\log P_o/P$ (absorbance) can be read directly off a scale.

The wavelength of radiation used in the measurement is usually the λ_{max} wavelength determined from an independent spectral curve. Use of this wavelength increases the selectivity and sensitivity of the determination. For established procedures, the most suitable wavelength is quoted in the literature; for new applications λ_{max} has to be determined by observing the absorbance of a test solution as the wavelength of the incident radiation is varied.

In some situations the test solution can contain more than one colored species (e.g., the reagent R may absorb as well as the complex). If the concentration of the interfering color is constant in all the standards and assays, the effect is corrected by the calibration graph. In most situations, however, the color intensity is variable and its effect has to be overcome by means of the multicomponent analysis system outlined in Section I,B, or by selective extraction of one species. Occasionally it is possible to selectively destroy one of the colored components.

With all instruments it is desirable to regularly calibrate both the wavelength scale and the photometric scale. A sample of didymium glass which has numerous sharp absorption bands is useful for checking the wavelength scale of spectrophotometers. For checking the accuracy of the photometric scale several standard solutions have been recommended including a solution containing 0.0400 gm of potassium chromate per liter in 0.05 molar potassium hydroxide. This solution should give a molar absorptivity of 482 at 370 mμ.

3. *Calculations*

By the Beer-Lambert law, the observed absorbance equals abc, and by using a standard solution the appropriate value of a can be calculated. With the cell thickness, b, also known calculations associated with test assays devolve into substitution in the equation $A = abc$.

Since this approach makes no allowance for "deviations from the law" the preferable technique is to use a series of standard solutions to prepare a calibration graph showing absorbance versus concentration. Even this approach only has real validity if the composition of the standards closely resembles the composition of the assay.

The optimum concentration range for a specific method and a specific instrument can be ascertained by preparing a plot of percent transmittance against the logarithm of the desired constituent concentration. If the range of concentrations examined is broad enough this graph will be observed to possess two curved segments joined by a linear section. The virtually linear segment of the "Ringbom" plot corresponds to the optimum concentration range.

Colorimetric methods have been developed for the determination of metals, nonmetals and organic compounds in an almost endless series of materials. The technique is widely used in the fields of clinical medicine, fats, fertilizers, foods, soaps and detergents, water analysis, mineral analysis, and metal analysis.

The major virtues include comparative simplicity, and reasonable precision when dealing with minor components of samples. With modern instrumentation and careful calibration, overall errors can be kept within the ± 1–2% range in most applications.

D. SPECIAL TECHNIQUES AND MODIFIED PROCEDURES

1. *Differential Techniques*

When colored solutions having absorbance values above 1 ($<10\%$ transmittance) are to be measured, the photometric error can be minimized by using a differential technique.

In the "transmittance ratio method" the 0% transmittance (T) setting is made with the phototube in total darkness and the 100% setting is made using a colored solution just slightly less concentrated than the most dilute solution to be measured. For example, with samples exhibiting about 1% transmission when measured against pure solvent as the reference material, optimum precision is obtained by using as the standard reference a solution possessing an absorbance of about 2 when measured against pure solvent. To allow for the small amount of light being transmitted and to permit reasonable detector response, instrument slit widths have to be increased.

The increased precision obtained by using this ratio method allows more concentrated solutions to be studied, and the final results have an accuracy comparable with gravimetric and titrimetric methods. As in all colorimetric procedures, the preparation of a calibration graph is recommended, since Beer's law is not necessarily followed.

For dilute solutions, increased precision can be achieved by means of the "trace-analysis method." In this modification of the conventional spectrophotometric technique, the 100% T setting is made as usual by having pure solvent (or a blank) in the cell. However, the instrument is adjusted to show 0% transmission when the light beam is passing through a reference solution which is only slightly more concentrated than the most concentrated unknown or standard to be examined.

The simplest physical interpretation of both these differential techniques is that the procedures expand the photometric scale.

In the "ultimate-precision method," two reference solutions are prepared, one more concentrated and the other more dilute than the standards or unknowns to be studied. These two reference solutions are then used to make the 0% and 100% transmittance settings.

For maximum precision the absorbance of the test solution (against pure solvent) should be about 0.43 (37% T) with the absorbances of the reference solutions being only slightly above and below this value.

This technique is so sensitive that the small differences in the optical path lengths in sets of cells can introduce significant errors. It is therefore necessary to make cell corrections to observed results. These corrections can be made by measuring the transmittance of several standard solutions in cells 1, 2, 3, etc. For each cell, the observed transmittance for each standard solution is plotted against the transmittance observed for the corresponding standard in cell 1 (considered as the reference cell). This gives a series of straight lines (if Beer's Law is obeyed) of differing slope. If cell 2 is used in any subsequent analysis, the appropriate cell correction is then obtained by using vertical projection to find the transmittance which would have been observed if cell 1 had been used.

By means of the ultimate precision method it is possible to determine the major component in a salt or alloy with an accuracy comparable to that obtained by any other technique.

2.   *Multicomponent Analysis*

Under the proper conditions it is possible to determine the concentration of two or more colored components of a solution.

From the absorbance-wavelength spectrophotometric curves for standard solutions of all the components, wavelengths are selected which represent maximum differences in absorbances; i.e., for each component a wavelength is chosen on the basis that this component has a high absorbance value in this region while the other components possess low values.

A calibration plot of absorbance versus concentration is then prepared for each component to test if the various species obey Beer's law at the selected wavelength.

The absorbance of the multicomponent system is measured at each of the selected wavelengths and the concentration of each component present is calculated using a series of simultaneous equations as outlined in Section I,B of this chapter.

3.   *Spectrophotometric Titrations*

In a spectrophotometric titration the transmittance or absorbance at a suitable wavelength is measured after the addition of increments of titrant. Usually the absorbance is plotted against the volume of titrant added and if the titrant is much more concentrated than the solution being titrated, two intersecting straight lines are obtained. The point of intersection represents the end point.

In principle the method may be used when either reactants or products absorb, but for practical purposes there should be a significant difference in the absorptivities of the various species.

The technique provides a means of locating the equivalence point in many titrations and is particularly advantageous in titrations at low concentrations (10^{-4} M or less). It can also be used to determine the stoichiometry of reactions occurring in solution and for the analysis of multicomponent colored systems. In the latter case the titrant is chosen so that it reacts with only one of the colored species.

IV.   INFRARED ABSORPTION SPECTROMETRY

A.   Vibrational Energy Changes

The term infrared spectrometry normally denotes the study of the absorption of radiation having wavelengths of 1 to 1000 μ. For convenience this range is usually divided into three regions. The "near infrared" is the region between 0.8 and 25 μ (12,500 to 4000 cm^{-1}); the "mid-infrared" is between 2.5 and 50 μ (4000 to 200 cm^{-1}); and the "far infrared" lies in the spectral region of 50 to 1000 μ (200 to 10 cm^{-1}).

Absorption spectra in the infrared originate in transitions between vibrational and rotational levels of a molecule present in its ground electronic state.

The absorption of far infrared radiation results in a change of rotational energy only. Pure rotational spectra are observed only with gases; in liquids and solids the rotational states are no longer well defined and any absorption produces a continuum rather than discrete lines. When radiations from the other infrared regions are absorbed, the main effect of rotational changes on the spectra of condensed phases is a broadening of the vibrational bands. Because the majority of studies of analytical interest involve condensed phase spectra and the shorter wavelength radiations, the discussion in this section will be concerned mainly with vibrational energy changes.

The atoms in a molecule are never stationary and it is a good approximation to treat the vibrating molecule as a number of point masses held together by Hooke's Law forces. In fact, the vibrational behavior of two atoms connected by a bond is similar to that of a pair of spheres connected by a spring. For small displacements, the restoring force is proportional to the displacement, and if such a system is put in motion, the vibrations are described by the law of simple harmonic motion.

With atomic systems the rate of vibration depends on the mass of the atoms and the strength of the bond connecting them. These parameters are related by an equation such as

$$\nu = \frac{1}{2\pi} \sqrt{k(M_a + M_b)/M_a M_b}$$

where ν is the appropriate stretching frequency, M_a and M_b are the masses of the two atoms in the chemical bond, and k is the force constant of the chemical bond.

The amplitude of the oscillation of the atom is of the order of 10^{-9} to 10^{-10} cm, while the vibrational frequencies are of the order of infrared radiation, namely 10^{12} to 10^{14} cps. Single, double, and triple bonds exhibit force constants of about 5×10^5, 1.0×10^6, and 1.5×10^6 dynes/cm, respectively.

N uncombined atoms, free to move in three dimensions, possess $3N$ translational degrees of freedom. When the N atoms are combined into a molecule, the total number of degrees of freedom remains $3N$. However, in a molecule three degrees of freedom are required to describe the translation of the center of gravity of the molecule. For a nonlinear molecule another three degrees are required to describe the rotation of the molecule since the rotation may be resolved about three perpendicular axes.

If it is assumed that there are only three types of motion, then: translational + vibrational + rotational degrees of freedom = $3N$, which means that there are in general, $3N - 6$ coordinates left to describe the motions of the atoms relative to each other for a fixed orientation of the molecule, i.e.,

vibrational degrees of freedom = $3N - 6$

Linear molecules require only two rotational coordinates and have $3N - 5$ vibrational degrees of freedom.

By classic mechanics it can be shown that the displacements of the masses from their mean positions are always the sum of the displacements of particular sets of vibrations. In these particular vibrations the masses move in straight lines and in phase. When the masses are in phase the nuclei pass through their mean positions and turning points simultaneously, so that the center of gravity of the molecule remains unaltered. The number of sets of permitted vibrations equals the number of vibrational degrees of freedom $(3N - 6)$ and these vibrations are known as the normal or fundamental modes of the molecule. A normal mode of vibration is, therefore, one where, in the absence of other normal modes, each nucleus executes simple harmonic oscillations in a straight line about its equilibrium position.

The nuclear displacements actually undergone when the molecule is carrying out only one of these normal modes may be described as the normal coordinate of that mode. Frequently a normal mode is localized largely in a group in the molecule, and it is then possible to describe the normal coordinate in terms of the stretching or bending of one or a few bonds.

Thus, some of the molecular vibrations are characteristic of the entire molecule ("fingerprint" vibrations) and others are associated with certain functional groups.

Vibrations can be classified into two major types—stretching and bending.

Stretching vibrations are those in which two bonded atoms continuously oscillate, changing the distance between them without altering the bond axis or bond angles.

Stretching vibrations are either isolated vibrations (e.g., —O—H, —C=O), or coupled vibrations (e.g., the methylene group or benzene molecule).

Some of the various types of vibration are shown schematically in Fig. 6.8.

Bending vibrations are characterized by a continuously changing angle between the bonds. They may be subclassified as wagging, rocking, twisting, or scissoring.

Wagging describes a vibration in which a nonlinear three-atom structural unit oscillates back and forth in the equilibrium plane formed by the atoms and their two bonds; while in rocking vibrations the structural unit oscillates back and forth out of the equilibrium plane. In scissoring, two nonbonded atoms move back and forth towards each other. Rotation of the structural unit around the bond which joins it to the rest of the molecule is termed twisting. These various modes are illustrated in Fig. 6.8.

Stretching and bending vibrations are quantized, and the total vibrational energy can only have values which satisfy the equation

$$\text{Total energy } E \text{ (ergs)} = \sum_j E_j = h\nu_1(\nu_1 + \tfrac{1}{2}) + h\nu_2(\nu_2 + \tfrac{1}{2}) + \cdots.$$

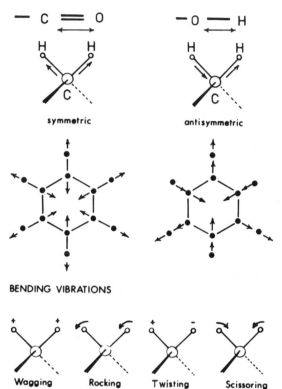

FIGURE 6.8   Examples of different modes of molecular vibration; $+$ indicates movement out of the plane of the paper, $-$ indicates movement back from the plane of the paper.

where ν_j is the frequency of the jth normal mode, and v_j is the vibrational quantum number which may take integral values only, 0, 1, 2, 3, etc.

When electromagnetic radiation of the same frequency as one of the normal modes of vibration is directed at the molecule, absorption may occur. In the process of absorption reinforcement occurs and the amplitude of the molecular vibration is increased. The excess energy released as the molecule falls back to the vibrational ground state is dissipated as heat.

In order that the vibrating molecule should interact with the oscillating electric field of the incident radiation and undergo a transition between two of its energy levels, the molecular electrical dipole moment must change its magnitude or its orientation during the motion. The intensity of the absorption band is determined by the magnitude of this dipole moment change.

Owing to the symmetry of the molecule some vibrations do not induce a change in the dipole moment and these are not detected in an infrared spectrum.

Many of the other possible modes of vibration in a polyatomic molecule can be extremely weak absorbers or degenerate; hence the number of fundamental modes observed in an infrared spectrum is usually less than the theoretical value of $(3 N - 6)$. On the other hand, in experimental studies, some molecules appear to exhibit more than the normal number of vibrational absorptions. The larger number can be attributed to "overtones" (multiples of normal frequencies, 2ν, 3ν, etc.), "combination tones" (i.e., $\nu_1 + \nu_2$, $\nu_3 + \nu_4$, etc.), and "difference tones" $(\nu_1 - \nu_2, \nu_6 - \nu_5$, etc). Usually these overtones and combination tones give rise to weak absorption bands. The appearance of these bands indicates that the motion of the atoms is not truly harmonic, since the transitions involved are not permitted if calculations are based on a harmonic oscillator model.

The large number of modes of vibration possessed by polyatomic molecules means that a large amount of information is contained in an infrared spectrum. The vibrational spectrum is a highly characteristic property of the molecule, and it has proved extremely useful for qualitative identification purposes and for the elucidation of structure.

B. INFRARED SPECTROMETRY

Infrared spectrometry is a widely used analytical technique and instrument makers offer a wide variety of instruments with varying degrees of sophistication.

In principle they all have a source, which consists of a hot body whose emanations are dispersed by a suitable monochromator unit. The intensity of selected frequencies is then measured by a suitable detector before and after passing through the sample. In most instruments, absorbance versus wavenumber spectral curves are automatically recorded, the movement of the recorder being synchronized with the changing frequency of incident radiation.

The most important characteristics of the infrared spectrometers are probably the nature of the optical system and the sensitivity of the detector–recorder system.

All species tend to absorb infrared radiations of some frequency or other; hence it is necessary to ensure that components of the apparatus do not absorb more incident radiation than the samples being studied. For various types of material, absorption is minimal below a certain wavelength known as the "cut-off" point, and some of the approximate values are listed in Table 6.5.

The most widely used segment of the infrared range is the "middle region"; hence the range of the majority of instruments extends to 25μ (400 cm^{-1}). It can be seen from Table 6.5 that the optics of the instruments, the prisms, sample cells, etc. should, therefore, be manufactured from materials such as KBr, CsBr, TlBr, etc. For many routine purposes, however, the limit imposed by NaCl components is not critical and quite a number of commercial instruments use this material.

TABLE 6.5

TRANSMISSION RANGE OF SOME MATERIALS USED IN INFRARED SPECTROSCOPY

Material	Range (μ)
Glass	up to 2.5
Quartz	up to 3.5
NaCl (rock salt)	up to 16.0
AgCl	up to 22.0
KBr	up to 25.0
CsBr	15 to 40.0
TlBr	15 to 40.0
CsI	15 to 50.0

All infrared spectrophotometers use mirror optics, since the reflection from most metallic surfaces is generally very good in the infrared region.

Two types of instrument are available, namely the single and double beam. In the double beam instrument, the radiation from the source is split into two exactly equivalent beams, one of which passes through the sample while the second beam traverses a reference cell. By means of a rotating, semicircular front-surfaced plane mirror, the beams are focused alternatively onto the entrance slit of the monochromator. When the radiant energies of the beams are equal, as in the absence of sample, the detector produces a dc signal which is not amplified by the ac amplifier circuit. When the intensity of the two beams is unequal due to absorption by the sample, an ac signal is produced (the frequency being controlled by the rate of rotation of the mirror). This signal is amplified and is used to drive a servo motor which moves an optical attenuator into or out of the reference beam to equalize the beam intensities. The position of the attenuator is a measure of the transmittance of the sample, and if coupled to a pen recorder its movements are a record of transmittance.

A schematic representation of the optical path in the optical-null type of double beam spectrophotometer is shown in Fig. 6.9.

The double beam system was introduced to facilitate the rapid recording of spectra covering a wide range. With this type of instrument it is also possible to compare two samples directly, giving "difference" spectra. This facilitates elimination of effects due to extraneous absorption, e.g., absorption by the solvent. Another important aspect is the automatic compensation for the absorption of infrared radiations by water vapor and carbon dioxide in the atmosphere.

There are technical problems associated with this type of instrument but these are usually compensated by appropriate adjustments to the circuit. For high quality instruments it is essential that the recorder pen should respond rapidly and sensitively to any small change in the absorbance of the sample.

The spectrum is scanned by rotating the dispersing medium about a vertical axis, and if suitably calibrated the recorder trace should represent an accurate plot of transmittance against wavelength (or wavenumber) of incident radiation.

The single beam instrument functions with one beam only, and the instrument measures directly the amount of energy transmitted by the sample and the spectrometer optics.

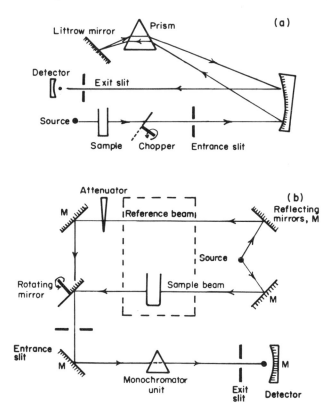

FIGURE 6.9 Schematic diagrams showing the optical path in (a) a single beam infrared spectrophotometer and (b) a double beam infrared spectrophotometer.

Usually, a "blank" measurement is made by inserting in the beam either an empty sample cell (for gases) or a cell containing the solvent (in solution studies). This provides a measure of the intensity of the incident radiation (P_o), and the position of the pen on the recorder is normally adjusted to a suitable position by varying the slit width and amplifier gain.

The value of the transmitted radiation (P) is measured immediately afterwards by substituting a sample cell for the blank. A single beam instrument yields more accurate transmittance values than a double beam instrument, and any malfunctions are more readily apparent. Hence it is well suited to quantitative studies. To ensure an accuracy of better than $\pm 3\%$ in the transmittance readings, the source, detector and amplifier units must be highly stable. In order to scan transmittance as a function of frequency, however, the slit width has to be varied to maintain a constant P_o value. This operation is not as convenient as double beam operation.

Since a single beam is preferable for quantitative analysis, and double beam operation is preferable for qualitative scanning, many commercial instruments are designed to operate by either mode.

The spectrophotometers are calibrated by recording the spectra of substances whose exact absorption wavelengths are known. The materials should have absorption bands which are sharp, easily recognized, and found at frequent intervals throughout the range of the spectrophotometer.

A thin film of polystyrene has a number of sharp absorption bands occurring at wavelengths between 3 and 15 μ, and this material is widely used to check the calibration of commercial instruments. Other suitable substances for checking more restricted ranges are the mercury arc (0.5 to 1.5 μ), water vapor (5 to 7 μ), ammonia (8 to 14 μ), and methyl alcohol (12 to 24 μ).

As indicated by this list, the infrared technique can be used to study solid, gaseous, or liquid samples. Infrared spectrometry is thus applicable to almost any sample, the procedure differing only in respect to the manner of sample preparation.

The sample cells for vapors and gases normally consist of a glass or metal case (several cm long) fitted with end "windows" constructed of some material which is transparent in the part of the infrared region chosen for investigation. The cell is fitted with inlet and outlet taps which permit it to be attached to any apparatus designed to facilitate the manipulating of gas flows and the measurement of pressures. The concentration of vapor present in a cell can be varied by changing the partial pressure of the vapor. The total pressure is preferably maintained near atmospheric pressure; hence, pure dry nitrogen gas is widely used as a diluent.

Where the initial gas samples are themselves very dilute, the sensitivity of the technique can be increased by increasing the length of the sample cell. The sampling areas in most instruments limit the length of sample cells to about 10 cm but longer path cells are available commercially. These cells employ multiple reflection within a reasonable volume of gas, and path lengths equivalent to 40 meters may be obtained.

The infrared absorption bands of most liquids and solids are quite intense; hence the thickness of the samples employed is quite small, e.g., between 20 and 100 μ.

Cells for use with liquids, therefore, consist of two flat polished "windows" separated either by lead or copper spacers (demountable cells) or by a micrometer screw arrangement (adjustable cell type). Two holes are provided for filling.

The thickness and an indication of the quality of a cell may be obtained by recording the interference pattern observed when the spectrum is scanned with the empty cell in the light beam. The better the quality of the cell, the greater is the ordinate separation between maxima and minima and the smoother is the curve. The distance between the maxima is used to calculate the cell thickness. An accurate determination of this value is important in quantitative studies, particularly if precise values of molar extinction coefficients are to be calculated.

For qualitative liquid phase spectra, a suitable sample may be obtained by squeezing some of the liquid between two plates of infrared transmitting material. The resulting capillary film of liquid is about 10 μ thick.

The solvent used to prepare solutions has to meet two specifications. First, it must dissolve the sample, and second, it must transmit infrared radiation in the region where the solute absorbs. Tabulations in the "Handbook of Analytical Chemistry" and infrared spectroscopy monographs list the absorption bands of common solvents. Two widely used solvents are carbon tetrachloride (for the 4000–1330 cm^{-1} range) and carbon disulfide (for the 1330–450 cm^{-1} range). It can be observed from the short list of solvent absorption bands recorded in Table 6.6 that these solvents have at least one absorption band within the spectral ranges recommended, and due allowance has to be made for this in interpreting the solute spectrum.

TABLE 6.6

SOLVENT INFRARED ABSORPTION BANDS[a]

| Solvent | Absorption Bands | |
	(μ)	(cm^{-1})
Chloroform	3.3–3.5	3030–2850
	4.1–4.3	2440–2330
	6.6–7.1	1520–1410
	10.6–15.0	943–666
Nujol oil	3.4–3.8	2940–2630
	6.8–7.4	1470–1350
Carbon disulfide	4.2–4.8	2380–2080
	6.1–7.1	1640–1410
	11.4–11.8	877–847
Cyclohexane	5.0–6.6	2000–1500
	12.5–20	800–500
Carbon tetrachloride	6.2–6.5	1610–1540
	7.8–8.3	1280–1200
	9.9–10.5	1010–950
	11.7–15.0	855–666

[a] Range 2–15 μ.

Other useful solvents include benzene, chloroform, dioxane, acetone, acetonitrile, methylene chloride, nitromethane, pyridine, dimethyl formamide, tetrachloroethylene, and dibromo- or tribromomethane. Many of these are of greatest value in quantitative studies where measurements are made at a few selected wavelengths only.

The concentration of solute required for use in a 0.1 mm cell is about 10% (i.e., 0.1 gm in 1 ml).

Besides absorbing infrared radiation, solvents can interact with the solute (e.g., through hydrogen bonding) causing shifts in the observed absorption spectrum. This effect has to be considered in interpreting spectral data in terms of qualitative composition. Conversely, infrared absorption has proved a convenient means of studying solute–solvent interaction.

For qualitative analysis it is often more convenient to examine the solid directly rather than form a solution. A solid may be examined in several ways, some of which are given below.

In the mull technique, finely ground solid is wetted with a suitable liquid which has only a few absorption bands in the infrared. Nujol oil, fluorinated hydrocarbons, and hexachlorobutadiene have been widely used for this purpose. The suspension is then placed between two NaCl or KBr plates. The function of the liquid is to surround the fine particles with a medium of high refractive index and so reduce light scattering.

Infrequently it is possible to obtain a solid sample in a layer thin enough to examine directly, e.g., a thin layer of plastic may be shaved off the surface of a solid specimen. Alternative methods of obtaining such a thin layer are evaporation and melting. In the evaporation method, a solution or suspension of the sample is evaporated on the surface of a suitable transparent plate. In the melting method, a small amount of sample is caused to melt and flow over the surface between two plates.

More widely used than these two methods, however, is the pressed disc technique. In this procedure, the solid sample is ground up with an excess (200 times) of powdered KBr. The mixture is then compressed under vacuum, using pressures of several tons/square inch, to form a clear transparent disc. The disc is then inserted in the sample beam of the spectrometer. The spectrum is usually sharper than that obtained by the mull technique, but the electric field of the positive and negative ions can modify some spectra. In addition, unless great care is taken, the discs exhibit bands due to absorbed water. Other metal halides or other substances can be used in place of KBr.

A useful modification makes use of the fact that under conditions of total reflection at a boundary between two media of different refractive indices, the radiation penetrates a short distance into the medium of lower refractive index. The reflected radiation, when examined spectrally, shows the absorption spectrum of the medium of lower refractive index. This is known as attenuated total reflection (ATR).

A special attachment is required for this technique which is of particular value for samples which are difficult to examine by the usual absorption technique. Simple units employ one reflection at the boundary between an infrared transmitting material (e.g., AgCl or germanium) and the sample; other units increase sensitivity by the use of multiple internal reflections.

C. APPLICATIONS

Infrared spectroscopy is a technique which has been found of value in a variety of fields and it has been applied to the study of an extremely wide range of materials.

In broad terms, it is used for qualitative analysis, quantitative analysis, and structure elucidation.

Qualitative analysis usually proceeds in two steps, identification of characteristic groups and confirmation of molecular structure by comparison with standard spectra.

The spectra recorded by the spectrometers normally contain a large number of significant peaks which can be identified in terms of frequency and intensity, the absorption intensity being described in terms of the symbols: s, strong, m, medium, w, weak, and v, variable. The broadening due to rotational changes is sometimes noted by the symbol b, but for interpretation purposes the absorption peaks are attributed solely to vibrational changes.

The frequency of a characteristic vibration in a molecule is controlled by the type of atomic groupings in the structure and the type of environment in which these groupings exist.

For example, in the molecule

the carbonyl ($-C{=}O$) stretching frequency depends on the nature of the substituents R_1 and R_2. When R_1 and R_2 are $-CH_2$ groups, the frequency of the ketonic carbonyl vibration is 1720–1740 cm^{-1}. Replacement of one $-CH_2$ by a phenyl group (C_6H_5) shifts the range to 1680–1700 cm^{-1}, while dual substitution to give a diaryl ketone moves the peak to the 1660–1670 cm^{-1} region. The $-C{=}O$ stretch vibration is observed in the spectra of many different types of compounds, including anhydrides, acyl halides, esters, aldehydes, ketones, and carboxylic acids, and strong peaks attributable to this vibration have been observed between 1670 and 1870 cm^{-1}. In fact, tables listing the correlation between absorption band ranges and characteristic groups show over forty separate entries for this carbonyl stretch in different environments.

TABLE 6.7

CHARACTERISTIC INFRARED ABSORPTION BANDS

Vibration	Group	Absorption wavenumber (cm^{-1})	Intensity
a. *Stretching modes*			
C—H	alkane (—CH$_3$—CH$_2$)	2962–2853	m-s
	alkene-vinyl	3095–3075	m
	-disubstituted,	3040–3010	m
	cis and trans	3040–3010	m
	alkyne	$\simeq$3300	s
	aromatic	$\simeq$3030	v
	aldehydes	2900–2820	w
		2775–2700	w
N—H	secondary amines	3500–3310	m
	secondary, bonded amides	3320–3140	m
	amine salts	3130–3030	m
O—H	alcohols and phenols		
	free O—H	3650–3590	v
	hydrogen bonded	3570–3450	v
	carboxylic acids	2700–2500	w
S—H	sulfur compounds	2600–2550	w
C=C	alkenes-nonconjugated	1680–1620	v
	-substituted	$\simeq$1667	m
C≡C	alkyne-disubstituted	2260–2190	v, w
C—N	aromatic amines	1340–1250	m
C=N	alkyl imines, oximes	1690–1640	s
C≡N	isocyanates	2275–2240	m
	alkyl nitriles	2260–2240	m
	aryl nitriles	2240–2220	m
C—O	alkyl ethers	1150–1060	m
C=O	sat. acyclic esters	1750–1735	s
	unsat. vinyl ester	1800–1770	s
	carbonates	1780–1740	s
	sat. aliph. aldehydes	1740–1720	s
	sat. aryl aldehyde	1715–1695	s
	sat. aliphatic carboxylic acids	1725–1700	s
	carboxylate anion	1610–1550	s
		1400–1300	s
C=S	organo sulfur compounds	1200–1050	w
—N=N—	azo compounds	1630–1575	v
b. *Bending modes*			
N—H	primary amines	1650–1590	s-m
	secondary amines	1650–1550	w
	amine salts	1600–1575	s
		$\simeq$1500	s
	primary amides	1620–1590	s
	secondary amides	1550–1510	s
C—H	alkanes-CH$_2$—	1485–1445	m
	alkenes-disubstituted	$\simeq$690	s
O—H	carboxylic acids	1320–1210	m
		950–900	s

164

Thus, in order to correlate absorption peaks with characteristic groups, it is desirable to have on hand data for the different types of vibrational groupings in different environments. Extensive tabulations of this kind are available (e.g., in the "Handbook" and in "Infrared Monographs") but considerable experience is required in the application of such data since the absorption peaks of different types of material tend to overlap.

The location of a few characteristic group infrared absorption regions are shown in Table 6.7. The primary aim of this list is to indicate the effect of environment, type of atoms involved, bond type, etc., on the frequency of vibration but many cases of overlap can also be observed.

With due care it is possible to assign the major peaks of an unknown spectrum. This may be sufficient to allow prediction of the type of compound involved. The unknown spectrum can then be compared with the standard spectra of materials of this type. Very extensive compilations of standard infrared spectra are available, both in "atlas" and "card" form, but as these are somewhat expensive only a limited number of laboratories tend to have complete sets. As a general rule, no two compounds have identical spectra; hence a similarity between the spectrum of the unknown and that of a standard provides fairly convincing proof of the nature of the unknown.

Characteristic group frequencies represent a powerful device for identification purposes, provided the influence of environment is always considered. The effect of environment can be attributed to many factors, among which can be listed conjugation effects, chelation, hydrogen bonding, ring strain, ionic forms, and electronegativity effects.

In structural studies, the position of the characteristic absorption peaks is used to elucidate the juxtaposition of the various groupings in the molecule. Peak shifts may indicate inter- and intramolecular bonding, solute–solvent interaction, isomeric transformations, etc., and accordingly infrared absorption is used in a wide variety of theoretical studies, involving most branches of chemistry.

Component and structural information obtained solely from the analysis of an absorption spectrum seldom provides enough data to make an absolute identification of an unknown compound, although identification is becoming, more feasible with the introduction of computer programs and other automated means of recording and comparing spectra.

In general it is necessary to make positive identification by combining the infrared radiation data with information provided by other methods such as elemental analysis, ultraviolet absorption, mass spectrometry, and nuclear magnetic resonance spectroscopy.

Quantitative studies in the infrared region are based on the Beer-Lambert law, $A = \varepsilon bc$. For a particular absorption band a calibration curve is constructed by plotting A against c for solutions of known concentration.

Two or three component mixtures can be analyzed in favorable cases by making absorbance readings at several wavelengths.

In principle, the procedure involved is identical with that used in ultraviolet and visible spectrometry. However, since the peaks are much sharper and subject to solvent interaction, deviations from Beer's law are more frequent.

A systematic error can be introduced in the determinations by the variable background absorption which is often encountered. The practical method of dealing with this is the base line method, which requires that a straight line be drawn across the base of the band such that it touches the two points tangentially.

The transmittance indicated by this line at the frequency corresponding to the absorption peak is taken as a measure of P_o, the incident radiation intensity. This is a reasonable approximation unless the neighboring bands arise from other components in the mixture. The transmittance at the peak maximum provides the value of the transmitted intensity (P) required to complete the calculation of the absorbance $(A = \log P_o/P)$.

The need to draw a baseline requires that at least a small region of the spectrum should be scanned for every sample measurement. In this respect it differs from the UV-visible technique where readings are made using a single wavelength setting.

For multicomponent mixtures it is sometimes not possible to find characteristic bands which are not overlapped by bands of the other components present. With weak absorption and overlapping bands, errors of $\pm 15\%$ are possible, but in favorable circumstances accuracies of a fraction of a percent have been achieved. Despite these limitations one important advantage of IR measurements is that it can be applied to problems which are very difficult or even impossible to solve by other analytical techniques, e.g., the estimation of one isomer in the presence of many others.

With samples in solution, the quantitative procedure is similar to the UV-visible technique and requires no elaboration here. Where solids are involved, however, some modifications are required. Using KBr discs two approaches can be adopted, viz., one can measure the thickness of the disc and use this as b in the basic equation or one can add a known quantity of a standard substance to each disc. Comparison of the absorbance of the unknown at a suitable wavelength with that of the standard (at its λ_{max}) permits a quantitative determination without the need for a thickness figure.

The analysis of gas mixtures for a single component can be carried out effectively by means of infrared absorption provided that the desired component has an absorption band which is located in a region where other gases do not absorb. For quantitative studies the effect of pressure has to be considered, but the method is ideally suited to monitoring the composition of gas streams of reasonably simple composition.

Qualitative and quantitative analyses represent only part of the list of possible applications of infrared spectroscopy. The technique has been widely used to study the structure of inorganic complexes, intermolecular hydrogen bonding, solute–solvent interaction, and molecular symmetry. With modifications, it has been used to study adsorption on solid surfaces, rotational isomerism, the effect of deuterium labeling, and the crystallinity of polymers. Physical and chemical processes have been monitored by this technique and multiple analyses performed on complex samples.

In short, the range of applications appears to be limited only by the ingenuity of the user.

V.   RAMAN AND MICROWAVE SPECTROSCOPY

A.   RAMAN SPECTROSCOPY

When monochromatic light is directed at a cell containing molecular species (in any physical form), the scattered emergent light contains frequencies other than the original frequency. The spectrum of the scattered radiation contains a series of lines which are displaced in frequency ($\pm\Delta\nu$) from the main exciting frequency (ν). These lines are characteristic of the scattering material and constitute its Raman spectrum.

Normally an intense monochromatic source of light from the visible or near ultraviolet region is employed as the incident radiation (e.g., the 4358 Å or 2537 Å line of the mercury spectrum), the radiations scattered at right angles to the initial beam being dispersed in a spectrograph capable of high resolution. If the sample is "Raman active" the resultant spectrogram consists of an extremely strong line corresponding to the frequency of the incident light (Rayleigh scattering) and lines very close to the initial line, the lines on the lower frequency side of the exciting frequency being the more intense. These are known as Stokes lines; the fainter lines on the high frequency side are termed Anti-Stokes lines.

The incident radiation possesses sufficient energy (ΔE_1) to cause electronic, vibrational and rotational transitions in the molecular species, and when an excited molecule returns to the electronic ground state, it does not necessarily return to the original rotational and/or vibrational level. The energy lost in the deactivation transition (ΔE_2), therefore, can differ from the energy of the incident radiation and any photons emitted have a new frequency (ν'). The rest of the energy ($\Delta E_1 - \Delta E_2$) is absorbed and can be accounted for in terms of a change in the vibrational or rotational state of the molecular species. Stokes lines represent the transition in which the final state involves a vibrational level of higher vibrational quantum number than the original state. Anti-Stokes lines (ν'') result from transitions where the final vibrational quantum number is smaller than the quantum number for the initial state.

The frequency shifts ($\nu - \nu'$ or $\nu'' - \nu$) are thus a measure of the energy required to cause vibrational and/or rotational changes. Raman spectroscopy complements infrared spectroscopy, the origin of the bands being the same in both cases. The shift in wave number represented by a Raman band equals the wavenumber of the infrared absorption band that corresponds to the same vibrational transition, but the intensity relationships are different.

For an infrared band to be absorbed the normal vibrational mode must produce a change in the electrical dipole moment of the molecule, and the intensity of an infrared absorption band depends on the magnitude of the change in the dipole moment during the vibration.

For a line to appear in the Raman spectrum, the vibration must change the polarizability of the molecule, and the intensity depends on the magnitude of the change of polarizability associated with the vibration. Hence frequencies related to the motion of strongly polar linkages lead to intense infrared spectra; the presence of highly polarizable linkages gives rise to intense Raman lines.

The symmetry of the molecule also plays a part in determining the intensity. With small molecules the less symmetric vibrations often yield intense infrared bands, while the symmetric vibrations often lead to strong Raman bands. In studies of molecules possessing a center of symmetry it is, therefore, desirable to use both methods.

The vibrational frequencies determined from the Raman spectrum of a substance may be used for identification purposes, the normal spectroscopic procedure of comparing the spectrum of the unknown with that of known compounds being used. The Raman spectrum of a mixture consists simply of the superimposed spectra of the components.

However, Raman spectra are difficult to obtain, and the analytical chemist is well advised to use infrared spectra for qualitative analysis under most circumstances.

The intrinsic weakness of the Raman effect necessitates the use of an intense source of monochromatic radiation. The intensity of the emitted radiation is proportional to the fourth power of the frequency of the exciting line, but fluorescence, photodecomposition, and other side effects increase as the wavelength is shortened. The widely used mercury blue line at 4358 Å is a good compromise, although in the near future, laser beams may become the most popular source unit.

Fluorescent materials must be absent from the samples examined by this technique and colored solutions tend to absorb the emissions. Most analytical applications are concerned with measurements on pure liquids and to a lesser extent, solutions. Measurements involving gases or solids are somewhat more complex.

Raman spectra are usually determined at high concentration levels; hence the sample must be very soluble in the chosen solvent. The solvent must yield a simple spectrum which does not mask the lines of the sample, and the most widely used solvents are those recommended for infrared studies, viz., carbon disulfide, carbon tetrachloride, chloroform, etc., with the important addition of water.

The spectrograph used to disperse the scattered radiation has to possess a high resolving power, and in much of the early work the spectra were recorded photographically. The modern trend, however, is to use photoelectric detection associated with chart recorders. The latter approach permits an accurate measurement of both line intensity and line position.

For quantitative studies the approach is similar to that used in other forms of emission spectroscopy. It is customary in these techniques to compare the intensity of the lines of interest with the intensity of some standard line. In Raman work, the solvent can act as the internal standard, e.g., the 459 cm^{-1} band of carbon tetrachloride has been widely used.

In the Raman technique there is a linear relationship between intensity and concentration which makes the identification of the major components of a mixture easier than in the infrared. Analysis of multicomponent mixtures is also often simpler by the Raman method, and one extensive application has been the qualitative analysis of mixtures of hydrocarbons. In oil companies, where a standard procedure can be employed, the technique has been widely adopted.

Another advantage of Raman spectroscopy is the ability to use aqueous solutions.

On the other hand, the equipment required is expensive and the experimental procedure is more complex than alternative procedures. For these reasons analytical applications may not expand rapidly although great interest is now being shown in the determination of molecular structures by this technique.

B. Microwave Spectroscopy

The microwave region of the spectrum is situated between the far infrared and radiofrequency regions and encompasses radiations having wavelengths ranging from a millimeter to several centimeters (frequency range 10^{10} to 3×10^{11} cycles per second). Absorption of this type of radiation causes changes in the rotational levels of gaseous polar molecules.

A vapor pressure of a few hundredths of a millimeter is all that is required and the spectra of molecules with dipole moments as small as 0.1 Debye units have been observed.

For a diatomic molecule the rotational spectrum yields the moment of inertia, since the possible energy levels of a diatomic rigid rotator are given by

$$E_r = J(J+1)h^2/8\pi^2 I = hcBJ(J+1)$$

where B is the rotational constant and is related to the moment of inertia (I) by $B = h/8\pi^2 cI$. J is the rotational quantum number and governs the angular momentum which has the value of $\sqrt{J(J+1)}h/2\pi$.

The spectrum of a rigid rotator consists of a series of equidistant lines separated from one another by 2 B cm^{-1}. From this distance I can be calculated and the internuclear distance follows immediately. Similarly, for nonlinear molecules, the principal moments of inertia lead directly to bond angles and internuclear distances.

Measurements in the microwave region have become possible only in quite recent times and most microwave spectrometers have been laboratory built. Since commercial instruments are now available an upsurge of interest in this field is predictable.

In a simple spectrometer the source is a klystron oscillator, the wavelength of the emitted radiation being varied by alteration of the voltage applied to the tube. The monochromatic radiation is directed towards a sample cell by means of a "wave guide." This is a hollow metal tube which plays a part similar to that of lenses and mirrors in optical spectroscopy. The absorption cell can be simply a length of wave guide fitted with mica windows and a sample inlet connected to a vacuum line. After passage through the sample cell, the radiation falls on a crystal detector. The output of the detector is amplified and displayed on an oscilloscope. By varying the voltage applied to the klystron, a frequency range can be scanned and any absorption of radiation appears as a dip in the oscilloscope trace. Auxiliary equipment is used to determine the frequency of the incident radiation and most instruments are sensitive to about 0.5 Mc/sec, i.e., 10^{-5} of a wavenumber.

This high resolution is an attractive feature for identifying and distinguishing between chemical substances. Microwave spectra, unlike almost all other spectra, show neither group frequencies nor any dependence on molecular constitution. The spectra are determined by the moments of inertia, and even the spectra of chemically similar molecules bear little resemblance to each other.

The sensitivity of the technique depends on the particular case since the absorption intensity depends on the molecular dipole moment, the rotational transition and the frequency. However, it would appear that substances present in amounts greater than 1% of the sample would be readily detected.

At the moment the application of this technique to qualitative analysis is limited by the absence of a large catalog of molecular absorption frequencies. The procedure is also limited to gaseous species having a measurable dipole moment.

Quantitative applications are subject to interference from the effect of pressure broadening. As the pressure of an absorbing gas sample is increased, the line width increases in proportion to pressure over a wide pressure range. The integrated absorption intensity also increases in direct proportion to the pressure, and consequently the peak height remains independent of pressure over this range. The addition of a foreign gas to an absorbing sample leads to similar behavior, the band half-width depending on the pressure of added gas for low partial pressures of the absorbing substance.

With low partial pressure of the absorbing substance the peak height varies with the mole fraction of the absorbing gas, and this has been used in the analysis of binary mixtures. For mixtures containing more than two substances, measurements are made at low microwave power levels and the integrated intensity levels are used for comparison purposes.

With further improvements in techniques and instruments microwave spectroscopy may find new applications such as monitoring gas mixtures or quantitatively following a tracer isotope in a chemical process.

At the moment, however, its major uses are associated with the determination of molecular structures, the determination of dipole moments, and the derivation of information about the mechanical properties of molecules.

VI.   TUTORIAL AND REVISION QUESTIONS

1. The manganese content of a 1.0000 gm sample of alloy steel, on oxidation to permanganic acid and dilution to 500 ml gave an absorbance reading of 0.68 when examined at a wavelength of 540 mμ in a spectrophotometer using a 2 cm cell.

   a. If the molar absorptivity of permanganic acid at 540 mμ is 3000, what is the percentage of manganese in the steel? (Atomic weight Mn $= 55$).

   b. In order to make the absorption measurement the chemist could have used either a 0.5, 1.0, 2.0, or 4.0 cm cell. Which of these cells would have yielded the smallest relative error, assuming that the error associated with reading a linear transmission scale is the major contributing factor.

   c. A calibration graph prepared from standard solutions of KMnO$_4$ was linear and passed through the origin.

   A calibration graph prepared from standard samples of mild steel containing varying amounts of manganese was also linear but the straight line did not pass through the origin.

   A calibration graph prepared from standard samples of alloy steels containing different amounts of Cr, V, Mo, and Mn gave a series of scattered points.

   Explain these three observations.

   d. After observing that the results obtained using the absorbance reading of 0.68 were higher than those obtained using a titrimetric procedure, the chemist selectively reduced the permanganate ion, and found that the colored "blank" had an absorbance of 0.16 using a 4 cm cell. What was the true manganese content of the steel?

   The steel contained 20% Cr; hence if the "blank" was due entirely to chromic acid, what is the absorptivity of chromic acid at 540 mμ?

e. Outline a procedure for determining colorimetrically both the Mn and Cr content of the steel (without selective reduction of the permanganate ion).

f. 20 ml of the oxidized alloy steel solution were titrated with a 0.001 M solution of sodium arsenite which selectively reduced the permanganate ion to the manganese (II) state.

Draw the curve which would have been obtained if the titration had been followed spectrophotometrically. The axes should be marked with appropriate numerical values.

g. Assuming that suitable standards are available, indicate how a modified spectrophotometric procedure could be used to analyze the Mn content of a special steel containing about 15% Mn.

2. Discuss the following topics:
   a. The determination of film thickness by X-ray absorption.
   b. The scope of X-ray absorption as a quantitative tool.
   c. The preparation of samples for infrared absorption studies.
   d. The basic circuitry for single beam and double beam spectrophotometers.

3. Using review articles, papers, and modern texts prepare a detailed essay on one or all of the following:
   a. Interference effects in atomic absorption spectroscopy.
   b. The use of nitrous oxide—acetylene flames in atomic absorption spectroscopy.
   c. Microwave spectroscopy as an analytical tool.
   d. Quantitative analysis by Raman spectroscopy.
   e. Automatic and semiautomatic methods of on-stream analysis based on the absorption of radiation.

4. As part of a survey of river pollution, samples of river water are to be regularly analyzed.
   a. Using the "Handbook of Analytical Chemistry" or other reference books, select colorimetric methods which might prove suitable for determining the concentration of Cu, Pb, Fe, and Zn (present in the parts per million range). Summarize the selected methods in the form of a table which gives the element, color forming reagent, experimental conditions, wavelength for color measurement, etc.
   b. If free to use alternative procedures, what technique would you use to study the heavy metal content of the water? Give reasons for your choice.
   c. Indicate how radiation absorption might be used to identify organic contaminants selectively extracted from the river water
      i. In terms of the "type" of compound present, e.g., aldehyde, aromatic compounds, etc.
      ii. In terms of "positive" identification of a particular compound.

What problems might be encountered in quantitatively determining the amount of each organic species present?

d. A new colorimetric method has to be devised for the determination of one of the pollutants. A literature reference indicates that the addition of reagent X causes the species to form an intensely colored compound, but no further details are listed. Outline the various steps which would be required to ascertain the optimum experimental conditions for the colorimetric determination of the pollutant using reagent X.

5. a. Explain why trace amounts of lead in zinc are readily determined by means of atomic absorption spectroscopy, but the determination of the same element by flame photometry is completely unsatisfactory.

b. Explain the energy transitions which occur within a molecule on the absorption of ultraviolet light.

c. Explain why the vibrational frequencies observed in a Raman spectrum differ in intensity from corresponding infrared absorption bands and indicate why it is advisable to use both techniques if possible in structural studies.

d. Explain why the absorption spectra obtained using ultraviolet or visible radiation have much broader peaks than infrared spectra.

e. Explain why solvents have to be carefully selected in radiation absorption studies.

6. Using spectra determined experimentally in the laboratory or spectra copied from reference books, try to identify
   i. the chromophoric groups responsible for the peaks in ultraviolet absorption spectra.
   ii. the atomic groupings responsible for major peaks in infrared spectra. (The spectra selected should be of simple compounds to facilitate the use of abbreviated lists of characteristic absorption bands.)

7. Write brief notes on:
   a. Dispersion of radiation by diffraction gratings.
   b. The generation of microwaves.
   c. Phototube characteristics.
   d. X-ray absorption edges.
   e. Matrix effects in X-ray absorption studies.
   f. Auxochromes.
   g. The effect of conjugation on the nature of ultraviolet spectra.
   h. The effect of environment on the frequency of infrared absorption peaks of characteristic atomic groupings.
   i. d-Orbital splitting due to ligand field effects.
   j. Pure rotational spectra.

8. a. The instruments used to measure the intensity of light absorption by colored solutions are sometimes subdivided into three classifications:

namely, filter absorptiometers, photometers, and spectrophotometers.

With the aid of books on instrumental analysis prepare a summary of the basic principles of the three types and comment on their suitability for routine analysis.

b. Comment on why radioisotopes are sometimes substituted for an X-ray generator in high frequency radiation absorption studies.

c. Discuss why infrared absorption spectrophotometers are more widely used to study vibrational changes than Ramán spectrographs.

9. Critically evaluate the validity of the statement: "In quantitative analysis based on the absorption of radiation, most of the errors are associated with the steps which precede the actual measurement of absorbance in a spectrophotometer, for example, sample preparation, dilution (including solute-solvent interactions and competing equilibria in sample solutions), careless selection and preparation of the standards used for calibration purposes, ..."

10. a. Infrared spectroscopy is widely used in all branches of chemistry. Briefly outline some of the applications of the technique in fields other than organic and analytical chemistry.

b. "If you want to know the elemental composition of a sample, use emission spectroscopy. If you want to know the structure of the sample, use absorption spectroscopy" Is this a fair generalization?

References

ATOMIC ABSORPTION

L. S. Birks, *Spectrochemical Analysis*, Wiley (Interscience), New York, 1959.

W. T. Elwell and J. A. F. Gidley, *Atomic Absorption Spectrophotometry*, 2nd Ed., Macmillan (Pergamon), New York, 1966.

H. A. Liebhafsky, H. G. Pfeiffer, E. H. Winslow and P. D. Zemany, *X-ray Absorption and Emission in Analytical Chemistry*, Wiley, New York, 1960.

J. Ramirez-Munoz, *Atomic Absorption Spectroscopy*, Elsevier, Amsterdam, 1968.

J. W. Robinson, *Atomic Absorption Spectroscopy*, Marcel Dekker, New York, 1966.

A. Zettner, *Advances in Clinical Chemistry*, Vol. 7 (H. Sobotka and C. P. Stewart, eds.), Academic Press, New York, 1964, pp. 1–62.

ELECTRONIC ABSORPTION SPECTROSCOPY

R. P. Bauman, *Absorption Spectroscopy*, Wiley, New York, 1962.

D. E. Boltz, ed., *Colorimetric Determination of Non-metals*, Wiley (Interscience), New York, 1958.

G. Charlot, *Colorimetric Determination of Elements*, Elsevier, Amsterdam, 1964.

A. E. Gillam and E. S. Stern, *Electronic Absorption Spectroscopy*, 2nd Ed., Arnold, London, 1957.

H. H. Jaffe and M. Orchin, *Theory and Applications of Ultraviolet Spectroscopy*, Wiley, New York, 1962.

M. G. Mellon (ed.), *Analytical Absorption Spectroscopy*, Wiley, New York, 1950.

E. B. Sandell, *Colorimetric Determination of Traces of Metals*, 3rd Ed., Wiley (Interscience), New York, 1959.

W. West (ed.), *Chemical Applications of Spectroscopy*, 2nd Ed., Wiley (Interscience) Publishers, New York, 1969.

INFRARED SPECTROSCOPY

L. J. Bellamy, *Infrared Spectra of Complex Molecules*, 2nd Ed., Methuen, London, 1958.

N. B. Colthup, L. H. Daly and S. Wiberly, *Introduction to Infrared and Raman Spectroscopy*, Academic Press, New York, 1964.

R. T. Conley, *Infrared Spectroscopy*, Allyn & Bacon, Boston, 1966.

W. Gordy, W. V. Smith and R. F. Trambarulo, *Microwave Spectroscopy*, Wiley, New York, 1953.

K. Nakamoto, *Infrared Spectra of Inorganic and Co-ordination Compounds*, Wiley, New York, 1963.

W. J. Potts, *Chemical Infrared Spectroscopy*, Vol. 1: Techniques, Wiley, New York, 1963.

7

INTERACTIONS WITH MAGNETIC FIELDS

I.	Basic Concepts	177
II.	Magnetic Susceptibility	183
	A. Methods of Determining Magnetic Susceptibility	183
	B. Inorganic Applications	185
	C. Applications to Organic Chemistry	187
	D. Structural Aspects of Coordination Compounds	187
III.	Nuclear Magnetic Resonance Spectroscopy	189
	A. NMR Spectrometry	189
	B. Shielding Effects	192
	C. Chemical Shifts	194
	D. Spin–Spin Coupling	196
	E. NMR with Nuclei Other Than Protons	200
IV.	Electron Spin Resonance Spectroscopy	201
	A. Hyperfine Splitting	203
	B. ESR Spectrometry	205
	C. Applications of ESR	207
V.	Mössbauer Spectroscopy	208
VI.	Mass Spectroscopy	212
	A. Principles of the Technique	212
	B. The Mass Spectrum	217
	C. Applications	219
VII.	Tutorial and Revision Problems	222
	References	223

I. BASIC CONCEPTS

The phenomenon of magnetism has been known since ancient times and early scientists observed that matter is either attracted or repelled by a magnetic field. However, the application of this phenomenon to analytical problems is of recent origin.

The four major techniques in which interaction of chemical species with an applied magnetic field forms an integral part of the basic process are magnetic susceptibility, nuclear magnetic resonance spectroscopy, electron spin resonance spectroscopy, and mass spectrometry. At the moment these techniques are used primarily for structure elucidation but each can be used for quantitative determinations.

The techniques are based on different aspects of magnetic interaction; hence, it is appropriate to begin this chapter with a brief revision of magnetic terms.

A magnetic system can be macroscopic or microscopic, but in each case it possesses a north pole and a south pole, separated by a definite distance. The two poles are equal in magnitude but opposite in character; together they constitute a magnetic dipole. Like poles repel, unlike poles attract, and under the influence of an external magnetic field a dipole tends to orient itself parallel to the applied field.

The strength of a magnet (its attractive or repulsive power) is measured by the number of unit poles to which each pole of the magnet is equivalent. A unit pole is defined as one which repels an equal and similar pole with a force of one dyne when separated in vacuo by a distance of one centimeter.

The force between two magnetic poles is governed by Coulomb's law which states that if two poles of strength m_1 and m_2 are separated by a distance r, the force between them is inversely proportional to the square of the distance, i.e.,

$$\text{Force} = km_1m_2/r^2$$

where k is a proportionality constant.

North and south poles are denoted by positive and negative signs respectively. Accordingly, a positive value of the force corresponds to repulsion, a negative value to attraction. In a vacuum, $k = 1$, and the value of the proportionality constant in air is also approximately unity. In other media, the magnitude of k can differ from this value, and the reciprocal of k is known as the magnetic permeability of the medium (μ).

The region surrounding a magnetized body, which is capable of inducing magnetism in other bodies, is termed a magnetic field. The intensity of the field is measured in terms of the force a unit pole would feel if placed in the field. Unit magnetic intensity exists at a point where the force on the unit pole is one dyne. Unit magnetic intensity was formerly called the "gauss"; the modern term is the oersted (Oe). A smaller unit is the gamma, γ, which is equivalent to 10^{-5} oersted.

The free path followed by a hypothetical, mobile unit pole, due to the forces acting upon it, is called the line of force. The number of lines of force per square centimeter is said to be numerically equal to the intensity of the field at that point and is termed the flux density. The unit of flux generally used for theoretical purposes is the maxwell which is defined as the flux through one square centimeter normal to a field of one oersted. 10^8 maxwells equals one weber.

The total number of lines emanating from the (north) pole face of a magnet is called the total magnetic flux, and for a pole of strength m, the total flux is $4\pi m$ maxwells.

The south pole of a magnet is located at some finite distance l from the north pole, and when the dipole comes under the influence of a magnetic field of strength H, it tends to turn and align itself with the applied field. The north pole experiences a force of $+mH$ and the south pole a force of $-mH$. These two equal and opposite forces constitute a couple, the turning moment of which is given by force times distance. If the dipole eventually makes an angle θ with the direction of the applied field, the distance between the two forces normal to the applied field is $l \sin \theta$.

$$\therefore \text{ Turning moment } M = mH \cdot l \sin \theta$$

$$= \mu H \sin \theta$$

The term $\mu\ (= ml)$ defines the magnetic moment and serves as a measure of the turning effect. It has units of dyne–cm per oersted or ergs per oersted. A more fundamental unit is the Bohr magneton (BM), which has a value of 0.927×10^{-20} ergs/oersted. The unit, often designated μ_B, is equal to $eh/4\pi mc$ where e is the charge on the electron, h is Planck's constant, m is the mass of the electron and c is the velocity of light. In studies of chemical compounds, the Bohr magneton is usually associated with a molar quantity of material; hence is defined as 1 BM $= 5585$ erg/oersted/mole.

Unmagnetized material, when placed in a uniform magnetic field, becomes magnetized through orientation of the randomly arranged dipoles in the material.

The intensity of magnetization (I) is measured by the pole strength (m) induced over unit area. Thus $I = m/A$, where A is the total area. By multiplying numerator and denominator by l, this relationship becomes $I = \mu/V$, i.e., the intensity of magnetization equals the magnetic moment per unit volume.

For a pole of strength, m, $4\pi m$ maxwells emanate from its surface. An induced intensity of I will, therefore, create $4\pi I$ unit lines of force across unit surface. Superimposed on this will be H lines of force derived from the applied magnetic field. The total number of lines of force across the unit surface is, therefore, given by $4\pi I + H$; this quantity is known as the magnetic induction B.

In a vacuum $B = H$; for a magnetic material of permeability μ, the magnetic induction B is given by $B = \mu H$.

The intensity of magnetization induced at any point is proportional to the strength of the applied field. That is,

$$I \propto H \quad \text{or} \quad I = \kappa H$$

where κ is a proportionality constant. The value of κ depends on the material being magnetized, and it is called the magnetic susceptibility per unit volume.

In qualitative terms, κ represents the extent to which a material is susceptible to induced magnetization.

$$B = \mu H = 4\pi I + H \quad \text{and} \quad I = \kappa H$$
$$\mu = 4\pi\kappa + 1$$
$$\kappa = (\mu - 1)/4\pi$$

The susceptibility per gram of material is called the mass or specific susceptibility (χ) and it is equal in magnitude to κ/ρ where ρ is the density of the material.

The atomic susceptibility χ_A and the molar susceptibility χ_M are simply defined as the susceptibility per gram-atom and per gram-mole respectively.

$$\chi_A = \chi \times \text{atomic weight}$$
$$\chi_M = \chi \times \text{molecular weight}$$

The ionic susceptibility is similarly defined as the susceptibility per gram ion.

According to Pascal, the molecular susceptibility χ_M of a compound can be expressed by

$$\chi_M = \Sigma_j \, N_j\chi_j + \lambda$$

where N_j is the number of atoms of the jth element in each molecule of the compound and χ_j is the atomic susceptibility of that element. The magnitude of λ, a correction constant, depends on the nature of the chemical bonds between the atoms.

For salts the following is assumed:

$$\chi_M = \chi_{\text{cation}} + \chi_{\text{anion}}$$

Typical values for atomic and ionic susceptibilities and constitutive correction factors (λ) are listed in the "Handbook of Analytical Chemistry" and in other reference books describing magnetic susceptibility.

For mixtures and solutions, the specific susceptibility (χ) is given by the sum of the susceptibilities of each component multiplied by the appropriate weight fraction (P).

$$\chi = P_1\chi_1 + P_2\chi_2 + \cdots P_n\chi_n$$

The magnitude of the specific susceptibility of different types of compounds or materials varies from small negative values to very large positive values. On this basis the species are classified into several types, the three major groupings being termed diamagnetic, paramagnetic and ferromagnetic.

Diamagnetic materials are those in which there is feeble repulsion between the material and the applied field. The specific susceptibility is very small and negative ($\simeq 1 \times 10^{-6}$) and there is very little change with temperature. Practically all organic and inorganic compounds with the exception of free radicals and compounds containing the ions of the transition elements exhibit diamagnetism.

With paramagnetic substances, the intensity of the induced magnetization is greater than the applied field in a vacuum, i.e., the specific susceptibility is positive and small ($\simeq 100 \times 10^{-6}$).

The susceptibility varies with temperature in accordance with the Curie law ($\chi \propto 1/T$) or Curie-Weiss law [$\chi \propto 1/(T + \theta)$]. Paramagnetism is exhibited by the salts and complexes of transition elements, some free radicals, and "odd" electron molecules. In precise work, the experimental susceptibility has to be corrected for the inherent diamagnetism. This is usually achieved by using Pascal's constants and the additivity law.

With metals such as iron, cobalt, nickel, and their alloys, there is an intense attraction of the substance towards the magnetic field. The specific susceptibility is positive and very large ($\simeq 1 \times 10^{2}$) and depends in a complex manner on the temperature and applied field strength.

The magnetic behavior of bulk matter can be explained in terms of the magnetic properties of the constituent molecules and atoms. In the ultimate, the magnetic properties can be associated with the structure of the constituent atoms.

For example, diamagnetism is caused by the orbital motion of electrons and is a universal property. However, it is only observed experimentally when all the electrons are paired. In the presence of an unpaired electron, the system contains a permanent magnetic dipole and is attracted by the magnetic field. This paramagnetic effect usually swamps the basic diamagnetism of the system. The intense attraction observed with ferromagnetic materials is attributed to the presence of domains or lattices containing electrons with parallel spins. Interaction between the dipoles is responsible for the high susceptibility values.

Most nuclei (including the proton) possess inherent magnetic fields, but the nuclear effects are too small to be observed in low magnetic fields, the susceptibility of nuclei being of the order of 10^{-10}. In an intense magnetic field, however, the nuclei tend to assume specific orientations.

An electron spinning around an axis behaves like a micromagnet and gives rise to a magnetic moment. Nothing is known about the shape of an electron or its charge distribution, but experimental results indicate that the spin magnetic moment of an electron (μ_s) may be defined by the relationship

$$\mu_s = 2\sqrt{s(s + 1)} \cdot eh/4\pi \, mc \text{ erg/oersted}$$

where s represents the spin angular momentum of the electron which has a value of $1/2$. μ_s, therefore, equals 1.6×10^{-20} erg/oersted or 1.73 Bohr magnetons, and in vectorial presentation has the opposite direction to s.

A magnetic moment attributable to spinning electrons is observed in species possessing one or more unpaired electrons. Such systems are attracted towards an applied field, i.e., they are paramagnetic. In most organic and inorganic compounds, however, electrons are paired and the magnetism due to the spin of one electron is cancelled out by that of another spinning in the opposite direction. The fact that these compounds are feebly repelled by an applied field can be attributed to the effect of the magnetic field on the orbital motion of the paired electrons.

An electron carrying a negative charge moving in a circular orbit is equivalent to a circular current. When subjected to a magnetic field, the system as a whole is repelled away from the magnetic field. The form of the orbit, motion in the orbit, and inclination of the orbit to magnetic lines of force all remain unaltered. What does result is a uniform precession of the orbit about the

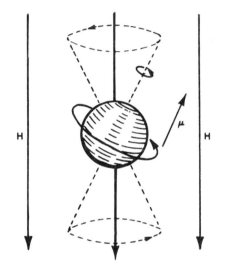

FIGURE 7.1   In a magnetic field rotating charge precesses.

direction of lines of force. The effect of the magnetic field is similar to that observed when a spinning top is pushed from a vertical position (cf. Fig. 7.1). The magnetic moment, μ_1, for the orbital motion is equal to l Bohr magnetons where l is the orbital quantum number of the electron.

A number of theoretical equations have been developed which relate the magnetic susceptibility to fundamental properties of the electrons. One approach for diamagnetism shows the atomic susceptibility to be proportional to the sum of the mean square radii of the various orbits in the atom. Another approach relates the susceptibility, in a complex manner, with the atomic number and all the orbital quantum numbers. For paramagnetic materials, the molar susceptibility has been shown to be proportional to the square of the permanent moment and inversely proportional to the absolute temperature, i.e., $\chi_M \propto \mu^2/T$.

The magnetic properties of nuclei are conveniently explained by assuming that the nuclear charge is spinning around an axis. The angular momentum possessed by such a nucleus can then be represented by a spin quantum number I.

Nuclei which have either an odd number of neutrons (^{13}C, ^{127}I) or an odd number of protons (^1H, ^{15}N, ^{19}F, ^{31}P, ^{11}B, ^{79}Br), but not both, exhibit half-integral spin quantum numbers. The circulating charge associated with these nuclei generates a magnetic field with which there is an associated nuclear magnetic moment, μ, along the axis of spin. Nuclei in which the number of protons and neutrons are both even (^{12}C, ^{16}O, ^{28}Si, ^{32}S) have no angular momentum ($I = 0$) and no magnetic properties.

The magnetic nuclei interact with an external magnetic field by assuming discrete orientations with corresponding energy levels. For a given nucleus there are $2I + 1$ possible levels or orientations ($I, I - 1, I - 2, \ldots, -I$), so that in the particular case where $I = 1/2$ (^1H, ^{15}N, ^{19}F and ^{31}P) there are two possible orientations. The spinning nucleus can align itself with the applied field ($E = -\mu H$) or against the applied field ($E = \mu H$). The energy difference between these two orientations is $2\mu H$ where H is the intensity of the applied magnetic field. The more general relationship for nuclei of other spin quantum numbers is $\Delta E = \mu H / I$.

The existence of such distinct energy states is extremely important since transition of nuclei between these states is the basic process in nuclear magnetic resonance spectroscopy.

II. MAGNETIC SUSCEPTIBILITY

A. METHODS OF DETERMINING MAGNETIC SUSCEPTIBILITY

A number of methods have been proposed for measuring magnetic susceptibilities and there are many variations of each basic method. In this section only two approaches will be considered, namely, the Gouy and the Faraday methods. Both these techniques subject the samples to magnetic fields of between 3000 and 15,000 oersteds, and both can be applied over a range of temperatures.

The major difference between these two methods lies in the nature of the magnetic field. In the Gouy method the sample is subjected to a uniform field while in the Faraday method the sample is exposed to a nonuniform field with a constant field gradient.

The basic components of the equipment used are shown in schematic form in Fig. 7.2. With the Gouy method, probably the more widely used technique, a long cylindrical sample is suspended from the arm of a sensitive balance such that one end lies in a region of strong uniform field and the other lies in a region of negligible field. The sample plus container is accurately weighed both in and out of the magnetic field. The difference in weight can then be related to the magnetic susceptibility of the sample. The method is particularly suitable

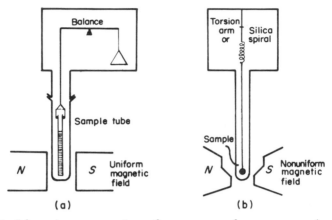

FIGURE 7.2 Schematic representation of two types of apparatus used to measure magnetic susceptibility; (a) Gouy method and (b) Faraday method.

for the measurement of dia- and paramagnetic susceptibilities. It cannot be used for ferromagnetic materials.

The samples studied can be solids, liquids, or solutions, but theory requires the sample to be in the form of a long homogeneous cylinder. Samples of metals, alloys, glass, polymers, etc., can be obtained in this physical form, and the dimensions required depend somewhat on the apparatus, but usually a length between 10–15 cm is adequate with diameters varying from a few milli-meters to a few centimeters. In the case of powdered solids, an approximation to cylindrical form can be achieved by packing the powder uniformly in a glass tube of uniform diameter up to some arbitrary mark. Solution samples are also held in tubes of uniform bore. The amount of sample required is of the order of 0.5 gm of solid or 5 ml of liquid.

The Gouy method provides a measure of volume susceptibility, and con-version to mass susceptibility requires an independent measurement of density. The results obtained generally have an accuracy of $\pm 1\%$.

The calculation of susceptibility values can be greatly simplified by de-termining the "tube constant"—a value obtained by measuring the effect of the magnetic field on a standard of known susceptibility when retained in a particular sample tube. Empty sample tubes possess some inherent susceptibility; hence the effect of the magnetic field on the weight of the tube must be ascertained and used as a correction in all subsequent measurements.

The Faraday method can handle ferromagnetic materials as well as dia- and paramagnetic compounds. In this technique, milligram amounts of powdered samples are compressed into tablets before being subjected to a nonuniform magnetic field. The latter is achieved by inclining the poles of the magnet or by using pole pieces of a special design. When a sample of mass m and sus-ceptibility χ is placed in such a nonuniform field, it becomes subjected to a force (f) given by

$$f = m\chi H(\delta H / \delta s)$$

where H is the strength of the nonuniform field possessing a gradient of $\delta H / \delta s$ in the direction s.

The magnitude of this force is measured accurately either by means of a silica spring or a torsion arm suspended from a fiber. The relationship between experimental readings and susceptibility is established by using standards of known susceptibility (e.g., KCl).

This technique can be used to study susceptibility changes associated with the adsorption of gases on solids or the thermal decomposition of solids. The weight changes (i.e., spring extensions) associated with the chemical changes are first observed in the absence of a magnetic field. The introduction of the field then gives a quantitative measure of any changes in magnetic properties.

The susceptibility of a sample can also be determined by observing the change in inductance which occurs when materials are introduced into the core of a solenoid. This principle is used in several of the commercial instruments designed for process control, e.g., the determination of carbon in steel using permeameters.

A number of quantitative determinations of commercial importance are based on susceptibility measurements. For example, oxygen is one of the few paramagnetic gases known, the other common constituents of air are diamagnetic. Hence magnetic susceptibility measurements provide a convenient means of analyzing for oxygen. Other instruments yield information on the composition of alloys and the particle size distribution of ferromagnetic catalysts.

B. Inorganic Applications

The majority of common applications are associated with paramagnetic materials.

The additivity law can be used, e.g., to determine the concentration and oxidation state of paramagnetic ions in solution, in solid mixtures, and in glass. Provided that there is no interaction between the solute and the solvent, results accurate to $\pm 1\%$ may be obtained.

To use the additivity law there should be no spin-spin interaction between adjacent ions. This can be tested by measuring the susceptibility at several temperatures. Where there are magnetic interactions, the variation of susceptibility with temperature is described by the Curie-Weiss law, ($\chi_m = C/T + \theta$), and the magnitude of the Weiss constant θ is an indication of the extent of any interactions.

The specific susceptibility of a salt in solution is given by $\chi_{cation} + \chi_{anion}/$ molecular weight of salt. If the percentage of salt present is $p\%$, then the weight fraction equals $p/100$ and the weight fraction of solvent is $(100 - p)/100$. The additivity equation then becomes

$$100 \, \chi_s{}^{\text{gm}} = p \cdot \frac{\chi_{\text{cation}} + \chi_{\text{anion}}}{\text{mol wt salt}} + (100 - p)\chi_{\text{solvent}}^{\text{gm}}$$

For diamagnetic materials appropriate values of the gram ionic susceptibilities χ_{cation} and χ_{anion} are listed in tabulations, and for aqueous solutions $\chi_{\text{H}_2\text{O}}^{\text{gm}}$ is taken as -0.720×10^{-6}.

In the case of a paramagnetic ion, χ_{cation} depends on the temperature and its effective magnetic moment μ_{eff}.

In many cases, χ_{cation} can be calculated from the relationship

$$\chi_{\text{cation}} = N\mu_{\text{eff}}{}^2/3kT$$

where N is Avogadro's number, k is the Boltzmann constant, and T is the absolute temperature.

The "effective Bohr magneton number," μ_{eff} can be defined by the theoretical equation

$$\mu_{\text{eff}} = \sqrt{3k\chi_m T/N\beta^2}$$

where β is the Bohr magneton.

From the experimental Curie law, $\chi_m T = \text{Constant } (C)$, and substituting numerical values for the constants in the basic equation one obtains the relationship

$$\mu_{\text{eff}} = 2.84\sqrt{C}$$

On the basis of classic theory,

$$\mu_{\text{eff}} = g\sqrt{J(J+1)} \ \text{BM units.}$$

Here g is the Lande splitting factor and J is the resultant angular momentum, which is a vector sum of L, the total angular momentum of the orbital motion of the electrons, and S, the corresponding spin angular momentum.

For atoms in which $L = 0$ (i.e., atoms in the S state) the magnetic moment is entirely due to electron spins. In this case $g = 2$ and $J = S$. The total spin momentum S equals the total number of unpaired electrons, n, times the spin quantum number for each electron which is $1/2$ quantum unit. Substitution yields the relationship

$$\mu_{\text{eff}} = \sqrt{n(n+2)}$$

This formula is applicable to some ions of the transition group of elements, but in many cases orbital contributions cause the observed moments to be greater than calculated.

Tabulations of μ_{eff} (both calculated and observed) are available in the literature.

Since ions of the same element present in different valency states possess different numbers of unpaired electrons, μ_{eff} varies with the valency state. This difference can be used to determine the proportion of each valency state present in a mixture.

The magnetic susceptibility method has also proved useful in studying the polymerization of paramagnetic ions in solution since the formation of the polymer usually converts the paramagnetic species into a diamagnetic compound.

This is a particular example of a more general field of application. Wherever a chemical reaction (e.g., adsorption, decomposition, oxidation, etc.) causes a paramagnetic species to lose magnetic susceptibility (or vice versa), the change in susceptibility can be used to follow the course of the reaction and/or the position at equilibrium.

C. APPLICATIONS TO ORGANIC CHEMISTRY

Many organic and organometallic compounds dissociate under certain conditions to yield free radicals which contain unpaired electrons. Longer lived' species can be studied by magnetic susceptibility measurements, shorter lived species by the electron spin resonance technique described later in this chapter.

Magnetic measurements have been used to study the course of oxidation–reduction reactions (e.g., the oxidation of styrene) and polymerization. The proportion of the keto and enol forms present in an equilibrium mixture has been determined by measuring the susceptibility of the mixture. Pascal's empirical constants are used to calculate the molar susceptibility for each form, and the amount of each present is ascertained by proportioning these values to equal the experimental value observed with the mixture.

With other organic compounds, comparing the experimental diamagnetic susceptibility with susceptibilities calculated from proposed structures provides a means of elucidating the bond structure of the compound.

In the biochemical field, magnetic measurements have been used for quantitative analysis. For example, the change in susceptibility when paramagnetic ferrohemoglobin is converted to diamagnetic carbon-monoxy ferrohemoglobin can be used to calculate the amount of hemoglobin present.

A number of workers have described magnetic titration procedures. Solutions of diamagnetic material (e.g., oxyhemoglobin) are treated with aliquots of a diamagnetic reagent solution (e.g., sodium dithionate), and the change in susceptibility is observed after each addition. Where the product is paramagnetic (e.g., hemoglobin in the above example) susceptibility changes rapidly up to the equivalence point and then flattens off.

D. STRUCTURAL ASPECTS OF COORDINATION COMPOUNDS

The major applications of magnetic susceptibility in recent years have involved studies of the complexes and compounds of the transition elements.

Magnetic measurements over a range of temperatures permit evaluation of the Weiss constant, and the magnetic susceptibility provides an estimate of the number of unpaired electrons; this information can then be related to structure.

For example, in the octahedral complexes of metal ions having d^n configurations, the n electrons can be placed in several different orbitals. In the valence bond approach, d^2sp^3 hybrid orbitals are used for bonding and the nonbonding d electrons are placed in the remaining d orbitals (d_{xy}, d_{xz}, and d_{yz}). After each level has taken up one electron (d^3) the next electron has to pair up, giving a lower number of free electrons ($n = 2$), unless the higher energy d orbitals of the next principal quantum number are used for bonding. Thus, for elements in the first transition series, two hybridization schemes can be proposed—$3d^24s4p^3$ or $4s4p^34d^2$—and the complexes using these schemes are referred to as inner and outer complexes respectively. The use of the outer orbitals yields bonds which are weaker and more polar than bonds formed from inner orbitals.

A d^5 ion such as Fe^{3+} can, therefore, form two series of complexes with distinctly different paramagnetic susceptibilities. With five unpaired electrons (as in the hydrated ion and outer complex compounds) μ_{eff} is observed to lie between 5.2 and 6.0 BM. With inner complexes (e.g., $Fe(CN)_6{}^{3-}$) only one electron is unpaired and μ_{eff} lies closer to 1.7.

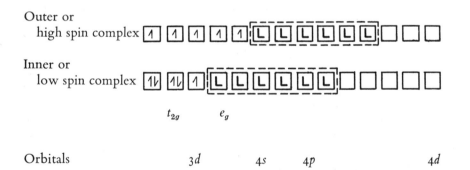

In the ligand field approach, each electron is placed into the lowest energy level that is available. Where the ligand field is weak (Δ is small), the number of unpaired electrons is a maximum. This is known as a spin-free or high spin complex, and these are the same as the outer complexes of the valence bond approach. With a strong ligand field (large Δ between t_{2g} and e_g) spin pairing is favored; this gives a spin–paired or low spin complex.

With ions having an even number of d electrons, e.g., Ni(II) with d^8, transition between the two forms results in a change from paramagnetism to diamagnetism.

III. NUCLEAR MAGNETIC RESONANCE SPECTROSCOPY

A. NMR SPECTROMETRY

Since its introduction less than two decades ago, nuclear magnetic resonance spectroscopy (NMR) has developed at a fantastic rate. In principle the technique is suitable for studying any nucleus which possesses angular momentum and an associated magnetic moment. The most important nuclei falling into this category are ^{1}H, ^{19}F, ^{31}P, and ^{11}B. Of these, the most extensive studies have been based on the proton and accordingly the discussion of this topic is based primarily on proton magnetic resonance (PMR).

A nucleus, possessing a spin quantum number (I) of 1/2 (such as ^{1}H, ^{19}F, ^{31}P) can align itself either with or against an applied magnetic field (cf. Section I). The energy difference between these two states (ΔE) is given by

$$\Delta E = \mu H_o / I = 2\mu H_o$$

where μ is the nuclear magnetic moment and H_o is the applied magnetic field.

If one replaces ΔE by the quantum relationship $h\nu$ the equation becomes

$$\nu = \mu H_o / hI$$

and multiplying both sides by 2π

$$2\pi\nu = 2\pi\mu H_o / hI.$$

The term $2\pi\mu/hI$ is related to characteristic properties of the nucleus and may be replaced by a new parameter, γ, known as the gyromagnetic ratio

$$\gamma = 2\pi\nu / H_o \quad \text{or} \quad \nu = \gamma H_o / 2\pi$$

These equations imply that under the influence of an oscillating field of frequency ν applied at right angles to a uniform magnetic field of strength H_o, a nucleus of gyromagnetic ratio γ should gain the energy required to shift from the low energy orientation to the high energy orientation.

The transition between the two states is called "magnetic resonance." That is, the nucleus resonates between two definite energy levels when subjected to the influence of the resonance frequency (with nuclei having spin quantum numbers other than 1/2, there are more than two possible orientations and energy levels). The absorption of energy by the nuclei causes a drop in the intensity of the incident oscillations which can be measured and used in the study of this phenomenon.

The gyromagnetic ratio defines the ratio of frequency to magnetic field required to achieve resonance conditions. For example, most work with protons is currently done using frequencies of 60×10^6 cps or 100×10^6 cps. The gyromagnetic ratio of the proton is 2.674×10^4 oersted^{-1} sec^{-1}. Hence the magnetic field strengths required to give proton resonance are 14,094 or

23,490 oersteds* respectively. The gyromagnetic ratios for other nuclei are ^{13}C (6.721 × 10^3), ^{31}P (1.082 × 10^4), ^{19}F (2.515 × 10^4), ^{11}B (2.573 × 10^4). To cause resonance with these nuclei a different combination of magnetic field and applied frequency is obviously required in each case.

For a proton, the difference in energy between the two nuclear orientations (ΔE) is extremely small, and the thermal energy at room temperature is sufficient to cause the nuclei to be almost equally divided between the states. In fact, the excess of nuclei present in the low energy state is only one or two in every million.

In the NMR experiment the absorption of radiation would cease once this minority had been excited if it were not for additional mechanisms which ensure that an excess population is maintained in the lower level. The two mechanisms responsible are called spin–lattice relaxation and spin–spin relaxation.

In spin–lattice relaxation, some of the excited nuclei relax back by losing thermal energy to other nuclei in the surrounding molecular environment. The energy is conserved within the system and appears as extra translational or rotational energy distributed around the lattice; the process is characterized by a spin–lattice relaxation time, T_1.

In spin–spin relaxation, a nucleus in the upper level transfers its energy to a neighboring nucleus by an exchange of spin. There is no net effect on the population distribution but it does limit the average time a nucleus spends in a given state. The two relaxation times, T_1 and T_2, can influence the absorption line width through the uncertainty principle ($\Delta E \cdot \Delta t = h/2\pi$), but with the nuclei most commonly encountered, T_1 and T_2 are sufficiently long to make this effect small when compared with the influence of instrumental factors.

The preceding discussion serves to define the basic elements required in a NMR spectrometer. These are shown in diagram form in Fig. 7.3 and can be listed as:

1. An electro- or permanent magnet with a strong stable homogeneous field to supply the principal part of the required field, H_0.
2. A set of Helmholtz coils fed with dc current from a sweep generator to superimpose on the main field of the magnet the additional field required to bring the total to the resonance condition.
3. A radiofrequency oscillator (frequency 60 or 100 Mc for proton studies) connected to a coil which transmits energy to the sample in a direction perpendicular to the magnetic field.
4. A radiofrequency receiver connected to a coil encircling the sample. This coil is mounted perpendicular to both the transmitting coil and the applied field.

* Most references on NMR tend to retain the older name gauss when describing magnetic field strengths.

5. A probe which serves primarily to hold the sample between the pole pieces. Commonly it also carries the coils of the sweep, transmitter and receiver circuit. The sample container is usually a 5 mm OD glass tube spun by an air driven turbine to average the magnetic field over the sample dimensions.
6. An amplification and recording unit attached to the radiofrequency detector.
7. An integrator unit to record the relative areas of the peaks in the spectrum.

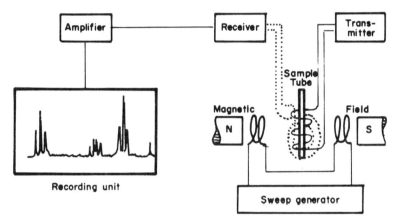

FIGURE 7.3   Schematic diagram of an NMR spectrometer.

The instrumental requirements for high resolution work are extremely stringent, and a commercial spectrometer contains an intricate collection of electronics. Both the frequency and magnetic field have to be regulated so that the variation in output is less than one part in 10^8 or 10^9.

The stabilization gear such as control loop circuitry is thus as much a part of the spectrometer as the basic components listed above. This aspect also contributes greatly to the high cost of NMR spectrometers. It is not practical to construct a high precision instrument for wide range use and each instrument is usually used to study one kind of nucleus only.

Some instruments are provided with interchangeable probes so that more than one nucleus can be examined with the basic instrument. Other instruments sacrifice resolution in order to achieve greater flexibility.

While theoretically it is possible to vary the applied frequency and maintain the magnetic field constant, it has been found preferable to design instruments based on the inverse procedure. For recording purposes, however, absorption is commonly recorded as a function of frequency. Hence where the magnetic field is varied to achieve resonance, the values in oersteds are converted to values in cycles per sec (Hertz). For the proton, one oersted is equivalent to 4260 cps.

For high resolution studies, the sample must be a liquid; it may be either a neat sample, a melt if the melting point is not too high, or a dilute solution. Solvents for NMR should have a high solvent power and give no resonance which overlaps the sample. Thus for proton studies, the ideal solvent contains no protons. Suitable solvents include carbon tetrachloride, carbon disulfide, and deuterated materials such as deuterated chloroform, benzene, water, acetone, acetonitrile, and dimethyl sulfoxide. The amount of material required to give a satisfactory NMR spectrum varies but is commonly of the order of 2 to 10% w/v. If the proton content is high, smaller concentrations are required. For this reason, pure liquid samples are often diluted with a suitable solvent.

To produce an NMR spectrum the sample is placed in the probe which is accurately positioned at the point of maximum homogeneity for the field. The sample is then subjected to the fixed applied radiofrequency field while additional magnetic field is supplied by means of the sweep coils until the resonance condition is attained. This condition is sensed by the detector and the amplified signal is fed to the recording unit.

The environment surrounding a nucleus influences the magnitude of the resonating frequency. The NMR spectrum of a sample, therefore, consists of a series of peaks, each group of peaks corresponding to nuclei in a particular type of environment. The intensity of the signals is directly proportional to the number of nuclei in each type of environment and the relative intensities of the different groups of peaks are thus often recorded on the NMR spectrum by means of the stepped trace produced by an integrator unit.

The spectrum which is obtained contains three types of information. The position and magnitude of the peaks indicate the molecular environment of particular nuclei, the number of resonating nuclei present in similar environments, and the nature of near neighbors. In short, the spectrum provides evidence which can be used to elucidate the structure of both simple and complex molecules. Proton magnetic resonance spectroscopy is particularly useful since hydrogen nuclei are major components of nearly all organic compounds. The subsequent sections are, therefore, restricted almost entirely to PMR.

B. Shielding Effects

A sample contains a large number of charged species (nuclei and electrons), each one of which is influenced by a magnetic field. While electron pairs normally produce no net magnetic field, in an applied magnetic field the circulation of the electrons generates a small local magnetic field proportional to, but opposing, the applied field. This phenomenon is called "diamagnetic shielding" because it shields the nucleus to some degree from the effects of the applied field. The nucleus thus finds itself in an effective field which is somewhat smaller than the applied field H_o,

$$H_{\text{eff}} = H_o - \sigma H_o$$

where σ is the shielding parameter. In order to attain the resonance condition, the applied field must be made greater than the field required in the absence of shielding. The magnitude of the increase required depends on the magnitude of σ; this in turn is related to the electron density around the proton. The more electronegative the neighbors, the more strongly are the shielding electrons pulled away. This results in less shielding and the proton goes into resonance at a lower field. If a near neighbor is electron-repelling, then the shielding magnetic field is more effective, so that H_o must be greater in order to achieve

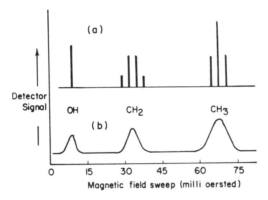

FIGURE 7.4   NMR spectrum of ethanol using a magnetic field of 14,100 oersted and a frequency of 60 Mc sec^{-1}; (a) "low" resolution pattern showing chemical shifts and (b) splitting pattern observed with higher resolution.

an effective field (H_{eff}) at the proton which equals the resonance field requirement.

To bring each kind of proton into resonance at a fixed frequency it is thus necessary to change H_o. As a result, the NMR spectrum shows an absorption peak for each kind of proton at some particular value of H_o. Figure 7.4a is a schematic representation of the spectrum of ethanol. In this molecule there are three different sets of hydrogen atoms, namely, hydroxyl, methylene, and methyl hydrogens present in the numerical ratio $1:2:3$. The low resolution spectrum shown in Fig. 7.4, therefore, contains three peaks with the peak areas showing a ratio of approximately $1:2:3$.

Additional shielding and deshielding effects arise from electronic circulations induced within molecules, such as about a π bond. This electronic motion creates a field which may act with or against the applied field, depending primarily on molecular geometry. The effect is therefore known as molecular magnetic anisotropy. In an aldehyde, e.g., the π electron clouds associated with the carbonyl group may be regarded as being in a plane normal to the direction of the applied field. Circulation of electrons in this plane generates a magnetic field which opposes the applied field but the lines of force are perpendicular rather than parallel to the bond axis. The net effect is that the aldehydic proton finds itself in an effective field which is enhanced by the induced field. A similar effect is observed in the ethylene molecule. An NMR spectrum of these two species, however, shows a distinct separation between the fields required to cause proton resonance in ethylene and in an aldehyde group. The marked difference can be attributed to the electronegativity of the oxygen atom; in other words, in the aldehyde group there is deshielding by both anisotropic and diamagnetic effects. Because the field due to the π-electron ring currents acts with the applied field, aromatic protons resonate at a much lower field than ethylenic protons.

With acetylenic protons the effect is the opposite. In this case the electron distribution in the triple bond is symmetric about the axis. If the axis is parallel to the applied field, electronic circulation perpendicular to the axis is induced and the resultant induced magnetic field opposes the applied field. The lines of force extend beyond the triple bond and the protons of acetylene are thus in a lower effective field.

Other factors, such as fields from associated molecules or solvents, tend to cause resonance shifts, but these effects are much smaller in magnitude.

The resonance equation should therefore be modified by substituting H_{eff} for H_o, i.e.,

$$\gamma = 2\pi\nu/H_{\text{eff}}$$

To obtain resonance operating with a fixed frequency one must thus change H_o as shown in Fig. 7.4.

Alternatively, if the field is considered to be fixed at H_o, then the protons find themselves in greater effective fields and a higher frequency is required to bring the various types of protons into resonance.

C. Chemical Shifts

The variation in the position of the resonance line with chemical structure is called the "chemical shift." It is impractical to express NMR shifts in absolute frequencies or fields, and they are measured relative to some standard. In proton magnetic resonance (PMR) spectra, the standards used include water, benzene, and cyclohexane, all of which have but a single resonance line. The more favored standard is currently tetramethyl silane, $Si(CH_3)_4$. Each proton in this molecule exists in an identical electronic environment which provides very high shielding. As a result, tetramethyl silane (TMS) exhibits a single sharp resonance line at a high applied field, well beyond the protons in most other organic compounds.

By expressing the shifts in dimensionless units which are independent of the magnitude of the field and the frequency, workers using different instruments can get the same answers. For this reason, the chemical shift (δ) is expressed

TABLE 7.1

STANDARDIZED CHEMICAL SHIFTS[a]

Type of protons	Compound	τ value (ppm)
Methyl	$Si(CH_3)_4$	10.000 (standard)
CH_3-C	$CH_3(CH_2)_4CH_3$	9.10
	CH_3CH_2OH	8.83
	CH_3CN	8.03
	CH_3COOH	7.93
	CH_3COCH_3	7.915
CH_3-X	CH_3SCH_3	7.942
	CH_3I	7.843
	CH_3OH	6.622
	CH_3NO_2	5.722
Methylene	CH_2	
$C-CH_2-C$	$CH_2{-}^{\cdot}CH_2$	9.78
	$CH_3(CH_2)_4CH_3$	8.75
$C-CH_2-X$	$CH_3CH_2NH_2$	7.22
	$(C_6H_5)_2CH_2$	7.129
	$i\text{-}C_3H_7CH_2Br$	6.74
	$(CH_3CH_2)_2O$	6.64
	CH_3CH_2OH	6.41
	$C_6H_5CH_2OH$	5.61
Olefin	$(CH_3)_2C{=}CH_2$	5.399
$>C{=}CH_2$	$(C_6H_5)_2C{=}CH_2$	4.60
	cyclo C_6H_{10}	4.43
	$C_2H_5OOCCH{=}CHCOOC_2H_5$	3.26
Acetylene		
$-C{\equiv}CH_2$	$HC{\equiv}CCH_2Cl$	7.60
	$HC{\equiv}CC_6H_5$	7.07
	$HC{\equiv}CCH_2OH$	5.82
Aldehyde	CH_3CHO	0.28
$-CH{=}O$	C_6H_5CHO	0.035
Aromatic	$C_6H_2(CH_3)_4$	3.26
	$C_6H_3(CH_3)_3$	3.36
	$C_6H_4(CH_3)_2$	3.0
	$C_6H_5CH_3$	2.905
	C_6H_6	2.734
	$C_6H_5{\cdot}CH_2Cl$	2.724
	$C_6H_5C{\equiv}CH$	2.678

[a] Solvent composition 99% CCl_4, 1% $Si(CH_3)_4$.

in parts per million from some reference point, the shift being defined by:

$$\delta_{ppm} = (H_s - H_r) \cdot 10^6 / H_r$$

where H_s and H_r refer to the field at resonance of the sample and the reference compound respectively.

Because field strength is linearly related to frequency, this equation can take the equivalent form of

$$\delta_{ppm} = 10^6 \cdot (\nu_s - \nu_r)/\nu_r$$

where ν_s and ν_r refer to the sample and reference resonance frequencies.

Chemical shifts are also commonly described in terms of tau values (τ). When tetramethyl silane is used as the reference compound, τ values are calculated from the relationship

$$\tau = 10 - \delta_{ppm}$$

Proton shielding values (expressed as τ) for various compounds are tabulated in the "Handbook of Analytical Chemistry," and other compilations are available in monographs dealing with this technique. A short list is given in Table 7.1.

Compounds having high tau values are said to be "up field" and in these compounds the protons are highly shielded. Compounds having tau values closer to zero lie "down field" and the protons in this case are deshielded.

Where possible the reference should be used as an internal standard by adding it to the sample. If there are problems of miscibility with TMS, other internal references may be substituted. Alternatively, if absolutely necessary, the reference material may be contained in a capillary tube inserted in the sample.

In a complex molecule, a number of effects combine to establish the value of the chemical shift for each type of proton. The exact value can also depend on the concentration of the species and the nature of the solvent, e.g., alcohols or amines are subject to hydrogen bonding effects. A correlation chart which indicates the general region of chemical shifts for different types of compounds is therefore useful for indicating the type of protonic groups which may be present. However, final interpretation of an NMR spectrum should be made through comparison with the spectra of closely related compounds, if this is possible.

D. Spin-Spin Coupling

In addition to chemical shifts, high resolution NMR spectra show further character arising from interactions transmitted by intervening electrons between nonequivalent nuclei. Each proton may be considered to be a spinning magnet which in an applied field is orientated with or against the field. The oriented spin of the nucleus gives an adjacent electron an opposite orientation. This orientating effect is transmitted down the chain and may ultimately reach a second nucleus. The spin orientations are characterized by different energies, and so create multiplets in the chemically shifted peaks. This effect is known as spin–spin splitting or spin–spin coupling. Coupling occurs through bonds by means of a slight unpairing of the bonding electrons. Propagation through multiple bonds is stronger than through single ones, and the coupling effect falls off rapidly as the number of intervening bonds increases.

The components of the multiplets are separated by a coupling constant J. The effect is a reciprocal one, and if nucleus X splits nucleus Y, then Y splits X and the value of J is the same for both.

Let us consider protons X and Y to be associated with two neighboring carbon atoms. The effective field at X is either decreased or enhanced by the local field generated by Y, depending on whether Y is orientated with or against the applied field. The resonance line for X is thus shifted down-field when the field of Y is aligned with the applied field, or up-field when the field generated by Y is aligned against the applied field. Since the distribution between the two orientations is essentially equal, a high resolution spectrum shows two distinct peaks separated by a distance J cps. In a similar way proton Y is split into a doublet.

The rigorous analysis of spin–spin multiplets is a problem in quantum mechanics but in many cases an approximate treatment suffices for interpretation of spectra.

For example, if the NMR spectrum of ethanol (Fig. 7.4) had been recorded on an instrument of higher resolving power, the CH_2 peak would be split into four segments and the CH_3 peak into three segments as shown in Fig. 7.4b.

This splitting of the peaks can be explained in a qualitative way in the following manner.

The spins associated with the two methylene protons can have the following configurations, each configuration being equally probable and having a certain energy associated with it.

$$\uparrow\ \uparrow,\ \uparrow\ \downarrow,\ \downarrow\ \uparrow,\ \downarrow\ \downarrow$$

The middle two sets with oppositely directed spins are equivalent. Two of the four possible perturbations of the methyl protons by the methylene protons are thus degenerate, giving rise to a triplet structure of component intensities $1:2:1$.

Similarly, the spins associated with the methyl protons can have the following configurations.

$$\uparrow\uparrow\uparrow,\ \uparrow\downarrow\uparrow,\ \uparrow\uparrow\downarrow,\ \downarrow\uparrow\uparrow,\ \uparrow\downarrow\downarrow,\ \downarrow\uparrow\downarrow,\ \downarrow\downarrow\uparrow,\ \downarrow\downarrow\downarrow\ .$$

In this group there are two states made up of three equivalent sets each and two states with no equivalent sets. Thus the methyl splits the methylene signal into a quartet of relative intensity ratios $1:3:3:1$.

These observations can be generalized by stating that a proton with n

equivalent protons on the neighboring carbon atom will be split by the n protons into $(n + 1)$ lines with relative subareas given by the coefficients of the binomial expansion $(1 + x)^n$.

This generalization also applies if the proton is subject to the influence of equivalent protons from two adjacent carbon atoms. For example, in propane, $CH_3CH_2CH_3$, the methylene protons are split by six equivalent methyl protons. This leads to a 7-membered multiplet with subareas in the ratio $1:6:15:20:15:6:1$.

When two sets of neighboring protons are not equivalent, the splitting pattern is a little different.

For example, in a substituted propane, the proton signal attributable to the central methylene group can be split into a quartet by the adjacent methyl group or split into a triplet by the substituent CH_2 group. Since the two sets of protons are not equivalent the final pattern is a multiplet of twelve lines, many of which are too weak to be observed in an NMR spectrum.

As the chemical shifts of two nonequivalent types of protons become more nearly the same, the symmetry of the splitting pattern is disturbed. In general the area of the inner peaks increases at the expense of the outer peaks, so that when the chemical shifts become identical both multiplets merge into a singlet. It can, therefore, be suggested that equivalent protons do not interact with each other to cause observable splitting.

Spin–spin interactions are independent of the strength of the applied magnetic field. Chemical shifts (cps) vary with the magnitude of the applied field. Hence if a set of peaks cannot be identified as a multiplet or as a collection of several individual peaks, resolution may be achieved by changing the strength of the applied field. Coupling constants rarely exceed 20 cps, whereas chemical shifts vary over 1000 cps. A few typical proton spin–spin coupling constant values are shown in Table 7.2. In multiplets the chemical shift is measured at the center of the system of peaks.

In simple, "first order" NMR spectra, the chemical shift (Δv cps) is much greater than the coupling constant (J cps), and a simple pattern of multiplets appears.

Departure from simple first order is shown in various ways, such as deviation from the predicted intensities and the presence of additional peaks. The exact form of the spectrum is governed somewhat by the magnitude of the ratio $\Delta v/J$. The interpretation of higher order spectra or spectra in which multiplets overlap requires a mathematical analysis which is too complex to consider here.

The preceding discussion should be sufficient, however, to facilitate interpretation of some simple spectra. To reiterate, the spectrum provides three kinds of information:

   1. Types of protons present—identified through their electronic environments as indicated by chemical shifts.

TABLE 7.2
PROTON SPIN–SPIN COUPLING CONSTANTS

Type	J_{cps}	Type	J_{cps}	
HC—C—C—CH	0	H　　　　H ＼＿／ (C=C ring)	5 mem	3–4
			6 mem	6–9
C (H,H)	10–15		7 mem	10–13
HC—OH	4–7	benzene ring H—ring—H	o	6–10
			m	1–3
			p	0–1
HC—CH (O)	2–3			
HC—CH	2–9	pyridine ring c,d,b,e,N,a	J_{ab}	5–6
			J_{bc}	7–9
			J_{ac}	1–2
			J_{bd}	1–2
C=HC—CH=C	10–13		J_{ad}	0–1
			J_{ae}	0–1
H,H C=C	6–14			
H C=C H	15–18	ring c,b,d,a,N	J_{ab}	2–2.5
			J_{bc}	2.5–3.5
			J_{ac}	1.5–2
H,H C=C	0–2	ring c,b,d,a,O	J_{ab}	1.5–2
			J_{bc}	3–4
			J_{ac}	0–1
			J_{ad}	1–2

2. Identity of neighboring protons as indicated by spin-spin splitting patterns.

3. Number of each type of proton present—determined on the basis that peak areas are proportional to the number of protons causing a given resonance line.

E. NMR with Nuclei Other Than Protons

By using an instrument which can be adjusted to provide a wide range of applied magnetic fields and/or a wide range of radio frequencies, it is possible to select conditions such that nuclei other than protons are brought into resonance. The only isotopes which are not observable are those with zero spin such as ^{4}He, ^{12}C, ^{16}O, etc. Of the remaining isotopes, the sensitivities vary significantly, due in part to the different relative natural abundances.

In practice, the operation is usually carried out at a fixed frequency with a variable magnetic field. Under these conditions, the only isotopes which possess a high relative natural abundance and a sensitivity of the same order as the proton are ^{11}B, ^{19}F, and ^{31}P. For this reason the majority of studies using nuclei other than protons have been restricted to these three.

The general principles involved in obtaining and studying the magnetic resonance spectra of these nuclei are similar to those discussed in proton studies, although some modifications are necessary.

For example, for spin-spin coupling, the general rule states that if n is the number of nuclei creating the multiplet and I is the spin quantum number of that nucleus, then the multiplicity is $(2nI + 1)$. The I values of ^{19}F and ^{31}P are 1/2, and the coupling effect is similar to that observed with protons; ^{11}B has a spin quantum number of 3/2; hence, the multiplets are much more complex.

Chemical shifts in fluorine compounds have been observed to depend sensitively on the solvent. Accordingly the compound CCl_3F (Freon-11) is employed both as the internal reference and as the solvent. Some of the shielding values are concentration dependent; hence, those measured at finite concentration are termed ϕ^* values while those obtained by extrapolation of ϕ^* to infinite dilution are termed ϕ values.

The ϕ value of CCl_3F is taken as zero, and increasing ϕ values represent increasing shielding. If ν_x represents the increase in frequency in cycles per second required to cause the CCl_3F peak to fall at the position previously occupied by peak x, then

$$\phi_x^* = \nu_x \times 10^6/\text{rf oscillator frequency}$$

Tabulations of fluorine shielding values are available; e.g., the "Handbook of Analytical Chemistry" has a short list.

Information related to other nuclei is accumulating rapidly and NMR studies are contributing highly significant information in the realms of fluorine, phosphorus, and boron chemistry.

The major application of NMR is undoubtedly structure determination. This technique has facilitated the elucidation of innumerable complex structures and is a major tool in most organic investigations. As more instruments suitable for observing nuclei other than protons become available, the range of inorganic applications can be expected to increase markedly.

Since the height of the peaks is proportional to the number of a particular type of nuclei present, NMR spectra can be, and have been, used for quantitative analysis. By observing the height of peaks corresponding to different types of nuclei, several determinations may be made from a single spectrum. The majority of spectra record proton resonance; hence, the majority of analytical applications involve hydrocarbons or natural and synthetic organic compounds. One interesting quantitative aspect is the determination of water in a wide range of materials.

In physical chemistry, NMR has found many other applications such as studies of crystal structure, phase transitions, molecular motions in solids, fast kinetics, solute-solvent interaction, and the determination of nuclear constants.

IV. ELECTRON SPIN RESONANCE SPECTROSCOPY

As indicated in Sections I and II, the magnetic field associated with a free electron is far greater than the field associated with nuclei, and chemical species with an odd number of electrons exhibit characteristic magnetic properties.

The magnetic moment, μ, of the magnetic field generated by the spinning action of an unpaired electron is proportional to e/m, where e is the charge of the electron and m its mass. For a free electron $\mu = \beta$ Bohr magnetons, but when an electron is present as an atomic particle the momentum and energy of the electron are also governed by quantum mechanical considerations. The magnetic moment of any electron is thus best described in the general form

$$\mu = -g\beta s$$

where s is the spin quantum number ($\pm 1/2$) and g is the spectroscopic (or Lande) splitting factor. For free electrons g equals 2.0023 and the value for nearly all free radicals is within one percent of this figure. For other systems g has a value close to two if there is a large energy difference between the excited electronic states and the ground state. When the overall electric field is not effectively spherical, the value of g depends upon the angle between the magnetic field and the principal axis.

When spinning electrons are placed in a magnetic field they can assume orientations aligned with or against the field. The difference in energy between these two states (ΔE) is given by

$$\Delta E = 2\mu H_o = 2g\beta s H_o = g\beta H_o$$

As in other spectroscopic systems, there is an electromagnetic radiation frequency which possesses an equivalent amount of energy, i.e.,

$$\Delta E = g\beta H_o = h\nu$$

$$\nu = 1.3997 g H_o \text{ Mc sec}^{-1}$$

For a free electron, the resonance frequency ν is thus about 9500 Mc sec^{-1} in a field of 3400 oersted or 2.8×10^{10} cps in a field of 10,000 oersted.

These frequencies correspond to wavelengths measured in centimeters, which means that free electrons under the influence of a strong magnetic field can be brought into resonance by microwaves.

As in NMR studies, the small difference in energy between the two states means that at room temperature the relative population of the two energy levels is very nearly the same.

By the Boltzmann equation, the ratio of the two populations (n_2/n_1) is given by

$$n_2/n_1 = \exp\left(-\Delta E/kT\right)$$

where k is the Boltzmann constant and T is the absolute temperature.

Thus at 298°K, using a magnetic field of 3400 oersted, there are 997 electrons in the high energy state for every 1000 electrons in the low energy state. With a field of 10,000 oersted the corresponding number in the higher state is 990.

Because the relative populations are so nearly equal, it is sometimes possible to cause the populations to become exactly equal. This is known as "saturation," and when it occurs it causes the signal absorption to decrease and the absorbed frequency band to widen. Two factors control saturation, the intensity of the microwave signal and the spin-lattice relaxation time. The latter is a measure of the interaction of the unpaired electron with its environment.

It can be observed from preceding equations that in order to cause a free electron to resonate between the two states, ν/H_o must equal 1.3997 g. When ν is fixed, as in most ESR spectrometers, then the magnetic field required to bring an electron into resonance is related to the magnitude of the splitting factor, g, as well as the magnitude of the microwave frequency. As noted previously, the value of g can vary if the electron is not subject to a spherical electrical field. Thus whenever the surrounding electric field is not uniform, the energy required for resonance depends on the orientation of the magnetic field with respect to a crystal axis. For example, with odd electron species possessing axial symmetry, the value of g depends on whether the magnetic field is aligned parallel or perpendicular to the symmetry axis. For symmetry lower than axial there can be three principal g components. If the system possesses a half integral value for its spin quantum number, sufficient residual spin degeneracy remains for transitions between spin states to be possible. However, with crystalline electric fields of low symmetry and integral spin quantum numbers, the spin degeneracy of a system can be completely removed in the absence of a magnetic field, and if this zero field splitting is sufficiently great, no ESR transitions are observed.

Rotation of a crystalline paramagnetic solid can thus sometimes cause changes in the field required to produce resonance conditions. Apart from these anisotropic effects, a large spin-orbit interaction can cause g to deviate considerably from a value of two.

When an unpaired electron spends an appreciable fraction of time near an atom having nonzero nuclear spin, the energy levels of the upper and lower states can be further split. This is known as hyperfine splitting.

A.   Hyperfine Splitting

The resonance frequency of an electron depends on the magnetic field at the electron, this field being composed of the applied field, H_o, and any local fields, H_n, due to the magnetic fields of nuclei or other effects. Since the local fields can be aligned with or against the applied field, the original two energy levels can be split.

For example, the proton has a nuclear spin, I, or $1/2$. The relative orientations of the nuclear magnetic moment and the electron magnetic moment can thus be represented as

$$\uparrow \qquad \uparrow \quad , \quad \uparrow \quad \downarrow \qquad , \qquad \downarrow \qquad \uparrow \quad , \quad \downarrow \qquad \downarrow$$

$$\cdot S\tfrac{1}{2} \quad I\tfrac{1}{2} \qquad S\tfrac{1}{2} \quad I - \tfrac{1}{2} \qquad S - \tfrac{1}{2}, \quad I\tfrac{1}{2} \qquad S - \tfrac{1}{2}, \quad I - \tfrac{1}{2}$$

In this case, one has a pair of energy levels symmetrically spaced about the position that a single level ($S\tfrac{1}{2}$ or $S - \tfrac{1}{2}$) would have occupied in the absence of hyperfine interaction; the intensity cf each line is also halved.

Because the nuclear moment remains fixed during electronic transitions, only two transitions are observed, namely, the transition between the two $I = +1/2$ states and the transition between the $I = -1/2$ states. The result of having a proton near an unpaired electron is thus a splitting of an ESR signal into a doublet. When the proton points in the same direction as the applied field, the electron finds the appropriate H at a lower value of the applied field H_o. A higher value of H_o is required when the proton magnetic moment opposes the field. The magnitude of the splitting, usually given in gauss, is called the hyperfine coupling constant, a. For the hydrogen atom a_H is about 508 oersted.

If an electron interacts with n protons, the number of hyperfine components can be shown by a similar argument to the above to be $(n + 1)$. Consider the case where $n = 2$. The proton fields may both oppose ($I = -1$) or act ($I = 1$) in the direction of the applied field. When the two proton fields are themselves opposed $I = 0$. The probability of the local field H_n being 1, 0, or -1 occurs in the ratio of 1:2:1. The ESR spectrum thus consists of a triplet with the central peak twice as large as its two flanking companions. The position of the central peak is similar to what would be observed in the absence of neighboring protons; the two-proton system is shown diagrammatically in Fig. 7.5.

With two nonequivalent protons, each reacts with the electron with a different coupling constant to produce hyperfine splitting. For example, consider an electron interacting with one proton with a coupling constant a_1,

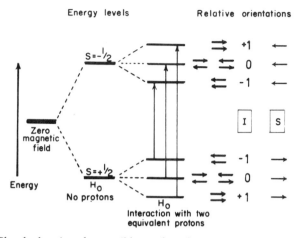

FIGURE 7.5   Sketch showing the possible modes of interaction of an unpaired electron with two equivalent protons.

and another proton with a coupling constant a_2, a_1 being much greater than a_2. In this combined field the doublet due to the interaction of the electron with the first proton is split into two doublets through the influence of the second proton (cf. Fig. 7.6).

In general, the number of hyperfine components from interaction with a nucleus of spin I is $(2I + 1)$. If the electron interacts with n chemically identical nuclei the number is $(2nI + 1)$. Taking the extreme case where there are m sets of nuclei, containing respectively n_1, n_2, $n_3 \ldots n_m$ nuclei, each chemically equivalent within one set, the maximum number of possible hyperfine

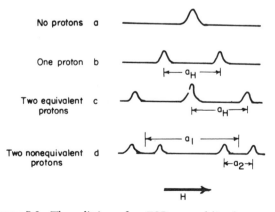

FIGURE 7.6   The splitting of an ESR spectral line by protons.

components is

$$\prod_{i=1}^{m} (2n_i I + 1)$$

The number of lines actually observed may be less due to accidental coincidence of energy levels and overlap of broad peaks.

Interaction of a single electron with a single nitrogen nucleus (^{14}N, $I = 1$) thus causes a split into three lines while interaction with a single manganese (II) nucleus (^{55}Mn, $I = 5/2$) can yield a six line spectrum. The presence of two equivalent ^{14}N nuclei produces a five line spectrum, two equivalent ^{55}Mn nuclei split the spectrum into eleven peaks.

B. ESR SPECTROMETRY

The basic components of an ESR spectrometer are shown diagrammatically in Fig. 7.7.

The source of radiation is a klystron tube, a typical unit yielding radiations of wavelength 3 cm or frequency 9×10^9 cps. A klystron may be tuned over a range of about $\pm 5 \times 10^{10}$ cps by varying the dimensions of a resonant cavity inside the tube, but after adjustment the source is usually held at a fixed frequency by means of a frequency control unit.

The microwaves are conducted to the sample and detecting crystal through a waveguide consisting in many cases of a hollow, rectangular brass tube. The insertion of a piece of resistive material in the waveguide decreases the power propagated, and by this means the power reaching the sample can be varied.

The sample is mounted in a resonant cavity in which a standing wave is set

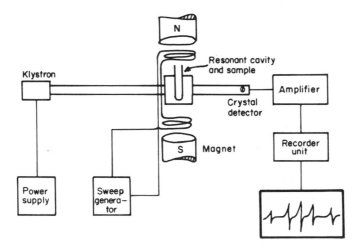

FIGURE 7.7 Diagrammatic representation of an electron spin resonance spectrometer.

up. A standing wave is composed of both a magnetic and an electric component, these being orientated at right angles to each other. Since it is the
magnetic component which interacts with the sample to cause spin resonance
and because the electrical field can also interact, the sample is usually located in
the cavity in a position which provides maximum rf magnetic field and
minimum electric field interaction.

The resonant cavity is made one or more half-wavelengths long (or in
diameter) with a hole at one end for coupling to the microwave system. With
rectangular cavities flat sample cells of about 0.25 mm thickness (volume
0.05 ml) are often used while with cylindrical cavities 3–5 mm ID tubing has
been used to make sample containers of capacity 0.15–0.5 ml.

The microwave radiation is usually detected by means of a silicon crystal
detector which converts the incident radiation into a direct current. To
eliminate "noise" from the detector, auxiliary electronic units are highly
desirable. The output of the detector is amplified and fed to a recording unit.
With strong signals the absorption lines may be displayed on an oscilloscope.
For weak signals, the field is modulated, amplified, and eventually recorded on
a chart as the first derivative of the absorption spectrum.

The resonant cavity and sample are mounted between the poles of an electromagnet capable of producing a highly stable, homogeneous magnetic field of
3400 oersted or greater. The ESR spectrum is recorded by slowly varying the
magnetic field through the resonance condition. This is achieved by varying
the current supplied to the magnet by the power supply. At resonance there is
a marked change in the amount of power reflected from the cavity.

For quantitative studies, the overall absorption of the sample is obtained by
integrating under the absorption curve. When a first derivative presentation
is involved the integration is preferably achieved with an electronic unit. In
applications where line widths are constant, peak heights may be used instead
of peak areas.

The integrated area is usually related to concentration by comparison with
a standard. The intensity of any absorption is proportional to the number of
unpaired spins in the sample, and is more or less independent of the source of
unpaired spins. The standard used may, therefore, be of different chemical
composition to the assay, although, as in all other forms of spectroscopy, the
best kind of standards are standard solutions or mixtures of the substance being
determined. In general, standards should be stable, the line widths should be
similar to those of the sample, and preferably the number of unpaired spins in
the standard should be close to the number in the sample.

In order to be able to accurately compare the intensity of the assay response
with that of the standard, many experimental precautions are required. The
volume of the standard in the rf field must be the same as that of the sample;
the size of the sample tubes and the wall thicknesses must be identical; and all
spectrometer adjustments must be the same for standard and sample. Sample

and reference should be similar in dielectric constant, in number of paramagnetic centers, and in microwave saturation characteristics. Many of these complications are minimized by permanently locating a reference substance within the cavity at some point remote from the sample hole. In turn one inserts a paramagnetic standard and the unknown sample. The ESR lines of the samples and the mounted reference should not overlap, so that the ESR spectrum in each case shows two sets of lines—one set due to the reference material, the other set due to the sample in the hole. One trace is used to relate the number of paramagnetic centers in the intermediate standard to the number in the mounted reference. The second trace relates the number of unpaired spins in the sample to the mounted reference. In this way standard and assay are compared with reasonable correction for many instrumental variables.

The type of paramagnetic material used as the standard can be a salt (e.g., $CuSO_4 \cdot 5H_2O$), a free radical in solid form or in solution [e.g., diphenylpicrylhydrazyl, DPPH or solutions of $K_2(SO_3)_2NO$], a transition metal ion highly diluted in a convenient host lattice (e.g., Cr^{3+} in Al_2O_3 or MgO) or various forms of carbon such as charred dextrose or coal.

For the mounted reference a synthetic ruby (Cr^{3+} in Al_2O_3) is quite convenient since the g value of the ruby resonance is anisotropic. The ruby signal can thus be shifted away from the sample signal by proper orientation of the ruby.

An alternative approach is to use a dual cavity technique, one cavity containing a standard, the other the sample. Each cavity is modulated at a different frequency so that the signals can be separated in tuned circuits and separately displayed.

C. Applications of ESR

By careful interpretation, an ESR spectrum can provide the following types of information:

1. The site of unpaired electron(s). The number of line components is used to elucidate the number and type of nuclei in the neighborhood of the odd electron.
2. Quantitative data. The relative intensities of the spectrum lines are of value in confirming the type of nuclei responsible for the splitting pattern. Summation of the intensities can be used to evaluate the total number of free electrons in the sample.
3. g values. The g value of a line or system may be measured approximately by comparing the position of the line with that of a reference of known g value, e.g., DPPH powder ($g = 2.0039$) or $K_2(SO_3)_2NO$ ($g = 2.0057$).
4. Field symmetry. If the electric field is not spherical, the spectrum is anisotropic, that is, rotation of the sample shifts the ESR spectrum.

The ESR techniques have proved of great value in studying substances which possess odd (unpaired) electrons. The classes of substances which have been investigated include:

1. Paramagnetic salts of the transition elements. With transition metal ions the valence state has a big influence on the nature of the spectrum.
2. Free radicals. The free radicals may be present as transient intermediates in chemical reactions or as semistable species produced by chemical, photochemical, electrochemical, thermal, or other means.
3. Defects in solids. These include impurities, trapped atoms, electrons or holes, and triplet states.

The most widespread use of ESR has been for the identification and measurement of radicals. The technique is extremely sensitive, the detection limit for commercial instruments being about 10^{-9} M, but this figure varies by several powers of ten depending on the experimental conditions being used.

A number of quantitative methods for the determination of paramagnetic ions have been published. For example, the analysis of vanadium in petroleum oils by ESR spectroscopy has been shown to be a rapid and convenient means of analysis and a continuous analyzer based on this principle has been designed.

In other published methods various classes of compounds are converted to free radicals prior to measurement. For example, polynuclear hydrocarbons like anthracene have been converted to radical cations by treating benzene solutions of the materials with activated silica-alumina catalyst.

Each year the number of direct and indirect quantitative applications increases but in many cases alternative procedures require simpler equipment and are less prone to error.

On the other hand, ESR has proved to be extremely useful in studying reaction mechanisms which proceed through free radical intermediates. The technique is also of great value in examining molecular triplet electronic states and has been used to elucidate the electronic structures of complexes of transition metal ions.

V.   MÖSSBAUER SPECTROSCOPY

One of the latest additions to the range of techniques available for studying molecular structure and molecular bonding is Mössbauer spectroscopy, sometimes known as nuclear radiation resonance or γ-ray nuclear resonance fluorescence. As indicated by these titles, Mössbauer spectroscopy is concerned with the emission and resonant reabsorption of γ rays, and the technique promises to become the inorganic counterpart of NMR.

When an isolated atom of mass m emits γ-radiation characteristic of the daughter atom, the atom recoils with conservation of momentum. Since the momentum of a γ-photon, of energy E_γ, is E_γ/c, the magnitude of the recoil energy is

$$E_R = 1/2mv^2 = 1/2(mv)^2/m = 1/2E_\gamma^2/mc^2$$

where v is the recoil velocity and c is the velocity of light.

Where there is recoil, the observed γ-ray energy differs from the excitation energy of the daughter nucleus by this amount (E_R).

When such γ rays fall on a sample of the daughter nucleus they are not absorbed, because the energy content is less than the quantum required to reach the excited state of the nucleus. A radiation detector placed behind the sample suffers no loss of signal.

The recoil loss can be reduced to an extremely small value by mounting the γ-ray emitter in a fixed position in a solid. In this way the mass term can be made almost as large as desired, a mass of about 10^8 atomic units reducing the recoil energy loss to about 10^{-9} eV. With E_R reduced to a negligible value, absorption by a daughter nucleus is more feasible.

Absorbance of recoilless emission due to nuclear radiation resonance occurs when the nuclei of the absorber are in a chemically similar environment to the emitter. The last condition is difficult to achieve in practice and in general there are small differences between E_γ and the excitation energy of the nuclei of the absorber, due to small differences in chemical environment. These can be adjusted, however, by imposing a Doppler effect on the source, by which means the γ-ray energy is increased, or decreased, slightly until resonance occurs. From the Doppler theory, the change in E_γ is given by $E_\gamma v/c$ where v is the velocity of the source (cm/sec).

In the Mössbauer technique the transmitted intensity of the emitted γ-radiation is determined as a function of the source or absorber velocity. A plot of intensity against velocity gives a spectrum which may possess one or more absorption peaks located at specific velocities (the resonance conditions).

The velocity required for resonance depends on the difference in the nuclear-energy levels of the source and the absorber, which in turn is determined by the nature and magnitude of the interaction between the nuclei and their chemical environment.

In chemical studies the Mössbauer technique is used to examine the nature of these interactions. A change in environment is indicated on a spectrum by a shift in the position of the absorption peak and/or through different splitting patterns. The experimental observations are reported as chemical shifts (δ), quadrupole splittings (Δ), and magnetic hyperfine Zeeman splittings.

The chemical shifts arise from the fact that the spacings of the nuclear energy levels (hence E_γ) depend minutely on the chemical environment of the nucleus.

The radius of the nucleus differs in the excited and the ground state by an amount $\delta r = r_{ex} - r_{gr}$, and it can be shown that the chemical shift (δ) is proportional to $\delta r/r$ times a term which represents the change in s electron density at the nucleus in going from the source to the absorber.

When $\delta r/r$ is positive, a positive chemical shift corresponds to an increase in s electron density, whereas if $\delta r/r$ is negative, a positive chemical shift corresponds to a decrease in s electron density; the opposite argument applies to negative shifts.

By setting up suitable calibration scales it is thus possible to deduce the relative s electron densities in a series of compounds. This in turn can be correlated with the type of chemical bonds present in the compounds.

Changes in the valency of an element change the magnitude of the chemical shift. Increased shielding of the nucleus from s electrons through introduction of electrons into d and f orbitals alters the chemical shift.

In summary, chemical shifts give information about:

1. The sign and magnitude of $\delta r/r$; Mössbauer spectroscopy is proving an unrivalled tool for elucidating the fine details of nuclear structure.
2. s electron density at the nucleus.

The s electron density effect can be used to examine:

1. Charge states
2. Effects of p, d and f electron shielding
3. Covalency effects
4. σ electron withdrawing power of electronegative groups
5. Deshielding effect of d_π back-bonding.

The nucleus is not necessarily spherical and any nuclear state with a spin $I > 1/2$ has a quadrupole moment, Q, which can align itself either with or across an electric field gradient, q. The excited state frequently possesses a spin value of $3/2$ and this leads to splitting of the nuclear substate into a doublet of separation Δ where

$$\Delta = \text{constant} \cdot Q \cdot q$$

As with the chemical shift, the quadrupole splitting, Δ, is the product of a nuclear factor (Q) and an extra nuclear factor (q).

Quadrupole splitting causes the absorption spectrum to split into two or more peaks and it is possible to estimate the field gradient q from the splitting.

This provides information about the site symmetries and field gradients within a crystal and gives details of imbalance between p and d electrons.

Quadrapole shifts have been used to give information about:

1. Site symmetries around the central atom in complex compounds.
2. The bonding and disposition of molecules in crystals.
3. Minute distortions from ideal symmetry.
4. High spin $\rightleftarrows$ low spin equilibria.

The third type of interaction that can be investigated by Mössbauer spectroscopy is the hyperfine Zeeman splitting of the nuclear energy levels in a magnetic field.

As in the case of unpaired extra nuclear electrons, each level of spin quantum number I splits into $(2I + 1)$ sublevels, the energies of these levels being given by

$$E = -\mu H m/I$$

where μ is the nuclear magnetic moment, m is an integer and H is the hyperfine magnetic field. It can be noted that, as in the case of chemical shifts and quadrupole interactions, the energy levels depend on both a nuclear factor (μ) and extra nuclear factor (H).

In Mössbauer spectroscopy one observes transitions between one of the sublevels of the excited state and one of the sublevels of the ground state. Transitions between these sublevels are governed by selection rules and for many Mössbauer nuclei the rules are represented by $\Delta m = 0, \pm1$.

The even spacing of the magnetic splitting is modified if quadrupole interactions are also present.

Hyperfine splitting provides information about:

1. Magnetic exchange interactions.
2. Magnitude and sign of the hyperfine magnetic field at selected sites within the crystal, e.g., the identification of site symmetries around magnetic nuclei in a range of alloys.
3. Curie point or Néel point of magnetic species.
4. Rates of fast electron transfer reactions (e.g., $Fe^{3+} + e \rightleftarrows Fe^{2+}$, half life $\simeq 10^{-7}$ sec).
5. Bonding in compounds like iron borides and phosphides.
6. Magnetic phenomena associated with particular types of solids.

The Mössbauer effect has been observed with over thirty elements, but so far information of chemical significance has been obtained for only half this number. Of the first thousand compounds studied, approximately half were compounds of iron and a third were compounds of tin. With the exception of potassium there are no Mössbauer elements having an atomic number < 26.

The information obtained by Mössbauer spectroscopy is of great importance in elucidating chemical structures and in increasing the understanding of chemical bonding and rate processes in fast reactions. Analytical applications may eventually prove to be of some importance in specialized cases.

The technique is best suited to studying solid powders and metal foils. The samples are mounted in a carrier which can be moved relative to the γ-radiation source. The primary detector is a γ spectrometer which is usually a scintillation counter fitted with a thin sodium iodide crystal to ensure that the soft or low energy radiations are detected efficiently in the presence of higher energy emissions. The γ-induced scintillation of the crystal is detected by a photomultiplier tube, the output of which is amplified and passed through a single channel pulse-height analyzer set at some desired energy peak.

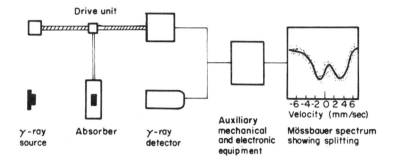

FIGURE 7.8   Schematic diagram showing the major components of a typical Mössbauer assembly.

Two general arrangements are possible. In the first, the source or absorber is moved at constant velocity, either in one or both directions, relative to each other, measurements being made at different velocities in turn.

The alternative method arranges for all velocities to be occupied for a predetermined or equal time each cycle. The spectrum built up is then displayed on a cathode ray oscilloscope.

Several different methods are available for moving the source and absorber relative to each other, and there are many methods for recording the detector counts. Hence in the diagrammatic representation of a typical assembly shown in Fig. 7.8, all the auxiliary operations are grouped together under the heading of auxiliary mechanical and electronic units.

VI.   MASS SPECTROMETRY

A.   PRINCIPLES OF THE TECHNIQUE

As indicated by the flow sheet (Fig. 7.9), in mass spectrometry a gaseous sample is bombarded with a beam of electrons to produce an ion molecule or

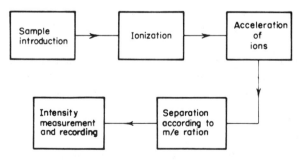

FIGURE 7.9   Flow sheet for the production of a mass spectrum.

ionic fragments of the original species. The resulting assortment of positively charged ions and fragments are then pulled out of the gas stream by means of a negatively charged accelerating electrode, and subsequently separated in accordance with their masses.

The mass distribution and relative abundance of the charged species are recorded. This provides the mass spectrum, which is a unique record representing the molecule from which it was formed.

The mass spectrum can be used to determine exact molecular formulas and molecular weights, and experienced mass spectrometrists use the spectra to characterize and completely determine the structure of the parent species. The technique is sensitive and specific; hence it can also be applied to qualitative and quantitative analysis.

1. The Formation of Ions

Since collision with other molecules or fragments can vitiate the separation of species on the basis of their relative masses, mass spectrometers operate under a high vacuum and use very small samples. The vacuum in the separation and recording section is preferably 10^{-7} torr or less, while in the ionization chamber it is usually about 10^{-5} torr. The sample is usually admitted into this chamber through a fine capillary leak, and gases or liquids having vapor pressures of at least 10^{-2} torr at ambient temperature can be fed in directly from a reservoir vessel. Samples must yield a vapor pressure of at least 10^{-7} mm Hg; hence less volatile compounds often require heating to provide a suitable pressure. Compounds of very low volatility are sometimes vaporized directly in the ionization chamber.

The amount of sample required is very small (e.g., about one micromole) and of this sample only about 0.01% reaches the detection unit; with such small samples the major problem is contamination. Unless extreme care is used it is easy to obtain a spectrum showing the remains of a previous sample, or the components of stopcock grease, or impurities in the sample itself. The latter is permissible in qualitative studies but undesirable in structure determinations. Removal of contaminants from the apparatus by a baking out procedure can be difficult and in some cases takes days.

In most instruments ionization is effected by bombarding the sample vapor with high energy electrons. Thermal electrons are generated by electrically heating a resistance filament. The electrons are then accelerated through the sample vapor by an anode maintained as some suitable potential, the energy of the electron beam being controlled by the magnitude of this potential.

The minimum energy required to produce a singly charged molecular ion is about 10 eV, and if a mass spectrometer is operated with an ionization beam of about this strength, the mass spectrum obtained consists almost entirely of a single peak corresponding to the mass of the original molecule.

$$\text{ABC} + e \rightarrow \text{ABC}^+ + 2e$$

Increasing the energy of the electron beam results in the formation of an excited charged species which tends to fragment. For most applications the electron beam is given an energy of 50–100 eV since this yields the most reproducible spectra. With some systems this potential is sufficient to eject a second electron to yield a doubly charged ion. Bombardment of the sample with electrons also produces some negative ions, but these are outnumbered by the positive ions by a factor of 10^2 to 10^4.

A positively charged plate repels the positive ions from the ionizing beam towards a series of negatively charged accelerating grids.

The total number of ions formed from the original material comprises the mass spectrum which is also known as the cracking pattern or fragmentation pattern. The pattern may include ions caused by dissociation of the single charged molecule, e.g.,

$$ABC^+ \rightarrow AB^+ + C \text{ or } BC^+ + A \text{ or } A^+ + BC, \text{ etc.,}$$

and ions formed by rearrangement or recombination of fragments, e.g., AC^+.

Theory predicts that the fragments from multiply charged ions should possess an energy and velocity in excess of that encountered with singly charged species. In most separation units high energy ions are discriminated against; hence these fragments do not usually make a notable contribution to the mass spectrum.

2.   *Ion Acceleration*

Most of the ions leaving the ion beam reach the first of the accelerator grids with a relatively small but variable kinetic energy. Acceleration of the particles is then achieved by application of potential differences ranging from 1 to 100 kV on succeeding focusing plates.

During their acceleration by the electric field all singly charged ions acquire energy E given by the equation

$$E = eV$$

where e represents the charge on the ion and V the applied potential.

The ions leaving the field have a kinetic energy equivalent to E, so that

$$E = eV = mv^2/2$$

where m is the mass of the ion and v is its velocity.

Algebraic rearrangement gives

$$v = \sqrt{2eV/m}$$

Besides accelerating the charged ions and fragments, the ion gun unit assists in focusing the ions into a fine beam. There is no convenient way in which the beam can be collimated, but it is possible to limit its angle of divergence by circular or slit diaphragms.

Since the ions vary slightly in their initial energy contents, they leave the accelerating unit with slightly variable energies. If this energy spread is reduced before the particles are dispersed by a magnetic field, a much higher resolution can be achieved in the separation stage. Hence in a high resolution instrument double focusing is sometimes used, that is, the accelerated ion beam is passed through a radial electrostatic field which transmits only those ions possessing a certain velocity into the magnetic separation unit.

3. *Separation According to m/e Ratios*

It can be seen from the preceding equation that ions having different m/e ratios travel with different velocities, and in time-of-flight spectrometers separation is based on these velocity differences.

In these instruments the accelerating potential V is applied intermittently. During each pulse all ions receive the same energy and are then allowed to drift for 40 to 200 cm in a field free region. Inside the flight tube pairs of horizontal and vertical deflection plates position the most intense part of the ion beam on the cathode of a magnetic electron multiplier unit.

The ion fragments separate according to their various m/e ratios as they drift from the source to the detector, the lighter ions arriving at the detector first because the ionic velocity is inversely proportional to the square root of the mass. The output of the detector is applied to the vertical deflection plates of a cathode ray oscillograph which has a horizontal time base synchronized with the accelerator pulses. The mass spectrum displayed on the oscilloscope screen can then be photographed or studied directly. A large proportion of mass spectrometers, however, use magnetic deflection to achieve ion mass analysis.

A particle in motion suffers a radiationless change of direction on passing through a magnetic field of flux H oersted. The charged particles experience a magnetic force $F (= Hev)$ which is equaled by a centrifugal force $F (= mv^2/r)$ where r is the radius of the circular path.

By equation of these two terms, and substituting for v, one obtains

$$r = mv/He = \frac{1}{H}\sqrt{2Vm/e}$$

From this equation it follows that the trajectories of ions of different mass differ, and hence an initial mixed group of ions can be resolved into beams which are homogeneous in their mass-to-charge ratio (m/e).

In a mass spectrometer the beam of charged ionic fragments is fed into the magnetic field at normal incidence to the boundary of the field. The mass spectrum is then magnetically scanned by varying the magnetic field in a precise manner, thereby causing the mass-separated ion beams to impinge upon a collector electrode in sequential order of their mass, after being bent through an angle of 60°, 90°, or 180°. Figure 7.10 shows the arrangement of the

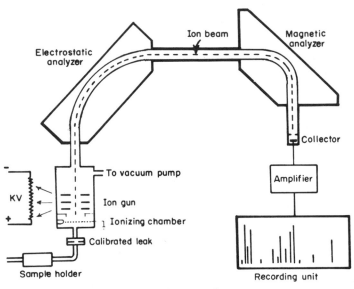

FIGURE 7.10   Schematic diagram of a mass spectrometer.

components of a typical mass spectrometer with magnetic scanning. This arrangement follows what is known as Neir–Johnson geometry.

It is also common to employ electrostatic scanning, in which system the magnetic field strength is held constant and the energy of the injected ions is varied in a controlled manner.

In a second type of instrument based on Mattauch–Herzog geometry all the ions are brought to focus on a plane. The detector in this instance can be a photographic plate which records all the ion beams simultaneously. That is, H and V are fixed and the instrument records the different m/e values through the differences in r values.

There are many other modifications such as spectrometers which use cycloidal focusing and units which separate ions by passing the beam through a series of grids pulsed with a high-frequency signal set to allow only one velocity to get through.

In single focus mass spectrometers the accelerating voltage is usually of the order of 2–4 kV with magnetic fields of up to 12,000 oersted. In double focusing instruments these operational parameters may be twice as great.

4. Recording

It has been noted that in time-of-flight spectrometers and Mattauch–Herzog design instruments recording is initially based on a photographic record. A strip chart record can be prepared, if desired, from the photographic plate by passing it through a recording densitometer.

The mass spectrum produced by most instruments, however, usually consists of a roll of chart paper which records the response of the detector as a function of the m/e ratios being sequentially brought to focus.

The ion beam currents which are collected and measured are of the order of 10^{-9} to 10^{-15} amp; hence amplifications of 4 or 5 orders of magnitude are required to feed recording units. For this reason, the collector electrode must be well shielded from stray ions and the amplifier must be stable and linear over a wide range.

In more advanced spectrometers, the detector signals are taped and fed into a computer for direct interpretation.

5. *Resolution*

One of the most important specifications of a mass spectrometer is its resolution. This is a measure of the instrument's ability to separate ions of adjacent mass. Two peaks of equal height are said to be just resolved if the valley between them is 10% of the peak height. The resolving power of the system at this point is defined as the difference in the masses of the two ions, e.g., one part in 2000 or 500 ppm, or in general terms, $M/\Delta M$ where M is the nominal mass of the pair of closely spaced peaks separated by ΔM mass units.

The degree of resolution required varies with the problem but it is often proposed that a resolution of at least one part in 10,000 is desirable for accurate formula assignment, and most current instruments have a resolving power of this order. For example, an instrument of lower resolving power would not separate N_2^+ (mass 28.0061) and CO^+ (mass 27.9949).

Some models are capable of achieving a resolution of one part in 40,000 while others are limited to one part in 200. Spectral scanning times and interpretational difficulty invariably increases markedly as the resolution goes up; hence, samples are usually run at the lowest resolution consistent with the type of data required.

For accurate mass measurements the spectrum of the sample and of a marker compound are recorded simultaneously. The marker, often a perfluoro compound, must give fragments of known composition which can be easily identified.

B. THE MASS SPECTRUM

When a molecular sample is subjected to an electron beam whose energy is only slightly greater than the ionization potential (e.g., 10–15 eV), the principal product is the molecular ion formed by the loss of one electron. The resultant mass spectrum is comparatively simple, with nearly all the ions appearing in the "parent peak." This parent peak gives the exact molecular weight of the sample.

Due to the natural abundance of the isotopes of constituent atoms, the spectrum also shows small peaks which correspond to molecular ions containing different isotopes. For example, with hydrocarbon compounds there is generally a small peak appearing one mass unit higher than the parent peak (the $M + 1$ peak) which is attributable to the small, but observable natural abundance of ^{13}C and for ^2H. Molecules containing chlorine and bromine yield significant $M + 2$ peaks since the isotopes of these elements differ by two mass units, and 24.2% and 49.5% respectively of the atoms are present as the heavier isotope. Isotope peaks sometimes prove very helpful in confirming proposed molecular weights.

With an electron beam energy of 50–70 eV, the molecular ion originally produced has considerable excess energy which needs to be rapidly delocalized. In most cases this is achieved by cleavage into fragments, one of which retains the charge. The parent peak in the spectrum is thus often quite small. A π-electron system more readily adjusts to the loss of an electron than a σ bond system; hence π bonded compounds tend to yield more stable molecular ions, that is, bigger parent peaks. With different classes of compounds the relative height of the parent peak tends to decrease in the order: aromatics > conjugated olefins > alicycles > unbranched hydrocarbons > ketones > amines > esters > carboxylic acids > branched hydrocarbons and alcohols.

The fragmentation pattern of a molecule can be related to its structure since the possibility of cleavage is related to both the bond strengths in the molecular ion and the stability of the fragments (which can be either stable molecules or radicals).

The largest peak in the spectrum is described as the "base peak" and other peak heights are measured and recorded as a percentage of the base peak height.

For identification purposes, the "Handbook of Analytical Chemistry" lists the mass and relative size of the parent peak, base peak, and next three major peaks for a large selection of organic compounds. This list has been extracted from one of the more extensive compilations now available.

The base peak often results from the loss of a small stable molecule such as CO, CO_2, H_2O, NH_3, H_2S, HCN, or ROH. With branched hydrocarbon compounds cleavage at branched carbon atoms is favored as a consequence of the relative stability of the carbonium ions produced. The influence of a hetero atom can be reasonably well predicted. In the primary ionization process, electronegative atoms (such as N and O) tend to lose one of their unshared electrons. Electrons from neighboring bonds are then strongly attracted to the hetero atom and this can result in cleavage at the β bond.

Many of the fragments observed, however, cannot be described as part of the original molecule. These fragments arise from the rearrangement of ionic species. In molecules containing a hetero atom the rearrangement is usually quite specific and the resultant peak can be quite intense. In hydrocarbons the rearrangements tend to be nonspecific and unpredictable, apart from the generalization that rearrangement is favored when a stable fragment will result.

The fragmentation pattern usually contains many peaks, the number being increased by the presence of isotopes of the constituent atoms. For example, the mass spectrum of *n*-butane contains 24 peaks due to ions with mass numbers (M) varying from 26 to 59. The spectrum of the larger molecule 1-phenyl-5-ethylbarbituric acid has 32 major peaks varying in mass number from 25 to 233, the parent peak appearing at $M232$. The base peak at $M119$ (100) is due to the fragment $C_6H_5N\cdot CO^+$. The next three largest peaks are $M55$ (82), $M70$ (68) and the parent peak $M232$ (58), where the numbers in parentheses represent the percentage height of the peak relative to the height of the base peak.

The interpretation of a mass spectrum requires an understanding of the ionization processes which occur before the ions reach the collector. From a study of the fragmentation pattern it is then necessary to predict how the fragments were formed.

Through the examination of innumerable spectra it is possible to generalize in respect to the behavior of different types of compounds. Lists are available of the type of fragments commonly lost, and atlases of standard spectra are fairly readily available. With these aids, an experienced operator can adequately interpret most simple experimental spectra.

With high resolution instruments the process is far more complex, and a computer is regularly used to unravel the multiple peaks and so provide a final interpretation.

C. Applications

The applications of the mass spectrometer can be subdivided into three broad divisions, the determination of basic physical data, the elucidation of structure, and chemical analysis.

1. *The Determination of Physical Data*

The mass spectrometer is unsurpassed by any other method for the determination of molecular weights. The simplest procedure involves measurement of the mass at which the "parent peak" occurs. This method is limited in range by the resolution of the instrument and in scope to those molecules which yield a stable molecular ion.

Since most of the atoms encountered by a chemist are polyisotopic, a careful study of a high resolution spectrum indicates peaks which differ from others by an integral number of mass units. By comparing the height of the M, $M + 1$, $M + 2$, etc., peaks, it is possible to calculate the number of each type of atom present. For example, one can determine the number of carbon, chlorine, bromine, silicon, sulfur, or metal atoms in the parent compound.

The masses of isotopes are known very precisely from mass spectrometry measurements and the relative abundance of the isotopes has been accurately determined by this technique. Thus exact atomic weights for all elements are now available, and the molecular weights of chemical species can be calculated to three decimal places.

Mass spectrometers are now available which have sufficient resolving power to permit the precise determination of a molecular weight to this order of accuracy. Thus the exact formula can be computed from a single measurement of the exact mass of the parent peak. For example, acetophenone (mol wt 120.157), benzamidine (mol wt 120.069), ethyl toluene (mol wt 120.094), and purine (mol wt 120.044) all yield a parent peak at mass number 120 in a low resolution instrument. On the other hand, in a high resolution instrument the mass numbers could be determined as the values shown in parentheses, and if the material was initially unidentified, the exact molecular weight would provide the desired identification. For example, a parent peak at 120.156 could only be assigned to acetophenone.

Closely associated with molecular weight determinations is isotope labeling. A convenient means of determining the mechanism of a reaction or the structure of a complex compound is to use a reactant containing a labeled atom and then find its exact location in a product. In many cases it is possible to use a stable heavy isotope and to locate the position of the labeled isotope by the change it causes in the mass spectrum. Some fragments appear at a different mass number due to the inclusion of the heavier isotope.

The ionization potential and the bond strengths of molecular species can be estimated from the electron energy required to produce ions of different types.

2.   *Structure Determination*

The mass spectrum provides a tool of enormous potential for the determination of molecular structure.

However, the present knowledge of molecular fragmentation processes is so incomplete, and the complete mass spectrum can be so complex, that it is not possible to determine the exact structure of large molecules by mass spectrometry alone. The additional evidence provided by NMR and infrared studies is absolutely necessary in most cases.

A systematic description of structure determination is beyond the scope of this book and interested readers are recommended to read one of the monographs available on this topic.

3.   *Analytical Applications*

The mass spectrum can be used to identify a compound, although the process is not necessarily easy. The empirical formula can be ascertained from a molecular weight determination while partial molecular formulas can be gained from studies of isotope distribution patterns. Final confirmation of the identity is based on the fragmentation pattern.

The fragmentation pattern is unique for each compound and atlases of mass spectra patterns are available. Identification therefore relies on comparison of the unknown spectrum with standard spectra until a match is obtained.

An interesting application of this qualitative aspect of mass spectrometry is the use of the technique to identify the products formed during thermal decomposition or catalytic cracking of parent compounds. Time-of-flight mass spectrometry has also been used extensively in studies of fast reactions.

Interpreting the mass spectrum of a mixture of compounds is extremely difficult but if the identities of the compounds in the sample mixture are known, quantitative analysis is fairly straightforward.

The general principle is based on the idea that each compound acts independently of all others present. The observed spectrum is thus the sum of the individual spectra and the contributions to each peak are additive for each compound present.

Thus if I_1, I_2 and I_3 are the total ion currents observed on the spectrum at mass numbers 1, 2, 3, and if the letters x, y, z are used to represent the different compounds present, then

$$I_1 = i_{1x}p_x + i_{1y}p_y + i_{1z}p_z +$$
$$I_2 = i_{2x}p_x + i_{2y}p_y + i_{2z}p_z +$$
$$I_3 = i_{3x}p_x + i_{3y}p_y + i_{3z}p_z + \cdots$$

where i_{1x} is the ion current at mass 1 due to component x at unit pressure and p_x is the actual partial pressure in the sample.

Values of i_{1x}, i_{2x} etc., can be obtained from the spectra of pure compound x at known pressures.

The total pressure used in the assay equals $p_x + p_y + p_z + \cdots$.

A set of simultaneous equations (at least as many as there are unknowns) can therefore be set up and solved for the unknown partial pressures. This is invariably done with the aid of a computer.

Using this procedure the complete proximate analysis of a petroleum fraction can be obtained in less than an hour and an accuracy of $\pm 0.5\%$ is claimed for each component.

Recently the determination of the mass spectra of each component of a complicated mixture has become possible by coupling an appropriate gas-liquid chromatography column directly to the spectrometer. On emerging from the column the carrier gas is removed by differential pumping through a porous tube, and the spectrum of each component is determined. The carrier gas is normally helium since it is more easily removed by differential pumping than any other gas (other than hydrogen) and it has a high ionization potential (22 eV). By operating the mass spectrometer at 20 eV only the sample is ionized and the slight increase in total ion current due to the onset of a peak can be detected. Immediately after the ions are produced they are pulled into the spectrometer for mass separation.

The main advantage of mass spectrometry over other physical methods (NMR, UV, IR spectroscopy, etc.) is the small amount of material required. The main limitations are imposed by sample volatility and a lack of understanding of the fragmentation involved.

VII. TUTORIAL AND REVISION QUESTIONS

1. Write brief notes on the following topics:
   a. Magnetic moment.
   b. Effect of temperature on paramagnetism.
   c. Shielding effects in NMR.
   d. Energy transitions in NMR.
   e. Generation of microwaves.
   f. Recoilless γ emission.
   g. The trajectory of an accelerated ion in a magnetic field.
   h. Isotope effects in mass spectrometry.
2. The term "chemical shifts" has been used in descriptions of both NMR spectroscopy and Mössbauer spectroscopy. Outline the origin of chemical shifts and indicate how these shifts are used to elucidate the chemical environment of atomic species.
3. Compare the principles of a time-of-flight mass spectrometer and a double-focusing magnetic field mass spectrometer.
4. From review articles or other sources prepare an essay on the analytical applications of magnetic susceptibility measurements.
5. It has been proposed that in the future the analyst will be required to describe the composition of a sample in terms of molecular nature rather than elemental composition. For example, if aluminum is present in an alloy it may be necessary to indicate the amount present as aluminum atoms, as aluminum-metal compounds, and as inclusions such as aluminum oxide.

   Evaluate the possibility of using one (or more) of the techniques described in this chapter for this purpose.
6. Describe the components of an ESR spectrometer and prepare a summary of the possible analytical applications of this technique.
7. a. In high resolution NMR spectroscopy the splitting of peaks is attributed to spin-spin coupling. What is spin-spin coupling and how does it cause multiplet peak formation?
   b. Predict what the high resolution spectra of the following compounds would look like. Indicate the approximate δ values for the different groups of equivalent protons in these compounds: acetic acid, acetone, ethyl benzene, ethyl acetate, acetaldehyde, and ethyl bromide.
   c. For ^{19}F, $I = 1/2$ and $\mu = 1.329 \times 10^{-23}$ erg oersted^{-1}. If the radio-frequency oscillator has a constant frequency of 60.0 megacycles (Mc) per sec, to what value must the applied magnetic field be adjusted for ^{19}F resonance?

8. While NMR spectroscopy has been used primarily to elucidate the structure of organic compounds, it has also been used for quantitative analysis. From review articles and monographs, prepare a summary of the main types of analytical applications.

9. A mass spectrum can be used for qualitative identification of a compound, structure determination, or quantitative analysis of mixtures.

Write an essay (e.g., 1000 words) on each of these aspects of mass spectrometry.

10. a. What do you understand by the resolution of a mass spectrometer?

  b. A mass spectrometer is to be used to analyze the products of a gaseous oxidation reaction, the heaviest product formed having a molecular weight of sixty.

  Assuming that the nature of the products is known, would a low resolution instrument be suitable for this analysis?

  If the fuel were of mixed composition and the products had first to be identified, why might one desire to use a high resolution instrument?

  c. If one of the reactants was tagged with a heavy isotope how would this be observed in the mass spectrum and how might this tagging assist in elucidating the mechanism of the reaction?

References

MAGNETOCHEMISTRY

L. N. Mulay, *Magnetic Susceptibility*, Wiley (Interscience), New York, 1963.

B. N. Figgis and J. Lavis, *The Magnetochemistry of Complex Compounds*, in *Modern Coordination Chemistry*, J. Lewis and R. G. Wilkins, (eds.), Wiley (Interscience), New York, 1960.

B. N. Figgis and J. Lewis, *Magnetochemistry of Coordination Compounds*, in *Progress in Inorganic Chemistry*; F. A. Cotton, (ed.), Wiley (Interscience), New York, 1964.

NMR AND ESR

M. Bersohn and J. C. Baird, *An Introduction to Electron Paramagnetic Resonance*, Benjamin, New York, 1966.

A. Carrington and A. D. McLachlan, *Introduction to Magnetic Resonance*, Harper, New York, 1966.

J. W. Elmsley, J. Feeney, and L. H. Sutcliffe, *High Resolution Nuclear Magnetic Resonance Spectroscopy*, Vols. I and II, Macmillan (Pergamon), London, 1966.

L. M. Jackman, *Applications of Nuclear Magnetic Resonance Spectroscopy in Organic Chemistry*, Macmillan (Pergamon), New York, 1959.

J. A. Pople, W. G. Schneider and H. J. Bernstein, *High Resolution Nuclear Magnetic Resonance*, McGraw-Hill, New York, 1959.

J. D. Roberts, *An Introduction to the Analysis of Spin-Spin Splitting in Nuclear Magnetic Resonance*, Benjamin, 1961.

Sadtler Research Laboratories N.M.R. Spectra Catalogue, Philadelphia, 1966.

Mass Spectrometry

A. J. Ahearn, *Mass Spectrometric Analysis of Solids*, Elsevier, Amsterdam, 1966.
J. H. Beynon, R. A. Saunders, and A. E. Williams, *The Mass Spectra of Organic Molecules*, Elsevier, Amsterdam, 1968.
J. H. Beynon and A. E. Williams, *Mass and Abundance Tables for Use in Mass Spectrometry*, Elsevier, Amsterdam, 1963.
K. Biemann, *Mass Spectrometry: Organic Chemical Applications*, McGraw-Hill, New York, 1962.
H. Budzikiewicz, C. Djerassi, and D. H. Williams, *Structural Elucidation of Natural Products by Mass Spectrometry*, Vols. 1 and 2, Holden Day, San Francisco, 1967.
R. W. Kiser, *Introduction to Mass Spectrometry and Its Applications*, Prentice-Hall, Englewood Cliffs, New Jersey, 1965.
F. W. McLafferty, *Interpretation of Mass Spectra. An Introduction*, Benjamin, New York, 1966.
F. W. McLafferty, (ed.), *Mass Spectrometry of Organic Ions*, Academic Press, New York, 1963.

Mössbauer Spectroscopy

R. H. Herber and V. I. Goldanski, *Chemical Applications of Mössbauer Spectroscopy*, Academic Press, New York, 1968.

8

REFLECTION, REFRACTION, AND ROTATION OF LIGHT

I. Physical Optics 226
    A. Wave Motion 226
    B. Interfacial Phenomena 228
    C. Polarization 229
    D. Geometrical Optics 231
II. Refractometry 234
III. Polarimetry and Optical Rotatory Dispersion 238
    A. Polarimetry 238
    B. Optical Rotatory Dispersion 241
    C. Circular Dichroism 243
IV. Light Scattering Photometry 244
    A. Factors Influencing Scattering 244
    B. Turbidimetry and Nephelometry 246
    C. Scattering by Macromolecules 247
V. Microscopy 249
    A. Optical Microscopy 249
    B. Electron Microscopy 253
VI. Revision and Tutorial Exercises 255
    References 256

An integral part of most of the techniques described in the two preceding chapters is the absorption of electromagnetic radiation.

Radiation can also be reflected, refracted, or rotated about its axis of propagation. These phenomena play a significant indirect role in many of the absorption techniques (e.g., in the dispersion of radiation) and warrant discussion for this reason. In addition, there are analytical techniques where these processes represent the basic principle on which the experimental procedures are based.

Techniques in this category are microscopy, refractometry, polarimetry and light scattering photometry.

An appropriate introduction to a discussion of these procedures is accordingly a brief revision of the physical laws associated with optical systems.

I. PHYSICAL OPTICS

A. WAVE MOTION

In the general treatment of emission and absorption, radiation was con-sidered as a source of photons (i.e., $E = h\nu$), but the majority of the phenomena of physical optics can be satisfactorily described by considering electromagnetic radiation simply as a wave motion defined by an equation such as

$$y = A \sin (\omega t + \alpha)$$

where y is the amplitude of the wave at any point in time, t, and A is a constant equal to the maximum amplitude. ω is the angular frequency ($\omega = 2\pi\nu$); α is a constant giving the phase of the wave, i.e., α describes the part of the cycle in which the wave disturbance exists when the timing starts.

Combination of two waves of the same frequency but of different amplitude gives a resultant wave described by the expression

$$y = A_1(\sin \omega t + \alpha_1) + A_2 \sin (\omega t + \alpha_2)$$

which can be reduced to

$$y = A_o \sin (\omega t + \theta)$$

where A_o is the maximum amplitude of the resultant wave and θ represents its phase angle. A_o has a maximum value when $\alpha_1 - \alpha_2$ equals $0°$, $360°$, $720°, \ldots$, and a minimum value when the phase difference equals $180°$ or $540°$.

The intensity of any radiation is a function of its maximum amplitude; hence, interference arising from superposition of two waves can lead to an increase or decrease in intensity, the controlling factor being the phase difference between the two wave motions.

When a great many wave trains of the same frequency but of random phasing are superposed, the amplitude can be shown by statistical analysis to increase as the square root of the number of wave trains. Intensity varies with the square of the amplitude; hence the superposition of a large number of randomly phased waves theoretically causes a linear increase in intensity. Radiation sources emit such random distributions, and accordingly one is justified in saying that the intensity of emission from an excited atomic species is on the average equal to the intensity from a single excited atom multiplied by the total number of atoms emitting.

A more complicated situation arises when wave trains of different frequency as well as different amplitudes and phasing are combined, e.g.,

$$y = A_1 \sin (\omega_1 t + \alpha_1) + A_2 \sin (\omega_2 t + \alpha_2) + \cdots .$$

Provided the velocity of the component waves is the same, the resultant moves with this velocity. The form of the resultant wave does not, however, exhibit the simple symmetry associated with the usual sine curve. For example, Fig. 8.1(a) shows the form of the resultant wave when a selected wave is superposed on a wave of similar frequency (but different phase) and Fig. 8.1(b) shows the form of the resultant wave when superposition involves a wave of different frequency and phase.

When an electromagnetic wave impinges on an atom, molecule, or ion, it induces a periodic electrical dipole, since the electrical vector of the radiation tends to cause electrons to move from their equilibrium positions about the

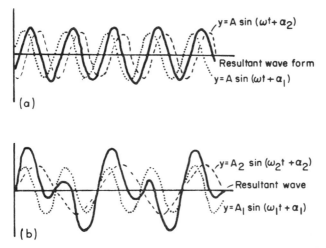

FIGURE 8.1 Diagram showing the effect of superposing wave trains having different phase angles; (a) same frequency and amplitude and (b) different frequency and amplitude.

nuclei. The energy associated with the electrical polarization is immediately released as radiation if the incident frequency is not one which is characteristically absorbed. This secondary radiation from the particle matter interferes with the incident radiation, and such interference is the basis of the phenomena known as refraction, reflection, and scattering.

The greater the polarizability of a particle, the greater the amplitude of the secondary waves, and the material is said to possess a larger refractive index. The secondary waves have the same frequency as the incident radiation, but they are plane polarized. That is, the electric vector, E, of the radiation vibrates in a single plane with the associated magnetic vector, H, vibrating in another plane at right angles to E. The plane of polarization is determined by the electrical vector of the incident wave and the direction of propagation of the secondary ray.

The refractive index of a medium, n, is defined by

$$n_v = c/v_v$$

where c is the velocity of light in a vacuum, v is the velocity in the medium of refractive index n, and the subscript ν establishes the fact that a particular velocity and index are associated with each frequency.

When the incident frequency is very much smaller than the frequency characteristically absorbed by the medium, n remains essentially constant at some positive value greater than unity; as the incident frequency approaches the absorption frequency, n becomes larger and larger. The behavior of the refractive index in the immediate vicinity of the absorption frequency is referred to as anomalous dispersion, whereas the gradual increase of n with ν observed at much lower frequencies is considered to be normal dispersion.

B.   INTERFACIAL PHENOMENA

When radiation passes from one type of transparent material to another it is partly reflected and partly transmitted. It retains its characteristic frequency in both media, but the velocity of propagation is different in each case. In general, the radiation also abruptly changes direction at the interface.

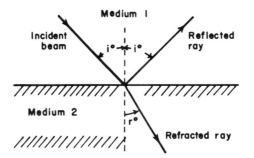

FIGURE 8.2   Interfacial phenomena. The incident beam, reflected beam, and transmitted beam are all in the same plane. The angle of incidence (i) equals the angle of reflection, but differs from the angle of refraction (r).

The interfacial effects associated with the incidence of some radiation passing from air on to a piece of transparent medium is shown schematically in Fig. 8.2.

The angle of refraction (r) differs from the angle of incidence (i), but the two are related by the equations

$$\sin i/\sin r = v_1/v_2 = n_2/n_1$$

The subscripts 1 and 2 refer to the two media involved; 1 represents the medium in which the radiation is initially traveling.

The angle of refraction cannot exceed a value known as the critical angle for each interface. Radiation is refracted at the critical angle, r_c, when the angle of incidence is 90°, i.e.,

$$\sin r_c = n_1/n_2$$

For vertical incidence ($i = 0°$) the ratio of the sines becomes indeterminate. In this case, the radiation is not deviated but merely undergoes a change of velocity on entering the second medium.

Reflection occurs whenever radiation is incident upon a boundary between two materials of differing refractive index. For a smooth surface the angle of incidence equals the angle of reflection; irregular surfaces give diffuse reflection. The reflected beam usually differs from the incident beam in total intensity, state of polarization and phase. In addition, if many frequencies are present, there are usually chromatic differences in the reflected radiation as well.

The intensity of the reflected radiation is related to the refractive indices of the two media involved. Thus for normal incidence, the reflectance ρ is given by the equation

$$\rho = I/I_o = (n_2 - n_1)^2/(n_2 + n_1)^2$$

where I and I_o are the reflected and incident intensities respectively.

When radiation falls obliquely on an interface, the reflectance varies with the angle of incidence. For angles of incidence up to about 60° it remains approximately constant. With larger angles reflectance increases quickly reaching the maximum value at 90°, the grazing incidence.

If the interface is small in extent, e.g., small particles suspended in some other material, the radiation is scattered rather than reflected. The origin of scattering, like reflection and refraction, is the induced secondary emission of particles in the path of the radiation. With larger particles, the lateral rays mutually cancel through destructive interference leaving only a refracted and a reflected beam. Where the particles have dimensions about equal to or smaller than the incident wavelengths, however, radiation is scattered randomly since there are fewer mutual interference effects.

C. POLARIZATION

The electromagnetic radiations produced by most sources are generally unpolarized, i.e., the electric and magnetic vectors of the waves occur at all orientations perpendicular to the direction of propagation. Contact with an interface usually causes the pattern to be altered. If the material medium causes the amplitude of the wave trains in particular directions to differ, it is said to polarize the beam. The radiation is said to be plane polarized if the vibration at any point is restricted to a straight line, elliptically polarized if restricted to an ellipse, and circularly polarized if restricted to a circle. An elliptical or circular vibration can always be resolved into two linear vibrations (i.e., plane polarized) at right angles with a suitable phase difference.

A system which transmits only one plane of vibration, e.g., a stack of glass plates or a Nicol prism, can be used to detect polarization by other media. When used in this way it is known as an analyzer.

For discussion purposes, the random orientation of waves can be considered to be resolved into two major beams vectorially at right angles to each other, and to the direction of radiation propagation. On this basis, an analyzer should transmit half the intensity of any incident beam, since it will be transparent to radiation polarized parallel to its characteristic vibrational planes but opaque to radiation polarized perpendicular to this direction.

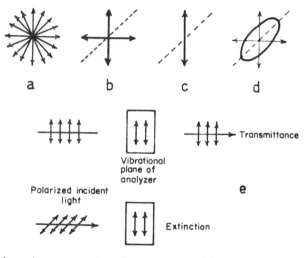

FIGURE 8.3   Schematic representation of some aspects of the polarization of light: (a) end view of an unresolved, unpolarized beam of radiation; (b) unpolarized beam resolved into planes perpendicular to each other; (c) single vibrational plane of an analyzer material; (d) elliptically polarized light; and (e) transmittance and extinction of polarized radiation by an analyzer.

The degree of polarization of a beam can, therefore, be ascertained by observing the position of the analyzer required to give maximum transmittance and by measuring the intensity of the incident and transmitted radiation.

These concepts are shown schematically in Fig. 8.3.

A number of crystalline materials preferentially absorb one vibrational plane of radiation. Vibrational components perpendicular to this plane are only slightly absorbed; these materials therefore behave like polarizers. The phenomenon is known as dichroism.

Except for the case of normal incidence, both reflected and refracted rays are partially polarized. The component vibrating parallel to the plane of incidence is selectively refracted; the component at right angles to this is more completely reflected. The reflected ray is entirely polarized when the angle of incidence is equal to the polarizing angle φ_p which is defined by the condition

$$\tan \varphi_p = n$$

where n is the refractive index of the reflecting surface.

Substances whose molecules or crystals lack a plane or center of symmetry are capable of rotating the plane of polarization of plane polarized radiation. Such substances are said to be optically active or are described as showing optical rotatory power. Different wavelengths are rotated by different amounts, a phenomenon known as rotatory dispersion.

Most crystalline substances exhibit optical anisotropy, i.e., one direction is not optically equivalent to another because of the differences in the spacing units in the lattice pattern and the number of atoms encountered. The speed with which a beam of polarized light passes through anisotropic crystals can therefore vary with the orientation of the vibrations with respect to the crystal. However, for each material there usually exists one particular direction along which the monochromatic beam can travel with the same velocity irrespective of the orientation of the polarization plane; this particular direction is known as the optic axis. By definition the velocity along the optic axis establishes the ordinary index of refraction.

On entering a crystal with a single optic axis incident radiation is resolved into two rays with mutually perpendicular planes of vibration. One vibration occurs in a plane formed by the optic axis and a perpendicular to one of the main faces of the crystal. This is known as the principal plane. The other vibration is in a plane at right angles to the principal plane, and its velocity equals that observed along the optic axis. This component is thus known as the ordinary ray. The component with vibrations in the principal plane is termed the extraordinary ray and has a velocity, hence index of refraction, that varies with the direction of propagation.

Thus, unless radiation traverses an anisotropic substance along the optic axis, what enters as a single ray emerges as two. The crystal resolves the incident radiation into beams that are transmitted at different angles because of their different indexes of refraction. This observation is termed double refraction or birefringence.

An important application of double refraction is in the construction of Nicol prisms which are used to both produce and analyze plane-polarized light. Nicol prisms are nearly always made from two triangular pieces of calcite crystals joined by a layer of Canada balsam. The incident ray is allowed to enter the unit at right angles to the optic axis, and is resolved into an ordinary and an extraordinary ray. At the layer of Canada balsam the ordinary ray is totally reflected, while the extraordinary ray is transmitted and passes through the other segment of the prism. The Nicol prism is restricted to use with visible light but a large number of experimental arrangements employ pairs of such prisms, one serving as a polarizer, the other as an analyzer.

D. GEOMETRICAL OPTICS

In optical systems, lenses, curved mirrors, and aperture devices are used to control the flow of electromagnetic energy from one point to another. Prisms and diffraction gratings are used to disperse polychromatic beams of radiation. The basic principles of these optical units can be described in terms of the phenomena listed in preceding sections.

For example, consider the dispersion of radiation by a 60° glass prism.

As shown in Fig. 8.4(a), a beam of monochromatic light AB traveling parallel to the base of the prism meets the face of the prism at an angle of 60°, that is, the angle of incidence is 30°. At the interface much of the beam is refracted, the angle of refraction being given by $\sin r = \sin i/n = \sin 30°/1.52$ since the refractive index of glass is about 1.52; solution of this equation gives a value for r of about 19°. The refracted beam BC then travels through the glass until it meets the glass–air interface at C.

At C the transmitted beam is again refracted. For the glass–air transition, the refractive index n^1 is the reciprocal of the index for an air–glass transition, i.e.,

$$n^1 = 1/n = 1/1.52 = 0.658$$

For the beam shown in Fig. 8.4(a), simple geometry shows that the angle of incidence to the glass–air interface is 41°, so that

$$\sin r = \sin 41°/0.658 = 0.962$$

This corresponds to an angle of 74°. The refracted ray leaving the prism therefore makes an angle of 74° with the normal to the face (i.e., 16° to the face).

Since the refractive index of a transparent medium varies with the frequency of the incident radiation, the angles of refraction also differ with frequency. Hence when a polychromatic beam follows the path AB, it is split into rays differing in refractive angle during passage through the prism, and the individual rays of different frequency ranges emerge at different angles to the prism face. The polychromatic radiation is said to be dispersed.

A slightly more complex argument is required to explain the dispersion of radiation by a diffraction grating.

Instead of a prism, let us consider the pattern of refraction in a transparent solid having a curved surface [cf. Fig. 8.4(b)].

A ray of light such as AB which is normal to the arc of the circle is not refracted and is transmitted directly through the medium, the path passing through the center of the circle of which the solid is a segment. Any other ray which is parallel to ray AB meets the surface at an angle of incidence other than 0°. The rays are therefore refracted in a manner similar to that described in detail for the glass prism. It can be shown that the emergent refracted rays all intersect at a point D, known as the focal point. In this respect, the circular segment bears some resemblance to a lens.

A lens is a transparent solid having two surfaces, at least one of which must be the curved boundary of a segment of a sphere. The other surface may be plane or spherical.

Lenses which cause parallel rays to converge [as shown in Fig. 8.4(b)] are called converging lenses; those which cause parallel rays to spread apart are called diverging lenses.

The most common type of lens is the double convex lens shown in Fig. 8.4(c). With this type of lens the refracted rays from parallel incident beams all pass through the focal point F. The distance CF (where C is the optical center of the lens) is known as the focal length, and the magnitude of this distance varies with the curvature of the two surfaces.

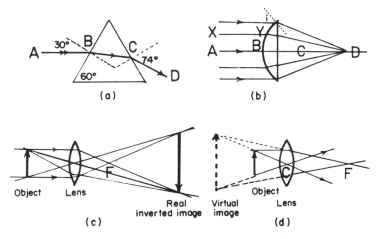

(a)   (b)

(c)   (d)

FIGURE 8.4 (a) Optical path of radiation passing through a 60° prism; (b) diffraction of parallel rays passing through a curved surface; (c) formation of a real image by a double convex lens; and (d) principle of a simple microscope.

If a source of light is placed at the focal point, the rays emerging from the lens are parallel to the principal axis.

When an object is placed in the path of parallel rays of incident radiation, the rays at the extremities of the object follow the geometric pattern outlined in Fig. 8.4(b) and pass through the focal point. Rays of minimum deviation pass through the optic center (C) and continue in a straight line. Other rays are refracted to different degrees because of the different angles of incidence. All the rays intersect in one plane and produce an inverted image which can be observed by placing a screen (e.g., a white card or a sheet of ground glass) in this plane.

Should the object be situated more than two focal lengths from the lens, the image is smaller than the original object. If the distance is less than two focal lengths, the image observed on the screen is larger. The images so observed are known as real images. At distances less than the focal length no images are observed on the screen, but an observer looking from the image side of the lens sees an enlarged upright virtual image of the object.

In this situation the converging lens is serving as a simple microscope; the geometry of this system is shown in Fig. 8.4(d). With a compound microscope, the real image produced by a converging lens of considerable magnifying power is further magnified by a second lens placed in an appropriate position.

The lenses used in instrumental optics are often of a different geometric form to the simple double convex lens discussed above. For example, instrument lenses may be plano-concave segments, or concavo-convex shapes, or combinations of several spherical forms.

The object of many of these complex forms is to overcome the optical aberrations observed with simple lenses.

The operation of a lens depends upon the refraction of light and the intersection of refracted rays at a common point. However, as indicated earlier, radiations of different wavelength are refracted to different extents. This unequal refraction of different colors means that for any single lens there is a different focal length for each color. Thus if an object illuminated by white light is magnified by a double convex lens, the image exhibits color fringes around the edges. This effect, known as chromatic aberration, can be minimized by using a diaphragm in front of the lens, so that only a small central portion is actually used. A more practical solution is to use a lens material possessing a low dispersing power. Alternatively, a combination of lenses is used to ensure that the various colors have practically the same focus.

Another important consideration is spherical aberration. When parallel rays of light enter a converging lens those which emerge near the periphery tend to come to focus at a point closer to the lens than those rays which emerge from the central portion of the lens. This aberration is sometimes remedied in part by using only the center part of the lens, but again the more practical answer is to use a combination of several different lenses.

Two other serious aberrations using simple lenses are distortion and curvature of the field. The image formed by a single lens does not lie in a perfectly plane field; hence in order to get a sharp image on a plane observing surface it is necessary to correct for the curvature of the field.

With mirror systems one is concerned with reflection instead of refraction, but a similar geometric approach can be used to describe their mode of operation.

The role of lenses and mirrors is to form an optical image at some desired plane, or focus radiations accurately at some desired point, or collimate (i.e., render parallel) the emissions from some small source. In all techniques which use ultraviolet and visible radiations the quality of the instrument design is usually determined by the quality of the optical system.

II. REFRACTOMETRY

Instruments capable of accurately measuring the refractive index of liquid phases are readily available and easy to operate. Accordingly, a number of analytical procedures based on the use of these instruments have been developed.

The refractive index is related to the number, charge and mass of the vibrating particles in the material through which the radiation is transmitted. Accordingly, for groups of similar compounds, the index has been found to vary with the density and molecular weight of the sample.

Thus refractive index, along with melting points, boiling points, and densities, can be used to characterize and identify liquid species.

In many cases the refractive index of binary mixtures varies linearly with the composition of the mixture, although it is always wise to check this additivity by the preparation of calibration curves. For complex mixtures, it is necessary to separate the mixture into single or binary fractions before making any measurements.

With modern refractometers the index can be measured with great precision, maximum precision ($\pm 0.2\%$) being achieved by using monochromatic radiation of known wavelength and careful temperature control. This degree of precision is highly desirable since in an homologous series of organic compounds the refractive index differs only slightly from compound to compound. A larger change in values is observed if dispersions are compared.

The partial dispersion D of a substance is defined as $D = n_2 - n_1$ where n_1 and n_2 are the indexes of refraction at specified wavelengths λ_1 and λ_2, λ_2 being the greater. The specific dispersion, S, is defined as $S = D/d \cdot 10^4$ where d is the density measured at the same temperature as the refractive indexes. Values for S and D are characteristic of the refracting substance and can be used to identify and characterize organic compounds, e.g., hydrocarbons present in successive cuts from a fractional distillation of petroleum.

Reliable data for the refractive indexes of high purity substances at specified temperatures and wavelengths are fairly readily available. The "Handbook of Analytical Chemistry" contains some tables while more extensive compilations are found in the "International Critical Tables" and "The Handbook of Chemistry."

The most popular instruments for refractometry are those which depend upon the critical angle method of measurement. They include the Abbe, dipping (immersion), and Pulfrich types, these instruments differing in the ranges covered and the type of light source employed. In general, when white light is used for illumination, color compensating prisms are included in the instrument.

The Abbe refractometer is widely used for measurements in the range of n equals 1.30–1.71 when an accuracy of ± 0.0001 is sufficient. The principal components of such an instrument are shown in Fig. 8.5.

A very thin layer of liquid is contained between an illuminating prism and a refracting prism. Light enters from below and grazing rays are refracted into the refracting prism. By scanning the upper edge of this prism with a telescope the angle of emergence of the critical rays can be ascertained. The prism is mounted so that it rotates about a central point in its surface, and rotation is

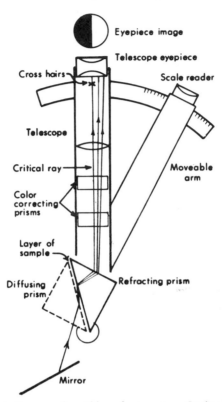

FIGURE 8.5   Schematic diagram of an Abbe refractometer. Omitted from this diagram are
the Amici color compensating prisms normally placed in the telescope tube above the
objective lens.

continued until the sharp edge of a light–dark boundary is centered on the cross
hairs in the telescope eyepiece. The refractive index reading is then taken from
the calibrated scale attached to the instrument. The light–dark boundary arises
from the fact that a segment of the prism is not illuminated by the rays refracted
in the sample. All nongrazing rays are refracted at angles smaller than the
critical angle. The grazing rays (i.e., angle of incidence equals 90°) are refracted
at the critical angle (r_c) and give rise to parallel critical rays which are coalesced
by a condensing lens.

For the critical rays

$$n_2/n_1 = 1/\sin r_c \quad \text{or} \quad n_1 = n_2 \sin r_c$$

where n_2 is the known refractive index of the refracting prism, n_1 is the refractive index of the sample.

By using three interchangeable sets of prisms and a monochromatic light
source a Precision Abbe refractometer can be used to cover the range of n equals
1.20 to 1.70 with a reading accuracy of ± 0.00003.

The dipping refractometer is quite similar to the regular Abbe except that the telescope tube is affixed to the refracting prism and there is no diffusion prism. For liquid samples the refracting prism is immersed almost completely and in the telescope the position of the light–dark border is read off an inbuilt scale. The instrument readings are converted to refractive indexes by reference to conversion tables.

In the Pulfrich instrument a horizontal beam of monochromatic radiation is directed along the surface of a refracting prism at the grazing angle. A sample reservoir is mounted on the horizontal surface of the prism, and a telescope is mounted at the side on a rotatable scale. The rays emerging from the refracting prism are reflected by a right angle prism and the telescope is moved until the light–dark boundary is focused on the cross-hairs of the eyepiece.

Other instruments use the displacement of images and interference effects to measure refractive indexes.

Automatic recording instruments have been designed for process control purposes, and great accuracy has been achieved with differential process refractometers. The differential type of equipment measures the difference in refractive index between a reference material and a sample from the process stream. One advantage of this approach is that the degree of temperature control required is slightly less rigid than is required for direct measurements.

The majority of liquids show a decrease in refractive index of about 0.00045 units per degree centigrade rise. Solids are generally much less sensitive. For a reading to be reliable to the fourth decimal place, liquid samples must, therefore, be thermostatted to $\pm 0.2°C$.

Although refractometry has been applied to the qualitative analysis of both inorganic and organic materials, in the vapor, liquid, and solid states, the most common application is for the identification of organic liquids.

The manufacturers of various refractometers generally provide tables for use in calculating the partial dispersion (D) from instrument scale readings, and this value, or the specific dispersion (S), is used for characterization purposes.

Alternatively, the density of the organic liquid (d) and its refractive index (n) can be used to calculate the molar refractivity M_D.

$$M_D = [(n^2 - 1)/(n^2 + 2)][m/d]$$

where m is the molecular weight.

The measured molecular refractivity of an unknown can then be compared with the theoretical molecular refractivity of various compounds.

The theoretical value is obtained by adding together listed values of the refractivities of atoms, correction terms being added for any unsaturated linkages, or structural systems. Values for the individual elements, structural units, and conjugation systems are available in a number of textbooks and reference handbooks.

Because of the additive nature of constituent refractivity values, characterization in terms of the type of compound can sometimes be achieved by plotting refractive indexes against density. For example, various types of hydrocarbons yield a series of lines of varying intercept, e.g., 1.0458 for paraffins and 1.1082 for naphthalenes.

With careful preparation of calibration curves, refractometry provides a comparatively simple means of analyzing binary mixtures. It has also proved useful in physiological chemistry, having been applied to the determination of total globulins, insoluble globulins, albumens, etc.

III. POLARIMETRY AND OPTICAL ROTATORY DISPERSION

A. POLARIMETRY

As a beam of polarized light passes through an anisotropic medium, interference between the ordinary and extraordinary electromagnetic rays leads to the production of elliptically, circularly, or plane-polarized radiation.

These different effects can be explained as follows. When two mutually perpendicular plane-polarized wave trains of the same frequency travel along the same line, the resultant is a vector that does not lie in a plane but traces out an ellipse or a circle. If the vibrations are out of phase by an angle between 0° and 90° the vector trace is an ellipse. A phase difference of 90° resolves either into an ellipse (if the wave heights differ) or a circle. The one exception is where the vibrations are in phase—in this case the vibrations can be resolved into a single wave that is plane polarized at an angle of 45° to the component vibrational planes. At greater phase differences figures repeat at intervals of +180°.

A linearly polarized beam passing through a medium can be conveniently considered to split into two circular components which are coherent and have an opposite sense of rotation. As light emerges from the medium, these two components add together to yield the original plane-polarized beam. However, if the medium is anisotropic the emergent beam is not identical with the original beam since in such a medium the refractive index for the right- and left-hand polarized light is different. That is, the velocity of propagation differs, giving a phase difference which causes the resultant beam to be rotated in its plane of polarization as it emerges from the medium.

A great many substances characteristically rotate the plane of polarized radiation. These substances are said to possess optical rotatory power.

Optical rotatory power has its origin in structural asymmetry such as exists in any substance that does not have either a plane or a center of symmetry. The asymmetry may be peculiar to the crystalline form of the substance or inherent in the structure of the molecules comprising the substance. In the latter case the optical activity is independent of the physical form of the material.

A typical optically active center is a carbon atom with four different substituents. Structural dissymmetry which results in a spatial left- and right-handedness can cause optical rotation, and compounds of these types come in right hand (D) and left hand (L) forms.

For example, in 3-phenyl-2-butanol there are two optically active carbon centers, and the compound exists in a D and L form (sometimes denoted as + and −).

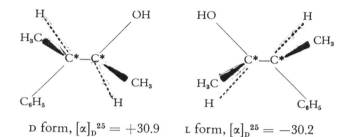

D form, $[\alpha]_D^{25} = +30.9$ L form, $[\alpha]_D^{25} = -30.2$

A heavy bar designates bonding coming out of the page; a dotted line represents a bond going to the back of the page. The asymmetric carbons are starred.

When equal amounts of D and L forms are mixed (racemic mixtures) there is no optical rotation because the activity of the two forms cancels out.

The extent to which the plane of polarized light is rotated varies with the nature of the optically active substance being studied, with the wavelength of the radiation, and with temperature. The specific rotation $[\alpha]$ is defined by the formula $[\alpha] = \alpha/dc$ where "α" is the angle, measured in degrees, through which the plane of light is rotated after passing through a cell d decimeters long containing a solution of concentration c (gm/100 ml). The wavelength of the light is commonly specified as the D line of a sodium vapor lamp, i.e., 5893 Å. To signify that a specific rotation has been measured with this radiation the symbol is given a subscript D. A superscript indicates the temperature of measurement (cf. 3-phenyl-2-butanol, $[\alpha]_D^{25} = +30.9$). In newer references rotation may be expressed as molar rotation $[\phi]$ rather than specific rotation ($[\phi] = [\alpha] \times$ molecular weight).

The measurement of optical rotation requires instruments of relatively simple design. The components of a polarimeter are shown schematically in Fig. 8.6.

Monochromatic light from a sodium lamp is polarized by a Nicol prism. The polarized radiation passes through the sample contained in a tube of known length, then through an analyzer Nicol prism to an eyepiece for visual observation. Since it is difficult to detect a position of complete extinction of light, most instruments incorporate another small Nicol prism mounted at an angle of a few degrees to the polarizer prism. The small prism covers half the total beam.

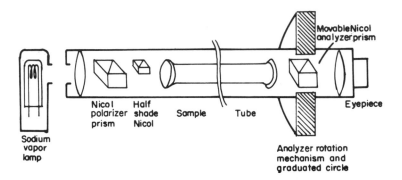

FIGURE 8.6 Diagrammatic sketch of a polarimeter. From "Fundamental Principles of Chemical Analysis," W. F. Pickering. Used with permission of Elsevier Publishing Co., Amsterdam, copyright © 1966.

With no sample in the beam, the analyzer is rotated until both halves of the beam are equal in power. That is, a uniform pale color covers the whole of the eyepiece. When the beam passing below the half-shade prism differs in angle of rotation from that passing through, the eyepiece shows distinct light and dark segments.

Insertion in the balanced system of a sample having optical activity causes a split field to appear and the analyzer has to be rotated to restore uniform illumination of both segments. The difference in angle observed on the graduated circle attached to the analyzer represents the optical rotation of the sample. Rotation is said to be dextra (+) if the required movement of the scale is clockwise to an observer looking toward the light source and levo (−) if counterclockwise.

Since visual polarimetric measurements are tiring on the eyes, precision tends to vary. To overcome this problem several models of photoelectric polarimeters have been developed.

If the specific rotation of the sample is known, the concentration of optically active substance in the solution can be calculated. Alternatively, if the concentration of material in the sample is known, the specific rotation can be calculated and used for identification purposes.

Extensive lists of specific rotation values are included in the "Handbook of Analytical Chemistry" and other handbooks and reference tables.

The technique can be applied to the determination of any optically active substance in the presence of nonactive species. Optical rotation is additive; hence readings can also be made in the presence of a known fixed amount of some other active species.

Quantitative methods of analysis by polarimetry are widespread, particularly for plant control in the pharmaceutical industry. The most extensive application, however, is in the sugar industry.

In the absence of other optically active substances, sucrose can be determined directly by measuring the angle of rotation, $[\alpha]_D^{20}$ being equal to $+66.5°$. If other active substances are present, it is necessary to measure the optical rotation before and after hydrolysis of the sugar solution. Sucrose is the only common sugar to undergo hydrolysis in the presence of acid and the resultant dextrose and levulose mixture has an optical rotation of $-20.2°$. The amount of sucrose present is calculated from the difference in rotation before and after inversion.

The optical rotatory power of quartz is almost identical with that of sucrose and this fact has been utilized in building instruments known as saccharimeters. By positioning a quartz wedge in the light path of a polarimeter, a variable path length through quartz can be obtained. In saccharimeters, the analyzer and polarizer prisms are fixed in orientation and the quartz wedge is moved into a point where it exactly compensates the rotation of the sugar solution. The position of the wedge is calibrated in terms of units known as sugar degrees (°S). The International Sugar Scale assigns 100°S to a pure sucrose solution of normal weight (26 gm in 100 ml) at 20° and 200 mm light path, using white light and a dichromate filter.

Virtually all sugar determinations are made with a saccharimeter. The instrument is first balanced with water to read zero °S and then the sample is placed in it. Rebalancing yields a figure in °S which corresponds to 0.26 gm/100 ml of sucrose per degree.

B. OPTICAL ROTATORY DISPERSION (ORD)

It has been noted in the previous section that a beam of plane polarized light passing through a medium may be considered to consist of two circularly polarized components moving in opposite directions. When the medium transmits the two components with unequal velocity, optical rotation is observed.

If in addition to unequal velocity there occurs unequal absorption of the left- and right-handed circularly polarized light, the emergent light is elliptically polarized. This unequal absorption is referred to as circular dichroism. The combined phenomenon of unequal velocity and unequal absorption is known as the "Cotton effect."

Experimentally the Cotton effect is most conveniently observed by measuring optical rotation at different wavelengths.

The refractive index of a medium varies with wavelength and in the absence of absorption the observed optical rotation at different wavelengths follows a smooth curve as shown as A in Fig. 8.7. Near an absorption band the refractive index changes rather abruptly, and as a consequence the Cotton effect gives rise to anomalous rotatory dispersion curves (cf. B in Fig. 8.7).

In an anomalous curve showing a single Cotton effect the curve possesses a maximum and a minimum. The mean of the wavelengths corresponding to

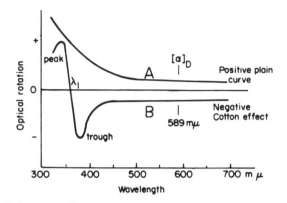

FIGURE 8.7   Optical rotatory dispersion curves. A. Typical plain curve. B. Typical Cotton effect due to an absorption band at λ_1. The sign of the curve is obtained by starting at the long wavelength end and observing the sign of the first extremum.

these two extrema usually corresponds approximately to the ultraviolet absorption maximum of the particular chromophore. To distinguish the extremum in the ORD curves the terms peak and trough are used and the vertical distance between the peak and the trough is called the amplitude. The horizontal distance (in mμ) between the two points represents the breadth of the Cotton effect curve. In articles, Cotton effects are described by recording the optical rotation at the highest wavelength measured, at 589 mμ, and at the shortest wavelength measured. In addition rotation values at each marked change of direction, such as peaks, troughs and shoulders, are listed in order of decreasing wavelength. Multiple Cotton effect curves possess two or more peaks with a corresponding number of troughs.

   Not all absorption bands of a given chromophore necessarily exhibit circular dichroism; hence the band which corresponds to zero rotation (λ_1) is conventionally referred to as an "optically active absorption band."

   Most of the results that are of particular interest are concerned with anomalous dispersion curves. To observe Cotton effect curves within the range of current instruments (700 to 250 mμ) the substances should show maximal absorption somewhere in this region. Furthermore, in order to locate experimentally the position of the peak and trough, the extinction coefficient should be quite low, e.g., $\varepsilon < 100$.

   Outside the region of an optically active absorption band, the relationship between optical rotatory power and wavelength is described by the Drude equation:

$$\alpha = k_1/\lambda^2 - \lambda_1{}^2 + k_2/\lambda^2 - \lambda_2{}^2 + \cdots = \sum_i k_i/\lambda^2 - \lambda_i{}^2$$

where k_i is a constant evaluated from experimental rotations and λ and λ_i are the wavelengths of the incident radiation and characteristic active absorption bands, respectively. For many substances one term of the equation is sufficient; the behavior of almost all others is well described by two terms. This equation has been employed to calculate λ_1 for systems where the anomalous dispersion lies outside the range of the measuring instrument.

In principle, the basic equipment required for optical rotatory dispersion studies is a polarimeter in which the sodium lamp has been replaced by a monochromator unit. In such a manual instrument the wavelength of incident radiation is selected manually, the analyzer is rotated by hand, and the angle of rotation is measured off the polarimeter scale. The ORD curve is obtained by plotting the series of individual points.

Today a number of companies manufacture automatic recording spectro-polarimeters, and the advent of these machines has markedly accelerated interest in this field.

The determination of a rotatory dispersion curve is highly desirable if a sensitive, quantitative polarimetric technique is to be established. By making polarimetric measurements, using a wavelength of incident radiation which corresponds to a peak or a trough, the response of the instrument to small changes in concentration is increased manyfold. In addition, if more than one optically active species is present, the ORD curves facilitate the selection of conditions which favor one of the components.

These aspects may lead to more widespread use of polarimetry in chemical analysis.

Small temperature variations appear to have little effect on the shape or position of ORD curves, but changes in the concentration or polarity of the solvent can affect the rotation appreciably.

Accordingly, for comparative studies experimental conditions should be standardized as much as possible. Two recommended solvents are methanol and dioxane, although hydrocarbon solvents, water, various alcohols, and chloroform have been employed from time to time.

To date the bulk of spectropolarimetric work has been directed towards problems of functional groups, structure, and conformation.

Functional groups which absorb light in the visible and ultraviolet region can give rise to Cotton effect curves when attached to an asymmetric molecule, and a change of location of such a functional group can have a marked effect on the ORD curve. As more information accumulates, it will become increasingly easier to recognize and locate functional groups from ORD studies.

By careful interpretation of ORD curves it is also possible to ascertain the spatial relationship of substituents around an asymmetric carbon atom.

These aspects are discussed in some detail in the monograph by Djerassi.

C. CIRCULAR DICHROISM

By studying the absorption of unpolarized light it is possible to determine a single value for the extinction coefficient of a material at a particular wavelength.

With polarized light, there are two extinction coefficients for each wave-length—one for the left-hand and one for the right-hand circularly polarized light.

The difference between these two, known as the dichroic absorption or extinction coefficient, is measurable and can be used in analytical studies. It is also possible to measure circular dichroism as a function of wavelength. In a typical instrument a scanning monochromator is followed by a polarizing prism and a crystal of ammonium dihydrogen phosphate. This is driven by an alternating voltage source to provide alternating circular polarization. The beam then passes through the sample to a photomultiplier sensor. The elec-tronics are arranged so that the difference in absorption between the left and right circularly polarized light is recorded directly as the wavelength range is scanned.

Circular dichroism peaks are quite isolated. They originate near the zero line and rise in a characteristic peak at defined wavelength ranges. A reversal in sign indicates reverse spatial arrangement and, therefore, the technique may be used to assign absolute configurations.

It has also been suggested that the technique may prove very useful for quantitatively analyzing complex molecules which are difficult by other methods.

IV. LIGHT SCATTERING PHOTOMETRY

A. FACTORS INFLUENCING SCATTERING

Isolated particles tend to scatter radiation if their major dimensions are less than one and a half times the wavelength of the incident light. Accordingly, particles having dimensions varying from 0.001 to 1 μ cause scattering of ultraviolet and visible light.

For suspended particles that are smaller than 0.1 λ, the scattering pattern in space is symmetrical and is termed Rayleigh scattering. Larger particles scatter more radiation overall, but relatively less in backward directions. The scattering caused by suspended particles is often termed the Tyndall effect.

When a beam of radiation of intensity I_o passes through a nonabsorbing medium that scatters light, the transmitted intensity I is given by the expression

$$I = I_o e^{-\tau b}, \quad \text{or} \quad \tau = (2.303/b) \log I_o/I$$

where τ is the turbidity and b the path length in the medium.

The intensity of light scattered at any particular angle is a function of the concentration of scattering particles, of their size, of their shape, of the difference in refractive indexes of the particle and the medium, and of the wavelength of light.

For particles which are small compared to the wavelength of the incident light, the scatter is proportional to the square of the effective radius of the particle and the total scatter from a number of particles is the sum of the individual scatterings. Hence the turbidity τ can be subdivided into a coefficient term and a concentration term. The intensity of scatter also varies inversely as the fourth power of the wavelength of the incident light.

At very low concentrations of suspended matter, the change in the intensity of the incident radiation is virtually undetectable, and measurement of the intensity of the scattered light becomes essential if reasonable sensitivity and precision is desired.

It can be shown that the intensity of scattered light (i_θ) at a distance r from the scattering particles is given by the equation

$$i_\theta = 2I_o\pi^2(dn/dc)^2(1 + \cos^2\theta)Mc/N_o\lambda^4r^2$$

where I_o is the intensity of the incident light of wavelength λ; (dn/dc) is the refractive index increment obtained from measurements of the refractive index, n, as a function of concentration, c; θ is the angle between the line to the detector and the direction of light propagation; M is the molecular weight and N_o is Avogadro's number. Using plane polarized light, the term $(1 + \cos^2\theta)$ is replaced by $2\sin^2\theta$.

Thus if the intensity of light scattered from a polarized beam is measured at a known angle θ and distance r from a known concentration c, the molecular weight can be calculated.

When concentrations of suspended matter are to be determined, the usual practice is to measure the intensity of scattering of the system at a given angle (e.g., 90°) and to read the results from calibration curves prepared from similar systems of known concentrations. This procedure is followed in the analytical techniques known as nephelometry and turbidimetry. On the other hand, if molecular weight or shape factors are to be calculated, it is necessary to determine the scattering at several known concentrations and usually at different angles.

Light scattering is a complex phenomenon and the experimental procedures are very prone to error.

The general principles of a light scattering photometer are shown in Fig. 8.8.

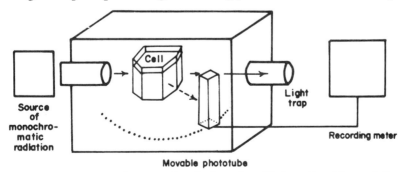

FIGURE 8.8 Apparatus for measuring the intensity of light scattered by a solution.

A source of high intensity monochromatic radiation is passed through a sample cell of such shape that the beams entering and leaving the cell pass through flat faces. Semioctagonal cells are useful since they allow measurements to be made at angles of 0°, 45°, 90°, and 135°. The detector is almost universally a photomultiplier tube which feeds some form of measuring meter. The detector may be fixed permanently at some angle to the incident beam or mounted on a circular disc to allow measurement at many angles.

Internal reflection of light must be kept to an absolute minimum; hence walls and nontransmitting surfaces are usually painted black. In addition a long narrow entrance support for the photomultiplier tube and a light trap for the exit radiation are generally provided.

B.   TURBIDIMETRY AND NEPHELOMETRY

The applications of turbidimetry and nephelometry are widely varied. Some determinations involve systems which are naturally turbid, e.g., river waters, while other systems require that the turbidity be developed in the laboratory.

In turbidimetric measurements, the intensity of the transmitted light is measured. The principle is identical with that used in colorimetry except that one measures the amount of light stopped by the suspended matter instead of measuring the amount absorbed by colored components. For this reason virtually any colorimeter or spectrophotometer can be used for turbidimetry measurements. For greatest sensitivity the wavelength of radiation used should not be strongly absorbed by any colored solutes present in the liquid phase. Apart from this consideration wavelength is not critical, although scattering is greatest with shorter wavelength radiations.

Turbidimetry lacks sensitivity at low concentrations of suspended matter, and under these conditions it is preferable to measure the amount of light scattered at right angles to the incident beam. Analyses based on the measurement of the intensity of scattered light are described as nephelometric determinations. Because of interparticle interference this technique is not reliable for dense suspensions. Both methods are relative, and are interpreted by using calibration curves.

Many colorimeters and spectrophotometers provide attachments for nephelometry, and specialized instruments are available commercially.

Smokes, fogs, aerosols, etc., can be studied quite readily by continuously drawing the sample through the scattering cell. Calibration curves are obtained with an aerosol of known particle size and physical properties.

Similarly, the amount of suspended matter in liquids can be readily estimated. Numerous applications are found in steam generating plants, beverage bottling industries, pulp and paper manufacture, water and sewerage treatment plants, pharmaceutical industries, and petroleum refineries.

The turbidity may be due to a single chemical species or a combination of several components, and the concentration ranges studied can vary from 0.1 to 100 ppm.

The range of applications of the technique has been extended by developing turbidity in a test sample through carefully controlled chemical reactions. For example, a widely used procedure is the determination of small amounts of sulfate ion through the formation of a fine suspension of barium sulfate.

In this type of analysis precise duplication of precipitation conditions is critical. There has to be close control of the temperature, the volume, the concentration of reagents, the rate of mixing, and the time of standing. The aim of these controls is to standardize the nucleation and growth patterns associated with precipitate formation. These aspects are discussed in more detail in Chapter 11. To prevent coagulation beyond a certain size, a stabilizer such as gelatin is often added.

The precipitation reaction used should be one which takes place rapidly and the precipitated particles should be small in size and of low solubility. In addition, the physical form of the precipitate should be reproducible in standards and assay solutions. The accuracy of the technique (typically $\pm 5\%$) is very frequently limited by the irreproducibility of the physical form of the precipitate.

Typical determinations using these methods cover a concentration range of 0.5 to 10 ppm. A table summarizing the conditions used in a number of the better known turbidimetric procedures is contained in the "Handbook of Analytical Chemistry" but more extensive compilations are to be found in reference books devoted to colorimetric analysis.

Turbidimetric titrations in which the turbidity of test solutions is measured after the addition of varying increments of reagent are of particular interest. At the equivalence point in the titration, turbidity reaches a maximum and accuracies of $\pm 2\%$ are claimed for this means of end-point detection.

The procedure has a number of advantages. It permits a study of low concentrations of materials and overcomes the difficulty associated with finding a suitable indicator for precipitation titrations. With slight modification it can be used to study the effect of variables (pH, temperature, etc.) on the nature of the precipitated material.

C. Scattering by Macromolecules

When dissolved in solvents of different refractive index, substances of high molecular weight effectively scatter incident radiation. The scattering is both angle and concentration dependent. For example, as the principal dimension of a particle exceeds one twentieth the wavelength of the incident radiation there is intramolecular interference between the scattering from different segments of the molecule. The net result is an asymmetrical pattern of scattering with reduced intensity in the backward direction ($\theta > 90°$). Intermolecular interference occurs on increasing the concentration.

In studying macromolecules, observations are thus ordinarily made as a function of angle and of concentration. With the larger particles, the angular dependence provides information as to shape.

If the dimensions of the molecule are much smaller than the wavelength of the incident light, the molecule acts as a single point light scatterer and the scattered intensity is equal in all directions.

In developing an equation to describe the intensity of light scattered by macromolecules present in a solvent composed of small molecules, it has been found desirable to add the term $n_o{}^2$ to the basic equation (n_o is the refractive index of the solvent). It has also been found convenient to divide the terms into groups which can be considered as constants related to the apparatus being used and the solvent system.

Thus the equation for light scattering by macromolecules can be written in the simple form

$$R_\theta = \kappa M c$$

where
$$R_\theta = i_\theta r^2/I_0(1 + \cos^2 \theta)$$

$$\kappa = 2\pi^2 n_o{}^2 (dn/dc)^2/N_o \lambda^4.$$

To determine molecular weights and basic information about size, light scattering data are taken at several concentrations and at one or more angles if the intensity varies with the angle θ.

$\kappa c/R_\theta$ is plotted against c and the intercept where c equals 0 provides a value for $1/M$.

If the plot varies with θ, the intercepts from the concentration plots are plotted against the trigonometric function of θ. Since the use of vertically polarized light is recommended in these studies, $(1 + \cos^2 \theta)$ in the R_θ equation is replaced by $2 \sin^2 \theta$ and the abscissa in the angular plot becomes $\sin^2 \theta/2$. The intercept of the angular plot (where $\sin^2 \theta = 0$) then gives the reciprocal of the weight average molecular weight.

Extrapolation of the experimental values has been found to be desirable because the behavior of polymers of low molecular weight tends to be nonideal and the scattering data are more consistent with an equation of the form

$$\kappa c/R_\theta = 1/MP_{(\theta)} + 2Bc$$

where B is a constant that measures intermolecular interaction at a fixed temperature and $P_{(\theta)}$ is a probability distribution function related to the size of the polymer molecule or the distribution of the molecular segments in solution. As θ approaches zero, $P_{(\theta)}$ goes to unity.

V. MICROSCOPY

A. OPTICAL MICROSCOPY

Microscopy is a comparatively old technique and is rarely included in chemical training programmes, yet there are many situations where microscopy is extremely useful. The technique provides a means of studying the physical properties of particles too small to be viewed by the human eye. If the particles are produced by microscale chemical reactions the physical size of the product can be related to the concentration of one of the reactants.

Chemical microscopy thus provides a means of studying minute amounts of material. It also provides a means of distinguishing between chemical species which are more readily characterized by physical properties than by chemical composition. For example, in assessing the quality of mineral beach sands, it is far more important to know the approximate proportions of the various minerals present (identified by microscopy) than to have a detailed knowledge of the Fe, Ti, Zr, Th content, since some of the elements may be present as minerals which are not amenable to processing.

Modern microscopes vary greatly in design and in the complexity of attachments, but each contains the following essential units:

1. Lens system.
2. Adjusting devices for focusing the lens system.
3. Stage on which to place specimens.
4. Source of illumination.

The basic components of a simple instrument are shown in Fig. 8.9.

The lens system consists of two components, an eyepiece lens or ocular and an objective lens system. The ocular lens commonly provides a magnification factor of $5\times$ or $10\times$ while the degree of magnification obtained with the objective depends on the focal length, but for chemical microscopy it is usually of the order of 10 to 50. The total enlargement in the microscope is the product of these two magnifications; and the chemist is advised to select a combination which provides the minimum magnification needed to examine the sample. All lenses are usually corrected for chromatic aberration and both the ocular and objective usually contain several lenses to give maximum clarity in the enlarged image.

Microscopes are fitted with rack and pinion devices and moveable tubes to allow the lens systems to be accurately focused on the sample.

For the examination of transparent materials or small semiopaque particles scattered over an area, the samples are ordinarily mounted on a glass microscope slide and illuminated from below. To prevent loss, the sample is usually covered with an optically flat cover glass of fixed dimensions. Large opaque specimens are normally illuminated from above by oblique reflection.

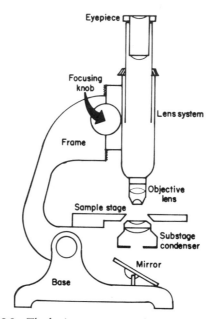

FIGURE 8.9   The basic components of an optical microscope.

The mounted samples are clipped in position on a stage which is often capable of being rotated or of being moved in transverse directions to bring different parts of the sample into the field of view of the objective.

Illumination is normally provided by white light reflected up into the lens system by a mirror. To provide an adequate cone of light of uniform intensity, most instruments have a condenser unit interposed between the mirror and the sample stage. The condenser unit contains a lens system and an adjustable diaphragm opening. In some studies colored light is substituted for the white light beam, and for the examination of anisotropic materials the use of polarized light is very advantageous. Polarizing microscopes are fitted with a polarizer prism below the stage and an analyzer prism located somewhere between the observer and the specimen.

While most microscopic studies are made visually, permanent records can be obtained by mounting a camera attachment on the eyepiece of the instrument.

A microscope is ideally suited for carrying out certain types of measurements with a fairly high degree of precision. Among the simpler applications are determinations of length and area. Some instruments are fitted with a graduated mechanical stage which allows the sample to be moved from left to right and from front to rear. A cross hair in the ocular is adjusted to coincide with one extremity of the sample, and the scale is noted. The stage is then moved until the cross hair reaches the other extremity of the dimension being measured.

The difference in scale readings (calibrated to 0.1 mm) yields the desired dimension.

For the precise measurement of very minute specimens an ocular micrometer is widely used. An ocular micrometer is a thin glass disc (bearing a scale) which is placed in the ocular by removing the eye lens. After replacement of the lens in the ocular and the ocular in the microscope, the scale visible in the eyepiece is calibrated against a glass stage micrometer held on the sample carrier. The dimensions of the ocular micrometer scale are calculated by counting the number of divisions of the scale which is equivalent to a known length of the stage micrometer image. When the object to be measured is substituted for the stage micrometer, its size can be calculated from the number of divisions of the ocular scale required to define the boundaries.

In certain types of work, e.g., blood counting, it is necessary to determine the number of fine particles suspended per unit of volume. In one method a small volume of the liquid is placed in a hollowed slide of known dimensions. The sample is covered with a cover glass, and the number of particles present in a region defined by ruled markings are counted and converted into the required count per unit volume by multiplication with an appropriate factor.

For mixtures of small particles of different color or shape, semiquantitative analysis can be achieved by counting the number of each type of particle present in a given area of the sample image.

The determination of the refractive index of a medium can be made by finding a combination of isotopic crystal and liquid in which the crystal is almost completely invisible. Thus crystals of known refractive index are added to a small amount of liquid placed in a hollowed slide and their image is observed under the microscope. When a crystal is surrounded by a liquid of different refractive index, it appears to be surrounded by a bright border or halo. The crystals employed for these determinations should not be very thick and the results can be accurate to ± 0.005.

A set of isotopic crystal standards allows one to ascertain the refractive index of liquids; a set of liquid standards makes it possible to determine the index of crystal samples.

More accurate results can be obtained using special accessories such as a stage refractometer.

Another physical property which can be determined under the microscope is the melting point or melting point range of a compound.

The microscopic determination of melting point requires the use of an electrically heated hot stage, designed so that the specimen can be observed by transmitted light. A few minute crystals are placed on the stage and the temperature is gradually increased. The sample is observed under the microscope and the melting temperature is read from a calibrated thermometer associated with the hot stage.

An indication of the lattice type of crystals or small particles can be obtained by studying the sample under polarized light with crossed Nicol prisms. Isotropic crystals (usually crystallized according to the cubic system) remain almost invisible at all orientations. Anisotropic crystals become invisible only when they have a particular orientation in respect to the incident beam. Should extinction result when an anisotropic crystal lies in a position parallel with the plane of polarization of one of the crossed Nicols, the compound is said to exhibit parallel extinction. Crystals which undergo extinction when in a position which is not parallel with either the polarizer or analyzer are said to exhibit oblique extinction.

Anisotropic crystals can be divided into two main groups. Uniaxial crystals embrace those which belong to either the tetragonal or hexagonal systems. Biaxial crystals embrace species which belong to the orthorhombic, monoclinic, and triclinic systems.

The division of compounds into these two types is achieved by studying the nature of the colored interference patterns which appear when the crystals are examined under cross Nicols using convergent polarized light. The appearance of these patterns has been widely used for the identification of minerals.

Apart from observing the physical properties of samples, microscopy can be used to follow chemical procedures performed with microgram quantities of sample and reagents.

The methods applied in chemical microscopy ultimately involve the formation of characteristic crystals which can be identified under a microscope.

The reactions used are generally of similar type to the classical procedures advocated for precipitation in a macro scale. However, the equipment has to be scaled down by an appropriate factor. Thus the surface of a microscope slide or a capillary tube replaces the beaker, a capillary pipette replaces the large burettes and pipettes of macro studies. Mixing and manipulation are observed under the microscope. With due patience and care almost all macrotechniques can be reproduced on the microscale.

The characteristic properties of the final precipitate are used for qualitative analysis, and with suitable calibration quantitative answers are derived from the amount of precipitate.

For example, the nature of metallic protective coatings may be ascertained by testing a minute amount of solution (derived from the coating) with a reagent which is capable of detecting several components. For example, potassium mercuric thiocyanate yields characteristic crystals with the cations of cadmium, copper, gold, lead, and zinc. The identification of some metals requires the use of more specific or selective reagents.

Procedures have been developed for the identification of most of the cations and anions encountered in simple inorganic systems.

It is not possible to attain a similar organized approach to organic compounds, but there are many areas in which crystalline derivatives are relatively readily obtained in a form suitable for characterization. For example, different types of sulfa drugs tend to yield crystals of different form with alcoholic picrolonic acid reagent.

Chapter 12 of Volume III,A of "Standard Methods of Chemical Analysis" (edited by Welcher, published by Van Nostrand, 1966) describes a number of the procedures used for the qualitative and quantitative analysis of inorganic and organic compounds. The "Handbook of Analytical Chemistry" (Table 6.62) provides a list of sources of general information useful to microscopists.

Microscopic procedures are especially suited for analytical problems in which only minute samples are available. Small samples require only small amounts of reagent which makes it feasible to use reagents which are extremely costly. In addition, work on a microscale is preferable when handling substances which are potentially toxic, explosive or highly combustible.

B. Electron Microscopy

The electron microscope is an optical device for obtaining enlarged images of small particles or small areas of large objects. The image-forming radiation is a beam of electrons; the lenses are electrostatic or magnetic fields.

In the optical column of a typical electron microscope one finds an electron gun, a condenser lens to control the illumination of the specimen, a specimen chamber, an objective lens, and a projector lens, the whole being maintained at a high vacuum.

The image is made visible either by means of a fluorescent screen or photographically.

The electron gun consists of a hot cathode and an anode. The potential difference (V) between these electrodes yields a stream of electrons moving with a velocity v, the magnitude of v being defined by the equation

$$\text{kinetic energy} = 1/2mv^2 = Ve$$

where m is the mass of the electron and e is its charge. Rearrangement gives

$$v = \sqrt{2Ve/m}$$

Associated with a moving stream of electrons is a wave motion whose effective wavelength (λ) can be calculated from the de Broglie equation, $\lambda = h/mv$, h being Planck's constant.

Combining these equations and substituting the appropriate values for the constants yields the relationship

$$\lambda = 12.3/\sqrt{V} \text{ Å}$$

In electron microscopes the accelerating voltage is measured in kilovolts; hence the effective wavelength of the electron beam is less than one Å, which is several orders of magnitude smaller than the wavelength of visible light.

This beam can be focused by means of a magnetic field arranged symmetrically about the axis along which the electrons travel. In magnetic lens systems the field can be derived from a permanent magnet, or from an electromagnet. Alternatively, the beam can be focused by an electrostatic lens, one design of which consists of three coaxial circular apertures. The outer two apertures are maintained at the same potential, the inner aperture is maintained at a considerably different potential. The focal lengths of the lenses are determined by the operating conditions.

To obtain a clear image of the specimen some degree of contrast is necessary. In light spectroscopy this contrast arises from differential absorption of the radiation by different portions of the sample. In an electron microscope, the main source of image contrast is differential scattering of electrons from different parts of the specimen. Thus in order to be observed, a specimen must be thin enough to transmit an appreciable fraction of the incident electron beam; it must be stable in the evacuated chamber; it must be stable upon electron bombardment; and it must display some degree of differential scattering.

Quite regularly the specimen is mounted upon a thin (100 Å) film of carbon or of a plastic material such as collodion or formvar.

Because of the extremely short wavelengths involved, the degree of resolution is very much greater than that achieved with visible light. The limit of resolution with an electron microscope is a few angstrom units; with visible light this limit is a thousand times larger.

The technique has been used in many areas of science and technology, since the size and shape of particles having dimensions ranging from ten microns to several angstroms can be determined with a degree of certainty which is directly proportional to the ratio of the size to resolution.

The identification of characteristic shapes can sometimes be used as a means of qualitative analysis, e.g., to distinguish between two different crystalline forms of some mineral.

The procedure visualizes very small particles; hence it can be applied to studies of colloids, or the nucleation and growth of crystals. Where particulate materials are dispersed in a polymer (e.g., fillers in rubber) the concentration can be estimated by a counting technique.

With the aid of selective staining or metallic coating the physical structure of many biological products and metallurgical materials has been ascertained.

In most electron microscope units it is also possible to study the electron diffraction patterns, and because of the low penetrating power of the electron beam this modification is particularly useful in studying the surfaces of materials.

When the impinging electron beam possesses sufficient energy to cause photoejection of inner orbital electrons from the sample atoms, characteristic X rays are emitted. Diffraction and measurement of these emissions is then used as a means of quantitatively determining the composition of small areas of a specimen. This technique, known as X-ray microprobe analysis, has been previously discussed in Chapter 5, Section II,C.

VI. REVISION AND TUTORIAL EXERCISES

1. Write brief notes on the following topics:
   a. Refraction of light
   b. Polarized light
   c. Dichroism
   d. Optical anisotropy
   e. Racemic mixtures
   f. Wave nature of electron beams.
2. Using introductory physics texts or books on instrumental analysis prepare diagrams and explanatory notes to illustrate that you understand the principles of the following:
   a. Reflection and collimation of light by mirrors
   b. Dispersion of polychromatic radiation by ruled gratings
   c. Role of different types of lenses in optical systems
3. a. Describe the determination of the refractive index of liquid samples using a grazing angle refractometer.
   b. Indicate how a microscope can be used to ascertain the refractive index of small crystals.
   c. From the literature, prepare a description of a method which would be suitable for determining the refractive index of a solid sample such as a thin slab of glass.
   d. Suggest reasons why quantitative analysis by refractometry is restricted to studies of binary mixtures.
4. a. Prepare simple diagrams which show the principle components of the following types of instrument:
      i. Polarimeter
      ii. Saccharimeter
      iii. Simple microscope
      iv. Turbidimeter
   b. Briefly explain the characteristics of an optically active substance.
   c. i. A sample of optically active material was found to rotate the plane of polarized light by $+12.5°$ when placed in a 20 cm tube in a standard polarimeter operating with a sodium lamp. Assuming that the material was sucrose ($[\alpha]_D^{20} = +66.5°$) what concentration of the sugar was present in the sample (gm/100 ml)?
      ii. What procedure could be used to confirm that the optically active material was sucrose?
      iii. What units would a sugar chemist tend to use in order to describe the optical rotation of a sucrose solution?
5. a. Write an essay on optical rotatory dispersion, including mention of the effect of wavelength on the optical rotation, the Cotton effect, and if possible, an indication of how ORD is used in structure determinations
      OR

     b. Write an essay on the use of optical rotatory dispersion in studies of inorganic complexes.

6. "Although nephelometry and turbidimetry have proved of value in isolated cases (e.g., the determination of small amounts of sulfate ions) their use should be restricted to situations where no alternative procedure is available." Critically evaluate the validity of this cynical statement.

7. a. Describe the principles involved in determining the size of small particles by
       i. Light scattering
      ii. Optical microscopy
     iii. Electron microscopy
   b. Comment briefly on the relative merits of the three techniques and indicate their range of greatest usefulness.
   c. If asked to monitor the density of the smog in an industrial city, what simple procedure would you propose for measuring the fluctuations in dispersed suspended matter?

8. a. Prepare a brief article (1000–2000 words) on chemical microscopy.
   b. Is there a case for replacing precipitation reactions with color-forming reactions in chemical microscopy?
   c. In view of the fact that studies of biological systems and natural products often require identification of extremely small amounts of sample, do you consider that greater efforts should be made to develop more microchemical procedures for organic compounds? Illustrate your answer with examples of existing applications, e.g., barbiturates, alkaloids.

References

A. A. Benedetti-Pichler, *Microtechnique of Inorganic Analysis*, Wiley, New York, 1942.
T. R. P. Gibbs, *Optical Methods of Chemical Analysis*, McGraw-Hill, New York, 1942.
H. A. Strobel, *Chemical Instrumentation*, Addison-Wesley, Reading, Massachusetts, 1960.

POLARIMETRY AND ORD

F. J. Bates et al., *Polarimetry, Saccharimetry and the Sugars*, Nat. Bur. Standards Circular C440, Washington, 1942.
P. Crabbé, *Optical Rotatory Dispersion and Circular Dichroism in Organic Chemistry*, Holden-Day, San Francisco, 1965.
C. Djerassi, *Optical Rotatory Dispersion*, McGraw-Hill, New York, 1960.
L. Velluz, M. Legrand, and M. Grossjean, *Optical Circular Dichroism*, Academic Press, New York, 1965.

LIGHT SCATTERING PHOTOMETRY

M. Bobtelsky, *Heterometry*, Elsevier, Amsterdam, 1960.
D. E. Boltz, (ed.), *Colorimetric Determination of Non-Metals*, Wiley (Interscience), New York, 1958.

H. C. van den Hulst, *Light Scattering by Small Particles*, Wiley, New York, 1957.

E. B. Sandell, *Colorimetric Determination of Traces of Metals*, 3rd Ed., Wiley (Interscience), New York, 1959.

F. D. Snell and C. T. Snell, *Colorimetric Methods of Analysis*, 3rd Ed., Van Nostrand, Princeton, New Jersey, Vols. I and II, 1949, Vol. IIA, 1959.

K. C. Stacey, *Light Scattering in Physical Chemistry*, Academic Press, New York, 1956.

MICROSCOPY

E. M. Chamot and C. W. Mason, *Handbook of Chemical Microscopy*, 3rd Ed., Wiley, New York, 1958.

G. L. Clark, *Encyclopedia of Microscopy*, Reinhold, New York, 1961.

C. E. Hall, *Introduction to Electron Microscopy*, McGraw-Hill, New York, 1953.

N. H. Hartshorne and A. Stuart, *Crystals and the Polarizing Microscope*, 3rd Ed., Arnold, London, 1960.

E. E. El-Hinnaur, *Methods in Chemical and Mineral Microscopy*, Elsevier, Amsterdam, 1966.

W. C. McCrone, *Fusion Methods in Chemical Microscopy*, Wiley (Interscience), New York, 1957.

H. F. Schaeffer, *Microscopy for Chemists*, Van Nostrand, Princeton, New Jersey, 1953.

B. M. Siegel (ed.), *Modern Developments in Electron Microscopy*, Academic Press, New York, 1964.

W. W. Wendlandt and H. G. Hecht, *Reflectance Spectroscopy*, Wiley (Interscience), New York, 1966.

9

STRUCTURE

I. Introduction 259
   A. Additive Effects 260
   B. The Determination of Molecular Weight 262
II. Elucidation of Structure by Means of Spectroscopic and Optical
   Methods 264
   A. Ultraviolet and Visible Spectroscopy 264
   B. Infrared Spectroscopy 268
   C. Nuclear Magnetic Resonance Spectroscopy 271
   D. Mass Spectrometry 277
   E. Joint Application of Physical Methods 284
   F. Optical Activity 285
III. X-Ray Diffraction 288
   A. Reflection of X Rays 288
   B. Crystal Systems 290
   C. X-Ray Powder Photographs 291
   D. Single Crystal Studies 293
   References 295

I. INTRODUCTION

Until recently, the role of the analytical chemist was to produce experimental procedures for the determination of the elemental composition of any type of material encountered. However, with the advent of chromatographic techniques, it has become possible to isolate and identify molecular species with relative ease, and interest in the molecular nature of the components of a sample has increased markedly. In some modern technological processes the actual structure of the molecular species has been found to be an important factor; hence one can predict that many future analytical chemists will be required to make structure determinations in addition to elemental analyses.

Accordingly, it was considered appropriate to include a brief introduction to the problem of structure determination in this analytical text. Probably the most positive arbiter in structural studies is X-ray diffraction, and the principles of this technique are outlined in Section III. However, since the mathematics associated with X-ray diffraction can be difficult, other physical methods have been widely used to provide partial or complete elucidation of structure.

Most of these techniques have been described in preceding chapters, and in this segment the aim is to indicate how these procedures may be applied to structure determination.

A.   ADDITIVE EFFECTS

Many physical properties of a compound are related to the molecular size of the molecule and the basic properties can usually be resolved into atomic and structural components. For example, the molar volume, M/D (where M is

TABLE 9.1

ATOMIC AND BOND PARACHORS

Atomic parachors	Bond parachors
	Single bond 0.0 (arbitrary)
C 4.8	Double bond 23.2
N 12.5,	Semi polar double bond −1.6
H 17.1	Triple bond 46.6
O 20.0	3-ring 16.7
P 37.7	4-ring 11.6
S 48.2	5-ring 8.5
Cl 54.3	6-ring 6.1
Br 68.0	
I 91.0	

the molecular weight and D is the density near the boiling point) of a liquid hydrocarbon, $C_xH_yO_z$, is often roughly equivalent to the sum of x carbon, y hydrogen, and z oxygen atomic volumes, plus a volume associated with any special bond groupings (e.g., the double bond if the molecule contains a carbonyl, C=O, group).

For liquids, wider agreement is obtained by using a property known as the parachor (P) which is defined by the expression

$$P = M\gamma^{\frac{1}{4}}/D - d$$

γ represents the surface tension of the liquid and d is a vapor density term.

Like molar volume, the parachor is an additive term; P values for different elements and bond types have been calculated by making measurements on liquids of known structures and some typical values are given in Table 9.1.

Tentative structures assigned to a new liquid compound may be verified by reference to tables of this type.

For example, consider the case where measurement of the density and surface tension of a liquid indicates a P value of 206.2. Elemental analysis shows the liquid to contain six carbon atoms and six hydrogen atoms.

To satisfy the tetravalency of carbon, these atoms must either be combined as an unsaturated chain containing one triple bond and two double bonds or as a cyclic compound with three double bonds. When the molecular P values for these two alternative structures are calculated using the data in Table 9.1, it is found that the value for the cyclic arrangement more closely resembles the experimental value.

For the chain structure $P_{calc} = (6 \times 4.8) + 6(17.1) + 46.6 + 2(23.2) = 224.4$. For a cyclic arrangement, $P_{calc} = (6 \times 4.8) + 6(17.1) + 3(23.2) + 6.1 = 207.1$.

It is, therefore, highly probable that the liquid is a cyclic hydrocarbon.

A P value can be combined with a refractive index measurement to yield another additive constant known as the refrachor (F).

$$F = -P \log (n_D{}^{20} - 1)$$

where $n_D{}^{20}$ is the refractive index of the liquid measured at 20° using the sodium D line.

Instead of calculating refrachor values, it is possible to use molar refractions (R) since these are also additive.

$$R = M(n^2 - 1)/D(n^2 + 2)$$

where M/D is the molar volume and n is the refractive index.

Molecular refraction measurements have proved useful in detecting conjugated double bonds in large molecules. When a conjugated system is present, the observed molecular refraction is higher than the value calculated by summation of the various atomic and bond refractions.

Tables of atomic and bond parachors, refrachors, and molar refractions are available in the literature.

It has been noted in Chapter 7 that the diamagnetic properties of a compound can be calculated by adding together tabulated values for atomic and structural magnetic susceptibilities (Pascal's constants).

These tabulated values can be applied to structural problems in a manner analogous to the use of the parachor. The diamagnetic susceptibility of the compound is determined by the Guoy method and this value is compared with the susceptibilities calculated using the listed values for the constituent elements and different bond groupings.

The advent of physical methods such as absorption spectroscopy and NMR has reduced the degree of reliance and interest in these additive procedures, but they do provide a convenient check on the conclusions reached in instrumental studies.

B. THE DETERMINATION OF MOLECULAR WEIGHT

A fundamental step in the determination of structure is the elucidation of the molecular size. Elemental analysis permits the calculation of an empirical formula; knowledge of the size of the molecular unit has to be ascertained by molecular weight determinations.

If the constituent groups of the molecule of a liquid species are known, its parachor can be calculated from tabulated values (cf. Table 9.1) and if the calculated value equals the observed value, the substance is a monomer. Similar arguments apply to calculations using other additive properties. On the other hand, if the calculated and observed values differ in magnitude, either the constitution of the molecule has been incorrectly assigned or the true molecular weight is a multiple of the formula weight.

The use of colligative properties for the determination of molecular weights is well known. For example, the vapor pressure of an ideal solution is proportional to the mole-fraction of the solvent present in the solution, i.e.,

$$p_o - p_s/p_o = n/N + n$$

where p_o is the vapor pressure of the solvent, p_s is the vapor pressure of a solution composed of n moles of solute in N moles of pure solvent.

Hence, if the decrease in vapor pressure brought about by dissolving a known weight of solute in a given weight of solvent is known, its molecular concentration can be determined. Knowing the weight of the substance and the solvent, it is an easy matter to calculate the molecular weight. Accordingly, the accurate measurement of solution vapor pressures may be proposed as a means of determining the molecular weight of a solute.

In many situations, it has been found more convenient to determine boiling points or freezing points. When the addition of a solute reduces the vapor pressure of a solvent, it increases the temperature required to make the vapor pressure equal to the superimposed pressure, i.e., it automatically elevates the boiling point.

It follows from Raoult's law (quoted above) that equimolecular weights of different substances should increase the boiling point of a solvent by the same amount, since they depress the vapor pressure by the same amount. The elevation of the boiling point of a solvent caused by the solution of one gram-molecule of any substance in 100 gm of solvent (known as the molecular elevation per 100 gm solvent) is, therefore, a constant (K) for that solvent. Typical values of K are water (5.2°), chloroform (38.8°), ether (21.1°), acetone (17.2°), benzene (25.7°), and ethyl alcohol (11.5°).

The molecular weight (M) of any solute can thus be calculated from the relationship

$$M = 100 \ WK/wT$$

where w is the weight of solute dissolved in W gram of solvent having a molecular elevation per 100 gm of K; the T represents the observed elevation in boiling point.

Lowering of the vapor pressure also causes a depression of the freezing point and for each solvent it is possible to determine a Molecular Depression Constant (per 100 gm solvent).

The same relationship can then be used to determine molecular weights except that in this case K represents the molecular depression constant and T represents the observed depression in freezing point. For a given solvent, the molecular depression constant is larger than the values recorded for molecular elevation (e.g., water, 18.5°; benzene, 51.2°; camphor, 400°); hence this approach is a little more sensitive.

In general, however, these two methods are not applicable to the measurement of macromolecular weights, chiefly because the effect observed is very small. On the other hand, another colligative property, osmotic pressure has been extensively used for this purpose.

If a solution obeys Raoult's law, the van't Hoff osmotic pressure equation has the form

$$\pi = c\,RT/M$$

where π is the osmotic pressure, c is the concentration in gm per liter of the solute of molecular weight M; R is the universal gas constant and T the absolute temperature. In these cases, a plot of π (observed) against c approximates a straight line of slope RT/M from which M can be calculated.

With solutions of macromolecular compounds, Raoult's law is only occasionally obeyed and in most cases dilute solutions follow a modified equation; the modification involves the addition of a term $BRTc^2$ to the right-hand side of the above relationship; B is a constant which varies with the solute–solvent system used. In this case, a plot of π/c against c yields a straight line, of slope RTB and intercept RT/M.

Thus by accurately measuring the osmotic pressure of a series of solutions, using one of the standard techniques, it is possible to ascertain the average molecular weight of the material. Molecular weights ranging from 45,000 to 2,000,000 have been determined in this way.

Another rapid and convenient method for determining molecular weight is based on the measurement of the viscosity of polymer solutions. The relationship between these two terms has been described in several different ways.

If η and η_0 represent the measured viscosities of the solution and solvent respectively, the specific viscosity, η_{sp}, of the solution is equal to $(\eta/\eta_0 - 1)$. The intrinsic viscosity of the solution, $[\eta]$, may be evaluated by plotting η_{sp}/c against concentration, c, and extrapolating to c equals zero. For a given polymer fraction the molecular weight then follows from one of the derived laws, e.g., the Standinger's law, which states that

$$\underset{c\to 0}{\text{Lim}}\ \eta_{sp}/c = [\eta] = KM$$

The value of the proportionality constant K is determined by using a compound of known molecular weight as a standard.

The use of light scattering as a means of determining the molecular weight of macromolecules has been discussed in Chapter 8, Section IV.

This method has been widely applied in the protein field for the evaluation of molecular weights ranging from 15,000 to 3,000,000.

For the measurement of very large molecular weights, a sedimentation technique is often preferred. A solution of the macromolecular compound is subjected to a centrifugal field of the order of 5000 to 10,000 g. To determine the molecular weight one either measures the solute concentration in the cell as a function of the distance from the axis of rotation at equilibrium, or one measures the rate of sedimentation. Suitable equations are available for relating these properties to the average molecular weight.

If a particular macromolecular compound is studied by several of the above techniques, the answers obtained usually agree within ± 5–10%. The molecular weights obtained are also generally similar to the values derived from the more accurate and more tedious procedure based on X-ray diffraction.

II. ELUCIDATION OF STRUCTURE BY MEANS OF SPECTROSCOPIC AND OPTICAL METHODS

Most of the physical methods discussed in the preceding chapters of this book can be used to provide some relevant information in regard to the structure of molecular species.

Since the principles of the various techniques have been described, the aim in this chapter is to indicate the role of these methods in structural studies.

A. Ultraviolet and Visible Spectroscopy

It has been indicated in Chapter 6, Section III,A that absorption of ultraviolet and visible light can be associated with atomic groupings known as chromophores, but there is no simple rule or set procedure for identifying a chromophore.

Many factors affect the character of an absorption spectrum, and positive identification of chromophoric groups requires comparison with a similar spectrum of known source. However, the search for a comparison spectrum can be simplified by considering initially the basic features of the spectrum of the unknown and by applying some general rules.

For example, a spectrum with many bands stretching into the visible region indicates the presence of either a long conjugated chromophoric group or a polycyclic aromatic chromophore. On the other hand, a system exhibiting only one band (or at most a few bands), at wavelengths less than 300 mμ, probably contains only two or three conjugated units.

The intensity of the bands can also be informative. Very low intensity absorption bands in the 270 to 350 mμ region with ε values of 10 to 100 are often the result of the $n \rightarrow \pi^*$ transition of ketones. The existence of absorption bands with ε values of 1000 to 10,000 generally corresponds to the presence of an aromatic system. Simple conjugated chromophores such as dienes and α,β unsaturated ketones have ε values of 10,000 to 20,000.

The changes in λ_{max} brought about by substitution in basic chromophoric groups is now sufficiently well documented to permit the enunciation of some general rules. There are many exceptions to the rules (e.g., due to distortion of the chromophore) but the rules outlined below do facilitate the selection of comparison spectra. This selection is aided further by a knowledge of the chemical behavior of the compound and the evidence provided by other physical methods such as infrared absorption spectroscopy and nuclear magnetic resonance spectrometry.

Briefly, the rules correlate the position of λ_{max} with the degree of substitution of a conjugated chromophore.

The latter can be calculated by summing a basic value and increments determined by the number and positions of any substituent alkyl groups in the chromophore.

For example, in order to estimate the position of the absorption maximum of an α, β unsaturated ketone, a base value of 215 mμ is assumed.

$$-\overset{|}{\underset{\delta}{C}}=\overset{|}{\underset{\gamma}{C}}-\overset{|}{\underset{\beta}{C}}=\overset{|}{\underset{\alpha}{C}}-C=O$$

For an α substituent 10 mμ are added; for each β substituent 12 mμ are added. For each alkyl group or ring residue in a γ or higher position 18 mμ are added. For each ring (six membered or less), to which the carbon-carbon double bond is exocyclic, 5 mμ are added. If the C=C and the C=O group are contained in a five-membered ring (cyclopentenones), 10 mμ are subtracted. When the carbon-carbon double bond is alone in the five-membered ring 5 mμ are added. The increment for a double bond which extends the conjugation is 30 mμ.

A similar empirical correlation exists for α, β unsaturated aldehydes, the value assigned to the parent aldehyde being 207 mμ.

The presence of auxochromes requires additional corrections to be made. These additive corrections vary in magnitude with the nature of the group and its structural position. Some representative values are given in Table 9.2.

TABLE 9.2

INCREMENTS IN λ_{max} OF THE UV SPECTRUM
OF α,β UNSATURATED
KETONES AND ALDEHYDES INDUCED BY AUXOCHROMES[a]

Auxochrome	Position	Increment (mμ)
—OH	α	35
	β	30
	δ	50
—OMe	α	35
	β	30
	γ	17
	δ	31
—OAc	α, β, δ	6
—Cl	α	15
	β	12
—Br	α	25
	β	30
—SAlk	β	85
—NR$_2$	β	95

[a] Solvent, ethanol.

The absorption values quoted in this example refer to systems which use 95 % ethyl alcohol as the solvent. When another solvent is used, a comparable value may be obtained by adding the following numerical corrections to the observed λ_{max}: methanol, 0 mμ; chloroform, +1 mμ; dioxan, +5 mμ; ether, +7 mμ; hexane and cyclohexane, +11 mμ; and water, −8 mμ.

To indicate how these rules can be applied to tentative structures, let us consider four different compounds.

1. Mesityl oxide—Me$_2$C=CHCOMe
   Parent value 215 mμ; 2 β substituents (Me$_2$), hence increment by 2 × 12, giving a calculated λ_{max} of 239 mμ. The spectrum of mesityl oxide shows maxima at 237 mμ (ε 11,700) and 310 mμ (ε 57).
2. Steroidal 1-enone—Structure I
   Parent value 215 mμ; 1 β substituent, hence calculated λ_{max} is 227 mμ.
3. Steroidal 4-enone—Structure II
   Parent value 215 mμ; 2 β substituents (increment 2 × 12) and C=C is exocyclic (add 5 mμ). Thus calculated λ_{max} is 244 mμ.
4. The trienone—Structure III
   Parent value 215 mμ; 1 β substituent (marked a, increment 12 mμ); extended conjugation (increment 2 × 30 mμ); 1 ω substituent (marked b, increment 18 mμ); the α, β double bond is exocyclic to ring A; hence allowance must be made for an exocyclic double bond (5 mμ). In this

case an additional 39 mμ has to be added to allow for the homoannular diene component. This gives a calculated λ_{max} of 349 mμ. The spectrum of this compound shows λ_{max} at 348 mμ (ε 11,000), 278 mμ (ε 3720) and 230 mμ (ε 18,000).

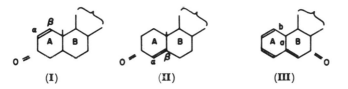

(I) (II) (III)

This last example draws attention to the fact that several different chromophoric groups may be responsible for the peaks in the sample spectrum.

If two or more chromophoric groups are present in a molecule and they are separated by two or more single bonds, the effect on the spectrum is usually additive. However, if two chromophoric systems are separated by only one single bond (a conjugated system) the high intensity π–π* transition is usually shifted to longer wavelengths with respect to the simple unconjugated chromophore.

Among the simple chromophoric groups, the absorption resulting from the $n \rightarrow \pi$* transitions of the carbonyl groups of aldehydes and ketones is among the most easily recognized. Conjugated dienes having alkyl substituents also lend themselves to numerical correlation.

For acyclic conjugated dienes and cyclic conjugated dienes contained in nonfused six-membered ring systems, a base value of 217 mμ (observed in butadiene) is used. To calculate the approximate λ_{max} of substituted dienes one adds 5 mμ for each acyclic alkyl substituent or ring residue and for each ring (6-membered or less) to which the diene system is exocyclic. The increment for substituent chlorine and bromine atoms is 17 mμ.

If the conjugated double bonds are contained in separate, but fused six-membered rings (a heteroannular diene) a base value of 214 mμ is used. When the conjugated double bonds are contained in the same ring (a homoannular diene), a base value of 253 mμ is used. A double bond extension (giving a triene) results in a 30 mμ increment in λ_{max}. The effect of auxochromes is much smaller than that observed with α, β unsaturated ketones.

As the number of double bonds in conjugation increases, the wavelength of maximum absorption encroaches on the visible region, the intensity increases, and a number of subsidiary bands appear.

Aromatic systems exhibit characteristic ultraviolet absorption patterns of a more complex character. Benzene, for example, has two main regions of absorption (around 200 mμ and in the 235–270 mμ region). The positions of both bands are moved to higher wavelength by auxochromic substituents, and rules have been prepared which permit prediction of the position of the principal band λ_{max} in various substituted benzene derivatives.

Since it is not the aim of this section to describe the interpretation of the ultraviolet spectra of all classes of compounds, the interested reader should consult one of the more comprehensive books listed in the bibliography if further details are desired.

Steric effects can influence the electronic spectra of organic molecules. For example, steric hindrance to coplanarity about a single bond gives rise to a marked decrease in intensity and may lead to a shift in the position of the absorption maximum. The absorption peaks of compounds containing extended conjugation are caused by π–π^* transitions that arise from the delocalized π electron systems present, and when the π electron system is prevented from achieving coplanarity, the degree of overlap of the π electron system is diminished.

Steric effects are often observed in isomer studies. Of a pair of geometric isomers, the *cis* form tends to be sterically hindered while the *trans* form generally achieves coplanarity of the π electron system. As a result, the more elongated arrangement (i.e., the *trans* form) absorbs at a longer wavelength and with a larger molar extinction coefficient than the *cis* isomer.

Information on stereochemical aspects of structures can therefore be revealed by spectral studies.

Collections of ultraviolet and visible absorption spectra have been initiated (see Bibliography), and modern guides to spectra eliminate time-consuming searches.

B. INFRARED SPECTROSCOPY

The infrared absorption spectrum provides a rich array of absorption bands. Many of these bands cannot be assigned accurately, but those that can provide a wealth of information about the functional groups present and the gross structural features of the molecule.

Examination of the infrared spectra of a multitude of compounds has enabled characteristic group frequencies to be assigned and charts of group frequencies to be compiled. The first step in dealing with the spectrum of an unknown, therefore, is to compare the position of observed absorption peaks with ranges shown on these correlation charts.

Almost every group encountered in organic structures has a characteristic absorption band or bands in the infrared region. Many inorganic groups also give rise to characteristic absorption bands, and the body of information available on inorganic and coordination compounds is growing rapidly.

The initial survey allows certain combinations of structures to be ruled out and some tentative assignments to be made. The tables corresponding to those groups which might be present are then consulted, since these contain more detailed information. In this regard, one should ensure that the bands under consideration are of the appropriate intensity for the structure suspected. Usually intensities are expressed subjectively as strong (s), medium (m), weak (w), and variable (v). The position of bands is usually given in cm^{-1}.

After gaining an indication of the types of functional groups present in the unknown molecule, positive identification of the structure is achieved by comparing the spectrum of the unknown with that of known compounds. Catalogs of infrared spectra are available, and card-indexing and other documentation processes facilitate the comparison procedure.

There are very few exceptions to the general rule that all nonidentical molecules yield different infrared spectra; the spectrum is a "fingerprint" of the molecule. Spectra differ particularly in the region 900 to 1350 cm^{-1}. This zone comprises a large number of bands, mostly of unknown origin, and complete coincidence here is an indication of identity. Enantiomorphous forms are exceptions to this rule since they give identical infrared spectra and need to be identified by optical rotation studies.

Characteristic peaks attributable to particular types of functional groups may be located by dividing the infrared spectrum into three regions.

In the frequency range 3600 to 2300 cm^{-1} one finds the stretching frequencies of O—H, N—H, S—H, P—H, and C—H groups. The stretching frequencies of triple bonds (C≡C, C≡N) and cumulated double bonds (X=Y=Z) are observed in the frequency range 2300 to 1900 cm^{-1}. Between 1700 and 1400 cm^{-1} are located the absorption peaks attributable to double bond stretching (C=C, C=O, C=N, N=O) and N—H bending.

Other stretching, bending and combination bands which are characteristic of different groups are found in the fingerprint region ($<$1400 cm^{-1}).

Some typical spectra and assignments are shown in Fig. 9.1.

The actual position of an absorption peak is related to the nature of the atoms involved and the environment. Hence the observed frequencies of a particular group are influenced by steric effects, electrical effects, inter-, and intra-molecular bonding (e.g., hydrogen bonding), etc. The correlation charts, therefore, show a broad region in which a particular absorption might be observed. Because of this range a peak in the unknown spectrum may be attributable to a number of functional groups. To resolve this overlap, one should study the correlation charts which have been prepared for each of the particular types of characteristic groups (e.g., C—H, O—H, N—H, R—H, triple bonds, C=O, etc.). These group charts list the position of other bands which should be present if the assignment is correct. They also tend to define more closely the source of the absorption, i.e., the probable environment of the vibrating group.

Figure 9.1 shows eight selected infrared spectra which illustrate some of the points made above. For example, spectra 3 and 5 both exhibit a characteristic C=O stretch absorption band at about 1720 cm^{-1}. On the other hand, the aldehyde (3) does not show the broad, hydrogen bonded OH stretch visible in the spectrum of *n*-butanoic acid (5) at about 3200 cm^{-1}. This latter peak is also absent from the spectrum of the carboxylate ion but in this case there is a very characteristic shift in the position of the carbonyl absorption peak. In the carboxylate ion, one can also assign peaks to both the symmetrical and

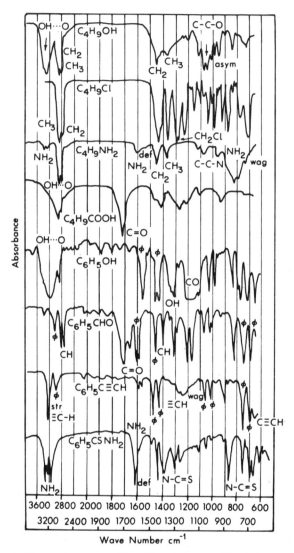

FIGURE 9.1 Typical infrared spectra. The abbreviations used include asym, asymmetric; sym, symmetric; str, stretch; def, deformations—scissors motion; wag, wagging; rk, rocking motion; ϕ phenyl group is origin of band; OH . . . CO implies interaction.

asymmetric stretching modes, viz., 1440 and 1580 cm^{-1}, respectively.

The existence of inter- and intramolecular bonding tends to broaden peaks and this is clearly shown by the broad OH bands observed at frequencies >2800 cm^{-1} in spectra 4, 5, and 8. The effect of hydrogen bonding may be minimized in spectrum 4 by examining a solution of phenol, rather than a solid sample. In this case there is a sharp peak at about 3600 cm^{-1} attributable to free —OH.

Spectra 1 and 6 show the characteristic twin peaks of the —NH$_2$ group but it can be observed that the position of the peaks is altered by the nature of the species with which the —NH$_2$ is associated.

Other spectra (e.g., 1–4) show bands which are attributable to vibrations within a ring structure (e.g., ϕ assignments). A dotted line implies interaction. Thus in spectra 4 and 5, the notation OH def . . . C—O str. could be added because some bands involve both OH deformation and C—O stretch.

If hydrogen bonding is suspected (OH . . . O) this point can often be clarified by varying the concentration of the sample in the chosen solvent or by changing the solvent.

An infrared spectrum does not commonly distinguish between a pure sample and an impure sample. In general, however, the spectrum of a pure sample has fairly sharp and well-resolved absorption bands, while the spectrum of a crude preparation (containing many different kinds of molecules) commonly displays broad and poorly resolved absorption bands.

Accordingly, in structural studies it is essential that the material studied be as pure as possible. It is also desirable that the spectrum be recorded using a variety of preparation techniques (e.g., KBr disc, mull, solution) to observe any inter- or intramolecular effects.

Raman spectra give information about molecular structure similar to that yielded by infrared spectra. Some groups have Raman lines, but do not show infrared absorption, and vice versa. The two spectra may therefore be regarded as complimentary, and since Raman frequencies tend to be less affected by structural changes in a more remote part of the molecule, the spectral bands can be assigned to groups with greater confidence.

C. Nuclear Magnetic Resonance Spectroscopy

A new approach to structural analysis is offered by NMR since it is primarily concerned with the study of proton patterns rather than carbon skeletons.

The more important factors necessary for the exact interpretation of a spectrum are line positions, line intensities, and the nature of spin-spin multiplets.

A number of effects combine to establish the value of the chemical shift of each type of proton, and in some cases the exact value can depend on the nature of the solvent and the concentration of the species.

On the other hand, correlation charts have been prepared from studies on known compounds and these can be used to identify the environment of the protons yielding a particular signal. The relative magnitude of the peak (given by an integrated signal) permits calculation of the number of protons occupying a similar type of environment. The spin-spin splitting patterns help to identify the structural character of neighboring protons. These aspects can be illustrated by considering the interpretation of a simple first order NMR spectrum. (In this context, the term "first order" means that the difference in chemical shifts $\Delta\nu$ is $\gg$ the coupling constant J.)

Figure 9.2(a) shows the NMR spectrum of a compound found by chemical analysis to be a monosubstituted aliphatic acid of empirical formula $C_3H_5O_2Cl$.

Three distinct signals can be observed at δ values 1.73, 4.47, and 11.22 ppm. The doublet at $\delta = 1.73$ has three times the intensity of the quartet visible at

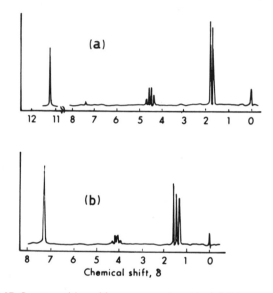

FIGURE 9.2   NMR Spectra: (a) α–chloro–propanoic acid, and (b) α–methyl benzylamine.

$\delta = 4.47$ (the least intense signal) which suggests that it might represent methyl protons (i.e., a CH_3 group).

In saturated hydrocarbon molecules, the CH_3 resonance appears in the 0.7 to 1.3 ppm range. However, the presence of a carbonyl group tends to deshield nearby protons, causing the resonance to occur at lower fields than in saturated hydrocarbons (i.e., δ is larger). The doublet at $\delta = 1.73$ may thus be considered in terms of position and intensity to represent a CH_3 group.

Provided $\Delta\nu \gg J$, the multiplicity of a peak is determined by the number of neighboring groups of equivalent nuclei; for protons, the multiplicity equals $n + 1$.

The doublet at $\delta = 1.73$ thus means that the methyl group has one proton neighbor.

The resonance observed far downfield ($\delta = 11.22$ ppm) is characteristic of an acidic proton. Its presence as a singlet also confirms the absence of neighboring protons; in intensity it equals the quartet observed at $\delta = 4.47$.

The proton resonating at $\delta = 4.47$ must have three equivalent proton neighbors if multiplicity equals $n + 1$. Thus this proton must be located on the carbon atom adjacent to the methyl group. The approximate chemical shift for a —CH—Cl group is $\delta = 4.0$, and allowing for the deshielding effect of the carbonyl group, the peak at $\delta = 4.5$ is undoubtedly attributable to this group.

The structure corresponding to spectrum (a), Fig. 9.2 is therefore

$$
\begin{array}{c}
\text{H} \quad \text{H} \quad \text{O} \\
| \quad\ | \quad\ || \\
\text{H—C—C—C—OH} \\
| \quad\ | \\
\text{H} \quad \text{Cl}
\end{array}
$$

This simplified approach is not applicable to all NMR spectra because of a number of less predictable factors. For example, the NMR spectrum of ethanol has been discussed in Chapter 7, Section III, and in this spectrum, the hydroxyl proton was shown to yield a singlet resonance peak. On the other hand, the simplified spin-spin coupling rules suggest that this signal should exist as a triplet, and the spectrum of highly purified ethanol actually displays a triplet, hydroxyl proton signal (coupling constant 5 cps). This apparent anomaly has been explained by invoking another theoretical concept.

The exact shape of the hydroxyl proton signal is considered to depend on the time that the proton spends on a given ethyl alcohol molecule. The hydroxyl hydrogen readily undergoes exchange with other hydrogen nuclei or ions in its vicinity; when this happens the replacement hydrogen does not necessarily have the same spin direction as that displaced. If the exchange occurs sufficiently fast (as in ordinary ethanol containing traces of acid or basic impurities) the neighboring CH nucleus experiences a "coupling field" which is averaged to zero. In other words, rapid chemical exchange causes spin decoupling.

The anticipated multiplicity is observed when the rate of chemical exchange is much slower than the frequency separation of the components of the multiplet (this occurs with very pure ethanol).

A broad absorption peak results when the rate of chemical exchange is of the same order of magnitude as the frequency separation of the components of the multiplet in the absence of exchange. It can be shown theoretically that for a coupling constant of J cps the coupled nucleus must exchange more rapidly than $J/2\pi$ times per sec for a multiplet to collapse to a singlet.

If two hydroxylic species are present, rapid chemical exchange usually occurs and instead of two resonance lines being observed for the different hydroxylic species present, only one signal appears. This resultant is located at an average intermediate position. When the exchange occurs between the test species and the solvent, the actual position of the resultant peak becomes concentration dependent.

The exchange process which gives rise to the averaging may be a physical exchange of nuclei (as in the exchange of OH protons) or merely an internal rearrangement, such as the rotation of a methyl group or the interconversion of the two chair forms of cyclohexane. If a single molecule contains both carboxyl and hydroxyl groups, chemical exchange usually causes the absorptions of these groups to lose their individuality.

The rate of chemical exchange increases with increasing temperature. Thus spin coupling can sometimes be observed by lowering the temperature of the sample. Conversely, spin decoupling may be observed on raising the sample temperature.

Coherent internuclear coupling can also be destroyed by means of an instrumental technique known as double irradiation or double resonance. Consider a sample containing two coupled nuclei which resonate at frequencies ν_1 and ν_2 in a given applied field. In the double resonance technique, both frequencies are applied simultaneously but the detector is tuned to receive only one signal, e.g., ν_1. The stirring of the nuclear spin caused by the second frequency destroys coherent coupling and the multiplet resonance of nucleus 1 collapses to a singlet. This technique is very useful for simplifying complex spectra.

Strong hydrogen bonding between an alcohol group and the solvent can cause the position of the hydroxyl absorption to shift downfield. In addition, the rate of chemical exchange is retarded, so spin-coupling of the hydroxyl proton to protons on an adjacent atom may be observed.

All protons associated with functional groups in which hydrogen atoms are attached to elements other than carbon (e.g., oxygen, nitrogen, and sulfur) are commonly considered to be active hydrogens. The NMR spectra of such protons depend on the extent of intermolecular hydrogen bonding and the rate of chemical exchange. The absorption positions of the protons depend on concentration, temperature and the nature of the solvent. The number of exchangeable protons present in the molecule can often be determined by adding an excess of deuterium oxide to a solution of the sample in carbon tetrachloride (or some other organic solvent). There is rapid exchange and the relative number of protons involved is indicated by the intensity of the water (or HDO) absorption signal that results from this equilibration process.

As mentioned in Chapter 7, additional shielding and deshielding of protons arises from electronic circulations induced within molecules. These anisotropic shielding effects are observed in the presence of multiple bonds and aromatic rings.

Let us now consider the NMR spectrum (Fig. 9.2(b)) of a species which preliminary tests indicate is an aromatic amine of empirical formula $C_8H_{11}N$.

Studies using different concentrations of sample show that the single line recorded in Fig. 9.2 at $\delta = 1.58$ ppm shifts with dilution. From this evidence it may be assumed that this peak represents active protons (associated with the

nitrogen atom) which exchange so rapidly that there are no spin–spin interactions. If this 1.58 ppm line is a singlet, then the other two peaks, of approximately equal intensity, which show small chemical shifts are probably members
of a doublet ($\delta = 1.38$ ppm). A doublet suggests a single proton on an adjacent
carbon atom.

Aromatic protons resonate far downfield; hence the peak in spectrum (b) at
$\delta = 7.30$ ppm can be attributed to the protons in a benzene ring.

The resonance signal at $\delta = 4.10$ is the least intense, and may be considered
to represent a single proton. It is split into a quartet; hence it has three neighboring protons on adjacent atoms. The peak is well downfield, which indicates
considerable deshielding, as might be expected if the proton were attached to a
carbon atom adjacent to the π electron system of the unsaturated ring.

Because of rapid exchange, the amine protons would not be expected to
split the proton signal; hence the quartet must be attributed to an adjacent
methyl group. This proposal is supported by the doublet nature of the peak at
$\delta = 1.38$ and the relative intensity of the signals.

Spectrum (b) therefore suggests that the aromatic amine must be α-methyl
benzylamine as follows.

In these two simple examples, it has not been necessary to invoke the use of
coupling constants for identification purposes or for elucidation of the spectral
pattern.

This simplified approach cannot be maintained when the difference in
chemical shifts, $\Delta\nu$, is no longer much greater than the magnitude of the coupling
constants J; and to distinguish between different types of NMR behavior, an
alphabetical nomenclature has been adopted. Thus the notation AX is used to
describe a system containing two nuclei whose chemical shift difference is large
compared with the magnitude of the coupling between them.

For nuclei close in chemical shift the letters AB (close in the alphabet) are used.
Close energy levels tend to repel each other; hence the A transitions move
slightly downfield, the B transitions move upfield. A second result of the small
chemical shift difference is perturbation of the intensities of the absorption lines.

The AB type systems tend to give complicated spectra which are usually
unrecognizable in terms of the simple splitting rules discussed in Chapter 7.
In these cases, to obtain values for the chemical shifts of the protons and the
coupling constants, mathematical analysis of the spectral data is required.

The nomenclature can be extended to include multiple nuclei systems.
Equivalent nuclei are assigned the same letter, e.g., ethyl bromide is an A_2B_3
system. Propyl iodide is described as an $A_2B_2C_3$ system. The A protons
(adjacent to iodine) are coupled to B protons, and the C protons are coupled
to the B protons. If the chemical shift values of some protons are very much

different from other interacting nuclei (A, B, ...) the dissimilar protons are described by the letters M, N ... and/or X, Y, ...

In the two examples discussed in preceding pages, Δv [$(\delta_2 - \delta_1)$ × frequency of sweep field] was much greater than J, and the spectra have been treated as AX_3 systems.

On the other hand, the NMR spectrum of pure ethanol shows a great multiplicity of peaks and this substance may be described as an AB_2C_3 system.

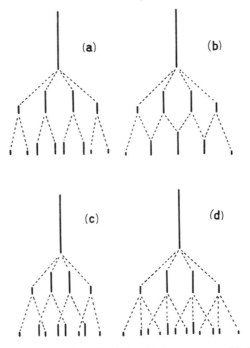

FIGURE 9.3   Schematic representation of NMR splitting patterns. (a) AB_2C_3 system with $J_{AB} < J_{BC}$; (b) AB_2C_3 system with $J_{AB} = J_{BC}$; (c) AB_2C_3 system with $J_{AB} > J_{BC}$; and (d) $A_2B_2C_3$ system with $J_{AB} < J_{BC}$.

The coupling constants have been determined as $J_{AB} = 5.0$ cps and $J_{BC} = 7.2$ cps, and in a high resolution instrument the signal attributable to methylene protons shows eight peaks, the two outer pairs being approximately one third the intensity of the inner quartet. This splitting pattern can be explained by proposing that the unperturbed line for the methylene protons is split into a quartet by the methyl protons ($J_{BC} = 7.2$ cps; relative intensities 1:3:3:1). Each component of this quartet is then split into a doublet ($J_{AB} = 5$ cps) by the hydroxyl proton. This is shown schematically in Fig. 9.3.

If in the above case J_{AB} had equaled J_{BC}, the splitting pattern would lead to overlap and a final quintet signal having relative intensities of approximately 1:4:6:4:1.

If J_{AB} had been greater than J_{BC}, then the secondary splittings of the quartet would have crossed over giving a different symmetry, for example, an octet having relative intensities of approximately $1:3:1:3:3:1:3:1$.

The picture is even more complicated in the case of an $A_2B_2C_3$ system such as a propyl halide. Here the methylene protons (B_2) are split into a quartet by the adjacent methyl protons. Each line of the quartet is then split into a triplet by the adjacent methylene group. The symmetry of the splitting is determined partly by the relative magnitude of the two coupling contents and partly by the fact that small peaks may not be observed.

Furthermore, as the magnitude of the coupling constants approaches the frequency separation of the proton signals from different nuclei, the relative intensity of the components of the multiplet differ from the simple pattern observed in A_nX_m systems where the ratios of the peak intensities are given by the coefficients of the expansion $(1 + x)^{m \text{ or } n}$ for nuclei A or X respectively.

Many systems of interacting nuclei do not yield a well-defined NMR pattern and it is not always possible to determine from the spectrum of a complex organic compound what types of interactions are occurring. Accordingly, it is sometimes better to use the evidence of other physical studies to propose a structure, and then check it by NMR.

A gradual transition in complexity occurs as one goes from a simple AMX system, through ABX to the ABC system. Complex systems can sometimes be resolved by changing the field strength since this effects the separation of peaks ($\Delta\nu$) but not the coupling constant J. Extensive mathematical treatment of the data derived from the complex spectra can also lead to elucidation of the structure. The other alternative is to compare the unknown spectrum with those of similar type compounds. Extensive catalogues of NMR spectra are now available.

D. Mass Spectrometry

A recorded mass spectrum shows the relative abundance of the positively charged particles formed on ionization of the sample in a suitable mass spectrometer.

In structural studies, these fragmentation patterns have to be correlated with the structure of the molecular species initially subjected to ionization.

The fragmentation pattern can be attributed to a number of factors as indicated below.

1. Molecular ions tend to fragment in a manner which leads to the formation of an even-electron carbonium ion and the elimination of a radical.

$$AB \xrightarrow{-e} [AB]^{\cdot +} \rightarrow A^+ + B^\cdot \quad \text{or} \quad A^\cdot + B^+$$

The relative stabilities of carbonium ions are tertiary > secondary > primary. Accordingly, hydrocarbons fragment preferably at the sites of branching. Thus in molecules containing side chains, decomposition can involve the elimination of methyl or ethyl radicals. With straight chain hydrocarbons, fragments can differ through the elimination of successive methylene groups. For this reason, the fragmentation patterns of isomeric hydrocarbons are vastly different in regard to the relative abundance of different fragments.

2. The fragmentation of some compounds tends to be dominated by the elimination of very stable neutral molecules, e.g., H_2O, HCN, CO, CO_2, NH_3, H_2S, ROH, etc.

Thus aliphatic alcohols usually decompose by elimination of water rather than a hydroxyl radical.

$$ROH \xrightarrow[-e]{} [ROH]^{.+} \xrightarrow{-H_2O} [M-H_2O]^{.+}$$

3. Isotope effects yield small peaks which differ by small units of mass from other larger peaks. Hydrocarbons contain, for example, about 1.1% of ^{13}C. A singly charged molecular ion or fragment therefore yields one peak due to the species containing the ^{12}C isotope and a second peak, one unit higher, attributable to the same ion containing ^{13}C. The abundance of this second peak should be $N \times 1.1\%$ of the abundance of the ion containing ^{12}C where N is the number of carbon atoms in that ion. Another small fraction of the peak may be attributable to a fragment containing ^{12}C and ^2H. The significance of the isotope effect varies with the natural abundance ratio of the various isotopic forms.

4. Occasionally fragmentation is accompanied by bond formation, i.e., an extra atom or group becomes attached to an incipient charged fragment. The migrating group is predominantly a hydrogen atom, and in most cases it can be proposed that the sample molecule undergoes rearrangement prior to cleavage.

In the majority of mass spectra, the ion occurring at highest mass may be assumed to be the molecular ion. (If reasonably intense, the peak may be accompanied by a small isotope peak, one mass unit heavier.) The m/e ratio of this molecular ion gives the molecular weight of the sample, an important starting point in structure elucidation.

The tendency of a compound to give a molecular ion increases with the thermodynamic stability of the species. Inherently unstable compounds, such as branched hydrocarbons, often do not give such an ion and it is desirable to check the validity of the assumption that the highest mass represents the molecule. One means of checking is to search for fragments immediately below the supposed molecular ion which correspond to $M-CH_3$ fragments (in hydrocarbon compounds containing methyl groups) or $M-H_2O$ fragments (in alcohols). Multiple loss of hydrogen atoms ($M-4$, $M-5$ ions, etc.) should be regarded as strong evidence that the heaviest unit does not correspond to the ionized molecular unit. Most aromatic compounds are sufficiently stable to give a molecular ion.

With high resolution instruments (yielding mass values to 3 or 4 decimal places) the molecular weight can be sufficient for partial identification since reference to Beynon's tables will indicate what combination of atoms ($C_xH_yO_z$ etc.) yields this exact mass. The lower the resolution of the instrument, the larger the number of empirical formula which might be responsible for the parent peak.

To take a simple example, a sample is known to contain only C, H, and O. Its mass spectrum contains a parent peak at mass 58, a base peak at mass 43, and significant peaks at mass 32, 28, 18, 17, 15, and 14.

Of these peaks, a number can be attributed to the background spectrum. Despite the extremely low pressures used in mass spectrometers, residual gas always gives a background which has to be subtracted from the observed spectrum. The background spectrum depends on many factors but there are generally peaks at 18 (H_2O^+), 28 (CO^+ and N_2^+), 32 (O_2^+) and 44 (CO_2^+) as well as small bumps due to hydrocarbon cracking if oil diffusion pumps are employed.

Subtracting this background from the unknown spectrum leaves peaks at mass 58, 43, 17, and 15, to be explained. Assuming mass 58 represents the molecular weight, Beynon's tables show that the only possible formulas are $C_2H_2O_2$ and C_3H_6O. The mass difference between the parent and base peak is 15 which corresponds to a methyl group (CH_3). The molecule therefore probably contains a methyl group which may be split off under bombardment. The peak at mass 43 could be due to $C_3H_7^+$ or CH_3CO^+. Since the maximum number of hydrogen atoms in the parent compound is 6 (in C_3H_6O), this restricts the source of the 43 peak to CH_3CO^+; consequently, the compound must be CH_3COCH_3.

The presence of a peak corresponding to $M–CH_3$ and the absence of peaks corresponding to $M–4H$ etc. tend to confirm that the peak at mass 58 is due to the molecular ion.

This type of argument can be applied to much larger and more complex molecules, and to assist in the elucidation of the composition and structure, a number of general rules have been evolved.

For example, once the molecular formula is known, the number of un-saturated sites can be calculated by means of the ring rule. The number of unsaturated sites (or double bond equivalents), R, is equal to the number of rings in the molecule, plus the number of double bonds plus twice the number of triple bonds.

For the molecule $C_aH_bN_cO_d$,

$$R = a + 1 + (c - b)/2$$

If the calculated value of R equals zero, the sample contains no rings or multiple bonds. A value of unity can indicate one ring or one double bond, etc.

In applying this equation to compounds containing other elements, mono-valent elements (e.g., Cl, Br, I) are treated as H, divalent elements may be effectively ignored, and trivalent elements (e.g., P) are treated as N. For pentavalent P three hydrogen atoms must be subtracted from b. Knowing the number of double bond equivalents helps in the prediction of plausible structures.

Further information can be ascertained by examining the fragmentation pattern. Where a wide range of reference spectra is available, the problem can sometimes be speedily solved by merely comparing the spectrum of the unknown with the standard spectra of all plausible compounds.

In the absence of reference spectra, the fragmentation pattern has to be subjected to some process of logical dissection.

The first step is to list all the major peaks, and all peaks near the parent peak. It is also helpful to list the relative intensities and the mass of the neutral fragments which must have been lost in order to give the various peaks.

TABLE 9.3

DATA ASSOCIATED WITH THE MASS SPECTRUM OF AN UNKNOWN COMPOUND

| Peak m/e | Relative intensity | Probable ion fragment | Neutral fragment lost | |
			m/e	Possible nature
58	38	$C_2H_2O_2$, C_3H_6O, C_4H_{10}, C_3H_8N etc.	—	—
57	10	C_2HO_2, C_3H_5O	1	H
43	11	C_3H_7, CH_3CO	15	CH_3
39	4		19	F
31	6	CH_2OH, OCH_3	27	C_2H_3
29	100	C_2H_5, OCH_3	29	C_2H_5, CHO
28	69	C_2H_4, CO, N_2	30	CH_2NH_2, NO
27	58	C_2H_3	31	CH_2OH, OCH_3
26	21	CN, C_2H_2	32	

By referring to Beynon's tables or other reference tables probable assignments can be made for each mass unit.

Unless there is evidence to the contrary, the heaviest unit is assumed to represent the molecular ion, and by applying the ring rule several alternative structures can be proposed. These probable structures usually represent different types of compounds. The fragmentation patterns of different classes of compounds are now reasonably well characterized; hence the pattern for each alternative structure can be predicted, and the predicted pattern can be compared with the experimental spectrum.

Consider the simple compound whose mass spectrum is summarized in Table 9.3. The heaviest ion is very likely to be the molecular ion since there is an M–CH_3 peak and no M–4 or M–5 peaks. Empirical formulas which correspond to this mass include $C_2H_2O_2$, C_3H_6O, C_4H_{10}, and C_3H_8N.

The molecular ion signal is relatively intense, and species which are known to give reasonably intense parent ions are C_1 to C_5 paraffin hydrocarbons, alkyl benzenes, aldehydes, ketones, and aromatic amines. Of these alternatives, the alkyl benzenes, and aromatic amines can be disregarded because the largest m/e

ratio is too small to account for these species. Since aliphatic amines do not give an intense parent peak, the compound C_3H_8N can be eliminated from further consideration.

Paraffin hydrocarbons generally give intense peaks with the general formula $C_nH_{2n+1}^+$ ($n \leq 5$), with additional peaks being observed at increments of 14 mass units (CH_2). In highly branched molecules, random rearrangements are common but not intense. Near each prominent peak there is a smaller peak one unit higher (for ^{13}C) and peaks one or two units lower corresponding to the loss of hydrogen atoms.

The spectrum quoted in Table 9.3 has many of the features of a paraffin spectrum except that there is no peak around $m/e = 15$, and there remains an unexplained peak at mass 31. Hence it is unlikely that the unknown is butane, C_4H_{10}.

One can now apply the ring rule to the two remaining probable compositions. For $C_2H_2O_2$, the ring rule gives $R = 2 + 1 - 1 = 2$. With two sites of unsaturation, and the formula $C_2H_2O_2$, the compound would have to be glyoxal OHC—CHO. For C_3H_6O, the ring rule indicates a single site of unsaturation; hence the compound could be CH_3COCH_3 (acetone); CH_2= CHCH$_2$OH (allyl alcohol); CH_3OCH=CH_2 (methyl vinyl ether); or $CH_3 \cdot CH_2 \cdot CHO$ (propionaldehyde).

Acetone can be eliminated as a possibility since its spectrum (discussed in the previous example) bears little similarity to this unknown. (See Fig. 9.4.)

This still leaves a number of alternatives, and in order to distinguish between these it is necessary to consider the general characteristics of the mass spectra observed with the different types of compounds.

Olefins tend to rupture at the C—C bond which is beta to the double bond, the charge remaining with the group containing the double bond. Thus peaks may be expected at m/e values of 41, 55, 69, etc., two units lower than for saturated compounds. In the example being considered, the chain length is insufficient to observe this effect.

With alcohols, the parent peak is small and the loss of water is a prevalent source of fragments, giving a significant M-18 peak. Generally the bond beta to the oxygen atom is cleaved giving prominent peaks at m/e ratios of 31 for primary alcohols (CH_2OH^+), at 45 for secondary alcohols (CH_3CHOH^+), and at 59 for tertiary alcohols ((CH_3)$_2COH^+$). In the unknown spectrum, the parent peak is of reasonable intensity and exhibits no M-18 line. Hence the possibility of the sample being allyl alcohol can be discounted.

In molecules containing heteroatoms (e.g., O, N, S) the most facile manner of ionization involves removal of one of the nonbonding electrons. This can be followed by electron transfer from adjacent groups and fission to give stable ions.

The carbonyl group, which is common to aldehydes, ketones, and esters, is instrumental in initiating two important types of cleavage, named α cleavage and β cleavage.

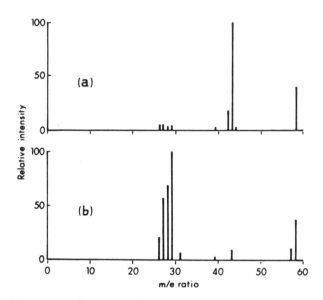

FIGURE 9.4   Mass spectra of two species having empirical formula C_3H_6O: (a) acetone, and (b) propionaldehyde.

An α cleavage can occur on either side of the carbonyl group.   The relatively stable acylium ions so formed are fairly pronounced in the spectra of ketones and esters in which the R and R′ groups contain only a few carbon atoms.   If the R and R′ groups consist of long alkyl chains, then the large increase in the number of bonds capable of being broken leads to many fragments and it is difficult to distinguish the acylium ion.   Similarly, with aldehydes higher than butyraldehyde, the $m/e = 29$ peak corresponds predominantly to $C_2H_5^+$.   With formaldehyde, acetaldehyde and propionaldehyde, m/e 29 is associated with the $(H—C\equiv O^+)$ acylium ion.   With these aldehydes one should also observe an M-1 peak due to the loss of a hydrogen atom.

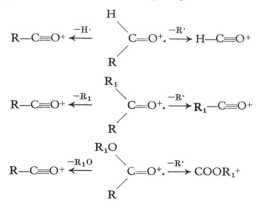

Where there is a choice between the loss of a hydrogen atom or an alkyl group, elimination of the latter predominates. Similarly if R and R_1 differ in length, elimination of the larger alkyl radical usually predominates.

During the cleavage process, it is also possible for the charge to be retained by the eliminated fragment, yielding ions corresponding to R^+, R_1^+, R_1O^+ etc.

If an alkyl group attached to a carbonyl moiety contains three or more carbon atoms in a chain with a hydrogen atom attached to the γ carbon, then β cleavage with γ hydrogen rearrangement becomes important.

$$R\cdot \; + \; \underset{\cdot CH_2}{\overset{H}{\diagdown}} C{=}\overset{+}{O} \quad \longleftarrow \quad \underset{R{-}CH_2}{\overset{H}{\diagdown}} C{=}\overset{+}{O} \quad \longrightarrow \quad R^+ \; + \; \underset{\cdot CH_2}{\overset{H}{\diagdown}} C{=}O$$

With hydrogen rearrangement, there is an intermediate ring structure, which leads to the following fragments.

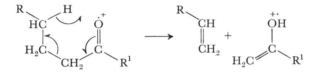

For aldehydes, $R^1 = H$; aldehydes also tend to eliminate small stable fragments such as CH_3, C_2H_4, C_2H_5, CO, and H_2O on ionization.

The important fragmentations of dialkyl ethers are in many ways similar to the systems discussed above. Ionization takes place predominantly by removal of a nonbonding electron from the oxygen atom. A radical may subsequently be eliminated by α cleavage. If the alkyl chain which is not involved in the α cleavage reaction contains two or more carbon atoms, there can be cleavage with hydrogen rearrangement to give an oxonium ion, e.g., $R'{-}CH{=}OH^+$.

Using this information, the nature of the unknown described in Table 9.3 can be restricted to a choice between methyl vinyl ether and propionaldehyde. The ether proposal fails to explain the peak at mass 28; the aldehyde proposal does not predict a peak at mass 26, although it appears ideal in all other aspects.

The matter would therefore have to be finally resolved by comparing the unknown spectrum with the spectra of these two highly probable species.

The two simple examples which have been discussed illustrate the type of thought processes involved in interpreting a mass spectrum.

The behavior of amines and amides is similar to that of ethers. The α cleavage process provides the most abundant ion in the spectra of primary, secondary and tertiary n-alkyl amines, and α mono- or disubstituted primary amines. In the spectra of α-branched secondary and tertiary amines, the base peak arises from the process of N—C cleavage with hydrogen rearrangement. The spectra of amides are a little more difficult to interpret.

Aromatic compounds behave quite differently from aliphatic compounds upon electron impact because the positive charge of the molecular ion can be localized to some extent in the π electron system of the aromatic nucleus.

Alkyl benzenes cleave mainly at the activated benzylic bond with charge retention by the aromatic ring. In addition, if the side chain is propyl or larger, β fission is also accompanied by rearrangement of a hydrogen atom.

Simple aromatic carbonyl compounds (C_6H_5COR) generally fragment with the loss of the R group as a neutral radical and the formation of the benzoyl cation ($C_6H_5C{\equiv}O^+$). Further decomposition of this to the phenyl cation and $C_4H_3^+$ is quite characteristic.

The mass spectrum of phenol contains an abundant molecular ion and pronounced M—CO and M—CHO peaks. In more complicated phenols, fragmentation occurs at the sites requiring least energy for bond rupture.

Aromatic ethers and amines tend to fragment through the loss of groups such as CO, CH_3 or —$CH_2{=}O$.

Nitro compounds decompose to a very large extent by simple elimination of the nitro group, although additional decomposition modes are sometimes very significant.

In general, the behavior induced by a single substituent (—OH, —COOR) in an aromatic ring is modified in a disubstituted ring, the nature of the modification depending on the nature of the disubstitution (e.g., ortho, meta, or para). Heterocyclic aromatic compounds behave rather similarly to their benzene analogs.

Frequently, the fragmentation pattern of a complex molecule can be assigned in many ways. In these cases, isotopic labeling or high resolution measurements must be employed to uncover the fragmentation mechanism and to achieve structure elucidation. The most common isotope employed in labeling work is deuterium, and by using different chemical procedures the isotope can be caused to replace hydrogen ions attached to different types of groups. Carbonyl groups are labeled with ^{18}O and nitrogen compounds with ^{15}N.

Rather than rely entirely on one technique for structure determination, it is often preferable to combine the information gleaned from several different approaches.

E.   Joint Application of Physical Methods

The elucidation of the structure of a compound using a range of techniques normally involves a fairly standard sequence of studies.

The first step is the determination of the molecular formula. This may be achieved by microanalysis, associated with a molecular weight determination. Alternatively this information may be derived from the mass spectrum.

From the molecular formula the number of double bond equivalents can be calculated, using the ring rule. Having determined that the molecule contains some unsaturated bonds an indication of the type of chromophore present can be provided by an ultraviolet absorption spectrum.

The presence (or absence) of certain functional groups may then be ascertained from an infrared spectrum.

Finally, the way in which these functional groups and remaining atoms are joined together can be deduced from the NMR spectrum and mass spectrum. The NMR spectrum clarifies the distribution of hydrogen atoms in the molecule while the fragmentation pattern facilitates location of the heteroatoms and provides confirmation of the overall structure.

F. OPTICAL ACTIVITY

In ascertaining the structure of optically active compounds, the measurement of molecular rotations $[\phi]$ can be of some value, the procedure being particularly useful in steroid studies.

It has been observed that if two similar asymmetric molecules (e.g., A and B) are changed in the same way (to give A′ and B′), then the molecular rotation differences ($M_{A'} - M_A$ and $M_{B'} - M_B$) are of the same sign. The effect of a particular structural change on the optical rotation decreases as the distance between the structural change and asymmetric center increases. Judicious application of these generalizations has facilitated the elucidation of the structure of several types of compounds.

For example, the molecular rotation of a steroid may be considered to be made up of the rotation of a fundamental structure (i.e., A or B) plus contributions from functional groups (Δ values). These Δ values are characteristic of the groups, their positions in the molecule, and their orientations. Provided the functional groups are not highly unsaturated and provided that the functional groups are separated by several saturated carbon atoms, the effects are additive.

This can be illustrated by considering the group of steroids represented by the formula

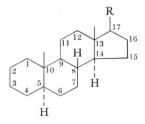

The molecular rotation of several members of this series and the Δ values observed with different substituents are recorded in Table 9.4.

For a similar series of steroids in which the rings are fused in a slightly different manner, both the $[\phi]_D$ values of the reference compounds and the Δ values for substituents are different in magnitude.

From tables such as Table 9.4, the molecular rotation of a steroid can be calculated. This value can then be compared with the experimental value and the structure adjusted until the two match.

TABLE 9.4
GROUP CONTRIBUTIONS TO MOLECULAR ROTATION[a]

Reference compound R	$[\phi]_D$	Position of substituent	Δ Values		
			ΔOH	ΔCO	ΔOAc
H	+5	1	+35(α)	+339	+79(α)
C_2H_5	+52		−17(β)		+27(β)
C_8H_{17}	+91	3	+5(α)	+71	+22(α)
COOH	+162		−2(β)		−31(β)
$(CH_2)_4$·COOH	+74	6	+55(α)	−113	+210(α)
			−50(β)		−110(β)

[a] α and β represent configurations; α groups are directed away from the reader, β groups are directed towards the reader.

For example, if chemical evidence indicates that a steroid possessing a substituent OH group is of the type shown above with R = C_2H_5, then the calculated molecular rotation would be +52 (reference value) + ΔOH. If, perchance, the observed molecular rotation were +35, ΔOH must equal (35 − 52) or −17 which corresponds to a β OH group in position 1.

It is also possible to use optical rotation measurements to identify the basic type of reference compound. Other molecular rotation studies permit the contributions of a double bond in various parts of a steroid molecule to be calculated.

Optical rotation measurements have been widely used in studies of the structure of carbohydrates. The extensive work in this field has led to the formulation of certain rules which correlate the configuration of carbohydrates with their optical rotations. Some of these rules (Hudson's isorotation rules) are summarized below.

The first rule states that the rotatory contribution of C_1 in a sugar molecule is influenced to only a minor extent by the rotation of the remainder of the molecule; in the D-series the α-anomer is always the more dextrorotatory. This rule is of general application and can be used to settle whether a C_1 epimeride belongs to the α or β series.

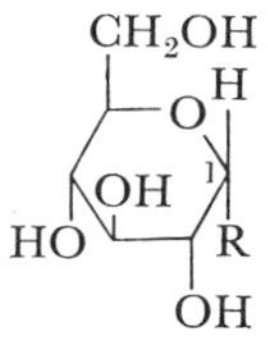

CH_2OH

α-D-Glucose

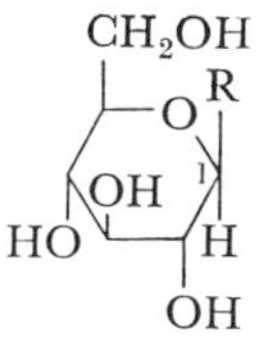

CH_2OH

β-D-Glucose

R	$[\phi]_D$	R	$[\phi]_D$
OH	+202	OH	+34
OMe	+309	OMe	−66
OEt	+314	OEt	−70
OPh	+463	OPh	−182

The second rule states that changes at position 1 affect to only a minor degree the rotation of the rest of the molecule.

Hudson's lactone rule states that if the lactone ring, as represented in the conventional projection formula, lies to the right of the molecule, the lactone is more dextrorotatory than the parent acid; if it lies to the left the lactone is more levorotatory.

A study of the molecular rotation of a lactone (γ or δ) and the parent acid can thus be used to provide information on the configuration of the hydroxyl group which participates in the lactone formation.

Other rules have been enunciated to cover the configurations of other carbon atoms in carbohydrates.

The technique, optical rotatory dispersion (ORD), has also been widely applied to structural problems.

Carbonyl groups in an asymmetric environment are subject to an induced asymmetric electron distribution within the group and a Cotton effect may be observed. In a large molecule like a steroid it has been possible to correlate the position of a keto-group in the molecule with the shape and size of the corresponding ORD curve. The correlation provides a method of locating the keto-group in a steroidal ketone of unknown constitution. A hydroxyl group which is oxidizable to a ketone may also be located in this manner.

The shape of the ORD curve is affected by stereochemical differences near to a keto-group and this effect can be used to confirm the type of steroid being investigated.

To determine the absolute configuration of a compound the ORD curve is compared with that of another compound (of known absolute configuration) in which the identical chromophore occurs in the same stereochemical and

conformational environment. Similar configurations in the known and unknown lead to curves which are superimposable.

It has been suggested that no other physical tool can approach rotatory
dispersion in its utility for settling stereochemical problems in the steroid field.

Other applications include studies of bicyclic ketones, triterpenoids, α-
haloketones, and to a lesser extent, monocyclic and aliphatic carbonyl compounds.

III. X-RAY DIFFRACTION

A. REFLECTION OF X RAYS

When a wave front of X rays falls on a set of atoms located in a plane, part of
the ray beam is transmitted and part is reflected by the individual atoms, as
shown schematically in Fig. 9.5(a).

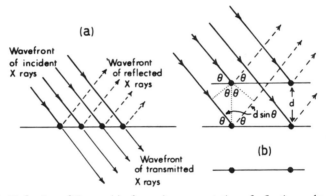

FIGURE 9.5   Reflection of X rays: (a) schematic representation of reflection and transmission;
and (b) the reflection of X rays by planes of atoms, showing the geometry associated with
Bragg's law.

In a crystal lattice the atoms are arranged in a systematic manner and can be
considered to exist in a series of planes, separated by finite distances.

Consider the simple case involving just two planes [Fig. 9.5(b)] where
incident X rays are being reflected from each atom. A study of Fig. 9.5(b)
shows that the path length from the impinging wave front to the reflecting
plane and thence to the scattered wave front is longer in the case of the lower
plane of atoms. The application of standard methods of geometry shows that
the additional distance involved is $2\,d \sin \theta$ where d is the distance between the
reflecting planes and θ is the angle made by the incident beam with the direction
of the reflecting plane.

To obtain a strong reflected beam the emergent rays must be in phase; hence
the relationship (Bragg's law)

$$2\,d \sin \theta = n\lambda$$

where n is an integer and λ is the wavelength of the incident radiation, needs to be satisfied.

It may be observed from this law that using monochromatic radiation of known wavelength, strong reflected rays will only be obtained at certain fixed values of d and θ. If the angle θ through which the beam is reflected is measured, then the distance d between the planes of atoms can be calculated.

On rotating a given crystal in all directions, reflected rays of differing intensity are observed at different angles of θ. Some of these can be attributed to different

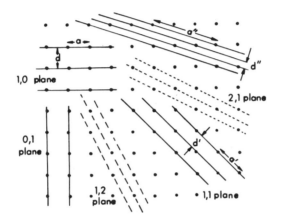

FIGURE 9.6 Two dimensional representation of different atomic planes in a simple cubic crystal lattice, showing the variations in interplanar distances d and reticular distances a.

values of n while others have to be attributed to the existence of several sets of reflecting planes within the crystal.

An indication of the source of different planes can be gained by considering a two dimensional representation of the simplest type of crystal pattern, namely the cubic arrangement. Figure 9.6 shows several different planes and it can be seen that in each set of planes there are differences in the distance between the atoms in each plane and in the distance between planes.

All crystals are made up of a repetition of some basic pattern, the nature of the basic pattern being due to the constituent atoms or ions settling in positions of minimum energy. The basic pattern is known as the unit cell.

A study of the diffraction pattern (that is, the angles through which radiation is reflected) can therefore be used to predict the arrangement of atoms within the crystal (i.e., the structure), and by substitution in Bragg's law the distance between the atoms can be calculated. The determination of the structure of a compound from its diffraction pattern is rendered difficult by the need to find a geometric shape which yields sets of planes of atoms whose interplanar distances correspond to the determined values for d.

B.   CRYSTAL SYSTEMS

A dimensional view of the various planes in a crystal is obtained by describing them in terms of nomenclatures such as the 010, 100, 111, etc. The numbers refer to the number of planes crossed in passing between individual atoms, in the directions a, b, and c.

Low nomenclature, i.e., low indexes indicate large interplanar distances with high reticular density. High indexes indicate small interplanar distances but low reticular densities, i.e., greater distances between atoms in the direction of the plane. Crystals tend to grow on planes with high reticular densities; hence exposed faces are faces of low indexes. This is sometimes known as the Law of Rational Indexes.

By convention, the indexes of a plane are given the general symbols h, k, l, and for a cubic unit cell it can be shown that

$$d_{hkl} = a/\sqrt{h^2 + k^2 + l^2}$$

where d is the interplanar distance determined from X-ray diffraction studies and a is the distance between the atoms. In this case, a represents the dimensions of the unit cubic cell; h, k, and l are small integers, so the ratio of d values should be $1:1/\sqrt{2}:1/\sqrt{3}:1/\sqrt{4}:1/\sqrt{5}$ etc. With complex cubic systems, e.g., face centered cubic, there are gaps in the series.

To calculate the interatomic distance, a, from a diffraction pattern, values of h, k, and l have to be chosen, mainly by inspection, so as to give consistent values of a for all observed values of d. The accuracy of the calculated value depends somewhat on whether the d values are calculated from large or small values of the angle θ. To obtain the best value of a, the calculated values are plotted against the appropriate values of $\cos^2 \theta$ or $1/2 \left(\cos^2 \theta / (\sin^2 \theta + \cos^2 \theta / \theta) \right)$. These plots are then extrapolated to zero to obtain the desired accurate value for the cell dimension.

The calculations become far more complex when the crystal unit cell has a geometric form which differs from a cube.

There are six basic unit cell types and the characteristics of these basic types are summarized in Table 9.5. The basic types can be further subdivided into thirty-two classes.

The type of equations which have to be solved in order to determine cell dimensions for cell types other than the cubic system can be indicated by taking just three examples.

In tetragonal systems

$$d_{hkl} = 1/\sqrt{(h^2 + k^2)/a^2 + l^2/c^2}$$

For a hexagonal unit cell

$$d_{hkl} = 1/\sqrt{4(h^2 + hk + k^2)/3a^2 + l^2/c^2}$$

When extended to rhombohedral systems, the mathematical equation is

$$d_{hkl} = a \Big/ \sqrt{(1+2\cos^3\alpha-3\cos^2\alpha)/(h^2+k^2+l^2)\sin^2\alpha+2(hk+kl+lh)(\cos^2\alpha-\cos\alpha)}$$

(The mathematical terms are defined in Table 9.5.)

The calculations involved in cubic, tetragonal, and hexagonal systems can be simplified by the use of special charts known as Bunn charts. In these simpler cases, the necessary d values can also be determined from powder photographs.

TABLE 9.5

CLASSIFICATION OF THE DIMENSIONS OF A UNIT CELL

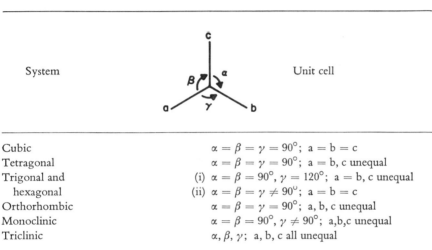

System		Unit cell
Cubic	$\alpha = \beta = \gamma = 90°$; a = b = c	
Tetragonal	$\alpha = \beta = \gamma = 90°$; a = b, c unequal	
Trigonal and	(i) $\alpha = \beta = 90°$, $\gamma = 120°$; a = b, c unequal	
hexagonal	(ii) $\alpha = \beta = \gamma \neq 90°$; a = b = c	
Orthorhombic	$\alpha = \beta = \gamma = 90°$; a, b, c unequal	
Monoclinic	$\alpha = \beta = 90°$, $\gamma \neq 90°$; a,b,c unequal	
Triclinic	α, β, γ; a, b, c all unequal	

To determine the cell constants of more complex systems, it is necessary to obtain single crystal rotation photographs and to use a computer for the calculations.

C. X-RAY POWDER PHOTOGRAPHS

Instead of rotating a single crystal of a compound to obtain a diffraction pattern, it is often more convenient to use a powdered specimen. With a powdered specimen, the small crystals of material are randomly arranged so that there are always some crystals present at the correct angle to the incident beam to give reflected rays from all the different planes of atoms present.

The X-ray diffraction pattern is readily obtained by using an X-ray powder camera or recording diffractometer.

Figure 9.7(a) shows the principle of the Debye-Scherrer circular powder camera which uses a thin cylindrical powder specimen set on the axis of the film cylinder. The specimen is rotated to increase the number of diffracting particles.

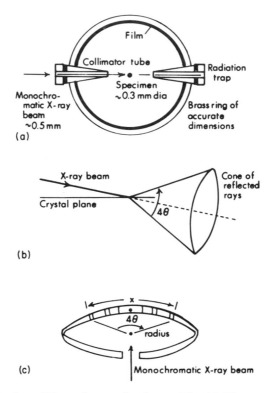

FIGURE 9.7   Diffraction of X rays by powdered materials. (a) Diagrammatic sketch of a powder camera; (b) cone of reflected X rays emanating from a crystal plane; and (c) geometry of an X-ray powder photograph.

Due to the random distribution of the particles, the incident X-ray beams are reflected from the various points in the crystal as cones of radiation [Fig. 9.7(b)]. By placing a film in a circle around the specimen, a photograph is obtained [Fig. 9.7(c)] which consists of a central spot corresponding to the center of the cone and a series of arcs, paired on either side of the center. These represent portions of the cones of reflected rays from the crystal planes. The powder photograph is a fingerprint of the d values of a crystal.

A study of the geometry of Fig. 9.7(c) shows that θ can be calculated from the distance between the corresponding arcs (x) on the photograph since

$$x/2\pi r = 4\theta/360$$

x and r are measured (for a given camera the radius r is known) and θ is calculated for each set of arcs. Knowing θ and λ of the X-radiation used, d values corresponding to each set of arcs can be calculated by substitution in Bragg's equation.

The d values and relative intensities of the arcs are then listed and can be used to identify the nature of the sample or to determine the crystal structure of the sample.

Instead of preparing powder photographs, an X-ray detector may be rotated in a circle around the specimen (i.e., a moveable detector replaces the film detector used in the camera system). The output of the detector is fed to a chart recorder which plots deflections against diffraction angle 2θ. The chart record can then be used for identification purposes or for calculation of d values.

The identification of a substance may be achieved by

a. Comparing the X-ray powder photograph or recorded diffraction pattern with reference films or spectra; or

b. Use of an ASTM index. The older index (the Hanawalt system) characterizes each powder pattern by the interplanar distances d_i and the relative intensities I_i of the three strongest lines in the pattern. The complete list of interplanar spacings, their relative intensities, and where known, the indexes of the crystallographic planes are given, together with the unit cell parameters of the compound. An alternative method, the FINK index has recently been provided for the same ASTM card file. The listing for the FINK index contains the inter-planar spacing values of the first eight strong lines in the pattern; or

c. Unit cell determination. An alternative approach to the data file method is to determine the unit cell parameters of the individual components of the specimen. A comprehensive determinative table has been prepared by the Donnays to enable a compound to be identified from the unit cell symmetry and dimensions.

D.   SINGLE CRYSTAL STUDIES

X rays of known, uniform wavelength can produce diffraction patterns from crystals which may be used to determine the spatial distribution of atoms in the crystal. A monochromatic, parallel, X-ray beam is allowed to fall onto a single crystal of the compound under investigation. The diffracted rays are then recorded on a photographic plate. Since crystals are three-dimensional, a series of photographs must be taken, using various orientations, in order to derive the spatial arrangement.

The diffraction pattern produced by a single crystal is formed by many separate diffractions, and the photographic record consists of a series of tiny spots. The intensity of the blackened parts of the photograph are measured individually by a photometric method. The intensities of the diffracted spots are important, since each is a function of both the number of electrons involved and the distribution of atoms in the plane which gives rise to the particular diffraction spot.

Through a complicated mathematical treatment involving a triple Fourier series integration it is possible to calculate, from the positions and intensities of the reflections, the electron density at various points within the crystal. These are extra-nuclear electron densities, since these electrons are responsible for the scattering of X rays by atoms. The treatment of the results involves many thousands of calculations but the use of computer programs has reduced the tedium of this task to a minimum.

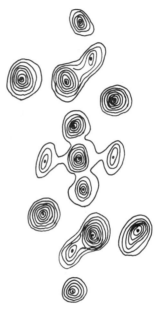

FIGURE 9.8   A typical electron density map computed from single crystal X-ray diffraction data.

The results are usually presented in the form of an electron density projection map in which the electron density within the three-dimensional crystal is projected on to a suitable plane, usually one face of the crystal. These maps, which consist of a series of contour lines, enable the positions of molecules within the crystal and atoms within the molecule to be ascertained. Figure 9.8 shows an electron density map computed from X-ray diffraction data.

From the projection map it is possible to determine intermolecular and interatomic distances, bond angles, and the dimensions of the unit cell.

The method has been used to elucidate some very complex structures, including macromolecules such as insulin, hemoglobin, and deoxyribonucleic acid (DNA).

At the moment, X-ray diffraction can be considered to be the final arbiter in any study of structure. The process can be much more complicated and time consuming than the other physical methods, but it is probably the only method capable of providing all the information necessary for a complete solution to a structural problem.

References

1. ULTRAVIOLET SPECTROSCOPY

I. Fleming and D. H. Williams, *Spectroscopic Methods in Organic Chemistry*, McGraw-Hill, London, 1966.

R. A. Friedel and M. Orchin, *Ultraviolet Spectra of Aromatic Compounds*, Wiley, New York, 1951.

H. H. Jaffé and M. Orchin, *Theory and Applications of Ultraviolet Spectroscopy*, Wiley, New York, 1962.

A. Gillam and E. S. Stern, *An Introduction to Electronic Absorption Spectroscopy in Organic Chemistry*, 2nd Ed., Arnold, London, 1957.

C. N. R. Rao, *Ultraviolet and Visible Spectroscopy*, Butterworth, London, 1961.

A. I. Scott, *Interpretation of the Ultraviolet Spectra of Natural Products*, Pergamon, Oxford, 1964.

CATALOGUES

H. M. Hershenson, *Ultraviolet and Visible Absorption Spectra*, Index for 1930–1954, Academic, New York, 1956; Index for 1955–59 (1961).

Organic Electronic Spectral Data, Vol. I (1946–52), M. J. Kamlet, (ed.); Vol. II (1953–55), H. E. Ungnade, (ed.); Vol. III (1956–57); Vol. IV (1958–59), J. P. Phillips and F. C. Nachod, (eds.), Wiley (Interscience), New York.

U.V. Atlas of Organic Compounds, Butterworth, London, 1965.

2. INFRARED SPECTROSCOPY

L. J. Bellamy, *The Infrared Spectra of Complex Molecules*, Methuen, London, 1958.

W. Brugel, *An Introduction to Infrared Spectroscopy*, Methuen, London, 1962.

N. B. Colthup, L. H. Daly, and S. E. Wiberley, *Introduction to Infrared and Raman Spectroscopy*, Academic Press, New York, 1964.

A. D. Cross, *Introduction to Practical Infrared Spectroscopy*, 2nd Ed., Butterworth, Lonond, 1964.

J. R. Dyer, *Applications of Absorption Spectroscopy of Organic Compounds*, Prentice-Hall, Englewood Cliffs, N.J., 1965.

K. Nakamoto, *Infrared Spectra of Inorganic and Co-ordination Compounds*, Wiley, New York, 1963.

K. Nakanishi, *Infrared Absorption Spectroscopy*, Holden-Day, San Francisco, 1962.

C. N. R. Rao, *Chemical Applications of Infrared Spectroscopy*, Academic Press, New York, 1963.

R. M. Silverstein and G. C. Bassler, *Spectrometric Identification of Organic Compounds*, 2nd Ed., Wiley, New York, 1967.

M. K. Wilson, *"Infrared and Raman Spectroscopy"* Chapter 3, Vol. 2, in F. C. Nachod and W. D. Phillips, *Determination of Organic Structures by Physical Methods*, Academic Press, New York, 1962.

CATALOGUES

H. M. Hershenson, *Infrared Absorption Spectra*, Academic Press, New York. Index for 1945–1957 (1959); Index for 1958–1962 (1964).

An Index of Published Infrared Spectra, H.M. Stationery Office, London, Vols. I and II, 1960.

Documentation of Molecular Spectroscopy, Butterworth, London.

3. NMR Spectroscopy

C. Banwell, *Fundamentals of Molecular Spectroscopy*, McGraw-Hill, London, 1966.

N. S. Bhacca and D. H. Williams, *Applications of N.M.R. Spectroscopy in Organic Chemistry*, Holden-Day, San Francisco, 1964.

J. C. D. Brand and G. Eglington, *Applications of Spectroscopy to Organic Chemistry*, Oldbourne, London, 1965.

J. W. Emsley, J. Feeney and L. H. Sutcliffe, *High Resolution Nuclear Magnetic Resonance Spectroscopy*, Vol. I, Pergamon, Oxford, 1965.

L. M. Jackman, *Applications of N.M.R. Spectroscopy in Organic Chemistry*, Pergamon, London, 1959.

J. A. Pople, W. G. Schneider and H. J. Berstein, *High Resolution Nuclear Magnetic Resonance*, McGraw-Hill, New York, 1959.

J. D. Roberts, *An Introduction to the Analysis of Spin-Spin Splitting in Nuclear Magnetic Resonance*, Benjamin, New York, 1961.

J. D. Roberts, *Nuclear Magnetic Resonance*, McGraw-Hill, New York, 1959.

K. Wiberg and B. J. Nist, *The Interpretation of NMR Spectra*, Benjamin, New York, 1962.

Catalogues

N.M.R. Spectra Catalogue. Vols. 1 and 2, Varian, Palo Alto, California.

N.M.R., NQR, EPR Current Literature Service, Butterworth, London and Washington, D.C.

4. Mass Spectrometry

J. H. Beynon, *Mass Spectrometry and Its Applications to Organic Chemistry*, Elsevier, Amsterdam, 1960.

K. Biemann, *Mass Spectrometry*, McGraw-Hill, New York, 1962.

H. Budzikiewicz, C. Djerassi and D. H. Williams, *Interpretation of Mass Spectra of Organic Compounds*, Holden-Day, San Francisco, 1964.

C. A. McDowell (ed.), *Mass Spectrometry*, McGraw-Hill, New York, 1963.

F. W. McLafferty (ed.), *Mass Spectrometry of Organic Ions*, Academic Press, New York, 1963.

Catalogues

J. H. Beynon and A. E. Williams, *Mass and Abundance Tables for Use in Mass Spectrometry*, Elsevier, Amsterdam, 1963.

Catalog of Mass Spectral Data, American Petroleum Institute Research Project 44, Carnegie Press, Pittsburgh, Pennsylvania.

J. Lederberg, *Computation of Molecular Formulas for Mass Spectrometry*, Holden-Day, San Francisco, 1964.

F. W. McLafferty, *Mass Spectral Correlations*, Am. Chem. Soc., Washington, D.C., 1963.

5. Optical Activity

E. A. Braude and F. C. Nachod, (eds.), *Determination of Organic Structures by Physical Methods*, Vol. I, Chapter 3, Academic Press, New York, 1955.

C. Djerassi, *Optical Rotatory Dispersion*, McGraw-Hill, New York, 1960.

W. Klyne, *The Chemistry of Steroids*, Methuen, London, 1957.

F. C. Nachod and W. D. Phillips, (eds.), *Determination of Organic Structures by Physical Methods*, Vol. II, Chapter 1, Academic Press, New York, 1962.

6. X-Ray Diffraction

C. W. Bunn, *Chemical Crystallography*, Oxford Univ. Press, London, 1945.

M. J. Buerger, *Elementary Crystallography*, Wiley, New York, 1956.

M. J. Buerger, *The Precession Method in X-ray Crystallography*, Wiley, New York, 1964.

H. P. Klug and L. E. Alexander, *X-ray Diffraction Procedures for Polycrystalline and Amorphous Materials*, Wiley, New York, 1954.

S. C. Nyburg, *X-ray Analysis of Organic Structures*, Academic Press, London, 1961.

H. S. Peiser, H. P. Rooksby and A. J. C. Wilson, *X-ray Diffraction by Polycrystalline Materials*, The Institute of Physics, London, 1955.

CATALOGUES

A.S.T.M. Card File and Index to Powder Diffraction File, A.S.T.M. Special Technical Publication 48-M2, American Society for Testing and Materials, Philadelphia, 1963.

Crystal Data Determinative Tables, J. D. H. Donnay and G. Donnay, 2nd Ed., ACA Monograph 5, Am. Crystallographic Assoc., 1963.

BASIC SOLUTION CHEMISTRY

I. Ions in Solution	299
II. Ion Association, Complex Formation, and Precipitation	303
III. Solvent Extraction and Adsorption	306
IV. Competing Equilibria	308
A. Selection of an Indicator in Titrimetric Analysis	309
B. Selective Precipitation and Masking	316
C. Solvent Extraction	323
D. Solid-Liquid Phase Equilibria	326
V. Effect of Solvent	328
VI. Tutorial Problems	330
References	333

A large proportion of the available analytical techniques initially require dissolution of the sample in some solvent, water being the solvent most commonly encountered in the chemical analysis of inorganic species.

The response of the analytical systems can be influenced by the interactions which can occur in the solvent phase; hence, it is desirable that analytical chemists should be familiar with the relevant aspects of solution chemistry.

I. IONS IN SOLUTION

The structure currently proposed for the water molecule implies the existence of four regions of charge residing at the corners of a tetrahedron. Two of these regions are filled by hydrogen atoms and carry a positive charge, the other two carry the lone pairs of oxygen which can participate in hydrogen bonds to two neighboring water molecules. Thus molecules of water are held together by tetrahedrally directed hydrogen bonds which are essentially electrostatic in character.

It can be calculated that the mutual electrostatic potential energy of two molecules in liquid water is approximately 3 kcal which at room temperature is several times the value of the thermal energy (i.e., kT) which governs the Brownian motion of the molecules and strives to maintain a random distribution.

The dipole of each water molecule thus has an appreciable influence over adjacent molecules. However, since hydrogen bonds are considered to be formed and broken several at a time, the structure of liquid water may be envisaged as short-lived clusters (of varying number) of hydrogen-bonded molecules surrounded by nonhydrogen-bonded molecules.

When an ion is introduced into water, the electrostatic attraction between the ion and the dipole of a water molecule can be many times greater than the attraction between two molecules of water. The presence of the ions accordingly has a profound effect on the structure of the surrounding solvent, and one can predict marked orientation of the water molecules which are in contact with ions.

Ions of opposite charge are attracted to each other mainly by electrostatic forces when the degree of separation is comparatively large, but as the interionic distance decreases, short range forces also come into play. Ions having an electronic structure resembling that of a rare-gas atom usually have repulsive short range forces; with other ions there can be reinforcement of electrostatic attraction by covalent bonding, and a new stable species may be formed in the solution.

According to Coulomb's law,

$$\text{Force} = e_1 e_2 / \varepsilon r^2$$

where e_1 and e_2 are the magnitudes of the charges, r is the distance between the charges, and ε is the dielectric constant of the medium.

Because of this attractive force, the concentration of positive ions is slightly higher in the neighborhood of a negative ion, and the concentration of negative ions is slightly higher in the neighborhood of a positive ion, than in the bulk of the solution. To allow for the energy of interaction of the electrical charges on the ions, an extra term is introduced into the thermodynamic equation describing the chemical potential of a solute (μ_{MX}). Thus for a nonelectrolyte

$$\mu_{MX} = \mu_{MX}{}^* + RT \ln m$$

where $\mu_{MX}{}^*$ is the potential the solute (MX) would have in a 1 molal ideal solution. For electrolytes, the molality (m) is replaced by activity (a)

$$a = \gamma m$$

where γ is the molal activity coefficient. If the solute MX dissociates into ions

For the ion M^+, $\mu_{M^+} = \mu_{M^+}{}^* + RT \ln m + RT \ln \gamma_+$
For the ion X^-, $\mu_{X^-} = \mu_{X^-}{}^* + RT \ln m + RT \ln \gamma_-$
$$\mu_{MX} = \mu_{M^+} + \mu_{X^-} = \mu_{M^+}{}^* + \mu_{X^-}{}^* + 2RT \ln m + 2RT \ln \gamma_{\pm}$$

where $\gamma_{\pm}$ represents the mean ionic activity coefficient for the 1:1 electrolyte (i.e., $\gamma_{\pm} = (\gamma_+ \gamma_-)^{1/2}$.

For a 1:1 electrolyte $a_{MX} = m^2\gamma_{\pm}^2$ while for a species such as M_xX_y

$$a_{M_xX_y} = x^x \cdot y^y \cdot m^{(x+y)} \cdot \gamma_{\pm}^{(x+y)}$$

where $\gamma_{\pm} = (\gamma_{+}^x \cdot \gamma_{-}^y)^{1/(x+y)}$.

It is not possible to determine experimentally values for γ_+ or γ_-, but several techniques are available for the determination of mean ionic activity, i.e., $\gamma_{\pm}$.

The tendency of an ion to surround itself with an atmosphere of oppositely charged ions is opposed by the thermal motions of the ions. On this premise, Debye and Hückel derived relationships which may be used to calculate the magnitude of $\gamma_{\pm}$ in extremely dilute solutions

$$\log \gamma_{\pm} = Az_+z_-I^{1/2}$$

where z_+ and z_- represent the charges on the ions.

$$A = (2\pi Ne^6)^{1/2}/2.303(10k\varepsilon T)^{3/2}$$

where $N =$ Avogadro's number, $e =$ charge of an electron, $k =$ Boltzmann constant, ε is dielectric constant of the solution, $T =$ absolute temperature; $I =$ ionic strength $= \frac{1}{2}\sum_i m_i z_i^2 = \frac{1}{2}(m_1 z_1^2 + m_2 z_2^2 + \cdots)$.   In   aqueous solution at 25°, $A = 0.509$.

A similar relationship applies to the activity coefficient for a single ion [(z_+z_-) being replaced by z^2], but in order to extend the theory to higher concentrations, Debye and Hückel introduced a term (a) which takes into account the finite size of the ions. The resulting expression is

$$\log \gamma_i = -\frac{Az^2I^{1/2}}{1 + BaI^{1/2}}$$

where B is another fundamental constant [$10^8 B = (8\pi Ne^2)^{1/2}/(10^3 k\varepsilon T)^{1/2}$] and a may be thought of as the mean effective diameter of the ions in solution or the distance of closest approach.

Since there is no independent method available for evaluating a, many workers prefer to use a modified equation such as

$$\log \gamma_i = -Az^2\left(\frac{I^{1/2}}{1 + I^{1/2}} - CI\right)$$

where the parameter C is given a value such as 0.3.

The concept of activity is extremely important to analytical chemists.

Techniques such as potentiometry provide readings which are directly related to the activity of the species (e.g., pH $= -\log a_{H^+}$); distribution processes are controlled by the magnitude of the activity terms in different phases; and the relevant terms in any equilibrium equation are the activity terms, e.g.,

$$A + B \rightleftharpoons C + D$$

$$K = \frac{a_C\, a_D}{a_A\, a_B} = \frac{[C][D]}{[A][B]} \cdot \frac{\gamma_C \gamma_D}{\gamma_A \gamma_B}$$

Reference to the empirical equations quoted in previous paragraphs indicates that if the species reacting are ions, the values of the activity coefficients will vary with the charge and size of the ion, the charge and concentration of all ions present in solution (ionic strength effect), the temperature, and the dielectric constant of the solvent. The latter can in turn differ from the bulk value due to interaction of the ions with the solvent. The actual concentration of a species present at equilibrium in a chemical reaction can accordingly vary with experimental conditions.

TABLE 10.1

ACTIVITY COEFFICIENT FOR IONS AT 25°C[a]

Ion	Ion size (Å)	Activity coefficient Ionic strength		
		0.002	0.02	0.2
K^+	3	0.964	0.899	0.76
OH^-	3.5	0.964	0.900	0.76
Na^+	4	0.964	0.902	0.78
H^+	9	0.967	0.914	0.83
Pb^{2+}	4.5	0.868	0.665	0.37
$Ca^{2+}, Cu^{2+}, Zn^{2+}$	6	0.870	0.675	0.40
Mg^{2+}	8	0.872	0.69	0.45
$Al^{3+}, Fe^{3+}, Cr^{3+}$	9	0.738	0.44	0.18
$Th^{4+}, Zr^{4+}, Ce^{4+}$	11	0.588	0.255	0.065

[a] Data taken from J. Kiellard, *J. Amer. Chem. Soc.* **59,** 1675 (1937). Reproduced by permission of the American Chemical Society.

In many analytical procedures the influence of changes in the magnitude of activity coefficients can be ignored, since terms may cancel or be duly compensated by the calibration procedures. Accordingly there is a tendency to use concentrations instead of activities in most calculation procedures. This simplification is justifiable in many situations, but should a chemist be developing a new procedure using calculations based on published fundamental data rather than actual experimental data, then due consideration must be given to the relative magnitudes of the activity coefficient terms.

For example, the solubility product of a sparingly soluble compound is generally quoted for a given temperature using experimental conditions where $\gamma_{\pm} \rightarrow 1$. The stability of metal complexes may be reported for similar conditions or for experimental conditions in which the ionic strength is maintained at some constant value, e.g., $I = 1.0\ M$. Calculations based on this tabulated solubility and stability data may indicate that a given precipitate should not dissolve in a certain concentration of complexing agent. In actual practice solution may occur because $\gamma_{\pm} \neq 1$ and different temperatures are used.

A greater appreciation of the possible significance of activity coefficients may be obtained by studying a few typical values. Table 10.1 records the calculated coefficient for a number of selected ions at three ionic strengths. It can be observed from this table that changing the ionic strength of the solutions from 0.002 to 0.2 changes the activity coefficient of univalent ions from about 0.96 to about 0.78. With divalent and trivalent cations the corresponding changes are about 0.87 to about 0.40, and 0.74 to 0.18, respectively.

More extensive compilations of activity coefficients are given in the paper from which the data recorded in Table 10.1 were taken and in the "Handbook of Analytical Chemistry."

II. ION ASSOCIATION, COMPLEX FORMATION AND PRECIPITATION

It has been noted earlier that when ions come close together the energy of mutual attraction is considerably greater than the thermal energy which strives to maintain a random distribution. The result is the formation of an ion-pair, an entity which persists through a number of collisions with solvent molecules. In an ion-pair, the ions may be separated by one or more solvent molecules, but the formation of the ion-pairs can be described by an equilibrium relationship similar to that used to describe complex formation. For example, in a solution of an electrolyte M_xX_y,

$$M^{y+} + X^{x-} \rightleftharpoons MX^{(y-x)+}$$

for which the thermodynamic association constant, K', is given by the equation

$$K' = \frac{\{MX^{(y-x)+}\}}{\{M^{y+}\}\{X^{x-}\}}$$

where braces enclose activities of the species.

If only those ions in actual contact are regarded as forming ion pairs, it has been argued that the association constant is given (to a first approximation) by the equation

$$\ln K = \ln K^\circ - z_+ z_- e^2 / a\varepsilon kT$$

where K° is the association constant of two uncharged particles in the solution. This equation indicates that $\ln K$ should vary inversely with temperature and with the dielectric constant of the solvent.

Analytical chemists must therefore take ion-pair formation into account when considering the effect of temperature and changes of solvent composition on the behavior of electrolyte solutions.

In a solution of a weak electrolyte there may be present free ions (with associated water molecules), ion-pairs and undissociated neutral molecules. In concentrated solutions of electrolytes, a large proportion of the water molecules can become associated with the ions, giving the effect of a reduced amount of "free" solvent for dispersion, and in these circumstances activity coefficients can have magnitudes of greater than unity (cf. Table 1.7 in "Handbook of Analytical Chemistry").

To this stage in the discussion of solution phenomena the possibility of short range forces which lead to a sharing of electrons has been ignored. However, with the ions of transition elements, e.g., water molecules are coordinated by the cation giving species such as $M(H_2O)_m^{n+}$. An increase in pH of the solution can result in the loss of one or more protons yielding $M(OH)_x(H_2O)_{m-x}^{(n-x)+}$. Simultaneously, or alternatively, water molecules can be replaced by other ligand groups.

The equilibria associated with the effects of short range forces may be summarized as follows:

$$M(H_2O)_m^{n+} \rightleftharpoons M(OH)_x(H_2O)_{m-x}^{(n-x)+} + xH^+ \tag{i}$$

In older nomenclature, the equilibrium constant for this equation was known as the hydrolysis constant but as an acid may be defined as any species (cation, neutral molecule, or anion) which can donate a proton, this is merely a special case of acid-base behavior. Accordingly the equilibrium constant may be expressed in the more familiar form, i.e., for the acid HB:

$$K_a = a_{H^+} \cdot a_{B^-} / a_{HB}$$

or for the acid H_2B,

$$K_1 = a_{H^+} \cdot a_{HB^-} / a_{HB}, \qquad K_2 = a_{H^+} \cdot a_{B^{2-}} / a_{HB^-}$$

since $H_2B \rightleftharpoons H^+ + HB^-$; $HB^- \rightleftharpoons H^+ + B^{2-}$; etc.

Because of the large charge to size ratio of the proton, several water molecules are strongly bound to this ion by electrostatic attraction. The predominant species is considered to be $H(H_2O)_4^+$ although there is evidence of other species with different degrees of aquation. For the sake of simplicity the symbol H^+ will be used to represent all the aquated proton species present in solution.

$$M(H_2O)_m^{n+} + L^{p-} \xrightarrow{K_1} M(H_2O)_{m-1}L^{(n-p)+} + H_2O \tag{ii}$$

$$M(H_2O)_{m-1}L^{(n-p)+} + L^{p-} \xrightarrow{K_2} M(H_2O)_{m-2}L_2^{(n-2p)+} \text{ etc.}$$

These are the general equations for stepwise formation of metal complexes, the ligand L being a neutral molecule ($p = 0$) or an anion ($p = -1, -2$, etc.). For the overall reaction (ignoring charges)

$$M + qL \rightleftharpoons ML_q$$

the equilibrium constant $\beta_q = a_{ML_q} / a_M \cdot (a_L)^q$.

Since M can be a proton, a competing reaction is that in which protons become associated with the ligand L.

Vast compilations of equilibrium constants for reactions of type (i) and (ii) have been prepared. The "Handbook of Analytical Chemistry" contains a number of tables but more comprehensive lists of data are contained in publications such as the "Stability Constants of Metal Ion Complexes" (Sillen and Martell, Chemical Society, London, 1964).

These data can be used to calculate the relative amounts of the different species present in solution over a range of experimental conditions (cf. Chapter 11). Such calculations can greatly facilitate the selection of suitable indicators for titrimetric techniques; can aid the selection of experimental conditions to ensure adequate masking of interfering ions; and can assist the selection of conditions for the solvent extraction of uncharged species, etc. In making these calculations, it is desirable to initially consider all the possible interactions between the various ions in solution, simplification by elimination of minor components being attempted only after consideration of the overall picture.

Because of the limited amount of tabulated information on ion-pair association constants, this particular aspect has often to be ignored even though experience may indicate that it should be considered.

$$M_{(aq)}^{n+} + X_{(aq)}^{(p-)} \rightleftharpoons MX_{(aq)}^{(n-p)+} \tag{iii}$$

Equation (iii) represents ion pairing as discussed in preceding sections.

The extreme case of ion pairing occurs when $n = p$ and the degree of attraction between the ions is sufficiently great to cause exclusion of water molecules on collision of ion pairs. This leads to $MX_{(s)}$, the nucleus of a crystal of ionic solid. Subsequent growth of the crystal based on attraction of the constituent ions to the surface of the solid leads to the formation of a visible precipitate.

$$xM_{(aq)}^{y+} + yX_{(aq)}^{x-} \rightleftharpoons M_xX_{y(s)} \tag{iv}$$

The activity of the solid phase is usually accepted as remaining constant, and in precipitation reactions such as described in equation (iv) equilibrium conditions are expressed in terms of the solubility product, K_{S_o} where $K_{S_o} = (a_M)^x(a_X)^y$.

Compilations of solubility products for a wide range of sparingly soluble compounds are included in the same reference books as recommended for reactions (i) and (ii).

A slightly more specialized case of equilibrium involving a solid and a liquid phase occurs in ion exchange reactions (v).

$$Ex^*A^+ + B^+ \rightleftharpoons Ex^*B^+ + A^+ \tag{v}$$

$$K = \frac{\{Ex^*B^+\}\{A^+\}}{\{Ex^*A^+\}\{B^+\}}$$

where braces represent activities and Ex^*A^+ represents the cation A^+ retained in the exchanger (Ex^*) phase.

While the exchange material is insoluble in the solvent, small amounts of solvent are associated with the exchange phase (on the surface and in the internal pores) and in many respects, the phase acts as though it were a concentrated solution of the electrolyte Ex^-A^+ or Ex^-B^+. For this reason it is difficult to propose values for the activity coefficients of species associated with the exchange material, and few equilibrium constants have been tabulated for this type of reaction.

A more realistic practical approach consists of determining experimentally the amount of each ion present in each phase at equilibrium. Substitution of these values in an equation of the same form yields a distribution coefficient which can be used to predict the influence of varying conditions in the solution phase on the amount of material exchanged.

For the distributions of solutes between an aqueous and an immiscible liquid phase, experimental distribution ratios are also considered of greater practical value than thermodynamic equilibrium constants.

III. SOLVENT EXTRACTION AND ADSORPTION

$$AB_{(aq)} \rightleftharpoons AB_{(org)}$$

For the distribution of the species AB between two immiscible solvent phases, equilibrium is described by the partition isotherm

$$p = (a_{AB})_{org}/(a_{AB})_{aq}$$

To overcome the difficulty of assigning numerical values to the activity coefficients for each phase, an alternative expression is often adopted.

$$\text{Distribution coefficient} = [AB]_{org}/[AB]_{aq}$$

This expression adequately describes the distribution of the species AB between the two phases, but it does not allow for secondary reactions which occur in the individual solvent phases. Thus the lower dielectric constant of the organic phase often favors polymerization of the extracted species. In the aqueous phase proton transfer and dissociation of complexes can occur. If one is therefore interested in the extraction of a particular element (A) irrespective of its chemical form, the relationship of interest is the

$$\text{Distribution Ratio } D = [A_T]_{org}/[A_T]_{aq}$$

where A_T represents the total concentration of A found to be present in the two phases by experiment.

The formation of a species which is extractable into the organic phase can involve one or more of the equilibrium processes described in preceding sections.

Covalent compounds and uncharged metal complexes both have a greater affinity for less polar solvents, and solution reactions which yield such products enhance extraction.

For example, iodide (I^-) ions are not extracted from aqueous solution into carbon tetrachloride. However, oxidation of the iodide to iodine leads to almost complete transfer of this compound into the organic layer. Similarly, hydrated metal ions can be rendered extractable by conversion into metal chelate species, such as the metal 8-hydroxyquinolates, e.g.,

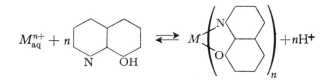

The reduction in polarity required to favor solubility in organic solvents can also be achieved by ion-pair formation. A suitable ion-pair may result if either the ligand associated with the cation, or the anion are bulky organic groups. A special case of ion-pair formation involves the substitution of organic solvent molecules for water molecules in the coordination sphere of the metal ion. The displacement of the coordinated water is favored by the presence of ligands such as halide ions or thiocyanate ions and by the reduction of the activity of the water through the addition of ions having strong polarizing field effects. Thus iron (III) is extracted into diethyl ether (Et_2O) from an aqueous solution which is $>2\ M$ in respect to HCl. The species extracted is probably the ion pair $[H^+Et_2O,\ FeCl_4^-(Et_2O)_2^-]$.

A solvated, uncharged species can also result from reaction of metal ions with organophosphorus compounds such as heptadecyl dihydrogen phosphate. These compounds are usually diluted with an inert solvent such as kerosene (in which they dimerize) and their behavior resembles that of a liquid ion exchanger, e.g.,

$$UO_2^{2+}{}_{(aq)} + 2(HX)_2 \rightleftarrows UO_2X_2 2HX_{(org)} + 2H^+$$

where $(HX)_2$ represents the dimerized organophosphoric acid.

The solute-solvent interactions responsible for the solubility of species in different liquids can range from direct coordination of the solvent to the solute, through dipole-dipole attractions to the weaker van der Waals dispersion effect.

It is equally difficult to define accurately the forces involved when a solid phase is brought in direct contact with a solution of ions.

When the solid phase is an ionic crystal or an ion exchange material, ions from the solution can be attracted to the phase interface by charged groups on the solid. With other solid adsorbents, adsorption can be attributed to the weaker electrostatic fields due to dipole-dipole interactions, or to dipole-induced dipole interactions, etc.

The actual amount of solute adsorbed on a solid is a complex function of the surface area of the solid, the number of active sites per unit area, the strength of the solute-solid bond, the frequency of collision at the interface (i.e., concentration of adsorbable species in the aqueous phase), temperature, polarity of the solvent, number of competing species, etc.

For this reason, the amount of material adsorbed per gram of solid (x/m) is usually related to the concentration (c) of adsorbable species in solutions by empirical relationships such as the Freundlich isotherm,

$$x/m = kc^{1/n}$$

or a Langmuir type of isotherm

$$x/m = k_1 k_2 c / 1 + k_1 c$$

where k, n, k_1, and k_2 have values ascertained by experiment.

The adsorption of an ion or strong dipole on the surface of a solid (forming a primary adsorbed layer) leaves the surface of the solid with some residual charge which can cause orientation of adjacent solvent molecules as well as attracting into this zone oppositely charged species. The composition of this diffuse secondary layer is accordingly different to that present in the bulk of the solution.

This phenomenon is of some significance in analytical procedures in which precipitation is involved. For example, an accepted procedure for the determination of silver ion involves titration with a standard solution containing thiocyanate ions. The fine precipitate of silver thiocyanate is capable of adsorbing significant amounts (e.g., 5%) of the remaining silver ions. This adsorbed material is not titrated unless its contact with the added thiocyanate is assured by vigorous shaking of the heterogeneous mixture.

Many coprecipitation effects may be attributed to the ions which become localized in the primary and secondary adsorption zones. For example, a solution may contain two cations, M_1 and M_2, each being capable of reacting with anion X to yield sparingly soluble products. Calculations based on the concentration of M_1 and M_2 initially present and the amount of X added may indicate that under the chosen conditions only the solubility product of $M_1 X$ should be exceeded. That is, theoretically only $M_1 X$ should precipitate. In practice it may be found that a large proportion of M_2 also appears in the isolated precipitate. A partial explanation for this observation can be based on the argument that the localized concentration of X in the surface zones associated with the precipitate $M_1 X$ is so great that the solubility product of $M_2 X$ is exceeded in this zone.

IV. COMPETING EQUILIBRIA

The preceding sections provide an indication of the types of chemical equilibria which may be encountered in studies involving a solution phase. In chemical analysis, the solution phase usually contains a number of ionic species and several types of chemical equilibria can be operating simultaneously. This situation arises because judicious use of competing reactions helps eliminate interference effects and facilitates the separation of the components of a mixture.

The relative importance of the competing equilibria can be estimated using the tables of equilibrium constants which are available, and such preliminary calculations can eliminate much of the trial and error associated with the development of new procedures.

In the calculations, two basic concepts have to be kept in mind.
1. The solution phase must be electrically neutral.
2. Chemical species may be simultaneously involved in several different equilibrium reactions.

In organizing the data prior to undertaking calculations, one should therefore prepare lists of
1. Charge balances, i.e.,

$$\Sigma \text{ cationic charges} = \Sigma \text{ anionic charges}$$

2. Mass balances, i.e., total conc. of component A added $= \Sigma$ conc. of derived species, e.g.,

$$[H_3A]_T = [H_3A] + [H_2A^-] + [HA^{2-}] + [A^{3-}]$$

3. Equilibrium systems. Initially all conceivable relevant equilibria should be listed. After due consideration of the relative magnitudes of the appropriate equilibrium constants, many can often be eliminated on the grounds that they do not significantly alter the final composition of the solution phase.

The mathematical problem then resolves into the simultaneous solution of all the remaining equilibrium equations. This process can be quite complex and solutions may be sought by introducing simplifying assumptions or by using graphical methods or through computer programming.

Comprehensive coverage of the first two approaches is contained in the monograph by Freiser and Fernando entitled "Ionic Equilibria in Analytical Chemistry," as well as in appropriate sections of text books on volumetric analysis and the "Treatise on Analytical Chemistry, Vol. 1."

In this particular chapter, only comparatively simple examples are considered since the major aim is to demonstrate the type of information which may be gleaned from preliminary calculations. To simplify the treatment further, concentrations ([ ]) and not activities have been substituted in the various equations.

A. SELECTION OF AN INDICATOR IN TITRIMETRIC ANALYSIS

The equivalence point in a titrimetric procedure corresponds to the complete conversion of one species into a new chemical form, e.g., $HA \rightarrow A^-$; $M^{n+} \rightarrow ML^{n+}$, $A^{n+} \rightarrow AX_{(s)}$, etc. In order to detect the point at which this chemical conversion is complete, indicators are usually added to the system.

Indicators are species which react with either the active compound in the assay solution or with the titrant being added. In nature, they resemble the species involved in the chemical reaction, e.g., indicators for acid-base titrations are weak acids or bases, indicators for complexometric titrations are capable of forming complexes with metal ions, etc. The unique property of indicators is that the reacted and unreacted form differ in color. A particular indicator is suitable for locating the equivalence point in a reaction if the conditions in the solution phase at the equivalence point of the titration correspond to the conditions required to convert most of the indicator material from one colored form to the other.

1. *Titration of an Acid (HA) with a Base B*

The following equilibrium systems have to be considered:

Dissociation of acid $HA \rightleftarrows H^+ + A^-$;

$$K_a = [H^+][A^-]/[HA]$$

Dissociation of solvent $H_2O \rightleftarrows H^+ + OH^-$;

$$K_w = [H^+][OH^-]$$

Dissociation of protonated titrant $HB^+ \rightleftarrows H^+ + B$;

$$K_a' = [H^+][B]/[HB^+]$$

Dissociation of indicator $HI_n \rightleftarrows H^+ + I_n^-$;

$$K_{I_n} = [H^+][I_n^-]/[HI_n]$$

It will be observed that $[H^+]$ is common to all these equilibria so that any change in pH influences the relative concentrations of all the species present.

Electroneutrality condition: $[H^+] + [HB^+] = [A^-] + [OH^-] + [I_n^-]$

Mass balances:
$$[HA]_T = [HA] + [A^-]$$
$$[HI_n]_T = [HI_n] + [I_n^-]$$
$$[B]_T = [B] + [HB^+]$$

a. *Indicator Action.* If $[HI_n] \geq 10[I_n^-]$ solution displays color of HI_n species (i.e., Color A).

If $[HI_n] \leq 0.1[I_n^-]$ solution displays color of I_n^- species (i.e., Color B).

Substituting in the appropriate equilibrium equation gives the relationships

$$K_{I_n}/[H^+] \leq 0.1 \text{—solution has Color A } (HI_n \text{ form})$$

$$K_{I_n}/[H^+] \geq 10 \text{—solution has Color B } (I_n^- \text{ form})$$

b. *Titration Curve.* If the pH of a solution is determined during the course of an acid-base titration, a curve similar to that shown in Fig. 10.1 is obtained.

Where excess acid is present, the major contributor to the $[H^+]$ is this acid. Similarly, in the presence of excess base, it is reaction of this base with the solvent which is mainly responsible for the pH. (By substituting mathematical values in the appropriate equilibrium equations it can be shown that the contributions of the indicator system and the dissociation of water are generally too small to warrant consideration.)

While the experimental determination of the titration curve is the best method of evaluating the equivalence point in the titration, good results can be obtained if a suitable visual indicator is available.

c. *Indicator Selection.* In considering an acid-base titration curve, it may be assumed that addition of base will be continued until the final concentration of excess base $([B]_f)$ equals the initial concentration of acid present $([HA]_i)$. For the reverse titration of base with acid, the converse situation may be assumed to apply and the argument re indicator requirements is then valid for both cases.

A suitable indicator for a titration system is one which retains its acid (HI_n) form until the concentration of excess acid has been reduced to at least 1% of its original value. At this stage, for strong acids $[H^+] = 10^{-2}[HA]_i$, and for weak acids' $[H^+] = 10^{-2} K_a$ since $[HA]/[A^-] \simeq 10^{-2}$.

The color of the HI_n form predominates when $K_{I_n}/[H^+] \leq 0.1$; hence a suitable indicator is one in which

$$K_{I_n} \leq 10^{-3} [HA]_i \text{ (strong acids) or } K_{I_n} \leq 10^{-3} K_a \text{ (weak acids)}$$

The second requirement of a suitable indicator is that it should display its basic color $(I_n^-$ form), when the concentration of excess base is about 1% of the original acid value, i.e., when $[\text{base}]_{\text{excess}} = 10^{-2}[HA]_i \,(= 10^{-2}[B]_f)$. At this stage, for a strong base,

$$[OH^-] = [\text{base}]_{\text{excess}} = 10^{-2}[HA]_i = K_w/[H^+]$$

or
$$[H^+] = 10^{-12}/[HA]_i$$

and for a weak base,

$$[H^+] = 10^2 K_a' \text{ since } [HB^+]/[B] \simeq 10^2$$

The basic indicator color is apparent when $K_{I_n} \geq 10[H^+]$; hence a suitable indicator is one for which

$$K_{I_n} \geq 10^{-11}/[HA]_i \text{ for strong bases}$$
$$K_{I_n} \geq 10^3 K_a' \text{ for weak bases}$$

From a list of indicator dissociation constants it is, therefore, possible to select indicators for any type of acid-base titration, the limits in respect to K_{I_n} values being defined in the preceding section. For example, for the titration of

0.01 M HCl with 0.01 M NaOH suitable indicators would be those whose K_{I_n} values lie between 10^{-5} ($10^{-3}[HA]_i$) and 10^{-9} ($10^{-11}/[HA]_i$).

For the titration of 0.01 M CH_3COOH ($K_a = 10^{-5}$) with 0.02 M NaOH, the limits for K_{I_n} are 10^{-8} ($10^{-3} K_a$) to 10^{-9}. For the titration of 0.1 M HCl with 0.1 M NH_3 ($K_a' = 10^{-9}$), the limits for K_{I_n} are K_{I_n} 10^{-4} to 10^{-6}.

Although the above limits have been derived from a consideration of monoprotic species, the same argument applies to polyprotic species provided that the equilibrium constants used refer to the predominant acid and base species present in the equivalence point region. Thus in the titration of sodium carbonate with 0.1 M hydrochloric acid, the species to be considered in selecting the indicator for the first equivalence point are the carbonate (excess base) and bicarbonate (excess acid) ions [K_a' for CO_3^{2-} is 4×10^{-11} (K_2 for H_2CO_3) and K_a for HCO_3^- is 4×10^{-7}]. For the second equivalence point the species to be considered are the bicarbonate ion (base) and excess hydrochloric acid.

Using the relationships derived for monoprotic acids, the indicator range for the second equivalence point (HCO_3^- to H_2CO_3) is $10^{-4} \geq K_{I_n} \geq 4 \times 10^{-4}$, while for the first equivalence point ($CO_3^{2-} \rightarrow HCO_3^-$) the calculated values are $4 \times 10^{-10} \geq K_{I_n} \geq 4 \times 10^{-8}$.

Since the inequations both appear absurd, it could be stated that no visible indicator system would accurately detect these two equivalence points. However, this bland statement requires some modification since the mathematics are based on the assumption that a colored species has to be present in tenfold excess to make its presence obvious (this ratio can be lower) and an accuracy of $\pm 1\%$ is desired. Thus, if one extends the allowed error to $\pm 5\%$ and uses an indicator which only requires a $[I_n^-]/[HI_n]$ change from 5 to 0.2 to give a distinct color change, the above inequations become $10^{-3} \geq K_{I_n} \geq 4 \times 10^{-5}$ and $4 \times 10^{-9} \geq K_{I_n} \geq 4 \times 10^{-9}$. In other words, within these newly defined limits, an indicator with $K_{I_n} = 4 \times 10^{-9}$ would detect the first equivalence point, while for the second end point the K_{I_n} of the indicator can lie between 10^{-3} and 4×10^{-5}.

Instead of listing indicators in terms of their K_{I_n} or pK_{I_n} ($= - \log K_{I_n}$) values, some reference books record the color change interval, e.g., methyl orange, 3.1–4.4. This interval records the highest pH at which the acid form color is distinctly discernible and the lowest pH at which the basic form color predominates. (As a rough approximation, the color change interval equals $pK_{I_n} \pm 1$.)

When an experimental titration curve (e.g., Fig. 10.1) is available for the reaction, suitable indicators can be selected by noting the position of indicator color-change intervals on the curve. The best indicators are those in which pK_{I_n} approximately equals the pH at equivalence point. Where the curve has an extended vertical section in the equivalence region, any indicator whose color-change interval lies wholly on this vertical segment can prove equally satisfactory.

In the absence of an experimental curve, an approximate curve can be prepared by calculation. Using the appropriate equilibrium relationships, pH values are calculated for the conditions predicted to exist at different stages of the titration. Full details of the calculations required are given in most texts devoted to Quantitative Chemistry.

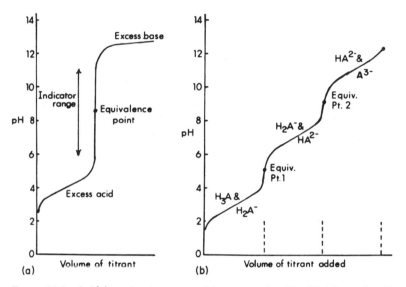

FIGURE 10.1   Acid–base titration curve: (a) monoprotic acid; (b) polyprotic acid.

A study of a series of experimental or calculated titration curves leads to a number of important conclusions.

1. In order to observe a distinct inflection on the curve in the vicinity of the equivalence point, the magnitude of the two controlling equilibrium constants (i.e., K_a and K_a') should differ by $>10^4$.
2. In order to obtain a near vertical inflection (corresponding to a pH change of >2 units in the region of $\pm 1\%$ of theoretical titration figure) the difference in equilibrium constants should be $>10^6$.
3. Where $K_a < 10^6 K_a'$, the inflection segment has a distinct slope and the use of visual indicators to detect the equivalence point introduces "indicator errors."
4. Indicators for any new system must be selected with care. A change in concentration (e.g., 0.1 to 0.001 M), a change in solvent (e.g., water to 50% aqueous ethanol), or a change in titrant (different K_a') may reduce the indicator interval (cf. Fig. 10.1) to such an extent that few available indicators have K_{I_n} which fall within the allowable limits for satisfactory equivalence point detection.

2.   Complexometric Titrations

In the preceding arguments, the term acid has been applied to any species which can donate a proton. A base is any species which can accept a proton. Hence,

$$HA \rightleftarrows H^+ + A^-$$
acid   conjugate base

and for a conjugate acid-base pair (HA, A^-) it can be shown that $K_a K_b = K_w$ where K_a is the dissociation constant of the acid (HA) and K_b is the basic dissociation constant for the base (A^-). This relationship is of value when using published data which record K_a and K_b. For example, the data source may give K_a values for acids (e.g., 2×10^{-5} for acetic acid at 25°) and K_b values for bases (e.g., 1×10^{-5} for ammonia solutions at 15°), while calculations require a knowledge of K_b for the acetate ion (not listed but equals $K_w/K_a = 5 \times 10^{-10}$) and K_a for the ammonium ion (equals K_w/K_b, i.e., 10^{-9}). It will be noted that two different temperatures have been quoted in this example. The K values vary with temperature; hence in accurate work the values to be used are those which correspond to the temperature of the analytical system being considered.

If, instead of the Lowry-Bronsted approach, one defines acids in terms of the Lewis theory, the titration of metal ions with a complexing agent becomes a special case of acid-base behavior. In the Lewis theory, neutralization is the formation of a coordinate covalent bond, e.g.,

$$H^+ + OH^- \rightleftarrows H_2O; \quad K_w = [H^+][OH^-]$$

$$M^{n+} + xL^{(a-)} \rightleftarrows ML_x^{(n-a)+}; \quad K_L = [M][L]^x/[ML_x]$$

$$aH^+ + L^{(a-)} \rightleftarrows H_aL; \quad K_a = [H^+]^a[L]/[H_aL]$$

In complexometric titrations, there may be several inflection points in the titration curve if $x > 1$ (cf. polybasic acids) and the shape of the curve becomes pH dependent when protons compete with metal ions for the ligand group L. A metallochromic indicator is a ligand (I_n) which reacts with the metal to give a colored complex which is less stable than the metal-titrant complex.

$$M^{n+} + yI_n^{b-} \rightleftarrows M(I_n)_y^{(n-yb)+}; \quad K_I = [M][I_n]^y/[MI_{ny}]$$
Color A   Color B

The pH of solutions to be titrated is usually controlled within narrow limits by the addition of buffer salts, and this introduces several more equilibrium reactions which should be listed in order to survey the overall picture. For example, a pH of 10 may be obtained using a mixture of ammonia and ammonium chloride

$$H^+ + NH_3 \rightleftarrows NH_4^+; \quad K_a' = [H^+][NH_3]/[NH_4^+]$$

but the metal ion may react with one of the buffer components, in this case forming metal ammines.

$$M^{n+} + pNH_3 \rightleftharpoons M(NH_3)_p^{n+}; \quad K = [M][NH_3]^p/[M(NH_3)_p]$$

In the titration, the metal ion is, therefore, converted from one coordinated form $[M(H_2O)_q^{n+}$ or $M(NH_3)_p^{n+}]$ to another (ML_x). During the conversion process, the concentration of hydrated metal ions changes and a titration curve may be prepared by plotting pM ($-\log[M]$) against volume of titrant (species L) added. Distinct inflection points are observed if the K values for successive steps vary by $>10^4$.

As in acid–base titrations, the preferred method of obtaining titration curves is by experiment. Indicators can then be selected by reference to these curves. The shape of titration curves can also be evaluated by calculation. The number of competing equilibria produce some mathematical problems which are best solved by the graphical methods described in Ringbom's monograph, "Complexation in Analytical Chemistry" (Interscience, 1963) or other modern books dealing with ionic equilibria.

For a sharp color transition to occur in the equivalence region it can be proposed that the ratio of $[I_n]/[MI_{n_y}]$ should be ≤ 0.1 just before equivalence point (e.g., reaction 99% complete) and ≥ 10 just after equivalence point (e.g., 1% excess titrant). One method of achieving this is to adjust conditions so that $[I_n] \simeq [MI_{n_y}]$ at the equivalence point.

Should the indicator form a 1:1 metal complex $[I_n]/[MI_n] = K_I/[M]$; and when $[I_n] = [MI_n]$, $[M] = K_I$. In this case, an ideal indicator would be one for which $pK_I = pM$ (at equivalence). The appropriate value for pM may be read off the titration curve.

At equivalence point, pM is controlled by the dissociation of ML_x and since $[ML_x] \simeq [M]_T$, $[M] \simeq (K_L[M]_T)^{1/(x+1)}$. Hence, in the absence of an experimental value for pM, it can be suggested that an ideal indicator would be one forming a 1:1 complex having a stability constant K_I of about $(K_L[M]_T)^{1/(x+1)}$.

After equivalence, $[M] = K_L[ML_x]/[L]^x$ and with a 1% excess of ligand $[L] \simeq 10^{-2} \cdot [M]_T$ and $[ML_x] \simeq [M]_T$. Under these conditions, $[M] \simeq 10^{2x}K_L[M]_T^{(1-x)}$. Since for a suitable indicator $[I_n]/[MI_n]$ should be ≥ 10 at this point, then $K_I/[M] \geq 10$. A suitable indicator for the reaction would thus be a compound which forms a 1:1 complex with the metal ion having a stability constant defined by

$$K_I > 10^{(2x+1)}K_L[M]_T^{(1-x)}$$

A suitable indicator must also retain the color of the indicator complex when 99% of the metal ion has reacted with the titrant. At this point $[M] \simeq 10^{-2}[M]_T$ if hydrated ions are being titrated. Should a weak complexing agent be in the assay solution, most of the untitrated metal ion will be in this complex form or associated with the indicator. Thus in the presence of ammonia, $[M]$ just before equivalence point is given by $[M] \simeq 10^{-2}[M]_T \cdot K/[NH_3]^p$ where K is the stability constant for the ammine complex.

For a suitable indicator, $[I_n]/[MI_n] \leq 0.1$, so that K_I should be $\leq 0.1\,[M]$. The upper limit for the stability of the indicator complex is thus given by either

$$K_I \leq 10^{-3}[M]_T \text{ or } \leq 10^{-3}[M]_T \cdot K/[NH_3]^p$$

In this simple example it has been assumed that the pH of the solution is such that competition of protons for the indicator and titrant ligand group can be ignored. When this assumption is not valid a more rigorous mathematical approach is required. Similarly, if the indicator complex has the formula MI_{n_y}, the stability requirements for the indicator are modified by an indicator concentration term. For example, when calculations are based on the measured pM at equivalence and it is assumed that $[MI_{n_y}] = [I_n]$ at this point, the indicator complex should have a stability constant given approximately by the expression

$$pK_I = pM - (\gamma - 1) \log [I_n]$$

B. SELECTIVE PRECIPITATION AND MASKING

A widely used procedure for overcoming interference effects in solutions is selective precipitation of one or more of the ions present. Where several of the ions tend to form sparingly soluble compounds, control of the precipitant concentration may be sufficient to allow isolation of a limited number of ions. In other situations, it is necessary to add a complexing reagent to react selectively with some of the ions and so prevent their interaction with the precipitant.

Since a majority of precipitants and complexing ligands are derived from weak acids, pH control is important since it provides a means of varying the concentrations of the reagent species over several powers of ten.

In predicting the experimental conditions required to yield selective precipitation one must therefore list the equilibria associated with the precipitation reactions, protonation reactions (including buffer systems) and complex formation.

1. *Precipitation from Simple Solutions*

In gravimetric analysis and removal of interferants by precipitation, the aim is to quantitatively remove some ion (e.g., cation A) from the solution. "Quantitatively" may be arbitrarily defined as 99.9% removal, hence the $[A]_f$ after precipitation should be $<10^{-3}[A]_{\text{initial}}$.

To ensure that the concentration of A reaches this limit, an excess of precipitant is usually added. In a solution saturated with the precipitate A_xX_y, equilibrium is described by the following (assuming activity coefficients $\simeq 1$):

$$K_{S_o} = (a_A)^x(a_X)^y$$
$$= [A]^x [X]^y$$

On the basis of this equation, the greater the excess of X, the more efficient should be the precipitation. Opposing this, however, is the reduction in magnitude of the activity coefficients which accompanies increases in ionic strength. This effectively increases the amount of solid in solution. The amount of excess should thus be restricted to the minimum required to ensure quantitative removal.

Consider the precipitation of calcium ions from a 0.1 M solution of $CaCl_2$ using oxalate ions as the precipitant.

$$K_{S_o} = [Ca^{2+}][C_2O_4^{2-}] = 10^{-9}$$

For quantitative precipitation, the product of the ionic concentrations must exceed K_{S_o} even when $[Ca^{2+}] < 10^{-3}[Ca^{2+}]_{initial}$, i.e., $< 10^{-4}$ in this case.

$$\therefore [C_2O_4^{2-}] \geq 10^{-9}/10^{-4} \geq 10^{-5} \text{ at equilibrium}$$

Since the oxalate ion is derived from a weak acid, the concentration of oxalate ions at equilibrium can be varied by altering the pH of the solution.

$$H_2C_2O_4 \rightleftarrows H^+ + HC_2O_4^-; \qquad K_1 = [H^+][HC_2O_4^-]/[H_2C_2O_4] = 6 \times 10^{-2}$$

$$HC_2O_4^- \rightleftarrows H^+ + C_2O_4^{2-}; \qquad K_2 = [H^+][C_2O_4^{2-}]/[HC_2O_4^-] = 6 \times 10^{-5}$$

$$[Ox]_T = [C_2O_4^{2-}] + [HC_2O_4^-] + [H_2C_2O_4]$$

At pH 3.2 ($[H^+] = 6 \times 10^{-4})[HC_2O_4^-]/[H_2C_2O_4] = 10^2$ (from K_1 equation) and $[C_2O_4^{2-}]/[HC_2O_4^-] = 0.1$ (from K_2 equation) $\therefore$ $[H_2C_2O_4]:[HC_2O_4^-]:[C_2O_4^{2-}] = 1:100:10$.

At pH > 3.2, $[H_2C_2O_4]$ will become an even smaller fraction of $[Ox]_T$ and its concentration may be neglected, i.e., calculations at pH > 3.2 can be based on the K_2 equation.

At pH 3.2, if $[C_2O_4^{2-}] \geq 10^{-5}$, $[HC_2O_4^-] \geq 10^{-4}$ and excess $[Ox]_T \geq 1.1 \cdot 10^{-4} M$.

At pH 4.2, if $[C_2O_4^{2-}] \geq 10^{-5}$, $[HC_2O_4^-] \geq 10^{-6}$ and $[Ox]_T \geq 2 \times 10^{-5} M$, etc.

At pH > 3.2, the amount of oxalate to be added need not therefore be much in excess of the stoichiometric requirement—e.g., 50.1 ml 0.1 M oxalate solution should quantitatively precipitate 50.0 ml 0.1 M calcium solution.

At pH 2.2, $[H_2C_2O_4]:[HC_2O_4^-]:[C_2O_4^{2-}] = 1:10:0.1$ so that $[C_2O_4^{2-}] = 0.1/11.1 [Ox]_T$, or in order that $[C_2O_4^{2-}] > 10^{-5}$, excess $[Ox]_T > 1.1 \times 10^{-3}$. That is, for 50 ml 0.1 M Ca^{2+}, > 51 ml 0.1 M oxalate solution is required.

At pH 1.2, $[H_2C_2O_4]:[HC_2O_4^-]:[C_2O_4^{2-}] = 1:1:10^{-3}$ or $[C_2O_4^{2-}] = 0.001/2.001[Ox]_T$.

For $[C_2O_4^{2-}] > 10^{-5}$, excess $[Ox]_T$ required is $> 2 \times 10^{-2}$ and in the above example > 75 ml 0.1 M oxalate solution would need to be added.

In order to obtain larger, purer crystals it is recommended that precipitation conditions be controlled so that initially only a few nuclei are formed, i.e., only a small fraction of the total material should initially precipitate.

In the above case this can be achieved by having the calcium solution so acid that no precipitate is formed when slightly more than the required stoichiometric amount of oxalate salt is added. If the pH of the solution is now slowly increased (e.g., by dropwise addition of ammonia or preferably by slow hydrolysis of urea) a point is reached when some slight cloudiness appears. With continuing increase in pH more and more oxalate ion becomes available for crystal growth and at pH > 3.2 precipitation should be quantitative.

In commercial samples, the assay solution may contain other ions which form precipitates or soluble complexes with the precipitating agent being added. The formation of complexes can be compensated by increasing the amount of excess precipitant by an appropriate amount. To overcome the problem of interfering precipitates one must either remove the interfering species in a preliminary step, use selective precipitation or resort to "masking."

2.   *Selective Precipitation*

If the relative magnitudes of the solubility products of the species capable of reacting with a given precipitant differ significantly, it is sometimes possible to precipitate the least soluble alone merely by control of the pH of the solution.

As an example, let us consider the precipitation of metal sulfides, using hydrogen sulfide as the precipitant. A saturated aqueous solution of hydrogen sulfide at 25° contains approximately 0.1 mole/liter. This dibasic acid dissociates in accordance with the equation

$$K_1 K_2 = 1 \times 10^{-22} = [H^+]^2 [S^{2-}]/[H_2S]$$

Therefore in a saturated solution, $[S^{2-}] \simeq 10^{-23}/[H^+]^2$, and accordingly at pH 1, $[S^{2-}] \simeq 10^{-21}$; pH 2, $[S^{2-}] = 10^{-19}$; etc.

Divalent metal ions form an insoluble sulfide when

$$[M^{2+}][S^{2-}] \geq K_{S_0}$$

and trivalent metal ions precipitate when

$$[M^{3+}]^2 [S^{2-}]^3 \geq K_{S_0}$$

and since $[S^{2-}]$ is directly related to pH, it is possible to prepare a graph such as Fig. 10.2 [pH versus—log K_{S_0} (i.e., pK_{S_0})] which indicates the maximum K_{S_0} a system can possess if a precipitate is to be formed at some selected pH. The single line represents the K_{S_0} values at which precipitation would commence in a saturated H_2S aqueous solution when $[M] = 0.1\ M$ and the broken line represents the corresponding values for when $[M] = 10^{-4}\ M$ (i.e., quantitative precipitation of solutions initially 0.1 M).

For example, for $[M^{2+}] = 0.1$ and pH 3, $K_{S_0} \leq [0.1][10^{-17}]$, i.e., at pH 5, $K_{S_0} \leq [0.1][10^{-13}]$ etc.

These graphs can be used to predict pH conditions suitable for the selective precipitation of ions. For example at pH 3, all divalent ions having pK_{S_0} values >21 will be quantitatively precipitated while divalent ions having $pK_{S_0} < 18$ should not precipitate at all. Alternatively, it may be suggested that if, for example, trivalent metal ions having $pK_{S_0} < 40$ are to be separated from

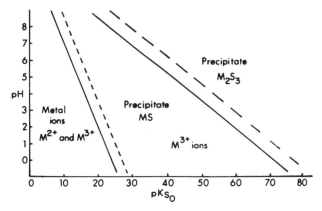

FIGURE 10.2   Graph showing the relationship between pH and the solubility product (K_{S_0}) for the precipitation of metal sulfides from a saturated aqueous solution of hydrogen sulfide: — limits for 0.1 M solutions of metal ions;   – – – limits for $10^{-4}\,M$ solution of metal ions.

trivalent metal ions having $pK_{S_0} > 50$, the pH of the solution should be adjusted to about 5.

Similar graphs can be prepared for the precipitation of metal hydroxides (Fig. 10.3). Decreasing the initial concentration of metal ion by a power of ten shifts the pH, pK_{S_0} plot along the pK_{S_0} scale by one unit. Thus, the series of parallel lines shown in Fig. 10.3 indicates the effect of metal ion concentration on the solubility product limits for precipitation at a given pH.

These simple graphs are very useful for predicting conditions for selective precipitation and the Handbooks and other specialized tabulations provide ample data for the preparation of similar graphs for other systems.

The predictions made using these graphs have to be treated with some caution and need to be proved by experiment since three factors have been ignored in their preparation. The factors not considered are (a) activity coefficients, (b) surface adsorption, and (c) complex formation.

Activity coefficients generally have numerical values of less than unity; hence this factor reduces the tendency of a species to precipitate. Opposing this factor is the higher concentration of precipitant species which occurs in the vicinity of the surface of any precipitate initially formed.

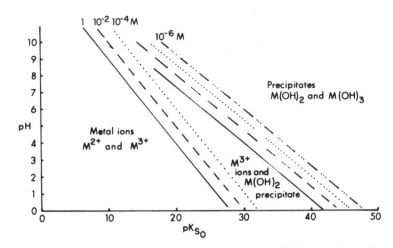

FIGURE 10.3   Graph showing the relationship between pH, solubility product and metal ion concentration for the precipitation of metal hydroxides.

Of greater interest is the effect of complex formation, since the formation of a complex effectively reduces the concentration of free ions available for reaction with the precipitant. The more stable the complex, the further the precipitation line is shifted along the pK_{S_o} scale. The differences in stability of different metal complexes can thus be used to enhance differences in solubility and so facilitate selective precipitation.

Complex formation and ion pairing are common phenomena in solution chemistry; hence in predicting precipitation conditions one must always consider initially all possible equilibria which may occur in solution and later eliminate those which can be shown by calculation to have a minimum effect on the overall reaction of interest.

In some situations complex formation enhances separations. Thus, ammonia is widely used to precipitate the hydrous oxides of iron, aluminium, chromium, etc., since the possibility of coprecipitating divalent metal ions such as copper, nickel, cobalt, zinc, etc., is reduced to a minimum by the tendency of these latter ions to form metal ammine complexes.

The use of a complexing agent to prevent precipitation or to suppress any characteristic reaction of a species in solution is known as "masking."

3.   Masking

Masking has been defined as a process for eliminating interference effects in which a substance is so transformed that certain of its reactions are prevented without any actual physical separation of the substance or its reaction products.

For example, the addition of citric or tartaric acid to a solution containing iron (III) salts prevents the iron from precipitating on the addition of ammonia.

The iron (III) has been masked by forming a soluble complex with the hydroxy-carboxylic acid.

Demasking is the process in which a substance is released from its masked form and regains its ability to enter certain reactions. Thus in determining the zinc content of an ore sample by titration with EDTA a preliminary step in one recommended procedure is the addition of cyanide ions which act as the masking agent for zinc and any other ions which form stable cyanide complexes. In the final evaluation step, the zinc has to be "demasked" prior to titration. One way of achieving this is through the addition of formaldehyde which is capable of causing the decomposition of zinc and cadmium complexes but not decomposition of the more stable complexes.

In predicting the best conditions for masking, using tabulated data, one is again interested in the relative importance of competing ionic equilibria.

As an example, let us consider the masking of iron (III) species by the addition of disodium ethylenediamine tetraacetate to an acid solution which is to be subsequently rendered alkaline by the addition of ammonia.

$$Fe^{3+} + 3OH^- \rightleftarrows Fe(OH)_3 \qquad K_{S_o} = [Fe^{3+}][OH^-]^3 = 10^{-38}$$

$$Fe^{3+} + Y^{4-} \rightleftarrows FeY^- \qquad K_c = \frac{[Fe^{3+}][Y^{4-}]}{[FeY^-]} = 10^{-25}$$

$$H^+ + Y^{4-} \rightleftarrows HY^{3-} \qquad K_1 = \frac{[H^+][Y^{4-}]}{[HY^{3-}]} = 5 \cdot 10^{-11}$$

$$H^+ + HY^{3-} \rightleftarrows H_2Y^{2-} \qquad K_3 = \frac{[H^+][HY^{3-}]}{[H_2Y^{2-}]} = 6 \cdot 10^{-7}$$

$$H^+ + NH_3 \rightleftarrows NH_4^+ \qquad K_a = \frac{[H^+][NH_3]}{[NH_4^+]} = 10^{-9}$$

In the above series, H_4Y represents ethylenediamine tetraacetic acid and less significant equilibria have been discarded.

The mathematical term of major interest in this system is $pFe(-\log[Fe^{3+}])$. If pFe for the principal reaction (precipitation) is represented by $(pFe)_p$ and pFe for the masking reaction (EDTA complex) is represented by $(pFe)_m$, the principal reaction will be favored if $(pFe)_p > (pFe)_m$. Alternatively, the relationship between these two terms can be described by the system proposed by Cheng.

Selectivity Ratio (SR) $= (pM_p)^2/pM_m$
Masking Ratio (MR) $= (pM_m)^2/pM_p$
The principal reaction tends to predominate if $SR > 7$
Masking tends to predominate if $MR > 7$.

It can be observed from the series of equations that both $(pFe)_p$ and $(pFe)_m$ vary with the pH of the solution since $[H^+]$ influences $[OH^-]$ and $[Y^{4-}]$. The value for $[H^+]$ is controlled to a certain extent by the $NH_4^+-NH_3$ buffer system.

Let us assume that masking is to be effective up to a pH of 10, that is, until the ratio $[NH_3]/[NH_4^+] \simeq 10$. At this pH, using the K_{S_0} relationship, $(pFe)_p = 26$. If masking is to be achieved at this pH, $(pFe)_m$ needs to be >26.

If $(pFe)_m > 26$, $[Fe^{3+}] < 10^{-26}$ and hence most of the iron (III) must be present as FeY^- if no precipitate is to be formed. Hence $[FeY^-] \simeq$ initial conc. iron (III).

From the K_c equation,

$$[Y^{4-}] = 10^{-25}[FeY^-]/[Fe^{3+}]$$

and if $[Fe^{3+}] < 10^{-26}$,

$$[Y^{4-}] \geq 10\ [FeY^-],\ \text{i.e.,}\ \geq 1\ \text{if}\ [FeY^-] \simeq 0.1$$

The $[Y^{4-}]$ present in solution is controlled by $[EDTA]_{total}$ and the pH.

At pH 10, from the K_4 equation $[Y^{4-}]/[HY^{3-}] = 0.5$ and from the K_3 equation $[HY^{3-}]/[H_2Y^{2-}] = 6000$. Hence $[Y^{4-}]:[HY^{3-}]:[H_2Y^{2-}] = 3000: 6000:1$ or $[Y^{4-}] = [3000/6001]\ [EDTA]_{total}$.

Thus to give $[Y^{4-}] \geq 10[FeY^-]$, concentration of EDTA must be $\geq 20[FeY^-]$, i.e., $\geq 2\ M$ if $[FeY^-] = 0.1$.

One may therefore state that if the precipitation of iron (III) as the hydroxide is to be prevented in a buffered solution of pH 10, the amount of excess masking agent, EDTA, to be added must be $\geq 20\ [\text{Iron (III)}]_{initial}$.

A repeat of the above calculations using pH 8 indicates that if $[EDTA] \geq 2 \times 10^{-3}\ [\text{Iron (III)}]_{initial}$ no precipitate should form, and the amount of EDTA salt to be added does not need to be much in excess of the stoichiometric amount required to convert $[\text{Iron (III)}]_{initial}$ to FeY^-.

From calculations such as these one can therefore obtain an indication of the amount of masking reagent required as well as predicting pH ranges over which masking would be effective. For example, to prevent precipitation of $Fe(OH)_3$ at pH 11, it can be calculated that $[EDTA]$ should be $\geq 10^4[FeY^-]$, a concentration which would be extremely difficult to achieve unless $[FeY^-]$ is $<10^{-4}\ M$. Hence, it may be stated that EDTA is an effective masking agent for this particular reaction only when pH < 10.

For masking, the principal reaction need not be a precipitation. It could be, e.g., the titration of a metal ion with a complexing agent. In this case $(pM)_p$ would be calculated from the appropriate equation for the dissociation of the metal complex, the concentration terms being substituted representing the experimental conditions which prevail in the vicinity of the equivalence point. Similarly, there are occasions when precipitation is used as the masking process.

The principles of selective precipitation and masking are particularly relevant to procedures in which inorganic species are to be separated by solvent extraction.

C. SOLVENT EXTRACTION

If a solute is added to a system of two immiscible liquids, it tends to distribute itself between the two phases. At equilibrium, the distribution can be expressed in the form of an equation such as

$$a_1/a_2 = \text{constant (at constant } T)$$

where a_1 and a_2 are the activities of the solute in phases 1 and 2 and the constant is known as the partition coefficient (in solvent extraction studies, phase 2 is usually taken as the aqueous phase), or with ideal dilute solutions activities can be replaced by concentrations and this yields the relationship

$$\text{Distribution coefficient } K_D = C_1/C_2$$

However, in most analytical applications, there is little justification for assuming that the ratio of activity coefficients remains constant. In addition, due to competing equilibria in one or both of the phases, the species being extracted may be present in more than one chemical form. Only one of these forms may be extractable; hence the quantity of greater practical significance is the

$$\text{Distribution Ratio, } D = \frac{\text{Total concentration in phase 1}}{\text{Total concentration in phase 2}}$$

D is not a constant, since its magnitude varies with the experimental conditions. D should be as large as possible if extraction into phase 1 is the desired aim, and for quantitative extraction D should be >100.

The difference between K_D and D becomes more apparent if we consider the extraction of some species, e.g., RHN, which is soluble in an organic solvent, and to a lesser extent soluble in water.

$$K_D = [\text{RHN}]_o/[\text{RHN}]_w = \frac{(\text{Wt RHN})_o}{V_o} \times \frac{V_w}{(\text{Wt RHN})_w}$$

where V_o and V_w represent the volumes of the organic and aqueous phases respectively.

The value of K_D at a given temperature should remain constant provided the same two solvents are used and activity effects can be neglected. The actual amount of RHN in the organic phase varies, however, with the phase volume ratio, i.e., V_o/V_w.

In the presence of acids or bases, RHN may gain or lose a proton to give the charged species RH_2N^+ or RN^-. Since the solubility of ionic species in organic solvents is small it may be assumed that the degree of extraction of these charged species is very small compared to the extraction of the uncharged species RHN.

In the aqueous phase, one can therefore expect to find RHN, RH_2N^+, and RN^-.

Distribution Ratio $D = [\text{Total conc. } RHN]_o/[\text{Total conc.}]_w$

$$= [RHN]_o/[RHN]_w + [RH_2N^+]_w + [RN^-]_w$$
$$= [RHN]_o/[RHN]_w\{1 + [RH_2N^+]/[RHN]_w$$
$$+ [RN^-]/[RHN]_w\}$$

In the aqueous phase the relative proportions of the various species of RHN are related in the following way:

For $RH_2N^+ \rightleftarrows H^+ + RHN$;

$$K_1 = \frac{[H^+][RHN]}{[RH_2N^+]}$$

and for $RHN \rightleftarrows H^+ + RN^-$;

$$K_2 = \frac{[H^+][RN^-]}{[RHN]}$$

Using these equilibrium relationships, the distribution ratio expression can be rewritten as

$$D = K_D\{1/(1 + [H^+]/K_1 + K_2/[H^+])\}$$

From this equation it can be observed that the magnitude of D varies with K_D, $[H^+]$ and K_1, K_2. Should the extracted species undergo some transformation in the organic phase (e.g., polymerization) the numerator has to be modified to allow for the equilibria occurring in this phase. The smaller the value of the term in braces, the larger the value of D, i.e., the fraction of the total RHN species in the organic phase is increased. Thus if values for K_1 and K_2 are known, the optimum pH for extraction can be calculated.

For example, if RHN = 8-hydroxyquinoline, $K_1 = 10^{-5}$ and $K_2 = 10^{-10}$ (at room temperature). Using these values a plot of $\phi(= \log \{ \})$ versus pH

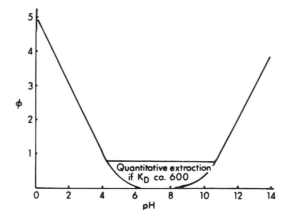

FIGURE 10.4   Plot of $\log \{1 + [H^+]/K_1 + K_2/[H^+]\}(\phi)$ versus pH for 8-hydroxyquinoline where K_1 equals 10^{-5} and K_2 equals 10^{-10}.

has been prepared (cf. Fig. 10.4) and it can be observed that a definite minimum occurs in the region of pH 6–9.

Since $\log D = \log K_D - \phi$, this minimum corresponds to the pH region in which maximum extraction occurs.

For a metal ion to be extracted into an organic solvent, it has first to be transformed into an extractable species (essentially uncharged). This conversion into an extractable species is impeded by the presence of complexing agents which form water soluble charged complexes with the metal ion.

Consider the case where M^{2+} reacts with RHN to form an extractable complex $M(RN)_2$.

$$K_D' = [M(RN)_2]_o/[M(RN)_2]_w$$

In order to determine D all possible equilibria have to be considered. For example, in the aqueous phase

$$M_{aq}^{2+} + 2RHN \rightleftharpoons M(RN)_2 + 2H^+; \quad K_c = [M(RN)_2][H^+]^2/[M^{2+}][RHN]^2$$

$$M_{aq}^{2+} \rightleftharpoons M(OH)_x^{(2-x)+} + xH^+; \quad K_h = [M(OH)_x][H^+]^x/[M^{2+}]$$

$$M^{2+} + y\,HL \rightleftharpoons ML_y^{(2-y)+} + yH^+; \quad K_L = [ML_y][H^+]^y/[M^{2+}][HL]^y$$

In addition, as discussed in the preceding section, $[RHN]_w$ will vary with K_D (for RHN) and pH.

If, for the sake of this example, it is assumed that the metal ion forms only one complex species with each ligand

$$D = \frac{[M(RN)_2]_0}{[M(RN)_2]_w + [M^{2+}] + [M(OH)_x] + [ML_y]}$$

$$= \frac{[M(RN)_2]_0}{[M(RN)_2]_w} \left(\frac{1}{1 + [M^{2+}]/[M(RN)_2] + \dfrac{K_h \cdot [M^{2+}]}{[H^+]^x[M(RN)_2]} + \dfrac{K_L[M^{2+}][HL]^y}{[H^+]^y[M(RN)_2]}} \right)$$

$$= K_D' \bigg/ \left\{ 1 + \frac{[H^+]^2}{K_c[RHN]^2} + \frac{K_h[H^+]^{2-x}}{K_c[RHN]^2} + \frac{K_L[H^+]^{2-y}[HL]^y}{K_c[RHN]^2} \right\}$$

$$= \frac{K_D' \cdot K_c[RHN]_w^2}{\{K_c[RHN]^2 + [H^+]^2 + K_h\,[H^+]^{2-x} + K_L[H^+]^{2-y}[HL]^y\}}$$

i.e., $D = \alpha \cdot K_D'$ where α is <1.

For quantitative extraction into the organic phase, D must be >100. Using the appropriate values for the various equilibrium constants it is, therefore, possible to calculate the pH conditions most suitable for the extraction of the species $M(RN)_2$, i.e., the conditions where $D \to K_D'$ or $\alpha \to 1$.

In solutions containing a number of metal ions selectivity is achieved by "masking"—i.e., the ligand HL is chosen so that the term $K_L[H^+]^{2-y}[HL]^y$ is $\gg K_c[RHN]^2$ for all the metal ions except that one species it is desired to extract.

Values of the formation constants K_c, K_L, etc., may be estimated from published lists of stability constants (the formation constant is the inverse of the stability constant) but the magnitude of the distribution coefficients (K_D, K_D') may have to be determined by experiment.

Unfortunately, the published data are restricted to systems in which the composition of the extracted species can be adequately defined. Many of the extraction systems of interest in analytical chemistry, however, do not fit into this category. For example, the species being extracted may be an ion pair in which one or both of the ions contains molecules of the organic solvent. With such systems, it is difficult to predict the effect of all the solution variables, and the optimum conditions for extraction are invariably determined by experiment.

D. SOLID-LIQUID PHASE EQUILIBRIA

In the distribution of species between a solid and a liquid phase, the magnitude of the equilibrium constant is a function of many variables, including the physical properties (particle size, surface characteristics) of the solid. Accordingly, any data required for calculation or prediction purposes usually has to be evaluated by experiment.

Once this basic information has been elucidated, semiquantitative prediction of the effect of variations in the solution composition can be attempted.

1. *Competition for the Solid Surface*

If a number of species (e.g., A, B, and C) are competing for the active adsorption sites on the surface of a solid, the position at equilibrium may be described mathematically by empirical equations such as

$$(x/m)_A = \frac{k_1[A]}{1 + k_1[A] + k_2[B] + k_3[C]} \; ; \; (x/m)_B = \frac{k_2[B]}{1 + k_1[A] + k_2[B] + k_3[C]}$$

where x/m represents the amount adsorbed per gram. After obtaining approximate values of k_1, k_2 and k_3 by determining the amount of A, B, and C adsorbed when the solution phase contains different concentrations of the species A, B, and C, the empirical equations can be used to predict what might happen in an analytical procedure.

Thus in precipitation titrations a common type of indicator is one which is adsorbed on to the surface of the solid phase formed during the titration, the adsorption leading to a color change on the solid or in the solution. The indicator (e.g., B) has to compete with the ions of the precipitate (e.g., A or C) for positions on the surface. Before equivalence point, the $[A]$ due to unprecipitated A will probably be $>[B]$, and after equivalence point $[C]$ (excess titrant) $> [B]$, but at equivalence the situation may be reversed, i.e., $[B] > [A]$ or $[C]$.

If k_2 for the indicator lies between k_1 and k_3 (for the species A and C), this change in relative concentrations in the vicinity of the equivalence point can result in the indicator being adsorbed $(k_1 > k_2 > k_3)$ or desorbed $(k_3 > k_2 > k_1)$. Should the indicator possess a different color in the sorbed and desorbed form, this transition near the equivalence point may be used to detect the equivalence point.

A suitable adsorption indicator is therefore a species which is either strongly adsorbed [high $(x/m)_B$ value] or nearly completely desorbed [very low $(x/m)_B$] when the concentrations of the reacting species (A and C) are approximately those to be found in solution at equivalence point (e.g., as calculated from Solubility Product data).

In other analytical situations, sorption on the surface of solids yields undesirable effects (e.g., adsorption maxima in polarography, "tailing" in partition chromatography) and these effects may be minimized by adding to the solution phase compounds which are preferentially adsorbed.

2. The Effect of Ligands in the Liquid Phase

The extent of adsorption on a solid surface is related to the equilibrium concentration of adsorbate in the liquid phase. Thus if this concentration is varied by some chemical means, adsorption may be increased or decreased.

Consider, for example, the adsorption (in accordance with a Langmuir type isotherm) of the un-ionized form of a monobasic acid (HA) by some solid (e.g., charcoal).

$$(x/m)_{HA} = \frac{k[HA]}{1 + k'[HA]} = \frac{k[H^+][HA]_T}{(K_1 + [H^+] + k'[H^+][HA]_T)}$$

where $K_1 = [H^+][A^-]/[HA]$ and $[HA] + [A^-] = [HA]_T$.

It can be observed from the derived equation that the extent of adsorption will vary not only with $[HA]_T$ but also with the magnitude of K_1 and the pH of the solution.

The influence of complex formation on adsorption can be illustrated by considering the cation exchange of metal M by a resin (in accordance with a Freundlich type isotherm) in the presence of a ligand HL which reacts with M to form a complex ML_y.

$$(x/m)_M = k[M]^{1/n} = k\left(\frac{K_L[M]_T}{([L]^y + K_L)}\right)^{1/n}$$

where

$$K_L = [M][L]^y/[ML_y]; \quad [M]_T = [M] + [ML_y]$$

$$[L] = K_1[HL]/[H^+] = K_1[HL]_T/(K_1 + [H^+])$$

and k and n are constants.

The amount of metal ion exchanged is therefore related to the total amount of metal ion in the aqueous phase, the stability of the metal complex, and the concentration of excess ligand.

With metals possessing similar charge and hydrated ionic radius, the experimental values of k and n will not differ significantly. Accordingly, if it is desired to separate these metal ions by ion exchange chromatography, it is necessary to add a complexing agent to the mobile phase. The agent selected should form complexes of differing stability with the ions present in the mixture so that separation can be enhanced by differences in K_L values.

Conversely, the magnitude of K_L can be estimated by determining $(x/m)_M$ in a series of studies in which the concentration of ligand L is varied from zero to a number of finite values.

V. EFFECT OF SOLVENT

In the preceding sections, the discussion has been confined almost solely to the behavior of ions in aqueous solution. However, the use of nonaqueous solvents is increasing; hence it is appropriate to conclude this chapter by considering the effect of a change of solvent.

In aqueous solution, interaction between the charged ions and the dipoles of water molecules results in orientation of the nearby solvent and close association of the ions with at least one layer of water. When water is replaced with a less polar solvent, solvation becomes less pronounced. At the same time, the attraction between oppositely charged species is inversely proportional to the dielectric constant of the medium; hence substitution of a solvent of lower dielectric constant results in greater attraction between ions, that is, increased ion pairing and decreased solubility for most ionic species.

On the other hand, the solubility of covalent or weakly polar compounds tends to increase as the polarity of the solvent decreases. An ancient adage states that "like dissolves like"; hence, polar solvents dissolve polar solids; aliphatic solvents are preferable for aliphatic compounds; and aromatic solvents are best for aromatic species. There are many exceptions to this simple rule since solution is a complex process, but it remains as a useful guide.

The variation in solubility brought about by a change in solvent can be used to advantage in analytical chemistry. For example, if an organic precipitant is insoluble in water, it may dissolve in alcohol or glacial acetic acid. On addition to an aqueous solution the probability of some of the reagent appearing again as a solid is high, and the final product is often contaminated with excess reagent, but the use of the two miscible solvents does facilitate the required initial contact between the reacting species.

In separations based on the precipitation of sparingly soluble compounds, the quantitative isolation of compounds which are moderately soluble in water may be obtained by adding some miscible solvent such as ethanol or acetone to the solution. Similarly, the addition of small amounts of these solvents sometimes enhances the sharpness of the indicator change in titration studies by reducing the degree of ionization of reaction products.

In adsorption studies, a change in solvent can result in increased selectivity. Adsorption on a solid is greatest for solutes having the least tendency to dissolve in the surrounding liquid, and in addition the solutes may have to compete with the solvent for the available adsorption sites.

Of greater interest than solubility effects are the changes induced in reactions where there is marked interaction between solute and solvent. For example, consider the solution of a Bronsted acid (i.e., proton donor) in a solvent.

$$HB_1 \quad + \quad B_2 \leftrightarrows HB_2^+ + B_1^-$$

Added acid   Solvent   Acid   Base

$$K = \frac{[HB_2][B_1]}{[HB_1][B_2]}$$

In aqueous solution, $[B_2]$ represents $[H_2O]$ and as the magnitude of this term remains virtually unchanged as a result of the interaction, it is incorporated with the equilibrium constant to give $K_a = K[B_2]$.

The K_a values, therefore, represent a measure of the extent of the interaction between the acid and water. For aqueous systems, it may be said that two bases, water, and the conjugate base B_1, are competing for the proton. If B_1 has a high affinity for protons, not only is there little tendency for protons to be transferred to the solvent, but any base B_1 formed by neutralization tends to remove protons from the solvent

$$B_1^- + H_2O \leftrightarrows HB_1 + OH^-$$

One result of this secondary reaction is a minimal change in $[H^+]$ during the neutralization of acids having K_a values of $<10^{-7}$. This means that visible indicators cannot be used to detect the equivalence point in titrations involving very weak acids or weak bases.

This problem can be overcome by using solvents other than water. For example, if B_1^- has a high affinity for protons, the conjugate acid HB_1 must dissociate to only a very slight extent in water. If a solvent having a greater affinity for protons is substituted for water (e.g., pyridine, morpholine, dimethylformamide, ethylene diamine, etc.) the transfer of protons to the solvent can become significant, yielding mainly HB_2^+ in the solution. This is equivalent to having a high concentration of H_3O^+ in water, that is, interaction of HB_1 with a basic solvent yields a solution which is similar to a strong acid solution in the water system. The protonated solvent HB_2^+ may then be titrated with a strong base (e.g., sodium-methoxide in benzene) using a visible indicator.

Weak bases (species which do not tend to accept protons from water) can be caused to fully protonate through solution in an acid solvent (e.g., glacial acetic acid).

$$B_1 \quad + \quad HB_2 \leftrightarrows HB_1^+ + B_2^-$$

Weak base   Acid solvent

The high concentration of solvent base (B_2^-) is equivalent to a high concentration of OH^- in the water system. The B_2^- can thus be titrated like a strong base using protonated solvent (e.g., $H_2B_2^+$) as titrant and a visible indicator to detect the equivalence point. Thus compounds such as primary, secondary, and tertiary amines; alkaloids; sulfonamides; and amino acids can be titrated with a solution of perchloric acid in glacial acetic acid if the bases are first dissolved in glacial acetic acid. Crystal violet is a suitable indicator in many cases.

Changing the nature of the solvent can thus profoundly influence the position of equilibrium on dissolution of a solute and this effect can facilitate quantitative determinations. Interest in nonaqueous titrations continues to increase, but the technique has been restricted mainly to acid–base systems. A nonaqueous solvent should be considered whenever a very weak acid or base is encountered, and ample guidance can be gained from the recent monographs on this field.

Less attention has been paid to the possible advantages of changing the solvent in oxidation–reduction and precipitation reactions, and this remains a promising field for future research. There is a growing volume of information on chemical reactions in nonaqueous solvents and this may be used as basic material in predicting the possible effect of a change of solvent on a chemical reaction.

VI.   TUTORIAL PROBLEMS

1. a. There are 56 mg of iron dissolved in 100 ml of molar hydrochloric acid. Calculate the approximate magnitude of the activity coefficient for the iron (III) in this solution.

    What would be the effect on the activity coefficient of diluting the solution one hundredfold?

   b. One gram of low grade ironstone (5.6% Fe) is fused with 5.3 gm Na_2CO_3. The melt is dissolved in 50 ml 4 M H_2SO_4 solution. What is the ionic strength of this solution and what would be the activity coefficient of the ferric ion in this case?

    The solution is then neutralized by the addition of 50 ml 3 M NH_3. At what pH might the iron begin to precipitate as the hydrous oxide, given that the solubility product for $Fe(OH)_3$ is 10^{-38} and assuming that concentrations can be substituted in the K_{S_0} equation? If activity terms are substituted for concentration, at what pH will precipitation commence?

   c. Calculate the pH of a 0.01 M HCl solution.

    If 100 ml aliquots of this acid are treated with 41, 82, or 123 mg of sodium acetate, what will be the pH of these aliquots? (K_a for acetic acid $= 10^{-5}$).

    Each aliquot is neutralized by the addition of 40 mg solid NaOH. What will be the pH of the final solutions (ignoring activity coefficients)? If an accurate pH meter was used to measure the pH, would one expect any differences in the readings? If so, why?

2. a. In aqueous solution oxalic acid can be present as $H_2C_2O_4$, $HC_2O_4^-$ and $C_2O_4^{2-}$. ($K_1 = 6 \times 10^{-2}$, $K_2 = 6 \times 10^{-5}$). Prepare a diagram showing the relative percentage of each species present as the pH is changed from 0 to 8.

   b. If 4.0 gm $Na_2C_2O_4$ are added to 100 ml of a solution which is 0.1 M in respect to calcium chloride and disodium ethylenediamine tetraacetate (buffered to pH 7), would a precipitate form? (pK_L for $Ca \cdot EDTA = 10^{-10}$; K_{S_0} for $CaC_2O_4 = 10^{-9}$).

   c. An acid solution contains the chlorides of As(III), Cu(II) and Cd(II) (0.1 M in respect to each). Using the data available in reference works, suggest an acidity at which As(III) could be separated from the divalent metal ions on saturation with hydrogen sulfide. At what maximum acidity would you predict that both Cu(II) and Cd(II) could be quantitatively precipitated as the sulfide?

   If a solution containing the Cu and Cd ions (about 0.01 M respectively) is adjusted to pH 5 and made 0.1 M in respect to cyanide ions before the admission of H_2S gas, what would happen after saturation of the solution with this reagent?

3. a. In studies of the distribution of solutes between water and chloroform the distribution coefficient for an organic precipitant (HR) was found to be 400 at room temperature. K_D for the metal compound MR_2 was found to be 700. Data tables indicate that K_a for HR equals 10^{-9}; $K_L = [M^{2+}][R^-]^2/[MR_2] = 10^{-12}$.

   If 50 ml of chloroform (containing 2.5×10^{-3} moles HR) are brought to equilibrium with 50 ml 0.01 M metal chloride solution, prepare a diagram showing the percentage of metal (M) present in the organic phase as a function of the pH in the aqueous phase.

   b. Suggest suitable conditions for the separation of metal M from metal N if the stability constant for NR_2 equals 10^{-8} (K_D equals 700).

   c. Outline a method for separating the nitrates of Ce(IV) from Th(IV). (Percentage extraction curves for the HNO_3 system are given in *Z. Anorg. Chem.*, **263** (1950) 146 and in reference works, e.g., Morrison and Freiser). Would the addition of some lithium nitrate to the aqueous phase influence the position of these distribution curves?

4. a. K_1 and K_2 for the acid H_2A are 10^{-3} and 10^{-10} respectively. Indicators are required to detect both equivalence points in the titration of an 0.1 M solution of this acid with 0.5 M potassium hydroxide. What properties should suitable indicator systems possess? From reference books suggest some appropriate indicators.

b. An organic dye is found to react with a number of divalent metal ions yielding 1:1 complexes (of different color) having stability constants of the order of 10^{-7}.

If this dye is to be used as a metallochromic indicator, how stable should be the complexes formed by the divalent metal ions with the titrant?

The metal ions form ammine complexes having stability constants ranging from 10^{-4} to 10^{-9}. Would the presence of an excess of ammonia (0.1 M) affect the suitability of the dye as an indicator for the titration of 0.01 M solutions of these divalent metal ions?

What information would be required to predict the best pH for the titrations?

5. a. In the titrimetric determination of the nickel content of steel, one gram samples are dissolved in mineral acid and some citric or tartaric acid is added before the solutions are adjusted to pH 8 by the addition of ammonia. Using data from reference books calculate the minimum weight of the separate hydroxy acids which should be present to prevent precipitation of hydrous iron oxide.

b. Discuss the statement: "Selective precipitation followed by filtration and washing is the most widely used procedure for separating macro (e.g., 0.5 gm) quantities of chemical species. Where the amount of material to be separated is present in micro quantities (e.g., mg), precipitation with an organic reagent followed by solvent extraction is often preferable. Both procedures are rendered more selective by the judicious use of masking agents."

6. a. The aluminum content of a commercial product was determined by three people. All precipitated the aluminum as the hydrous oxide but precipitation conditions varied. The novice chemist merely adjusted the pH after solution of the initial sample—hence precipitation occurred in a solution having a high concentration of electrolytes (0.1 M) and the final result was 5% high. Chemist No. 2 diluted the solution tenfold before precipitation and his results were only 1.5% high. The senior chemist reprecipitated the initial product and reduced his error to 0.5%. Assuming the errors are due to cation adsorption in accordance with a Freundlich isotherm $(x/m = kc^{1/n})$ use the first two results to calculate approximate values for k and n and suggest a possible value for the concentration of adsorbable species in the reprecipitation solution.

b. The major impurity in the solution forms a hexammine complex for which $K_L = 10^{-14}$. When the pH of the assay solutions was adjusted by adding ammonia until there was an excess equivalent to a 0.01 M solution, the results obtained by the three chemists were much closer. Assuming that only the hydrated metal ion was adsorbed, what percentage of contamination would be expected in each result? Would you expect the experimental values to be as low as predicted?

7. a. The "Handbook of Analytical Chemistry" lists some pK_a values for acids present in alcohol-water mixtures. Assuming that these data can be treated in the same way as pure aqueous solutions, prepare a series of titration curves for at least two acid systems to show the effect of adding alcohol on the shape of the curve when the titrant is KOH dissolved in the same solvent. Suggest visual indicator systems which may prove suitable for detecting the equivalence point in the various solvent mixtures.

   b. Table 2.21 in the Handbook lists some masking agents which have been used in chemical analysis.

      i. For any selected element (e.g., Ni) arrange the masking agents in order of the stability of the ion complexes formed by the metal ion with the masking agent.

      ii. Rearrange the table in terms of masking agents, i.e., list the elements masked by a particular reagent.

      Indicate how these two approaches would facilitate the selection of a specific masking agent for a particular reaction.

8. a. Using nonaqueous solvents, amino acids can be determined either through their basic or acidic properties. Outline suitable methods for the analysis of such products (use the literature to give specific examples).

   b. Write an essay discussing the possibility of using nonaqueous solvents for determinations other than acid-base reactions.

References

A. Albert and E. P. Serjeant, *Ionization Constants of Acids and Bases*, Methuen, London, 1962.

L. F. Andrieth and J. Kleinberg, *Non Aqueous Solvents*, Wiley, New York, 1953.

J. Butler, *Ionic Equilibrium*, Addison-Wesley, Reading, Massachusetts, 1964.

G. M. Fleck, *Equilibria in Solution*, Holt, London, 1966.

H. Freiser and Q. Fernando, *Ionic Equilibria in Analytical Chemistry*, Wiley, New York, 1963.

I. M. Kolthoff and P. J. Elving (eds.), *Treatise on Analytical Chemistry*, Wiley (Interscience), New York, 1968.

J. Kucharský and L. Safárik, *Titrations in Non-aqueous Solvents*, Elsevier, Amsterdam, 1965.

L. Meites (ed.), *Handbook of Analytical Chemistry*, McGraw-Hill, New York, 1963.

G. H. Nancollas, *Interactions in Electrolyte Solutions*, Elsevier, Amsterdam, 1966.

A. Ringbom, *Complexation in Analytical Chemistry*, Wiley (Interscience), New York, 1963.

L. G. Sillen and A. E. Martell, *Stability Constants of Metal Ion Complexes*, Chem. Soc., London, 1964.

H. H. Sisler, *Chemistry in Non-aqueous Solvents*, Reinhold, New York, 1961.

G. Swarzenbach, *Complexometric Titrations*, Methuen, London, 1957.

O. Tomicek, *Chemical Indicators*, Butterworth, London, 1951.

11

IONIC REACTIONS—TITRIMETRIC ANALYSIS

	I. Introduction	335
	II. Acid–Base Titrations	338
	A. Calculation of pH Changes	338
	B. Sources of Error	343
	C. Titration of Polybasic Acids	345
	D. Titration of Amino Acids	348
	III. Nonaqueous Titrations	350
	A. Classification of Solvents	350
	B. Solvent Selection and Titration Curves	353
	C. Typical Applications	356
	IV. Complexometric Titrations	358
	A. Lability and Stability of Complexes	358
	B. Chelometric Titrations	360
	C. pH Effects	361
	D. Solution Variables and Indicators	362
	E. Sources of Error	364
	F. Components of a Complexometric Titrimetric Procedure	365
	V. Precipitation Titrations	367
	A. pM Changes during Titration	367
	B. Indicators for Precipitation Titrations	369
	C. Applications	373
	VI. Oxidation–Reduction Titrations	374
	A. Oxidation and Reduction	374
	B. Standard and Formal Reduction Potentials	376
	C. Free Energy Changes	377
	D. Potential Changes	381
	E. Redox Indicators	383
	F. Selected Applications	384
	VII. Tutorial Questions and Problems	385
	References	389

I. INTRODUCTION

It will be apparent from the preceding chapter that reactions in aqueous solution invariably involve ions, and that the position of the system at equilibrium is an important factor. Since many analytical procedures are based on solution reactions, a general knowledge of solution chemistry is invaluable to an analytical chemist.

The two analytical techniques which make the greatest use of ionic reactions are titrimetric and gravimetric analyses. These procedures are sometimes combined under the general heading of classical chemical analysis and are widely used for routine analysis as well as for monitoring the efficiency of instrumental procedures.

This chapter is concerned solely with the various subclassifications of titrimetric procedures, while Chapter 12 discusses the principles of gravimetric analysis.

Another important application of solution reactions is the formation of stable colored species suitable for use in colorimetric determinations. The many aspects of colorimetry have been discussed in Chapter 6.

Some ionic reactions can be induced by supplying electrons from an external source, e.g., a battery. Such reactions are the basis of a wide range of electroanalytical techniques and these are described in Chapter 13.

A more recent development is chemical analysis based on the measurement of the rate of a solution reaction. This kinetic approach marks a distinct change in philosophy, since previous generations of analytical chemists considered only systems at equilibrium to be of value, and any slow processes were generally discarded in surveys of potential procedures.

As chemical advances often involve a combination of old and new ideas, it seemed appropriate to include a section on the kinetic approach in the same chapter as gravimetric procedures (Chapter 12).

In titrimetric methods of analysis the volume of a solution of known composition required to react quantitatively with the assay species is measured accurately. The titrant preferably reacts in a selective manner so that preliminary separation procedures are unnecessary.

The titrimetric procedures are commonly subdivided in terms of the type of chemical reaction involved; hence there are four major categories: acid-base, complex formation, oxidation-reduction, and precipitation titrations. The increased use of solvents other than water has introduced another category, known as nonaqueous titration procedures.

For all categories, the basic reaction should satisfy a number of general requirements, the more important being:

1. The reaction between the two species must proceed to completion at an extremely fast rate.
2. The reaction should proceed in a stoichiometric manner, i.e., the precise nature of the products should be known and should not alter with changes in experimental conditions.
3. The position of complete reaction should be detectable, i.e., an indicator should be available to permit detection of the equivalence point in the titration.

4. It must be possible to determine, with certainty, the exact concentration of the added solution. For this purpose, a supply of standard chemicals or materials is highly desirable.

It is an experimental fact that there is a rapid change in the concentration of some component of a reaction in the vicinity of the equivalence point.

To detect this equivalence point it is necessary to have available some instrumental technique which shows a sharp meter response with such changes in conditions. Alternatively, the equivalence point may be observed through the addition of another chemical species (designated an indicator) which responds to the rapid change in concentration by changing some visible physical property, e.g., color, solubility, fluorescence, etc.

As outlined in the previous chapter, the selection of a suitable indicator for a given titration requires a knowledge of the equilibrium conditions in the solution at all stages of the titration, a knowledge of the mechanism of indicator behavior, and an understanding of solvent effects.

Solvent effects are sometimes responsible for a lack of stoichiometry in the titration reaction. In some systems, reaction produces a range of products, the proportion of each varying with experimental conditions. Such reactions can only be used in titrimetric procedures if the experimental parameters can be standardized to a sufficient extent to ensure that the overall process is adequately described by one, single, balanced equation.

Apart from disturbing stoichiometric considerations, competing equilibria can cause a diminution in the magnitude of the concentration change associated with the equivalence point. This can be responsible for indistinct, erroneous, or false end points in a titration. In cases where the competitive effect is not so marked, the influence is limited to a more restricted choice of indicators.

The ability of an indicator system to accurately detect the equivalence point in a titration is preferably confirmed by analyzing a series of standard materials. Standards therefore play two roles: they permit evaluation of the titrant concentration and they facilitate evaluation of the suitability of different methods of end-point detection.

The specification that a reaction should be fast is essentially a matter of convenience. Slower reactions can be used in titrimetry, provided that one allows sufficient time for reaction to go to completion after each individual addition of titrant. This procedure is time-consuming and tedious, and in most cases attempts are made to speed up the rate of reaction by increasing the temperature or by the addition of a suitable catalyst.

In principle, titrimetry is probably the simplest analytical technique and despite the upsurge of interest in instrumental methods is still used extensively. There are fields of study in which the technique is superior to all alternatives, while successful application to other systems requires rigid adherence to some established experimental procedure. In most systems accuracies better than $\pm 1\%$ are readily attainable.

Detailed descriptions of titrimetric methods are recorded in a large number of reference books (a selected number are listed in Chapter 1) and these books generally provide methods appropriate to most of the applications commonly encountered.

The aim of this chapter is, therefore, to outline the chemical principles involved in different types of titrations, so that a new procedure can be evolved should the literature search prove fruitless.

II.   ACID-BASE TITRATIONS

A.   CALCULATION OF pH CHANGES

By the Bronsted-Lowry theory, an acid is any species capable of donating a proton and a base is any species capable of accepting a proton.   The basic equation for an acid-base reaction may thus be written

$$HA + B \rightleftharpoons HB^+ + A^-$$

Acid   Base   Conjugate   Conjugate
                      acid   base

To be suitable for use in a titrimetric procedure, base B and HA must react rapidly and the chemical equilibrium should lie far to the right, i.e., there should be an appreciable free energy change.

If this condition is met then in the vicinity of equivalence point there is usually a marked change in hydrogen ion concentration in the solution.

Prior to the equivalence point, the solution contains excess acid and the hydrogen ion concentration is related to the position of the equilibrium:

$$HA + S \rightleftharpoons HS^+ + A^-$$

Acid   Solvent   Conjugate base

where $S = H_2O$, $K_a = [H_3O^+][A^-]/[HA]$ or $[H_3O^+] = K_a[HA]/[A^-]$.

The ratio $[HA]/[A^-]$ varies during the course of the titration and K_a values vary with the nature of the acid involved and the temperature.

Figure 11.1 shows the effect of different K_a values on the pH $(-\log [H^+])$ changes which occur during titration.

The calculation of the pH of solutions of acids, bases, weak acids, weak bases, and buffer solutions is an integral part of many introductory chemistry courses; but in analytical usage allowance has to be made for the change in concentrations which occurs during the course of a titration.

A number of acids ionize completely when dissolved in water (represented by $K_a = \infty$), and the concentration of hydrogen ions in solution can then be taken as numerically equal to the concentration of acid present.   In a titration this acid concentration varies through neutralization and dilution.

Consider the case of 100 ml 0.1 M HCl treated with 20 ml 0.1 M NaOH.

After reaction 80 ml of the original acid will not have been neutralized, and the solution volume will have increased by 20 ml. The acid concentration at this stage is, therefore, equal to $0.1 \times 80/120$.

In general terms, where $K_a = \infty$,

$$[H^+] = \text{(Volume of acid unreacted)(initial molarity)}/\text{Total volume of solution}$$

Thus, by considering the volume of acid remaining and the total volume of solution, one can readily calculate the pH of the solution as a function of the volume of titrant added.

On the other hand, many of the acids encountered in chemical analysis have acid dissociation values $<10^{-1}$ and calculations are then based on the equilibrium equation. Such solutions contain measurable amounts of acid and conjugate base and are accordingly buffered solutions. The pH of buffer solutions are virtually independent of the total volume, but in the calculation of titration curves, the concentrations of acid and conjugate base substituted into the equilibrium equation must be relevant to the conditions in solution at that particular point during the titration.

For example, if 20 ml 0.2 M NaOH are added to 100 ml 0.1 M CH$_3$COOH ($K_a = 10^{-5}$), 40 ml of the original acid are neutralized to form a corresponding amount of acetate ion.

$$\therefore [CH_3COOH] = 60 \times 0.1/120 = 0.05$$

and
$$[CH_3COO^-] = 20 \times 0.2/120 \text{ or } 40 \times 0.1/120 = 0.03^{\cdot}$$

Substitution into the acid dissociation equation yields

$$[H^+] = K_a[CH_3COOH]/[CH_3COO^-] = 10^{-5} \times 0.05/0.03^{\cdot}$$

$$= 1.5 \times 10^{-5}$$

$$pH = 4.82$$

The calculations can be applied equally well to all species which are classified as acids or bases by the Bronsted definition. For example, in the titration of ammonia by hydrochloric acid, the solution contains ammonium and chloride ions. The ammonium ion is a weak acid and the method used to calculate the pH is the same as that used for a solution of acetic acid. Similarly, at the equivalence point in the titration of acetic acid with sodium hydroxide, the solution of sodium acetate present can be treated simply as a solution of acetate ions, a weak base having a K_b value of about 10^{-9}.

In solutions containing free base, e.g., after equivalence point in an acid-base titration, the pH is controlled by the amount and type of base present. In aqueous solution, bases tend to react with the solvent in accordance with the equation

$$B + H_2O \rightleftarrows HB^+ + OH^-;$$

the position of equilibrium being described by the relationship

$$K_b = [HB^+][OH^-]/[B]$$
$$[OH^-] = K_w/[H^+] = K_b[B]/[HB^+]$$

In a titration the ratio $[B]/[HB^+]$ increases as excess base is added and the pH of the resulting solution becomes a function of the stage of titration and the magnitude of K_b.

The effect of K_b on the pH changes is shown in Fig. 11.1.

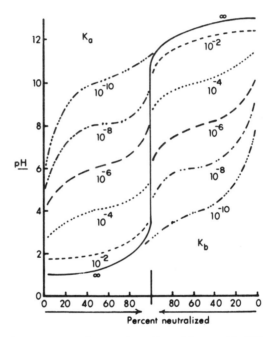

FIGURE 11.1   Titration curves for 0.10 M acids and bases with different dissociation constants. The dotted base curves, on the right hand side, have been plotted in the reverse direction to yield the equivalent of titration curves for a wide range of combinations.

The hydroxides of sodium and potassium ionize completely in water ($K_b = \infty$) and with these bases the concentration of hydroxyl ions is numerically equal to the concentration of base present. That is, in a titration where there is excess strong base

$$[OH^-] = (\text{Vol. of base unreacted})(\text{initial molarity})/\text{Total volume}$$

For a solution containing a weak base and no significant amount of the conjugate acid (e.g., a solution of sodium acetate) an approximate pH value can be calculated by assuming that reaction with the solvent produces equal amounts of HB^+ and OH^-, i.e., $[HB^+] = [OH^-]$. Substitution in the basic equation then yields the relationship

$$[OH^-]^2 = K_b[B]$$

Similarly, a solution containing a weak acid (and no conjugate base) can be assumed to react with the solvent to give equal amounts of H^+ and A^-. The approximate pH of the solution can then be calculated using the equation

$$[H^+]^2 = K_a[HA]$$

All calculated pH values tend to differ from experimental values because concentrations rather than activities are usually substituted into the equations. For this reason, a certain degree of simplification and approximation is permissable in this context.

In analytical chemistry, one is concerned mainly with establishing the form of a titration curve. For titrations in which there is a single proton transfer, the total titration curve normally resembles the shape obtained by bringing together appropriate K_a, K_b segments from Fig. 11.1.

For example, in the titration of a strong acid ($K_a = \infty$) with a strong base ($K_b = \infty$), combination of the appropriate ∞ segments indicates a large (7 pH unit) change in the hydrogen concentration in the vicinity of equivalence point. Such a change would be readily detected experimentally using a pH meter, and a range of indicators would meet the selectivity requirements outlined in Chapter 10. On the other hand, if the acid being titrated has a K_a value of 10^{-6}, the vertical segment is reduced in size to two pH units. This would be readily detected instrumentally but the number of suitable indicators would be extremely limited. Should an attempt be made to titrate this acid with a base having a K_b value of 10^{-4}, it can be observed from Fig. 11.1 that the resultant titration curve would have no vertical segment; selection of a suitable visual indicator would be impossible; instrumental detection of the equivalence point would be difficult and liable to be in error.

One may generalize by stating that as the values of K_a and/or K_b for the reactants become numerically smaller, the change in pH in the vicinity of equivalence point is smaller and indicator selection is more critical.

In evaluating the potential of any proposed acid-base titration procedure, the first step is therefore to ascertain (from the literature or by experiment) the K_a and K_b values for the two reactants. Where there is a choice of titrant, the aim should be to select the combination which produces the largest value for the product $K_a K_b$, since maximizing this factor leads to an optimum pH change at equivalence point.

The discussion to this stage has considered only monoprotic species and conditions where one acid or base is responsible for the hydrogen ion concentration.

Let us now consider the calculations associated with systems in which a weak acid interacts with a weak base, instead of the solvent.

For example, at the equivalence point in the titration of acetic acid with ammonia, the solution contains equivalent amounts of ammonium ions and acetate ions. Collision between these ions tends to result in proton transfer

$$NH_4^+ + Ac^- \rightleftharpoons NH_3 + HAc$$

This equation contains two conjugate acid-base pairs, and the equilibrium relationships between the components of each pair have to be satisfied simultaneously. These may be written as

$$K_1 = [NH_3][H^+]/[NH_4^+] \quad \text{and} \quad K_2 = [Ac^-][H^+]/[HAc]$$

Multiplication yields

$$K_1 K_2 = [H^+]^2[NH_3][Ac^-]/[NH_4^+][HAc]$$

In the situation where equal amounts of NH_4^+ and Ac^- are initially present, stoichiometric considerations dictate that the amount of NH_3 formed must equal the amount of HAc formed, leaving equal amounts of the two ionic species. It may, therefore, be stated that $[NH_4^+] = [Ac^-]$ and $[NH_3] = [HAc]$. Substitution in the equilibrium equation gives $K_1 K_2 = [H^+]^2$ or $[H^+] = \sqrt{K_1 K_2}$ or using exponential symbolism ($pX = -\log X$)

$$pH = \tfrac{1}{2}pK_1 + \tfrac{1}{2}pK_2$$

An identical approach may be used to calculate the pH of solutions containing acid salts derived from polybasic acids. Consider a solution of sodium bicarbonate, $NaHCO_3$. The bicarbonate ion is amphiprotic, that is, it can donate or accept protons.

$$HCO_3^- + HCO_3^- \rightleftharpoons H_2CO_3 + CO_3^{2-}$$

The equilibrium constants for the two conjugate acid-base systems involved in this equation correspond to the first and second dissociation constants of the parent acid, H_2CO_3.

$$K_1 = [H^+][HCO_3^-]/[H_2CO_3] \quad \text{and} \quad K_2 = [H^+][CO_3^{2-}]/[HCO_3^-]$$

By multiplication of these two equations and acceptance of the fact that the amount of H_2CO_3 formed equals the amount of CO_3^{2-} formed, it can be shown that $[H^+] = \sqrt{K_1 K_2}$ as before.

For a tribasic acid three dissociation constants have to be considered, namely

$$K_1 = [H^+][H_2A^-]/[H_3A]$$
$$K_2 = [H^+][HA^{2-}]/[H_2A^-]$$
$$K_3 = [H^+][A^{3-}]/[HA^{2-}]$$

If $K_1 > 10^4 K_2 > 10^4 K_3$, the species H_3A, H_2A^-, and HA^{2-} can be treated as independent acid species.

Using the reasoning adopted for the dibasic acid, it can be shown that a solution of the salt NaH_2A has a hydrogen concentration given by the equation

$[H^+] = \sqrt{K_1 K_2}$ since the predominant reaction is

$$H_2A^- + H_2A^- \rightleftarrows H_3A + HA^{2-}$$

On the other hand, a solution of Na_2HA can be shown to have a hydrogen concentration given by $[H^+] = \sqrt{K_2 K_3}$; the proton transfer process being represented by

$$HA^{2-} + HA^{2-} \rightleftarrows H_2A^- + A^{3-}$$

In the titration of a tribasic acid with a strong base, the removal of the first proton is equivalent to titrating a weak monobasic acid whose K_a value equals K_1.

At the first equivalence point, the solution contains predominantly H_2A^- ions and $[H^+] = \sqrt{K_1 K_2}$.

As further base is added, a proton is removed from this acid H_2A^- and the second stage of the titration is equivalent to titrating a weak monobasic acid whose K_a value equals K_2.

At the second equivalence point, the resultant solution containing HA^{2-} has a $[H^+]$ equal to $\sqrt{K_2 K_3}$.

Continued titration causes pH changes similar to that predictable for a simple titration in which the K_a of the acid is numerically equal to K_3 for the tribasic acid.

When one puts these three segments together to form a complete titration curve, the resultant has several inflection points as shown in Fig. 10.1(b). The larger the difference in magnitude between the constants, K_1 and K_2 or K_2 and K_3, the greater the pH change in the vicinity of the equivalence points.

For phosphoric acid $K_1 = 5.9 \times 10^{-3}$, $K_2 = 6.2 \times 10^{-8}$, and $K_3 = 4.8 \times 10^{-13}$.

These values are sufficiently different in magnitude for the titration curve to have distinct inflection points [as indicated in Fig. 10.1(b)]. It is thus possible to nominate indicators suitable for the detection of the first and second equivalence points.

On the other hand, the individual dissociation constants for the tetrabasic species, ethylenediaminetetraacetic acid (H_4Y) are quoted as $K_1 = 10^{-2}$, $K_2 = 2 \times 10^{-3}$, $K_3 = 6.9 \times 10^{-7}$, and $K_4 = 5.5 \times 10^{-11}$, and an experimental titration curve for this acid does not display an inflection point corresponding to the formation of the species H_3Y^-, but jumps are observed at points corresponding to the formation of solutions containing predominantly H_2Y^{2-} and HY^{3-}.

B. SOURCES OF ERROR

Errors in titrimetric analysis can be attributed to

1. Inaccurate standardization of the titrant.
2. Use of an inappropriate indicator system.

    3. Operator difficulty in distinguishing indicator changes.
    4. Competing solution equilibria.

The strength of any titrant solution should be checked regularly against standard compounds of known composition, purity, and stability. For standardizing acid solutions the compounds recommended include tris (hydroxymethyl) amino methane, $(HOCH_2)_3CNH_2$; mercury (II) oxide, HgO; potassium bicarbonate, $KHCO_3$; sodium carbonate, Na_2CO_3; borax, $Na_2B_4O_7 \cdot 10 H_2O$. For checking alkaline compounds the recommended primary standards include benzoic acid, C_6H_5COOH and potassium hydrogen o-phthalate, o-$C_6H_4(COOK)(COOH)$.

In titration experiments most analysts prefer to use internal indicators, but it is essential that an appropriate indicator be chosen.

Tabulations of indicators and their properties appear in the "Handbook of Analytical Chemistry" and in most reference books which discuss titrimetry. From these lists, potentially useful indicators can be selected for a particular application, but the suitability should always be checked by experiment.

Multiple determinations of the species of interest are carried out using different selected indicators. The results should then be treated statistically to evaluate the relative merits of the indicators in detecting the true equivalence point; an example of this procedure is given in Chapter 3, Section II.

A statistical approach using several different analysts can also be used to trace any errors arising from operator bias.

Continual difficulty in detecting end points or persistent large errors are indicative of competing reactions in the assay solution.

The two basic reactions in the titration process are

$$H^+ + OH^- \rightleftarrows H_2O$$

$$H^+ + I_n^- \rightleftarrows HI_n$$

Potential interferants are, therefore, those ions which tend to react with one or more of the species listed in the equations or which tend to act as a source of H^+ or OH^- ions.

For example, if an acid solution contains iron (III), aluminum (III), and chromium (III) ions, the addition of standard base initially causes the pH to rise due to removal of H^+. Above pH 2, however, base is consumed in producing the hydrous oxides of the metal ions, and the titer and pH changes bear no relationship to the original acidity of the solution. It is also possible for foreign ions in the assay solution to react with the indicator species, effectively removing it, e.g., the reaction may involve formation of a metal-indicator complex or precipitate. Alternatively, the indicator may be oxidized or reduced, thus eliminating the traditional color transition.

Added salts can act as sources of H^+ or OH^-, thus vitiating the stoichiometry of any titration reaction; e.g., the addition of a salt of a weak acid may raise the pH sufficiently to mask the indicator color change interval

$$NaA \rightleftarrows Na^+ + A^-; \qquad A^- + H_2O \rightleftarrows HA + OH^-$$

Conversely the salts of weak bases can lower the pH of alkaline solutions

$$NH_4^+ + H_2O \rightleftarrows NH_3 + H_3O^+$$

$$NH_4^+ + OH^- \rightleftarrows NH_3 + H_2O$$

A more important source of secondary protons are protonated anions. If the original solution contains a polybasic acid, the protonated anions are a product of the titration reaction,

$$H_xA + OH^- \rightleftarrows H_{(x-1)}A^- + H_2O$$

and these acid anions compete with the species being titrated for the added base. Reaction with the stronger acid (larger K_a) is favored, and if the K_a values of the acidic species in solution differ by more than a factor of a thousand, reaction with one species goes to virtual completion before reaction with the weaker acid becomes significant. Thus with polybasic acids such as carbonic acid and phosphoric acid, the titration curves show several distinct inflection points, and by suitable choice of indicator it is possible to stop a titration at any selected stage of the reaction. On the other hand, in the titration of sulfuric acid solutions, removal of protons from the species H_2SO_4 and HSO_4^- proceeds simultaneously and the distinct break in pH required for end-point detection occurs only after the original acid has been converted to sulfate ions and water.

C. Titration of Polybasic Acids

In the titration of polybasic acids with strong base, or the titration of poly-acidic bases with strong acid, it is essential to ascertain the character of the overall titration curve since from this curve one can predict the feasibility of using visual indicators and deduct the stoichiometry of the reaction(s) associated with distinct pH changes.

In favorable cases, mixtures of two species can be analyzed by merely using two indicators; e.g., Fig. 10.1(b) resembles the titration curves obtained on titration of 0.1 M phosphoric acid with 0.1 M sodium hydroxide. Methyl orange changes color in the region of equivalence point 1 (solution containing predominantly $H_2PO_4^-$) and phenolphthalein changes color in the pH range associated with equivalence point 2 (solution containing mainly HPO_4^{2-}). Figure 10.1(a) resembles the curve obtained on titration of 0.1 M hydrochloric acid with a strong base. From the shape of this curve both methyl orange and phenolphthalein would be suitable indicators for equivalence point detection.

Let us now consider a solution which is approximately 0.1 M in respect to both hydrochloric and phosphoric acids.

If an aliquot of this solution were to be titrated with base using methyl orange indicator, a color change would be observed when the following reactions were complete.

$$HCl + OH^- \rightarrow H_2O + Cl^-$$

$$H_3PO_4 + OH^- \rightarrow H_2O + H_2PO_4^-$$

Let the volume of titrant required be x ml.

Now if a second aliquot is titrated with the same base solution, but phenolphthalein is substituted as the indicator, a color change will not be observed until the following reactions have proceeded to completion.

$$HCl + OH^- \rightarrow H_2O + Cl^-$$

$$H_3PO_4 + 2\,OH^- \rightarrow 2\,H_2O + HPO_4^{2-}$$

Let the volume of titrant required be y ml. Comparison of these two equations shows that the difference in titer $(y - x$ ml$)$ can be attributed to the reaction

$$H_2PO_4^- + OH^- \rightarrow HPO_4^{2-} + H_2O$$

The volume of base required to remove one proton from the phosphate species is thus $(y - x)$ ml. The same volume will be required for the reaction

$$H_3PO_4 + OH^- \rightarrow H_2PO_4^-$$

Knowing the molarity of the titrant, the amount of phosphoric acid in the original solution can be calculated.

In the last equation, one mole of H_3PO_4 reacts with one mole of OH^-; in the titration, z mole H_3PO_4 (in the aliquot treated) reacts with $(y - x)$ ml of titrant. If the molarity of the titrant is m, then $(y - x)$ ml of solution contains $(y - x)m$ 10^{-3} moles base and $z = (y - x)m$ 10^{-3}.

The concentration of phosphoric acid originally present equals the (No. of mole H_3PO_4 in aliquot) $\times$ (10^3/volume of aliquot) which simplifies to $(y - x)$ m/aliquot volume.

The concentration of hydrochloric acid present can also be calculated, because if $(y - x)$ ml of base are required to convert the H_3PO_4 present to $H_2PO_4^-$, then $x - (y - x)$ ml of titrant must have been required to neutralize the HCl present.

The number of moles of HCl present in each aliquot is, therefore, equal to $(2x - y)m$ 10^{-3} or the molarity of the solution in respect to HCl is $(2x - y)m$/aliquot volume.

By means of equilibrium calculations similar to those outlined in the previous chapter, it can be shown that phosphate species differing in structure by more than one proton do not exist to any significant extent at equilibrium. For example, if phosphoric acid is added to a solution of K_2HPO_4, then at equilibrium the solution will contain either H_3PO_4 and $H_2PO_4^-$ ions (if excess acid added) or $H_2PO_4^-$ and HPO_4^{2-} (if excess salt present). The addition of excess hydrochloric acid to a solution of dipotassium hydrogen phosphate would result in a final solution containing excess HCl and an amount of H_3PO_4 equivalent to the amount of phosphate salt originally present. The addition of excess sodium hydroxide to the same salt solution would yield as final products free OH^- and PO_4^{3-}.

The above situation allows several other types of mixtures to be analyzed by using two different indicators.

For example, a mixture of sodium hydroxide and trisodium phosphate may be titrated with a standard acid solution using phenolphthalein indicator. This indicator changes color when the following reactions are complete:

$$OH^- + H^+ \rightarrow H_2O$$

$$PO_4^{3-} + H^+ \rightarrow HPO_4^{2-}$$

Repetition of the titration using methyl orange requires more acid, since in this case the hydroxyl ions are neutralized and the phosphate group gains two protons before any color change is observed.

By calculations similar to those outlined above, the amount of each component in the mixture may be ascertained.

In this particular case, separate aliquots are not essential. In the titration of bases with acids, phenolphthalein changes from red to colorless at the end point. Thus after neutralization of the free OH^- and formation of HPO_4^{2-}, methyl orange can be added to the colorless solution. Titration can then be continued until this indicator changes from yellow to red. The volume of titrant required is determined by the reaction

$$HPO_4^{2-} + H^+ \rightarrow H_2PO_4^-$$

The same two indicators can be used to analyze a mixture of H_3PO_4 and $H_2PO_4^-$. In this case, titration of an aliquot using methyl orange indicator and a basic titrant gives a titer (v ml) which corresponds to the reaction

$$H_3PO_4 + OH^- \rightarrow H_2PO_4^-$$

Repetition using phenolphthalein indicator gives a titer (w ml) corresponding to the reaction

$$H_3PO_4 + 2\ OH^- \rightarrow HPO_4^{2-}$$

$$H_2PO_4^- + OH^- \rightarrow HPO_4^{2-}$$

The volume of titrant required for the dihydrogen phosphate reaction is thus ($w - 2v$) ml, and simple calculations give the amount of each component present.

Other mixtures commonly analyzed by a two indicator technique are mixtures of carbonates and bicarbonates or mixtures of carbonates and free hydroxyl ions. The treatment of these systems is directly analogous to that outlined above for the phosphoric acid system.

The bases most widely used as titrants in the determination of acid species are the hydroxides of sodium and potassium. These compounds are prone to absorbing carbon dioxide from the air, and unless great care is exercised in the preparation and storage of the solutions, the final titrant contains a small, but significant, amount of carbonate ion. This can lead to titration errors. For example, titration of a strong acid (HCl, $HClO_4$, HNO_3, or H_2SO_4) solution with sodium hydroxide should give the same titer (within one drop) irrespective of whether methyl orange or phenolphthalein is used as indicator. However, if the titrant contains carbonate contamination, the titers observed with the two indicators differ, the magnitude of the difference being directly related to the degree of contamination.

As mentioned previously, it is desirable to optimize the magnitude of the term K_aK_b for the acid and base systems involved. In most titrations the selected titrant has a K_a (or K_b if it is a base) value of greater than unity, and titrations generally become possible when the other equilibrium constant has a magnitude greater than 10^{-8}.

This limitation excludes many well-known weak acids and bases.

One solution is to select a new solvent for the reaction, since K_a and K_b values reflect the degree of interaction between the species and the solvent. This leads to nonaqueous titrations, and these are discussed in Section III.

An alternative approach, which is applicable to a limited number of systems, is to increase the ease of proton transfer through complex formation.

Probably the best example of this is the determination of borates present in alkaline media.

Titration of the alkaline borate solution with standard acid using methyl orange indicator yields an end point when all free base has been neutralized and the borate has been transformed into boric acid.

$$Na_2B_4O_7 + 2\,H^+ + 5\,H_2O \rightarrow 4\,H_3BO_3 + 2\,Na^+$$

Boric acid cannot be directly backtitrated with standard base solution. However, if a sugar such as mannitol is added, the resultant boric acid–sugar complex behaves like a weak monobasic acid and can be successfully titrated using phenolphthalein as indicator.

D. TITRATION OF AMINO ACIDS

The determination of amino acids and other amine, amide type compounds is of vital interest to pharmaceutical chemists. To satisfactorily titrate most of these compounds it is necessary to use a nonaqueous solvent, but there are a few which have dissociation constants which are sufficiently large to allow their determination in aqueous media.

Amino acids may be titrated with either strong acids or strong bases, since even the simplest amino acid contains at least one acidic and one basic group. A typical example is alanine, $CH_3CH(NH_2)COOH$, in which the α amine group acts as a base and the carboxyl group behaves as an acid.

In aqueous solution amino acids tend to undergo internal proton transfer to yield an internally ionized molecule, commonly referred to as a "zwitterion" or "dipolar ion."

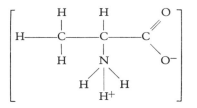

The zwitterion may be protonated or deprotonated according to the equilibria

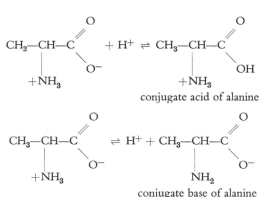

conjugate acid of alanine

conjugate base of alanine

and the pH of an aqueous solution of this amino acid is given by the expression

$$pH = \tfrac{1}{2}(pK_1 + pK_2) = \tfrac{1}{2}(2.35 + 9.87) = 6.11$$

where K_1 is the ionization constant of the conjugate acid of alanine and K_2 is the ionization constant of alanine itself.

If such a solution is titrated with a standard acid solution, there is a drop in pH, the pH halfway to the equivalence point being equal to pK_1 (2.35). Near equivalence point the rate of change of pH with the volume added increases to give a small inflection or curve. The change is not sufficient to be detected by a visual indicator but it can be observed if the pH changes are monitored with an appropriate meter.

Titration of the alanine solution with a standard base causes a gradual rise in pH, the pH halfway to equivalence point being equal to pK_2 (9.87). Near equivalence point the rate of change of pH per titrant increment again increases significantly.

Amino acids containing more than one carboxyl or amine group yield titration curves which display several inflection points [cf. Fig. 10.1(b)].

It can be observed that an amino acid can exist either as a positively charged ion, a neutral molecule, or as a negatively charged ion. Because amino acids differ in respect to the magnitude of their dissociation constants (K_1, K_2, etc.) each acid tends to require a different pH to attain electrical neutrality. Conversely, at any selected pH value, some amino acids are present as positively charged ions while others exist as neutral zwitterions or negatively charged ions. This behavior facilitates the separation of mixtures of amino acids by techniques such as electrophoresis and ion exchange where charge is an important consideration.

The pH at which the amino acid exists principally as the neutral zwitterion (or where the number of cations equals the number of anions) is called the "isoelectric point."

III. NONAQUEOUS TITRATIONS

A. CLASSIFICATION OF 'SOLVENTS

The strength of an acid, HB, in a given solvent, S, is defined in terms of the extent to which the reaction $HB + S = SH^+ + B^-$ proceeds. This reaction is a combination of two steps, ionization and dissociation:

$$HB + S \rightleftarrows \underset{\substack{\text{ion} \\ \text{pair}}}{SH^+B^-} \rightleftarrows SH^+ + B^-$$

$$\underset{\text{ionization}}{K_i} \qquad \underset{\text{dissociation}}{K_d}$$

The overall dissociation constant, K_a, is equal to the product of the ionization constant K_i and the dissociation constant K_d, and the activity of the solvent.

$$K_a = a_{SH^+} \times a_{B^-}/a_{HB} = K_i \cdot K_d \cdot a_S$$

where a_{HB}, a_{SH^+}, etc., represent activities.

The magnitude of the ionization constant depends on the relative basic strengths of B and S, while the degree of dissociation depends on the charge type of the members of the ion pair and the dielectric constant of the solvent.

The force of attraction (F) between two particles of charge q_1 and q_2 respectively, is given by Coulomb's law,

$$F = q_1 q_2/Dr^2$$

where r is the distance which separates the two charged particles and D is the dielectric constant of the medium in which the particles exist.

For solvents of high dielectric constant, such as water ($D = 78.5$ at 25°C), the force of attraction between the components of an ion-pair is relatively small and the dissociation step is virtually complete. However, in solvents having low dielectric constants there is considerable ion-pairing, together with the formation of larger ion aggregates. Some typical dielectric constant values are recorded in Table 11.1.

The acid-solvent reaction involves two conjugate acid-base systems:

$$HB \rightleftarrows H^+ + B^-; \quad K_{HB}, \text{ and } SH^+ \rightleftarrows H^+ + S; \quad K_{SH}$$

In terms of these equilibria, $K_a = a_S (K_{HB}/K_{SH})$. Hence for a given acid (where K_{HB} is fixed) the "strength" of the acid depends on the magnitude of K_{SH}.

TABLE 11.1

CLASSIFICATION OF SOLVENTS

Solvent classification	Name	Autoprotolysis constant	Dielectric constant
AMPHIPROTIC			
Protogenic	Acetic acid	14.45	6.4
	Formic acid	6.2	58.0
Intermediate	Water	14.0	78.5
	Methanol	16.7	31.0
	Ethanol	19.1	24.2
	Isopropanol		18.3
Protophilic	Ammonia (liquid)	33($-50°$C)	17.0
	Aniline		6.9
	Dimethylformamide		34.8
	Ethylenediamine	15.3	14.2
	Pyridine		12.3
APROTIC	Acetone		20.7
	Acetonitrile		37.5
	Benzene		2.3
	Chloroform		4.8
	Chlorobenzene		5.8
	1,4-Dioxane		2.2
	Methyl ethyl ketone		18.5
	Methyl isobutyl ketone		13.1

Solvents have been classified in terms of their acid-base properties; e.g., acidic or protogenic solvents are those capable of readily donating protons. By definition they are themselves acids, and examples of this class are formic acid and glacial acetic acid.

Similarly, basic or protophilic solvents are those which possess a strong tendency to accept protons. Solvents in this class include liquid ammonia, aniline, dimethylformamide, ethylenediamine, and pyridine.

These two definitions are not exclusive since many solvents can act as both proton acceptors and proton donors. Hence it is perhaps preferable to subdivide solvents as either aprotic or amphiprotic.

An aprotic or inert solvent displays no detectable acid or base properties. Benzene, chloroform, and carbon tetrachloride are typical aprotic solvents.

An amphiprotic solvent is one capable of acting as either an acid or base. There are gradations among amphiprototic solvents, ranging from predominantly acidic to predominantly basic. In the intermediate group (represented by water and the lower aliphatic alcohols), neither property predominates.

All amphiprotic solvents undergo self-ionization or autoprotolysis as illustrated by the systems:

$$H_2O + H_2O \rightleftarrows H_3O^+ + OH^-$$

$$CH_3OH + CH_3OH \rightleftarrows CH_3OH_2^+ + CH_3O^-$$

$$NH_3 + NH_3 \rightleftarrows NH_4^+ + NH_2^-$$

$$CH_3COOH + CH_3COOH \rightleftarrows CH_3COOH_2^+ + CH_3COO^-$$

or in general form

$$SH + SH \rightleftarrows SH_2^+ + S^-$$

where SH_2^+ is the solvated proton or "lyonium ion," S^- is the "lyate ion."

The autoprotolysis equilibrium constant K_s is defined by the relationship

$$K_s = (SH_2^+)(S^-)$$

The strongest acid which can exist in any solvent, SH, is the lyonium ion SH_2^+. For example, perchloric, hydrochloric, and nitric acids are stronger acids than H_3O^+; hence when placed in water, the reaction

$$HB + H_2O \rightleftarrows H_3O^+ + B^-$$

proceeds almost completely to the right. These acids are said to be leveled to the strength of the solvated proton.

In glacial acetic acid all acids are leveled to the strength of $CH_3COOH_2^+$; in formic acid all acids are leveled to the strength of $HCOOH_2^+$. In these solvents, the equilibrium between the mineral acids and the solvent does not necessarily proceed completely to the right and an order of acid strength becomes apparent, the order being perchloric > hydrochloric > nitric acid.

Solvents are termed differentiating solvents if they distinguish between the behavior of a series of acids. Thus glacial acetic acid differentiates the acid strengths of perchloric, hydrochloric, and nitric acids; water does not. Water, on the other hand, successfully differentiates between a mineral acid and the weaker acetic acid. A more basic solvent would tend to exert a greater leveling effect and so mask any innate differences.

The interaction of a base, *B*, with a solvent requires that the solvent possess acidic properties. A base leveling effect is encountered and different solvents differentiate between different groups of bases. The strongest base which can exist in a solvent is the lyate ion, S^-.

A strongly acidic solvent levels the strength of basic solutes to that of the lyate ion but it differentiates the strength of less acidic solutes.

Conversely, a strongly basic solvent levels the strength of acidic solutes to that of the lyonium ion and differentiates the strength of less basic solutes.

These generalizations may be expressed in more practical terms.

To make a weak base appear strong, use a strongly acidic solvent; to make a weak acid appear strong, use a strongly basic solvent.

This is the theme which is adopted in nonaqueous titrations. By varying the type of solvent used it has become possible to neutralize by titration a wide range of organic bases or acids and to determine directly the acidic and basic functional groups. In addition to their importance in organic analysis, non-aqueous acid-base titrations are used extensively in pharmaceutical analysis for the determination of constituents present in antihistamines, antibiotics, and sulfonamides.

B. Solvent Selection and Titration Curves

The following considerations are pertinent in the choice of a solvent for a specific, nonaqueous, acid-base titration.

1. The solvent should permit a large change in the solvated proton concentration near the equivalence point.
2. The solvent should preferably have a high dielectric constant to facilitate potentiometric detection of the equivalence point.
3. The substance to be titrated must be soluble, either in the solvent or in an excess of the titrant.
4. The products of the titration should be soluble in the solvent. If a precipitate is formed, it must be crystalline and not gelatinous.
5. The solvent should not introduce interfering side reactions.
6. The solvent should be inexpensive and easily purified.

The titrant selected must react rapidly and quantitatively with the substance being analyzed. It needs to be stable, i.e., it should not show any appreciable change in titer or color on standing.

The end point in many titrations can be detected visually using organic dyes as indicators but the selection of suitable indicators is somewhat difficult, and in practice the progress of an acid-base titration is best followed potentiometrically (cf. Chapter 13).

Acid-base titration curves in protolytic solvents having dielectric constants above about 25 can be closely approximated by means of the equations developed for aqueous solutions. Naturally it is essential to use the values of K_a or K_b appropriate to the solvent under consideration and to replace K_w with K_s.

A small number of nonaqueous K_a values are listed in the "Handbook of Analytical Chemistry" but the list is not sufficiently extensive to permit many calculations to be performed.

Thus titration curves for most systems have to be determined experimentally. In protogenic solvents, a glass electrode tends to change its potential more or less proportionally with pSH_2^+ and this change is detected by measuring the potential difference between this electrode and a reference electrode (such as a saturated calomel or silver–silver chloride electrode) placed in the test solution. The potential difference is plotted against volume of titrant added to give the required titration curve.

In highly basic solvents the glass electrode does not function as an indicator electrode and an antimony electrode is used instead. The design and construction of suitable reference and indicator electrodes for nonaqueous potentiometric titrations poses some problems. For example, the presence of water molecules within the glass membrane is essential to the mechanism of operation of a glass electrode, and this water can be leached out by the nonaqueous solvents to which the electrode is exposed.

The EMF readings obtained in nonaqueous media cannot be related to the pH system used in aqueous solutions; hence visual indicator selection cannot be based on the indicator lists prepared for aqueous studies.

Since most of the solvents have low dielectric constants, acid–base indicators tend to exist as ion-pairs, the ion-pair equilibria being complicated by the variations in the charge types exhibited by the indicators in their different forms. Frequently, the color changes are not the same as in water; probably because of the ion-pairing and other structural (electronic) changes in the indicator molecule caused by the solvent environment. Some indicators successively pass through a wide range of color shades (cf. Inset, Fig. 11.2).

To determine the indicator or color change which best signifies the equivalence point, a potentiometric titration should be run with the indicator present. The indicator which gives a sharp color change nearest to the potentiometric equivalence point may then be used in future titrations of the same kind.

The determination of two or more components in a mixture can be achieved if the solvent is capable of differentiating between all the species present. The ability of a protogenic or protophilic solvent to differentiate between species can be modified by adding inert solvent, since the presence of the inert solvent decreases the amphiprotic character of both acidic and basic solvents. Inert solvents yield large EMF changes near the equivalence point and are the preferred solvent type for successive titrations of acids or bases in mixtures. Unfortunately, solubility problems limit the number of possible applications of these aprotic solvents.

Figure 11.2 shows the type of titration curve obtained when a mixture of strong, weak, and very weak acids were titrated using methylisobutyl ketone as solvent.

As shown in this figure, each component in a mixture gives rise to a step in the titration curve. This step occurs at a potential characteristic of that individual

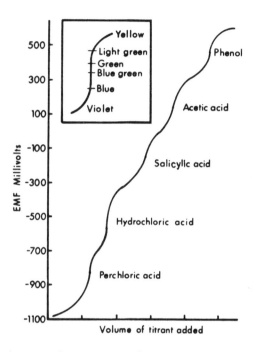

FIGURE 11.2   Titration curves in nonaqueous solvents. The main curve shows the resolution obtained when a mixture of acids in methylisobutyl ketone was titrated with 0.2 M tetrabutylammonium hydroxide in isopropanol. (After D. B. Bruss and G. E. A. Wyld, *Anal. Chem.*, **29** (1957) 232.) The inset shows the color changes undergone by methyl violet during the titration of 0.1 M potassium hydrogen phthalate in glacial acetic acid with acetous perchloric acid. Copyright (1957) by the American Chemical Society. Reprinted by permission of the copyright owner.

component. The midpoint potential for a given component is defined as being the potential observed when exactly one half of the constituent has been titrated (represented by the symbol HNP).

In a solvent of low dielectric constant the midpoint potential is both concentration- and salt-dependent. Under constant conditions, however, a linear relationship between midpoint potentials and pK_a values in water can be established for compounds of any given class.

$$\Delta HNP = (a - b)(pK_a)$$

a and *b* are constants which vary in magnitude with the solvent involved and class of compounds being studied.

The value of ΔHNP for a basic compound is defined as the difference in midpoint potential of the compound in question and the half-neutralization potential of 1,3–diphenylguanidine. For an acidic compound, ΔHNP is defined as the difference between the half-neutralization potential of the acid in question and the half-neutralization potential of benzoic acid.

Approximate ΔHNP values for series of monofunctional amines, amides, aliphatic, and *m*- and *p*-aromatic carboxylic acids, and hydroxyaromatic compounds can be calculated for a series of solvents using the data quoted in Table 3.22 of the "Handbook of Analytical Chemistry."

C.   TYPICAL APPLICATIONS

The apparatus required for nonaqueous titrations is usually quite simple, e.g., a pipette, a conventional burette, flasks or beakers and an electrometric potentiometer with indicating and reference electrode. Some basic solvents tend to absorb carbon dioxide from the air, and in these cases some additional precautions such as operating in a closed system are desirable.

For the titration of weakly basic substances the solvent most widely used is glacial acetic acid. If solubility permits, aprotic solvents are substituted, since they tend to give sharper end points with visual indicators.

The titrant employed is perchloric acid dissolved in either glacial acetic acid or dioxan, and a suitable visual indicator is often crystal violet since it undergoes a series of color changes (cf. Fig. 11.2 inset).

Despite the change of solvent the reaction involved is still a simple proton transfer. For example the titration of an amine of general formula $R_1 R_2 R_3 N$ (where R_2 and R_3 may be H atoms) is described by the equation:

$$R_1 - \overset{\displaystyle R_2}{\underset{\displaystyle R_3}{N}} + HClO_4 \longrightarrow \left[R_1 - \overset{\displaystyle R_2}{\underset{\displaystyle R_3}{NH}} \right]^+ ClO_4^-$$

The compounds amenable to titration include primary, secondary and tertiary amines, alkaloids, sulfonamides, purines, pyrazolones, and amino acids.

Many of the amino acids are insufficiently soluble in glacial acetic acid to permit direct-titration. Their perchlorates, however, are much more soluble and this permits a back-titration technique to be used. The amino acids are dissolved in an excess of the titrant, and the excess is then titrated with a standard solution of sodium acetate in glacial acetic acid.

Many metal salts of organic acids are soluble in glacial acetic acid and a large number are dissociated into their constituent ions. The anions readily accept protons under these conditions; hence the salts from which they are derived can be titrated as bases.

Bases like pyridine and triethylamine have been successfully titrated in carbon tetrachloride.

For the direct titration of acidic substances a number of solvents have been employed. These include dimethylformamide, *n*-butylamine, ethylene diamine, pyridine, and morpholine. For differential titrations of acid mixtures acetone and acetonitrile have been used. The titrant is commonly a solution of potassium methoxide in a benzene-methanol mixture. Methylisobutyl ketone, which does not undergo the usual type of autoprotolysis reaction, differentiates between a wide range of compounds (cf. Fig. 11.2). The titrant usually associated with this solvent is a solution of tetrabutylammonium hydroxide in isopropanol.

Lithium aluminum hydride has been employed as a basic titrant in the determination of alcohols and phenols. Lithium aluminum amide has been used as the titrant in determinations of alcohols, phenols, aldehydes, ketones, and esters. These titrants have to be used in oven dried glassware and the tetrahydrofuran which is used as the solvent needs purifying every few days since it readily forms a peroxide.

For the titration of acidic substances in dimethylformamide or butylamine, thymol blue and azo violet have proved to be satisfactory visual indicators.

The acid species which have been titrated range from mineral acids to carboxylic acids, from phenols to enols, from imides to sulfonamides.

The reactions involved are typified by the titration of medicinal sulfonamides.

The compounds are dissolved in a basic solvent (e.g., dimethylformamide) and titrated with potassium methoxide (in benzene/methanol) using thymol blue as indicator. The end point corresponds to completion of the reaction,

$$RSO_2NHR + CH_3OK \rightarrow RSO_2NKR + CH_3OH$$

Lewis acids, such as BCl_3 and $SnCl_4$, can also be titrated in nonaqueous media.

The majority of nonaqueous titrations involve acid-base reactions. Extension to oxidation–reduction reactions is hindered by the tendency of the solvents to interact with the reagents. However, one interesting reaction which comes into this category is the determination of water using the Karl Fischer Reagent (KFR). This reagent is composed of a mixture of iodine and sulfur dioxide in pyridine. Samples containing milligram amounts of water are normally suspended or dissolved in anhydrous methanol before being titrated with KFR until either visual observation or electrometric detection indicates the presence of free iodine.

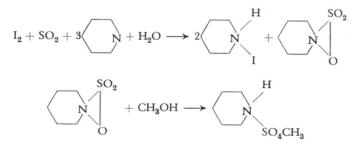

This procedure is particularly useful for determining the moisture content of oils, inorganic salts, and a wide range of commercial products. It is also invaluable for organic functional group analysis since it can be adapted to study any reaction in which water is a product. The scope of this special titrimetric procedure is fully described in the monograph "Aquametry" written by Mitchell and Smith.

IV. COMPLEXOMETRIC TITRATIONS

A. LABILITY AND STABILITY OF COMPLEXES

In the past decade, the formation of stable metal-complex ions, by titration of metal ion solutions with a standard solution of a complexing agent, has become increasingly important as a method of volumetric analysis.

For a reaction to be suitable for such a titration it must fulfill the same requirements as all other volumetric methods. That is, the complex-formation reaction must be rapid, must proceed according to well-defined stoichiometry, and must possess the desired characteristics for the application of the various modes of end-point detection.

The formation or dissociation of many complex ions is characterized by a slow rate of reaction. Such complex ions are called inert or nonlabile complexes, and among the metal ions whose complexes frequently exhibit this effect are chromium (III), cobalt (III), and platinum (IV). Those complexes which undergo ligand replacement within a few minutes at room temperature and 0.1 M reactant concentrations are arbitrarily termed labile. The group of metals which characteristically form labile complexes include cobalt (II), copper, lead, bismuth, silver, cadmium, nickel, zinc, mercury, and aluminum.

Kinetic studies indicate that the factors which influence the lability of metal complexes are the electronic structure, charge, and size of the central metal ion. Complexes which are extremely stable in a thermodynamic sense (with overall stability constants in the range 10^{20} to 10^{60}) can be inert. Thus it is important to distinguish carefully between lability and instability. Instability is decided by the difference between the free energies of the reactants and the products. Lability refers to the kinetics of the process. For example, cobalt (III) ammine complexes like $[Co(NH_3)_6]^{3+}$ are energetically unstable in acid solution but resist decomposition in such conditions for several days at room temperature.

It follows from this discussion that metal ions which form nonlabile complexes are not readily determined by means of a complexometric titration.

An equally important problem in complexometric titrations is stoichiometry of the reaction.

Most metal ions of analytical interest have several orbitals available for bond formation. For example, four ammonia molecules can become coordinated around a central copper ion to yield the square plane copper tetrammine cation.

Similarly, reaction between hexaaquochromium (III) ions and excess cyanide ions yields the stable hexacyanochromate (III) anion. However, it is not necessary that all available orbitals be filled simultaneously. Thus the replacement of water molecules in the hydrated copper ion proceeds in a stepwise manner, according to the following equilibria.

$$Cu(H_2O)_4^{2+} + NH_3 \rightleftarrows Cu(NH_3)(H_2O)_3^{2+} + H_2O$$

$$Cu(NH_3)(H_2O)_3 + NH_3 \rightleftarrows Cu(NH_3)_2(H_2O)_2^{2+} + H_2O$$

$$Cu(NH_3)_2(H_2O)_2 + NH_3 \rightleftarrows Cu(NH_3)_3(H_2O)^{2+} + H_2O$$

$$Cu(NH_3)_3(H_2O) + NH_3 \rightleftarrows Cu(NH_3)_4^{2+} + H_2O$$

A stepwise formation constant is associated with each equilibrium and is designated by K_n. In the present example, the subscript n is the integral value which corresponds to the addition of the nth ammonia ligand to the complex containing $(n-1)$ ammonia molecules. Thus

$$K_1 = [Cu(NH_3)(H_2O)_3^{2+}]/[Cu(H_2O)_4^{2+}][NH_3] = 10^{4.1}$$

or

$$K_3 = [Cu(NH_3)_3(H_2O)^{2+}]/[Cu(NH_3)_2(H_2O)_2^{2+}][NH_3] = 10^{2.9}$$

Overall formation constants are also used to describe the equilibria. Overall constants are designated by β_n where the subscript n gives the total number of ligands added to the original ion. For example, β_4 refers to the addition of four ammonia molecules to $Cu(H_2O)_4^{2+}$, and the pertinent equilibrium expression is

$$\beta_4 = [Cu(NH_3)_4^{2+}]/[Cu(H_2O)_4^{2+}][NH_3]^4 = 10^{12.6}$$

The relationship between stepwise formation constants and the overall formation constants is a simple one:

$$\beta_1 = K_1; \quad \beta_2 = K_1K_2; \quad \beta_3 = K_1K_2K_3; \quad \beta_n = K_1K_2 \ldots K_n$$

In this example, an ammonia molecule occupies only one coordination site; hence it is known as a monodentate ligand. Other typical monodentate ligands are chloride, bromide, iodide, cyanide, thiocyanate, and hydroxyl ions.

The relative stabilities of the various complexes formed by metal ions with most monodentate ligands are such that several complexes coexist in solution at any given concentration of ligand. With a mixture of complexes, no simple stoichiometric ratio of ligand to metal prevails. Even when exact stoichiometric quantities of metal ion and ligand have been mixed, the concentration of none of the species changes sufficiently to serve as an end-point signal. For these reasons complexes involving monodentate ligands are generally unsuitable for complexometric titrations.

The exceptions to this general rule arise when one member of a family of complex ions has unusual stability relative to other members of the family.

Fortunately, extensive compilations of stepwise and overall formation constants of metal complexes are available (e.g., L. G. Sillen and A. E. Martell "Stability Constants of Metal Ion Complexes," Chemical Society, London, 1964) and it is a reasonably simple matter to evaluate the relative amounts of each member of the complex family as a function of the ligand concentration. From the overall distribution graph [mole fraction of complex species versus log (ligand)] the probability of one species predominating can be ascertained. The stability data can also be used to calculate the concentration of hydrated metal ion, or its negative logarithm, $pM (= -\log[M])$ in the presence of different equilibrium concentrations of ligand. This can be replotted in the form of a titration curve (pM versus moles ligand added) to ascertain if there is a sharp change in pM near some point of well-defined stoichiometry.

With the advent of multidentate ligands the need for systematically evaluating the potential of a range of these types of systems has decreased and the number of analytical procedures based on the use of monodentate ligands is small. The more common methods are listed in Table 3.40 of the "Handbook of Analytical Chemistry." These include the determination of bromide, chloride, iodide, cyanide, and thiocyanate ions by titration with solutions of silver (I) or mercury (II) ions and the determination of copper (II), mercury (II), and nickel (II) ions through titration with a standard solution of potassium cyanide.

The end points in some of these titrations are located by a change in turbidity; in other systems there is a color change induced by the addition of indicators.

B.   CHELOMETRIC TITRATIONS

The full promise of complexometric titrations has only been realized with the introduction of multidentate complexing agents. For example, the compound triethylenetetraamine (trien) has four nitrogen atoms, linked by ethylene bridges, in a single molecule. This molecule can occupy the four coordination sites of a copper ion in one step.

$$Cu(H_2O)_4{}^{2+} + \text{trien} \rightleftarrows Cu(\text{trien})^{2+}$$

The complex is very stable ($K = 10^{20.4}$) and under ordinary conditions the stoichiometry of complex formation is 1:1. The reaction proceeds rapidly and both the ligand and complex are water soluble.

Thus, providing a suitable indicator is available, trien meets all the requirements desired in a titrant for copper ions.

Only a few metal ions such as copper, cobalt, nickel, zinc, cadmium, and mercury (II) form stable complexes with nitrogen ligands such as ammonia and trien. Other metal ions are better complexed with ligands which contain oxygen atoms as the electron donors. Chelating agents which contain both nitrogen and oxygen donor groups are particularly effective in forming metal chelates, and the introduction of amino polycarboxylic acids such as ethylene-diaminetetraacetic acid (EDTA) has been responsible for a remarkable increase in interest in chelometric titrations. To date, titration procedures have been developed for the determination of over sixty elements.

The term chelon has been proposed to designate the class of reagents that form stable, soluble, 1:1 chelates (chelonates) with metal ions and that consequently may be employed as titrants for metal ions. The best known chelon is EDTA,

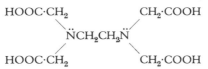

This reagent is potentially a sexidentate ligand, since it can coordinate with a metal ion through its two nitrogens and four carboxyl groups. In other cases, it behaves as a quinquedentate or quadridentate ligand, having one or two of its carboxyl groups free of strong interaction with the metal ion.

The abbreviation H_4Y is the usual representation for EDTA which is a tetraprotic acid; Y^{4-} is used to represent the ethylenediaminetetraacetate ion, the species which coordinates with the metal ion. For the general reaction

$$M^{n+} + Y^{4-} \rightleftarrows MY^{(n-4)+}, \qquad K_{abs} = [MY^{(n-4)+}]/[M^{n+}][Y^{4-}]$$

where K_{abs} is called the absolute stability constant or absolute formation constant.

A number of formation constants of metal chelonates are tabulated in summary form in Table 1.19 of the "Handbook of Analytical Chemistry." More extensive lists are found in the compilation by Martell and Sillen.

C. pH EFFECTS

The concentration of Y^{4-} present in solution is a function of pH. The four stepwise or successive acid dissociation constants of the acid H_4Y have magnitudes of $K_1 = 10^{-2}$, $K_2 = 2.16 \times 10^{-3}$, $K_3 = 6.92 \times 10^{-7}$, and $K_4 = 5.5 \times 10^{-11}$; at any particular pH the distribution of EDTA species may be calculated from the acid ionization constants. At pH > 12 most of the EDTA exists as the tetraanion Y^{4-}; around pH 8 the predominant species is HY^{3-}; while a solution containing mainly H_2Y^{2-} ions has a pH of 4.4.

Most metal-EDTA titrations are performed in neutral or alkaline solutions where the predominant species are H_2Y^{2-} and HY^{3-}. In a titration, protons, and metal ions compete for the reactive chelon ion and the net reaction may be visualized as

$$M^{n+} + H_2Y^{2-} \rightleftarrows MY^{(n-4)+} + 2H^+ \text{ or}$$

$$M^{n+} + HY^{3-} \rightleftarrows MY^{(n-4)+} + H^+$$

depending on the exact solution pH.

The protons liberated in these reactions could change the pH sufficiently to cause a significant amount of back reaction, i.e., metal complex dissociation in the presence of acid. To prevent this, titration systems usually contain an excess of an innocuous buffer system.

Maintaining a constant pH assists end–point detection since the change in pM at the equivalence point remains similar from assay to assay.

At any pH, only a fraction of any excess EDTA (i.e., uncomplexed with metal ion and represented by $[Y]_T$) is present as the species Y^{4-}.

$$[Y]_T = [Y^{4-}] + [HY^{3-}] + [H_2Y^{2-}] + [H_3Y^-] + [H_4Y]$$

The concentration of protonated species present can be expressed in terms of $[Y^{4-}]$ and $[H^+]$ by using the appropriate acid ionization constant formula, e.g.

$$[HY^{3-}] = [H^+][Y^{4-}]/K_4$$

and thence,

$$[H_2Y^{2-}] = [H^+][HY^{3-}]/K_3 = [H^+][H^+][Y^{4-}]/K_3K_4$$
$$[H_3Y^-] = [H^+][H_2Y^{2-}]/K_2 = [H^+]^3[Y^{4-}]/K_2K_3K_4$$
$$[H_4Y] = [H^+][H_3Y^-]/K_1 = [H^+]^4[Y^{4-}]/K_1K_2K_3K_4$$

Substituting these terms in the mass balance equation and factoring out $[Y^{4-}]$ on the right hand side gives

$$[Y]_T = [Y^{4-}]\left\{1 + \frac{[H^+]}{K_4} + \frac{[H^+]^2}{K_3K_4} + \frac{[H^+]^3}{K_2K_3K_4} + \frac{[H^+]^4}{K_1K_2K_3K_4}\right\}$$

The portion enclosed in braces is a function of $[H^+]$ only (or pH), since the K values are constant for a given chelon. This function is given the symbol α, so one may write

$$[Y]_T = [Y^{4-}]\alpha \quad \text{or} \quad [Y^{4-}] = [Y]_T/\alpha$$

It is obvious from the above that values for α can be calculated at any desired pH value for any chelon whose ionization constants are known (pK values for a number of common chelons are included in Table 1.18 in the Handbook). If the chelon is to be used over a range of pH values it is advisable to prepare a graph of log α versus pH.

Substitution of $[Y]_T/\alpha$ for $[Y^{4-}]$ in the absolute stability constant expression yields

$$K_{abs} = [MY^{(n-4)}]\alpha/[M^{n+}][Y]_T$$
$$[MY^{(n-4)}]/[M^{n+}][Y]_T = K_{abs}/\alpha = K_{eff}$$

The K_{eff} is called the effective or conditional stability constant; K_{eff} varies with pH and it is useful because it shows the tendency of a metal species to form the metal chelonate at the pH in question.

D.   SOLUTION VARIABLES AND INDICATORS

In calculating the titration curve for a chelometric titration it is considered advisable to use K_{eff} values rather than K_{abs} values. If K_{eff} is numerically greater than 10^7, the titration curve normally exhibits a sharp break in the value of pM in the vicinity of the equivalence point.

An estimate of feasibility can be made without calculating the titration curve. It has been proposed that the smallest value of K_{eff} which permits a feasible titration is represented by an equation such as

$$\log K_{eff} = T - \log C \, (= \log K_{abs} - \log \alpha)$$

where T is a titration factor (about 5 for visual titrations or 4 for potentiometric titrations) and C is the initial concentration of metal ion to be titrated.

This empirical equation can also be used to find the minimum value of $\log \alpha$ (thus pH) at which a particular titration is feasible.

While extremely useful, this simplified approach fails to allow for the presence of secondary or auxiliary complexing agents in the solution. For example, with metal ions that hydrolyze readily it may be necessary to add complexing ligands to prevent precipitation of the metal hydroxide. Thus the use of ammonia-ammonium ion buffers not only provides pH values >7 but also prevents precipitation of the hydrous oxides of Cu, Cd, Zn, Ni, Co, etc., through the formation of soluble metal ammines. In other cases, ligands such as cyanide ions are deliberately added to act as masking agents.

Interaction of the auxiliary complexing agents with the metal ion causes a further decrease in the magnitude of K_{eff}. Where the stability constants for all the complexes involved are known, then the effect of the complexes on the titration can be calculated. However, since many of the ligands used are unidentate, the proportion of the various complexes formed is usually strongly dependent on the ligand concentration and this may not be known exactly. Hence, often the magnitude of complexing effects can be estimated only in an empirical way.

The wide applicability of complexometric titrations and the preference of analysts for visual means of end-point detection has stimulated the development of a wide range of metallochromic indicators. An extensive list of such indicators and their properties is available in the "Handbook of Analytical Chemistry."

Basically, the metallochromic indicators are colored organic compounds which themselves form chelates with metal ions; the chelate must have a different color from the free indicator and the effective stability should be such that the indicator releases the metal ion to the titrant at a pM value very close to that of the equivalence point.

Metallochromic indicators usually show significant acid-base indicator behavior; hence the color changes observed on chelation can be pH dependent. The protons also compete with the metal ions for the indicator species; hence it is desirable to calculate a K_{eff} value for the indicator, using available acid dissociation constants and chelate formation constants.

As a general rule, the effective stability of the metal-indicator complex must be smaller than the effective stability of the metal-titrant complex.

Some metal ions form more stable complexes with the indicator and under such circumstances direct titration is not possible since the indicator tends to retain its complex color in the presence of excess titrant. This difficulty may be overcome by selecting another indicator or by using a back-titration technique.

In the latter technique, an excess of titrant is first added to the metal ion solution. The excess is then titrated with a standard solution of some other metal ion (e.g., Mg^{2+} or Zn^{2+}) using a suitable indicator such as Eriochrome Black T.

Trace impurities in the test solution can also block the indicator through the formation of stable indicator complexes. The answer in this case is normally a masking agent for the impurities.

Even where blocking of the indicator does not occur, impurities in the solution can cause titration errors by consuming titrant. Chelons react with a wide range of metal ions over a wide range of pH. Thus, selectivity is minimal and can usually only be achieved through the judicious use of auxiliary ligands or masking agents. The added ligands must form either sparingly soluble compounds or stable complexes with all metal ions except the species it is desired to titrate.

E. Sources of Error

Complexometric titrations have greatly expanded the range of applications of titrimetric procedures. They are simple to perform but erroneous results can be obtained if all the following factors are not considered.

1. The pH range selected for titration should lead to a sharp change in pM at equivalence point.
2. At the selected pH the uncomplexed indicator should exhibit a color which is distinctly different from the color of the metal-indicator complex. The effective stability of this complex should be less than the effective stability of the metal-chelon compound.
3. The amount of buffer solution present should be sufficient to minimize pH changes but a large excess should be avoided since components of the buffer may act as ligands.
4. Trace impurities and interfering ions should be effectively masked before the addition of the indicator and titrant.
5. The indicator molecule should not undergo any chemical change. In alkaline medium, certain metal ions, notably Mn(II), may catalyze the air-oxidation of an added metal indicator and thereby destroy its activity. This difficulty can be obviated by the addition of a reducing agent such as ascorbic acid to the solution. Some indicators are appreciably unstable in solution; hence stock solutions should be freshly prepared at short intervals. For occasional use, deterioration can be minimized by diluting the indicator 1:100 to 1:400 with a finely ground solid such as NaCl or sucrose. A small amount of solid mixture is added to the assay solution just prior to titration.

6. The titration reaction should be rapid. Chelonate formation with a number of metal ions is measurably slow, and the direct titration of readily hydrolyzable cations (e.g., aluminum and zirconium) presents difficulties, such as dragging end points and low results. Careful adjustment of pH and the use of a back-titration technique can obviate these problems.

F. Components of a Complexometric Titrimetric Procedure

Since the number of published applications of chelometric titrations are measured in thousands, a generalized statement in regard to the selection of experimental conditions is not feasible.

While EDTA is the chelon which has been most widely used, other amino-polycarboxylic acids have proved to possess distinct advantages for particular types of determinations. Besides metallo-chromic indicators, equivalence points have been detected using redox indicators, metallo-fluorescent indicators, metal-specific indicators, chemiluminescent indicators and turbidity indicators. Instrumental methods such as potentiometric measurement of pM changes, photometric detection of color changes, and amperometric titrations have proved applicable to many systems.

An extensive reference work entitled "Analytical Applications of EDTA," editor F. J. Welcher, is available in most libraries; the "Handbook of Analytical Chemistry" provides extensive lists of indicators and a summary of methods for the determination of inorganic cations by visual titration. In addition there are lists of masking agents and summaries of EDTA procedures involving masking.

To amplify the above discussion and to illustrate the number of steps which can be involved in the chelometric determination of an element present in a commercial product, let us examine the determination of zinc in an ore.

Determination of Zinc in Zinc Ore

Zinc may be determined by direct titration with the disodium salt of EDTA. The tetrabasic acid, H_4Y, has limited solubility in water; hence its disodium salt is usually used as the source of the chelon Y^{4-}.

In alkaline solution there is sharp change in pZn at the equivalence point and this change can be observed using eriochrome black T as indicator.

Eriochrome black T,

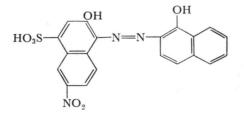

forms a red, slightly dissociated compound with zinc ions. The metal chelate is formed with this molecule by loss of the hydrogen ions from the phenolic —OH groups and the formation of bonds between the metal ions and the oxygen atoms as well as the azo group.

The molecule can be represented in abbreviated form as H_3I_n, the extent of ionization and color depending on pH.

$$H_3I_n \rightarrow H^+ + H_2I_n^- \xrightleftharpoons{\hspace{2cm}} H^+ + HI_n^{2-} \xrightleftharpoons{\hspace{2cm}} H^+ + I_n^{3-}$$

$$\text{red} \quad K_2 = 5 \times 10^{-7} \qquad\qquad \text{blue} \quad K_3 = 2.5 \times 10^{-12} \qquad\qquad \text{orange}$$

The sharpest color change obtainable in this system is from red (Zn-indicator complex) to blue (due to HI_n^{2-}); hence the most desirable pH in the solution is that which causes the indicator to be predominantly in the form $H_2I_n^{2-}$.

This requirement is met if the pH lies somewhere between 7 and 11. On the other hand, the pH influences the value of α for both the Zn-chelon ($K_{abs} = 3.2 \times 10^{16}$) and Zn-indicator complexes. (For EDTA, $\alpha \simeq 2000$ at pH 7, $\alpha \simeq 0.5$ at pH 10.) Hence for sharper end points the pH is preferably on the higher side, e.g., near to 10, since $K_{eff} = K_{abs}/\alpha$.

In this pH range, zinc ions tend to precipitate as the hydrous oxide unless present as some soluble complex. Zinc forms an ammine, and accordingly precipitation can be avoided by using ammonia to raise the pH of the test solution.

The indicator is not very stable in solution and is preferably added as a solid. The amount added should be kept to a minimum as any excess tends to obscure the end point. To facilitate handling of the small amount of indicator, it is preferably added in the form of a 1:500 dilution in finely ground NaCl.

At pH 10 a variety of different metals form stable complexes with EDTA; hence in order to make the titration specific for zinc, the interfering metals must be removed or masked.

The ore is first treated with concentrated nitric acid, the excess acid being removed by evaporation on a water bath.

The dry residue is taken up in hot water, and an excess of ammonia solution added. This procedure precipitates Fe(III) and most of the Al, Pb and Bi ions present. The hot solution is filtered to remove the hydrous oxides.

The filtrate may contain the ions of Mg, Ca, Sr, Ba, and Mn(II) as well as the ammines of Zn, Cd, Cu, Co, Ni, and Ag. The pH of the solution should be about 10. By adding potassium cyanide solution the ammine complexes are converted to the more stable cyanide complexes. These metals are effectively masked by this procedure and it is possible to titrate the alkaline earth metals [and Mn(II) if present] by adding eriochrome black T indicator, and EDTA solution until the solution changes in color from red to blue. If Mn(II) is present a few crystals of ascorbic acid should be added before the indicator.

This preliminary titration masks the alkaline earths in the form of EDTA complexes. At this stage the zinc present in the sample exists as a cyanide complex.

The cyanides of zinc and cadmium can be decomposed by adding formaldehyde. The metal ions react with the indicator to form the red complex and the solution changes color.

The red solution is now titrated with a standard EDTA solution until the color again changes from red to blue. The volume required in this titration is a measure of the amount of zinc and cadmium present in the sample.

In many zinc ores the amount of cadmium present is so small that it may be neglected. Where this is not permissible, the cadmium content has to be determined by a suitable method in a separate sample, and the amount of cadmium subtracted from the combined Zn + Cd figure obtained by EDTA titration.

V.   PRECIPITATION TITRATIONS

A.   pM CHANGES DURING TITRATION

The formation of a sparingly soluble compound can be used as the basic reaction for a titration procedure.

Almost all the established precipitation methods can be described by the general equation

$$A_{(aq)} + B_{(aq)} \rightleftharpoons AB_{(s)}$$

for which the appropriate equilibrium expression is

$$K_{sp} = [A][B].$$

(Ionic charges have been omitted for simplicity.)

During the course of such a titration there is a marked change in pA and pB in the vicinity of the equivalence point. In any solution in equilibrium with the solid, there is also a distinct relationship between pA and pB, since by taking the negative logarithm of both sides of the solubility product equation, one obtains

$$-\log K_{sp} = -\log [A] - \log [B],$$

or in exponential form

$$pA + pB = pK_{sp} \; (\text{cf. pH} + \text{pOH} = pK_w)$$

The magnitude of the pA change in the vicinity of equivalence point varies with the magnitude of K_{sp} and the initial concentration of reactant A. As with other titration systems, the larger the change in pA, the wider the range of suitable indicators.

Unfortunately, the range of indicators available for detecting equivalence points in this type of titration is limited. In addition, solid phases are prone to adsorb ions from the solution and this can lead to fading end points or false end points. In other systems, the rate of precipitate formation and growth is too slow for titration purposes. The more dilute the solution, the more marked is this effect.

For these reasons there are relatively few recommended precipitation titration procedures. In fact, most of the better known applications are associated with the sparingly soluble compounds of silver.

Titration curves for precipitation reactions can be constructed, and in form they are entirely analogous to those for acid–base titrations.

Let V_A ml of a solution of concentration C_A in respect to A be titrated with a solution of concentration C_B in respect to B. Let the volume of titrant added at any point be represented by V_B.

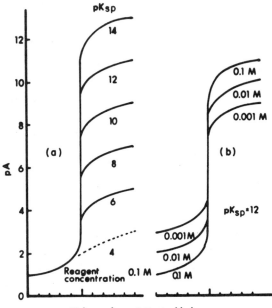

FIGURE 11.3   Titration curves for the precipitation titration, $A + B \rightarrow AB$: (a) effect of pK_{sp} on shape of curve. Concentration of titrant and titrand, 0.1 M. Numbers on the curves are the assumed values of pK_{sp}; and (b) effect of reagent concentration on shape of curve. pK_{sp} is assumed equal to 12.

Up to equivalence point, reactant A will be present in excess, the concentration after V_B ml of titrant have been added being given by

$$[A] = \frac{V_A C_A - V_B C_B}{V_A + V_B}$$

The precipitate formed will also tend to dissociate in accordance with the solubility product equation. This will contribute a further small amount of A to the solution. The amount of A arising from this source is equal to the amount of B simultaneously formed; $[B] = K_{sp}/[A]$; hence before the equivalence point,

$$[A] = \frac{V_A C_A - V_B C_B}{V_A + V_B} + \frac{K_{sp}}{[A]}$$

After the equivalence point, reagent B will be present in excess, and accordingly beyond the equivalence point

$$[B] = \frac{V_B C_B - V_A C_A}{V_A + V_B} + \frac{K_{sp}}{[B]}$$

$$[B] = K_{sp}/[A]$$

At the equivalence point $[A] = [B] = \sqrt{K_{sp}}$. In using these equations the solubility of the precipitate can often be ignored. The correction terms $(K_{sp}/[A]$ or $K_{sp}/[B])$ are necessary only if K_{sp} is rather large, or the solutions are quite dilute, or the point being considered is near the equivalence point.

The slope of the curve at the equivalence point, which determines the relative precision of the titration, depends upon both the value of K_{sp} and the concentrations of the solutions employed. Figure 11.3(a) shows some calculated titration curves for the titration of 0.1 M solutions of A with reagents that form precipitates having solubility products ranging from 10^{-6} to 10^{-16}. Figure 11.3(b) demonstrates the effect of reagent concentration on the structure of the curve.

B. Indicators for Precipitation Titrations

The indicators used to detect the equivalence point in precipitation titrations can be subdivided into three classes: (1) colored precipitate formers; (2) specific color forming reagents; and (3) adsorption indicators.

Before considering these indicator types in more detail, it is desirable to briefly consider the mechanism of precipitate formation, in particular, those aspects which influence indicator behavior.

On adding titrant B to a solution of A, crystal nuclei tend to form when the product $[A][B]$ exceeds the value of K_{sp}. The number of nuclei formed is related to the degree of instantaneous supersaturation achieved on mixing the two solutions. It is, therefore, a function of the concentration of the reacting species and the initial rate of mixing. Subsequent additions of titrant usually lead to growth of the original nuclei.

Slow mixing of dilute solutions may, therefore, be expected to yield a relatively small number of largish particles. Rapid mixing of more concentrated solutions generally leads to a large number of smallish particles—often of colloidal size. Although the weight of precipitate obtained in each case should be the same, the surface area of the colloidal material is many orders of magnitude greater than that observed with larger particles.

This surface area is important, because the behavior of indicators is influenced by phenomena associated with the interface, and the magnitude of the effect can be related to the surface area.

Any precipitate in contact with its mother liquor has a primary adsorbed layer on its surface which consists mainly of the constituent ion which is present in excess. Hence in the titration of A with B, the primary layer on the precipitate AB is composed of species A before equivalence point and species B after equivalence point. Other species in solution, e.g., C, can compete for positions in this layer, the order of preference depending on the relative magnitude of K_{sp} for the compound AC.

Where AC is only sparingly soluble, coprecipitation occurs; this leads to titration errors. Alternatively, if species C reacts with B to form a precipitate BC which is reasonably soluble, precipitation may be delayed until virtually all the initial A has been precipitated as AB. This can lead to a double step in the titration curve.

1. End Point Marked by the Formation of a Colored Precipitate

Should compound BC be colored, the appearance of the second phase can be observed, and in some cases used to detect the equivalence point in the titration of A.

The best known example of this is the use of potassium chromate as indicator in the titration of chloride ions by silver ions.

$$Ag^+ + Cl^- \rightleftharpoons AgCl_{(s)}; \quad K_{sp} = 10^{-10}$$

$$2Ag^+ + CrO_4{}^{2-} \rightleftharpoons Ag_2CrO_{4(s)(red)}; \quad K_{sp} = 2 \times 10^{-12}$$

Silver chloride is less soluble than silver chromate; hence when silver ions are added to a solution containing a large concentration of chloride ions and a small concentration of chromate ions, silver chloride is first precipitated. The red silver chromate appears when the excess silver ion concentration becomes large enough to exceed the value of K_{sp} for silver chromate.

Normally a chromate concentration of about $0.005\ M$ is used. To form a precipitate with this concentration, the silver ion concentration needs to be $2 \times 10^{-5}\ M$, since,

$$[Ag^+]^2 = K_{sp}/[CrO_4{}^{2-}] = 2 \times 10^{-12}/5 \times 10^{-3} = 4 \times 10^{-10}$$

At the equivalence point in the titration of Cl^- with Ag^+,

$$[Ag^+] = [Cl^-] = 1 \times 10^{-5} \quad \text{since} \quad [Ag^+][Cl^-] = 10^{-10}$$

The appearance of the red precipitate therefore occurs after the equivalence point. Assuming the total volume of the solution at equivalence to be 100 ml, the additional number of mole of Ag^+ required to give a concentration of $2 \times 10^{-5}\ M$ is $2 \times 10^{-5} \times 100/1000 = 2 \times 10^{-6}$ or 2×10^{-3} mmole.

If the original solution contained 2.50 mmole of chloride ion, the stoichiometric quantity of silver ion required to reach equivalence point is 2.50 mmole. To form the silver chromate an additional 0.002 mmole of titrant are required, and the percentage error is thus $0.002 \times 100/2.50$ or 0.08%.

In practice the error may be larger than this value because the calculation does not take into account the amount of silver ion that is used to form an observable amount of Ag_2CrO_4. Nor does it take into account the fact that silver ions form the primary adsorbed layer on the precipitate of AgCl. It is therefore best to correct for the end-point error by determining an indicator blank.

Use of this particular indicator is also limited to solutions with pH values from about 6 to 10. In more alkaline solutions silver oxide precipitates. In acid solutions, the chromate ion concentration is greatly reduced through the formation of hydrogen chromate and dichromate ions.

$$2H^+ + 2CrO_4{}^{2-} \rightleftarrows 2HCrO_4{}^- \rightleftarrows Cr_2O_7{}^{2-} + H_2O$$

2. Formation of a Colored Complex

In the preceding example, a slight excess of titrant is observed through the formation of a colored precipitate. The same result is achieved if the titrant reacts with the indicator to form a colored complex. For example, in titrations involving the thiocyanate ion, iron (III) salts are used as indicators and the presence of excess thiocyanate ions is indicated by the appearance of the red iron (III) thiocyanate complex.

In direct titrations of silver ions with thiocyanate ions, in nitric acid media, a premature end point is often observed. This can be attributed to silver ions being adsorbed on the surface of the precipitate. The effect can be overcome by vigorous stirring of the mixture near the end point.

3. Adsorption Indicators

Due to the change in the nature of the species present in the primary adsorbed layer before and after equivalence point, colloidal precipitates change the sign of their charge in the vicinity of equivalence point. For example AgCl in the presence of excess Ag^+ is positive; in the presence of excess Cl^- it is negative.

The charge of the primary adsorbed layer tends to attract ions of opposite sign from the solution to form a diffuse secondary adsorption layer.

Many organic dyes exist in ionic form and can become part of this secondary layer. In certain cases, the dye undergoes an abrupt color change in the process of being adsorbed. If the dye is adsorbed more strongly on one side of the equivalence point than the other, the abrupt color change can serve as a sensitive end-point detector in titrations.

One important family of adsorption indicators is derived from fluorescein. Fluorescein is a weak organic acid which may be represented by the symbol

HFlu. In alkaline solution it exists predominantly as the fluoresceinate ion, Flu^-.

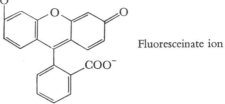

Fluoresceinate ion

This material is frequently used as an adsorption indicator in the titration of chloride ions with silver nitrate.

When added to the neutral or slightly basic solution present in the titration flask, the fluoresceinate ion is not adsorbed by colloidal silver chloride as long as chloride ions are present in excess. However, when silver ions are in excess the Flu^- ion is attracted to the surface of the positively charged particles $[(AgCl)Ag^+:Flu^-]$. The resultant aggregate is pink and the color is sufficiently intense to serve as a visual indicator.

The indicator imparts a yellow-green color to the solution and the color change on adsorption is believed to be due to a distortion or bending of the fluoresceinate structure when it is attracted to a precipitate particle.

The amount of color appearing is a function of the surface area of the solid phase; hence it is desirable to maintain this at a maximum. For this reason the particles should be kept as small as possible, and if necessary coagulation can be minimized by adding dextrin as a protective colloid.

Fluorescein, like other adsorption indicators, is a weak acid; hence pH control is important. Below pH7 the fraction of the indicator present as the anion is too small for observable color changes to occur. On the other hand, derivatives of this material, which are stronger acids, can be used in solutions of lower pH. For example, dichlorofluorescein can be used in the pH range 4 to 10.

It is preferable that the indicator ion be of opposite sign to the reactive ion added as titrant. Adsorption of the indicator does not then occur until excess titrant is present. This eliminates the possibility of indicator being trapped in growing particles.

The degree to which different indicator ions are adsorbed varies considerably, and an indicator must be chosen that is not too strongly or too weakly adsorbed. If the species is too strongly adsorbed it tends to displace the primary adsorbed ion before equivalence point is reached. If the indicator is weakly adsorbed, the end point is observed well after equivalence point. The ideal situation occurs when adsorption starts just before equivalence point and increases rapidly at equivalence point.

It has been observed that there is a relationship between the relative tendency of indicator ions to be adsorbed and the solubility of the salt formed by the reacting metal ion and the indicator.

4. Instrumental Methods

The number of visual and adsorption indicators suitable for precipitation titrations is rather limited, and many extensions of this technique rely on instrumental procedures.

For example, the titration of a mixture of halides is preferably followed potentiometrically, a silver electrode being used to follow the concentration of silver ions in solution.

With organic precipitants, equivalence points have been located using a spectrophotometer. The color intensity or turbidity of the assay solution is measured and plotted against the volume of titrant added. A change in gradient often occurs in the vicinity of equivalence point.

C. APPLICATIONS

The substances commonly determined by this technique include silver, mercury, and zinc ions; the halides, fluoride, chloride, bromide, and iodide; sulfates and arsenates.

All the procedures are subject to interference from other species capable of forming sparingly soluble compounds or complexes with the titrant.

In general principle, the number of applications could be expanded manyfold since extensive lists of methods of gravimetric analysis based on the formation and isolation of sparingly soluble compounds exist in the literature.

However, in practice few of these precipitation reactions meet all the criteria outlined in Section I,A; e.g., the rate of reaction should be fast. In many precipitation reactions, the rate of precipitation or attainment of solubility equilibrium is distinctly slow. This is of little concern in gravimetric analysis where solutions can be allowed to stand for hours if necessary before filtration. On the other hand, in a titration, rapid equilibrium is necessary if changes in concentration are to be followed accurately and closely. In some cases the problem of a slow reaction has been overcome by using a back-titration technique. An excess of precipitant is added and the formation of the sparingly soluble compound is allowed to proceed to completion. The amount of excess precipitant is then determined by titration with a standard solution of another reagent.

A second requirement is that any titration reaction should proceed in an exact stoichiometric manner. Few precipitates of analytical interest are formed from complex aqueous solutions with the high degree of purity and definite composition necessary for a successful titration. In gravimetric analysis, purification steps are integral parts of most published procedures, since all precipitates tend to adsorb considerable amounts of extraneous ions. In a titration it is not possible, in most cases, to overcome all the complications introduced by precipitate sorption.

The third requirement for titrimetry is availability of indicator systems. To date, the number of visual indicators available for precipitation systems is extremely limited, but instrumental procedures have proved satisfactory in most cases where the available visual indicators fail to meet the need.

In summation, the wider application of this technique is limited primarily by the tendency of precipitation reactions to proceed at a slow rate to produce nonstoichiometric compounds.

VI. OXIDATION–REDUCTION TITRATIONS

A. OXIDATION AND REDUCTION

Oxidation-reduction methods of analysis, commonly called redox methods, probably rank as the most widely used volumetric analytical procedures. The "Handbook of Analytical Chemistry" provides a summary of methods suitable for the determination of over seventy inorganic ionic species, and this number can be expanded by consulting volumetric reference books. Oxidation methods are also applicable to a large number of organic reactions. Redox reactions in ionic systems are generally quite rapid, and reactants can be titrated directly to the equivalence point, but for the determination of most organic compounds it is usually preferable to add an excess of titrant, followed by back titration.

A wide variety of visual indicators and electrometric methods are available for detecting the equivalence point in this type of titration, and suitable standards are available for standardizing all the common titrants. For oxidation purposes, the list of titrants includes solutions of chlorine, bromine, and iodine; the salts of Ce(IV), and the oxyanions, OCl_2^{2-}, BrO_3^-, IO_3^-, ClO_3^-, $Cr_2O_7^{2-}$, and MnO_4^-. A list of typical reducing agents would include the salts of Cr(II), Fe(II), Sn(II), and Ti(III); and compounds such as sodium arsenite, sodium thiosulfate, and hydrazine.

To understand the role of these reagents it is desirable to revise our understanding of oxidation and reduction.

Ions are formed from atoms by the loss or gain of electrons and for each atom there is a spontaneous force which tends to cause donation or acceptance of electrons.

$$A \pm ne \rightleftarrows A^{n-} \text{ or } A^{n+}$$

Similarly, ions tend to lose or gain electrons to yield either the original element or a species of different charge. This transfer of electrons is known as oxidation-reduction.

Oxidation is defined as the loss of electrons by an atom, molecule, or ion, and reduction is the gain of electrons by such particles.

Thus in a redox reaction there is a transfer of electrons from one reactant species to another. A flow of electrons is equivalent to an electrical current and accordingly the tendency of species to transfer electrons can be measured in electrical units. The potential or electromotive force of an oxidation-reduction system is therefore measured in volts (symbol E).

For the reduction process represented by the equation

$$A^{n+} + ne \rightleftharpoons A$$

it has been shown that

$$E = E^\circ + (RT/nF) \ln (a_A n^+/a_A)$$

In this equation, known as the Nernst equation, E is the potential tending to induce the ion A^{n+} to accept electrons and E° is known as the standard reduction potential for this equilibrium system. The symbol E° represents the value of E obtained when the logarithmic term equals zero. The symbol R is the universal gas constant, T the temperature in degrees absolute, n the number of electrons transferred, and F is the Faraday, the quantity of electricity associated with one mole of electrons (96, 493 coulombs); $a_A n^+$ is the activity of the species being reduced. For dilute solutions and simple calculations, the concentration in gram ions per liter is usually substituted for this term; a_A is the activity of the reduced species. By convention, where this species is in the elemental form it is considered to be in its standard state and to have an activity of unity. Thus for the system

$$Ag^+ + e \rightleftharpoons Ag, \quad E = E^\circ + (RT/F) \ln [Ag^+]; \quad E^\circ = 0.800 \text{ V and for}$$
$$Cu^{2+} + 2e \rightleftharpoons Cu, \quad E = E^\circ + (RT/2F) \ln [Cu^{2+}]; \quad E^\circ = 0.337 \text{ V}$$

If the reduction process involves the formation of an ion of lower valency, the activity of this reduced species cannot be taken as unity and the simplified equation for calculation purposes becomes

$$E = E^\circ + \frac{RT}{nF} \ln \frac{[\text{Ox}]}{[\text{Red}]}$$

where [Ox] and [Red] represent the equilibrium concentrations of the oxidized and reduced form of the species A.

Thus for the reaction

$$Fe^{3+} + e \rightleftharpoons Fe^{2+}, \quad E = E^\circ + \frac{RT}{F} \ln [Fe^{3+}]/[Fe^{2+}]; \quad E^\circ = 0.771 \text{ V}$$

The mechanism of many redox reactions is complex, the reaction proceeding in steps through unstable intermediates. The exact nature of the reaction determining the potential at an electrode surface is then usually unknown, and it is only possible to write a stoichiometric equation covering the overall process.

For example, the stoichiometric equation for the reduction of permanganate ions may be written as

$$MnO_4^- + 8H^+ + 5e \rightleftharpoons Mn^{2+} + 4H_2O$$

By analogy with the preceding example, the appropriate form of the Nernst equation should be

$$E = E^\circ + (RT/5F) \ln ([MnO_4^-][H^+]^8)/[Mn^{2+}]$$

In practice, calculations of the potential using such an equation generally prove to be in error, although in many cases the error is small and the potentials calculated are of value in predicting the feasibility of a proposed titration. In all redox systems involving hydrogen ions, the potential is normally found to depend on the hydrogen ion concentration, but the power dependence may differ significantly from the term derived from a balanced equation. The same remark can sometimes apply to the power dependence of the oxidized and/or reduced form of the reactant.

B. STANDARD AND FORMAL REDUCTION POTENTIALS

The calculated values of E derived from the Nernst equation provide some quantitative measure of the tendency of a system to accept or donate electrons.

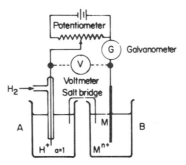

FIGURE 11.4 Schematic representation of apparatus suitable for measuring electrode potentials.

Experimentally, however, one can only measure the difference in potential between two systems. Values of E° have therefore been obtained by taking one system as an arbitrary standard and subsequently setting up an electrochemical cell (discussed in more detail in Chapter 13) to measure the potential of other systems relative to this standard.

The system chosen as the arbitrary standard for comparison purposes was

$$H^+ + e \rightleftarrows \tfrac{1}{2}H_2$$

and the standard reduction potential, E°, assigned to this system was zero. The reference electrode system is known as a hydrogen electrode, and consists of hydrogen at atmospheric pressure bubbling over a platinum electrode immersed in a solution having unit activity in respect to hydrogen ions.

The experimental setup is shown schematically in Fig. 11.4. Vessel A represents a hydrogen electrode with an arbitrary electrode potential of zero. Vessel B contains a reducible species M^{n+} in contact with its reduced form M. If system B is in its standard state (i.e., the activity of M^{n+} equals unity), then the observed potential (V) between the two electrodes may be recorded as the standard reduction potential for the system M^{n+}/M. The reading of V may be positive or negative, a positive value indicating that the system in vessel B has a greater tendency to accept electrons than H^+.

In measuring the potential difference, current must not flow through the system; hence in practice the voltmeter (V) is replaced by a potentiometer which applies an opposing EMF of equal magnitude. To provide a complete electrical circuit, the solutions in A and B have to be in contact. This is achieved by means of a salt bridge which may be a piece of filter paper moistened with an electrolyte solution or a plug of agar gel saturated with electrolyte.

For analytical applications, formal potentials are somewhat more meaningful than standard potentials. The formal potential of a system is the potential observed against a standard hydrogen electrode when the reactants and products are at one formal concentration (i.e., one gram formula weight per liter) and the concentrations of any other constituents of the solution are carefully specified. Formal potentials partially compensate for activity effects and errors resulting from side reactions. A formal potential is usually denoted by the symbol $E^{\circ\prime}$.

A selected list of standard and formal reduction potentials is given in Table 11.2. Since the system described represents only half of the electrochemical cell required to make measurements, the ionic equations are known as half-cell reactions.

It may be observed from Table 11.2 that formal potentials can differ significantly from the standard potential. This can be attributed in part to activity effects but very often a major cause is complex formation. This effect may be observed by comparing the formal potential of complex ions with the standard potential of the parent metal ion.

A study of this short table should also emphasize that it is desirable to know the end product of a reaction, since for a given oxidant the recorded potential can vary with the number of electrons transferred (cf. MnO_4^- to MnO_2 and MnO_4^- to Mn^{2+}). For systems in which hydrogen ions are part of the stoichiometric equation, it is usual practice to measure the potential in solutions of unit hydrogen ion activity.

Where the redox couple involves two ions of the same element, e.g., Fe^{3+} and Fe^{2+}, the electrode used in vessel B is usually inert, e.g., platinum, and serves merely as an electrical conductor.

With many systems, particularly the complex type involving hydrogen ions, standard potentials cannot be ascertained directly from galvanic cell measurements. In such situations, the appropriate values are derived indirectly, generally from thermodynamic considerations.

C.   Free Energy Changes

Energy is the driving force in all chemical transformations, and the sign of the electrode potential is preferably considered in terms of the free energy change involved.

TABLE 11.2
Selected Examples of Standard and Formal Reduction Potentials[a]

Half-cell reaction	$E°$ or $E°'$ (volts)
$S_2O_8^{2-} + 2e \rightleftarrows 2SO_4^{2-}$	2.01
$Ce^{4+} + e \rightleftarrows Ce^{3+}$ $(1F\ HClO_4)$	1.70
$Ce^{4+} + e \rightleftarrows Ce^{3+}$ $(1F\ HNO_3)$	1.61
$Ce^{4+} + e \rightleftarrows Ce^{3+}$ $(1F\ H_2SO_4)$	1.44
$MnO_4^- + 4H^+ + 3e \rightleftarrows MnO_2 + 2H_2O$	1.695
$MnO_4^- + 8H^+ + 5e \rightleftarrows Mn^{2+} + 4H_2O$	1.51
$MnO_2 + 4H^+ + 2e \rightleftarrows Mn^{2+} + 2H_2O$	1.23
$Cr_2O_7^{2-} + 14H^+ + 6e \rightleftarrows 2Cr^{3+} + 7H_2O$	1.33
$Br_2(aq) + 2e \rightleftarrows 2Br^-$	1.087
$Br_3^- + 2e \rightleftarrows 3Br^-$	1.05
$VO_2^+ + 2H^+ + e \rightleftarrows VO^{2+} + H_2O$	1.00
$Hg^{2+} + 2e \rightleftarrows Hg$	0.854
$Hg_2^{2+} + 2e \rightleftarrows 2Hg$	0.789
$Ag^+ + e \rightleftarrows Ag$	0.7995
$Fe^{3+} + e \rightleftarrows Fe^{2+}$	0.771
$Fe^{3+} + e \rightleftarrows Fe^{2+}$ $(1F\ HCl)$	0.70
$Fe^{3+} + e \rightleftarrows Fe^{2+}$ $(1F\ H_2SO_4)$	0.68
$Fe^{3+} + e \rightleftarrows Fe^{2+}$ $(0.5F\ H_3PO_4, 1F\ H_2SO_4)$	0.61
$I_2(aq) + 2e \rightleftarrows 2I^-$	0.6197
$I_3^- + 2e \rightleftarrows 3I^-$	0.5355
$VO^{2+} + 2H^+ + e \rightleftarrows V^{3+} + H_2O$	0.361
$Cu^{2+} + 2e \rightleftarrows Cu$	0.337
$Cu^{2+} + e \rightleftarrows Cu^+$	0.153
$AgCl + e \rightleftarrows Ag + Cl^-$	0.2222
$AgBr + e \rightleftarrows Ag + Br^-$	0.073
$AgI + e \rightleftarrows Ag + I^-$	−0.151
$Hg_2Cl_{2(s)} + 2K^+ + 2e \rightleftarrows 2Hg + 2KCl$ (sat)[Sat. calomel electrode]	0.2415
$Sn^{4+} + 2e \rightleftarrows Sn^{2+}$	0.154
$2H^+ + 2e \rightleftarrows H_2$	0.0000
$Pb^{2+} + 2e \rightleftarrows Pb$	−0.126
$N_2 + 5H^+ + 4e \rightleftarrows N_2H_5^+$	−0.23
$V^{3+} + e \rightleftarrows V^{2+}$	−0.255
$Ti^{3+} + e \rightleftarrows Ti^{2+}$	−0.37
$2CO_2 + 2H^+ + 2e \rightleftarrows H_2C_2O_4$	−0.49
$Zn^{2+} + 2e \rightleftarrows Zn$	−0.763
$Zn(NH_3)_4^{2+} + 2e \rightleftarrows Zn + 4NH_3$	−1.04
$Cd^{2+} + 2e \rightleftarrows Cd$	−0.403
$Cd(CN)_4^{2-} + 2e \rightleftarrows Cd + 4CN^-$	−1.09
$Al^{3+} + 3e \rightleftarrows Al$	−1.66
$Mg^{2+} + 2e \rightleftarrows Mg$	−2.37
$Na^+ + e \rightleftarrows Na$	−2.714
$K^+ + e \rightleftarrows K$	−2.925
$Li^+ + e \rightleftarrows Li$	−3.045

[a] Based on data from the "Handbook of Analytical Chemistry," edited by L. Meites. Used with permission of McGraw-Hill Book Company; copyright © 1963, McGraw-Hill Book Company.

For an electrolytic cell, the free energy change ΔG is equal to $-nFE$. A positive value of E thus leads to a negative value for ΔG, and reactions involving a decrease in free energy are considered to be spontaneous. Thus, if $E°$ for the system in vessel B is positive, then, theoretically, the oxidized form of B can be reduced by hydrogen.

Now if the systems in vessels A and B in Fig. 11.4 are both half-cells in their standard states, the observed value of V equals $(E_B - E_A)$. It is often more convenient to use a system of known potential in place of the hydrogen electrode. The systems used have to maintain a constant potential during the period of measurement and are known as reference electrodes; typical examples are the saturated calomel electrode and the silver–silver chloride electrode (see Chapter 13).

With the potential of the system in A fixed through the use of a reference electrode, any change in the potential of system B is indicated by changes in the voltmeter readings. This is the fundamental principle of the analytical technique known as potentiometric titrations.

The potential of the system in B is given by the equation

$$E = E_B° + \frac{RT}{nF} \ln \frac{[Ox]}{[Red]}$$

In the titration of a reducing agent with an oxidizing agent the proportion of the species B in the oxidized form continually increases during the titration and the calculated or observed value of E also increases. After the equivalence point, the solution contains only the oxidized form of species B, and hence system B ceases to contribute significantly to the potential of the solution. However, during the titration an equivalent amount of the reduced form of the titrant (e.g., species C) is formed and excess titrant completes the requirement for a new electrical equilibrium involving system C. The continued addition of excess oxidant accordingly results in further increased readings on the voltmeter. A plot of E versus the milliliters titrant added is known as a potentiometric titration curve.

Consider the titration of $0.1\ M$ iron (II) sulfate with $0.1\ M$ cerium (IV) sulfate in $1\ F\ H_2SO_4$. Prior to the equivalence point, both iron (II) and iron (III) ions are present and the EMF developed can be attributed to the Fe^{3+}/Fe^{2+} half-cell reaction. The ratio $Fe^{3+}:Fe^{2+}$ increases during the titration and the calculated EMF accordingly changes from

$$< E°_{Fe^{3+}/Fe^{2+}} \text{ to } > E°_{Fe^{3+}/Fe^{2+}}$$

as the equivalence point is approached. This can be seen by considering just a few points; e.g., after 10% of the initial iron (II) has been oxidized, the ratio of $[Ox]/[Red]$ equals one ninth and the half-cell potential is given by

$$E = 0.68 + (2.3\ RT/F) \log 0.11$$

After 50% of the iron (II) has been oxidized, the ratio [Ox]/[Red] equals one and $E = E°$.

After 75% oxidation, [Ox]/[Red] = 3, and

$$E = 0.68 + (2.3\,RT/F)\log 3$$

After the equivalence point, the solution contains both cerium (III) and cerium (IV) ions and the EMF can be attributed to the Ce^{4+}/Ce^{3+} half-cell reaction.

Accordingly,

$$E = 1.44 + (2.3\,RT/F)\log [Ce^{4+}]/[Ce^{3+}]$$

At the equivalence point, the iron (III) and cerium (III) ions formed in the titration tend to interact,

$$Fe^{3+} + Ce^{3+} \rightleftarrows Fe^{2+} + Ce^{4+}$$

Since the oxidized and reduced forms of both species are now present, both half-cell reactions contribute to the EMF of the solution. Summation of the two appropriate Nernst equations yields the relationship

$$2E_e = E°_{Fe^{3+}/Fe^{2+}} + E°_{Ce^{4+}/Ce^{3+}} + \frac{RT}{F}\ln\frac{[Fe^{3+}][Ce^{4+}]}{[Fe^{2+}][Ce^{3+}]}$$

From the equilibrium equation for the reaction, it can be observed that if the initial amounts of Fe^{3+} and Ce^{3+} are equal (the condition at equivalence point) then the amounts remaining after reaction will be equal, and equivalent amounts of Fe^{2+} and Ce^{4+} will be formed. Therefore $[Fe^{2+}] = [Ce^{4+}]$ and $[Fe^{3+}] = [Ce^{3+}]$ and substitution of these values in the summed equations leads to a zero value for the logarithmic term, thus,

$$E_e = (0.68 + 1.44)/2 = 1.06\text{ V}$$

In reactions where the values of n differ for the respective Nernst equations, the individual equations have to be multiplied by appropriate factors to give a common RT/nF value, and it can be shown that, for the general redox reaction

$$a\,Ox_1 + b\,Red_2 \rightleftarrows a\,Red_1 + b\,Ox_2$$

the potential at equivalence is given by

$$E_e = \frac{bE°_1 + aE°_2}{a + b}$$

Calculations based on the Nernst equation are only reasonably valid when both redox systems are reversible. As in acid-base studies, the preferable method of ascertaining the form of the titration curve is by experiment. If the assay solution contains several species which can react with the titrant, several distinct inflection points will be observed, provided that there is a significant difference in the magnitude of the respective formal potentials (e.g., >0.3 V).

The magnitude of the potential jump in the vicinity of equivalence point and the steepness of this curve (which influences indicator precision) are related to the difference in formal potentials for the two systems involved. It is also a function of the equilibrium constant for the overall reaction.

The general redox reaction may be subdivided into two half-cell reactions:

$$Ox_1 + be \rightleftarrows Red_1; \quad \Delta G° = -E_1° \, b \, F$$
$$Ox_2 + ae \rightleftarrows Red_2; \quad \Delta G° = -E_2° \, a \, F$$

Multiplying the first equation by a, and the second by b yields the equations

$$a \, Ox_1 + ab \, e \rightleftarrows a \, Red_1; \quad \Delta G° = -aE_1° \, b \, F$$
$$b \, Ox_2 + ba \, e \rightleftarrows b \, Red_2; \quad \Delta G° = -bE_2° \, a \, F$$

Subtracting and simplifying gives

$$a \, Ox_1 + b \, Red_2 \rightleftarrows a \, Red_1 + b \, Ox_2; \quad \Delta G° = -abF(E_1° - E_2°)$$

But for an equilibrium reaction

$$\Delta G° = -2.303 \, RT \log K$$

Therefore $\log K = (ab)F/2.3 \, RT \cdot (E_1° - E_2°)$ which reduces to

$$\log K = (ab)(E_1° - E_2°)/0.059 \text{ at } 25°$$

(ab) represents the total number of electrons transferred in the overall reaction. If K is large ($>10^{10}$), interaction of the reaction products yields insignificant amounts of the species Red_2, and the concentration, $[Red_2]$, can be calculated from the amount of initial reductant remaining. Thus after 99.9% of the original species has been oxidized

$$E = E_2° + (2.3 \, RT/aF) \log 99.9/0.1$$
$$\simeq E_2° + 0.18/a \text{ V}$$

With an 0.1% excess of titrant, the solution potential approximately equals

$$E = E_1° - 0.18/b \text{ V}$$

Hence the potential jump in the vicinity of equivalence point is about $0.18(a + b)/ab$ V less than the difference in formal potentials.

D. POTENTIAL CHANGES

Figure 11.5 shows the variation in potential with the fraction of reagent present in the oxidized form for several redox couples. A titration curve is obtained by combination of two of these curves. To illustrate the magnitude of the potential jump likely to occur in a titration, the curves for systems with formal reduction potentials greater than 1.0 V have been plotted to the right of systems of lower formal potentials.

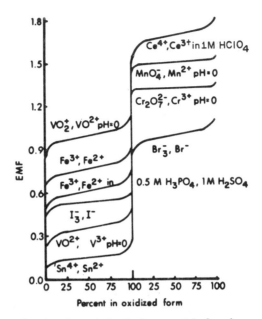

FIGURE 11.5   Curves showing the variation in the potential of a redox couple with variation in the amount present in the oxidized form.

The shape of a curve depends upon the value of n, the number of electrons involved in the half-cell reaction. The larger the value of n, the flatter is the curve. The slope is least near the midpoints (equal amounts of oxidized and reduced form), and a redox couple is said to be poised in this region.

The change in potential at the equivalence point, in the titration of a reducing agent, depends on the oxidant used. Hence a larger jump is observed if cerium (IV) ion is used instead of the dichromate ion. The potential change can also be increased by complexing one component of the couple being oxidized. For example, in the titration of iron (II) with potassium dichromate, it is recommended that a sulfuric acid–phosphoric acid media be used, since this reduces the formal potential of the Fe^{3+}/Fe^{2+} system by approximately 150 mV.

The calculation of equilibrium constants and the preparation of titration curves permits one to predict the feasibility of a redox titration.

Unfortunately a large number of potentially useful reactions proceed at a very slow rate. This effect is particularly marked with reduction processes which involve protons and the transfer of more than one electron.

In many cases, the rate can be accelerated to an acceptable level by the addition of a catalyst or by increasing the temperature of the reaction (see Chapter 12).

Apart from kinetic considerations, the major difficulty associated with redox titrations is the lack of specificity. In the ideal situation, the titration solution should contain only one species with a marked desire to accept electrons, and the titrant should be the sole source of electrons. If more than one electron donor or acceptor is present, there should be a marked difference in formal potential for the systems so that a stepped titration curve may be observed. These idealistic conditions are not always satisfied, and in commercial analysis it is possible for the assay solution to contain significant amounts of reagents (e.g., organic matter, nitric acid, sulfide ions, etc.) which consume titrant without causing any marked effect on the shape of the titration curve or the behavior of a visible indicator.

E. REDOX INDICATORS

As with acid-base titrations, it is relatively easy to ascertain the shape and nature of a redox titration curve by experiment, and it is always considered wise to check the suitability of a visual indicator against the potentiometric result.

Visual indicators can be divided into two main classes: (1) specific color formers, and (2) redox indicators.

The specific color group can be subdivided into three smaller groups:

1. Some titrants are intensely colored (e.g., $KMnO_4$, I_3^-) and the slightest excess can be detected by its color.
2. The intensity of this color can be intensified by having in solution a specific reagent which forms a complex of even higher absorptivity. For example, minute traces of iodine yield intense blue solutions in the presence of starch, and the presence of thiocyanate ions aids the detection of the first small excess of iron (III) salts.
3. Where the presence of the specific reagent can interfere in the titration, it may be employed externally by using a spot plate. Drops of the titrant solution are removed with a dropper and placed on drops of the test reagent held in depressions on an indented tile. This process is tedious and external indicators are now adopted only as a last resort.

A redox indicator is a substance which can accept or donate electrons and whose oxidized and reduced forms differ in color. The fundamental reaction may be represented by the half-reaction

$$Ox + ne \rightleftarrows Red$$

$$\text{(color A)} \qquad \text{(color B)}.$$

The half-reaction for many indicators also involves protons, but as the amount of indicator present is small, the quantity of hydrogen produced or liberated by the indicator does not appreciably alter the pH of the solution. Hence for the purpose of discussion it is permissible to use the simple equation given above.

For the above reaction, the appropriate Nernst equation may be written as

$$E = E^\circ_{I_n} + (0.059/n) \log [\text{Ox}]/[\text{Red}]$$

If one assumes that the intensity of the indicator color is proportional to its concentration, then the equation may be rewritten as

$$E = E^\circ_{I_n} + (0.059/n) \log \{\text{Intensity of A}/\text{Intensity of B}\}$$

As in the case of acid-base indicators, the solution will appear to be color A if the intensity of A is more than ten times that of B. Conversely, color B will predominate when the ratio of color intensity is less than 1/10.

Substituting these ratios in the Nernst equation leads to the conclusion that color A is seen when $E \geq E^\circ_{I_n} + 0.059/n$; and color B is seen when $E \leq E^\circ_{I_n} - 0.059/n$.

At intermediate values of E the color will be a mixture of A and B. The color change interval for a redox indicator is accordingly described approximately by the expression $E^\circ_{I_n} \pm 0.059/n$.

Since the indicator is present in small amount, the value of E is controlled by the potential of the solution being titrated. A suitable indicator for any redox titration is thus any species whose color change interval lies entirely within the potential jump observed in the vicinity of equivalence point.

Indicators are available which possess standard potentials ranging from 0.26 to 1.31. Some of the indicator reactions involve hydrogen ions, and so the potentials at which they change color tends to depend on pH. Other indicators are destructively oxidized if maintained for any time in the presence of excess oxidant, and these compounds are used primarily in the titration of reducing agents, since the oxidant is present in excess only after equivalence point.

Potentiometric Detection of the End Point

Where the difference between the formal potentials for the titrant and titrate system lies between 0.20 and 0.40 V, it is preferable to follow the progress of the redox titration by means of potentiometric measurement techniques.

For routine analysis, potentiometric measurements possess a further advantage. The potential change near the equivalence point can be used to stop the addition of titrant or activate a recorder. Automatic titrations are thus achieved with reasonably simple apparatus.

The potentiometric approach is also preferable in titrations of multicomponent systems.

F. SELECTED APPLICATIONS

In selecting a titrant for a particular application, several factors have to be considered besides the potential jump at equivalence point.

The most important considerations are interference effects and kinetic problems.

Interference can begin with dissolution of the sample in acid. If hydrochloric acid is used, potassium permanganate is an unsuitable titrant because of the tendency of chloride ions to reduce the oxidant. The use of nitric acid introduces a competitive oxidant which may attack the indicator or react with any reducing agents introduced.

A preliminary step in many determinations is prereduction or preoxidation of the element of interest. Failure to destroy excess of the reagent used for this purpose can vitiate the results of any subsequent titration.

Slow rates of reaction can sometimes be overcome by the introduction of catalysts and by the application of heat. With the catalyst approach it is necessary to ensure that the stoichiometry of the overall reaction is not disturbed. The application of heat is generally safe, except when it accelerates interference reactions to a similar degree or when it causes losses of steam volatile products.

Because it acts as its own indicator, one of the most popular titrants is potassium permanganate. It has been used in the titration of arsenious acid (catalyst ICl), oxalic acid [autocatalyzed by Mn(II)], iron (II), manganese (II), molybdenum (III), hydrogen peroxide, nitrites, antimony (II), organic compounds, selenate ions, and lower valency vanadium compounds.

Iodine is a weaker oxidant, but has proved very useful in titrations of sulfide ions, sulfites, thiosulfates, tin (II), copper (I), and arsenic (III). The end point in the titrations is readily detected by means of the intense blue complex formed by iodine with starch.

The most widely used standard reductant is probably sodium thiosulfate. In most applications of this reagent, oxidants are allowed to react with iodide ions to produce free iodine which is subsequently titrated with the thiosulfate solution.

Other titrants have particular advantages over these three in particular situations. For example, potassium dichromate is not reduced by hydrochloric acid, potassium iodate solution is more stable than iodine solutions, etc.

By varying the titrant, and by using indirect procedures, redox methods can be used to examine any material which contains a species capable of accepting or donating an electron.

VII. TUTORIAL QUESTIONS AND PROBLEMS

1. a. Discuss the factors which influence the strength of an acid.
   b. Explain why nonaqueous solvents can facilitate the determination of weak acids and weak bases.
   c. What is "autoprotolysis?"
   d. What properties should a reaction possess to be suitable for use in titrimetry?
   e. Derive a relationship between K_a for an acid and K_b of its conjugate base.

f. If the two dissociation constants of the acid H_2A are K_1 and K_2, what formula is used to calculate the pH of a solution of $NaHA$?

g. List the reagents commonly used to standardize base solutions and summarize any precautions needed in their use.

h. Both sodium carbonate and borax are recommended as primary standards for acid solutions. What precautions are needed in the preparation of the pure primary standards?

i. Prepare a summary of methods (in equation form) for the analysis of mixtures, which utilize two indicators.

j. What is a "zwitterion" and what effect has pH on the charge of an amino acid?

k. Outline what you understand by the terms "differentiating solvent" and "leveling effect."

l. Prepare a list of the various types of compounds which are commonly determined by nonaqueous titration. What factors might limit the extension of this list to a broader range of materials?

m. Make a brief summary of the properties and applications of the basic titrants used in nonaqueous titrations.

n. Provide an explanation for the many colors exhibited by crystal violet when used as an indicator in nonaqueous solvents.

2. a. An analyst is given a bottle of clear solution and is told that it is 0.1 M in respect to a tribasic acid H_3A.

How could the analyst check the concentration of the acid? Would he need to know the magnitude of the acid dissociation constants K_1, K_2, and K_3? If so, how might they be determined?

b. Assuming that the values of K_1, K_2, and K_3 are 10^{-2}, 10^{-6}, and 10^{-10} respectively, calculate the pH changes which occur on the addition of 0.1 molar sodium hydroxide to a 100 ml aliquot of the H_3A solution.

Plot the approximate titration curve and use literature lists to select suitable indicators for the detection of the first and second equivalence points.

c. Are there any regions in which the pH is buffered? What is a buffer? What is the buffer capacity?

d. If in the study of a new sample of the acid, the titration to the first equivalence point was found to be less than the volume of base required to pass from the first to second equivalence point, what conclusions could be drawn about the composition of the new sample?

e. A laboratory assistant adds 10 mmoles of the salt Na_2HA to 500 ml of 0.1 M H_3A solution. What will be the final pH? What volume of 0.1 M NaOH will be required to neutralize a 50 ml aliquot of the mixture to the first, second and third equivalence points?

3. Comment on the statement: "Acid-base titrations in nonaqueous media are now far more important as commercial methods of analysis than titrations in aqueous media."

4. a. Explain the meaning of the following terms: monodentate ligand, chelon, metallochromic indicator, lability, stepwise formation constants, and overall formation constant.

   b. The ethylenediamine tetraacetate ion forms stable complexes with a large number of metal ions. Outline two of the procedures used to increase selectivity.

   c. Abstract from reference sources, summaries of the EDTA titration procedures used for the direct titration of calcium, copper, lead, and nickel salts and the indirect titration of sulfates and phosphates.

   d. Table 3.38 of the "Handbook of Analytical Chemistry" provides a list of the masking agents commonly used in EDTA titration procedures. Use this table to prepare a list of suitable reagents for the masking of Al^{3+}, Cd^{2+}, Fe^{3+}, Pb^{2+}, Ti^{4+}, and Zr^{4+}. What other elements would be simultaneously masked by the various reagents proposed?

   e. For some applications a combination of masking agents is used—why might this lead to indistinct end points and titration errors?

   f. Explain the meaning of the term "demasking" and quote some examples of this procedure.

5. a. The titration of metal ions by monodentate ligands is usually not feasible. Why?

   b. Write down the chemical equations for the primary titration reaction and the indicator reaction in the determination of Ni^{2+} by titration with KCN.

      Repeat for the determination of Ag^+ by titration with the same reagent.

   c. What is the weight of KCN present per liter of a plating solution, if a 50 ml aliquot requires 17.5 ml of $0.5\,M$ silver solution for titration to the appearance of a slight turbidity?

   d. When a 100 ml aliquot of a solution containing potassium di-hydrogen phosphate and magnesium sulfate was passed through a bed of cation-exchanger present in the hydrogen form, the resultant acid solution required 24.8 ml $0.2\,M$ NaOH for neutralization using methyl orange indicator and a further 5.2 ml of titrant using phenolphthalein as indicator. What volume of $0.1\,M$ EDTA would be required to titrate the magnesium content of this aliquot?

      If the phosphate content was determined indirectly, what volume of $0.1\,M$ EDTA would be required in this determination?

6. a. Explain why the pH of the solution is an important factor in selecting an indicator for a chelometric titration.

b. The concept of an effective stability constant is useful in chelometric titrations. Why?

c. If the titration factor, T, for the visual detection of equivalence point is taken as 5, what is the smallest value of K_{eff} which would permit a feasible titration when the initial concentration of metal ion is 0.15 M, 0.03 M, or 0.006 M?

  Would this calculation be valid when the solution contains species which form weak complexes with the metal ion involved?

d. It is desired to titrate an 0.05 M solution of Cu^{2+} with EDTA using a visual indicator. What is the lowest pH at which the titration could be performed? K for Cu EDTA $\simeq 10^{18}$.

e. Outline some of the factors which can cause errors in complexo-metric titrations.

7. a. Prepare a titration curve by calculating the appropriate pM values, for the titration of:

  i. 0.1 M NaCl with 0.2 M AgNO$_3$
    K_{sp} for AgCl $= 10^{-10}$

  ii. 0.01 M K$_2$CrO$_4$ with 0.1 M AgNO$_3$
    K_{sp} for Ag$_2$CrO$_4 = 10^{-12}$

  iii. 0.001 M AgNO$_3$ with 0.005 M KCNS
    K_{sp} for AgCNS $= 10^{-12}$

b. If 50 ml of a solution which was 0.08 M in respect to KCl and 0.005 M in respect to K$_2$CrO$_4$ were titrated with 0.1 M AgNO$_3$, what volume of titrant would be required to precipitate the first trace of red silver chromate?

c. In one particular area sea water is being contaminated by the discharge of sulfuric acid from an industrial plant. Suggest titrimetric methods for determining the total acidity, the chloride content and the sulfate content of the water.

8. a. Prepare a brief dissertation on the mechanism of precipitation indicators, covering all three classes (specific color reactions, colored precipitate formation and adsorption indicators) and quote examples of each type.

b. Why is it desirable to add a permanent colloid like dextrin to solutions when using an adsorption indicator?

c. What factors inhibit the wider acceptability of precipitation titration procedures?

9. a. Explain why formal potentials are of more practical value than standard reduction potentials.

b. Describe the basic properties an organic compound should possess if it is to be used as a visual redox indicator.

c. Using the Nernst equation calculate the potential changes associated with the titration of 0.01 M iron (II) solutions (in M H$_2$SO$_4$) with cerium (IV) sulfate, potassium dichromate, and potassium permanganate solutions of equal molarity.

From a list of redox indicators, nominate suitable indicators for the visual detection of the equivalence point in these reactions.

Would the number of suitable indicators be increased or unchanged if the iron solution were treated with sodium fluoride or phosphoric acid?

d. In neutral solution, arsenic (III) is quantitatively oxidized to arsenic (V) by a solution of potassium triiodide (KI_3). In acid medium, arsenic (V) oxidizes iodide ions to free iodine. Explain this pH effect.

e. Enumerate the possible sources of error in a redox titrimetric procedure.

10. Standard quantitative text books quote few examples of the use of redox methods for the determination of organic compounds. Using Ashworth's book on organic analysis or other reference books, prepare a brief article on this topic which might be included in a future text.

References

M. R. F. Ashworth, *Titrimetric Organic Analysis*, Wiley (Interscience), New York, 1964–65.

M. Bobtelsky, "*Heterometry*," Elsevier, Amsterdam, 1960.

I. Gyenes, *Titrations in Non-Aqueous Media*, Iliffe, London, 1967.

I. M. Kolthoff and P. J. Elving (eds.), *Treatise on Analytical Chemistry*, Vol. 1, Wiley (Interscience), New York, 1959.

I. M. Kolthoff and N. H. Furman, *Potentiometric Titrations*, Wiley, New York, 1931.

I. M. Kolthoff and N. H. Furman, *Volumetric Analysis, I*, Wiley, New York, 1928.

I. M. Kolthoff and V. Stenger, *Volumetric Analysis, II*, Wiley (Interscience), New York, 1947.

I. M. Kolthoff and R. Belcher, *Volumetric Analysis, III*, Wiley (Interscience), New York, 1957.

J. Kucharsky and L. Safarik, *Titrations in Non-Aqueous Solvents*, Elsevier, Amsterdam, 1964.

J. Mitchell and D. M. Smith, *Aquametry*, Wiley (Interscience), New York, 1948.

A. Ringbom, *Complexation in Analytical Chemistry*, Wiley (Interscience), New York, 1963.

G. Swarzenbach and H. Flaschka, *Complexometric Titrations*, Methuen, London, 1968.

O. Tomicek, *Chemical Indicators*, Butterworth, London, 1951.

F. J. Welcher, *The Analytical Uses of Ethylene Diamine Tetraacetic Acid*, Van Nostrand, Princeton, 1958.

C. L. Wilson and D. W. Wilson (eds.), *Comprehensive Analytical Chemistry*, Vol. 1B, Elsevier, Amsterdam, 1960.

12

IONIC REACTIONS—
SELECTIVE PROCEDURES AND KINETICS

I. Selectivity through Preoxidation or Reduction 391
   A. Thermodynamic Prediction 391
   B. Methods of Preoxidation and Reduction 393
II. Organic Reagents 394
   A. Applications . , ... 394
   B. Factors Influencing Selectivity 395
III. Precipitate Formation 399
   A. Kinetics of Precipitate Formation 399
   B. Properties of Precipitates 403
   C. Gravimetric Analysis 405
IV. Analytical Methods Based on Solution Kinetics 407
   A. An Introduction to Solution Kinetics 407
   B. Kinetic Methods for Catalyzed Reactions. 416
   C. Kinetic Methods Involving Uncatalyzed Reactions 420
V. Literature Assignments 424
   References 426

I. SELECTIVITY THROUGH PREOXIDATION OR REDUCTION

A. THERMODYNAMIC PREDICTION

A change in the oxidation state of an element can often facilitate separation procedures.

In the discussion of selective precipitation (Chapter 10, Section IV,B) it was noted that the pH at which a metal hydroxide precipitates is related to both the magnitude of the solubility product and the valency of the metal ion involved.

Thus, iron (III) hydroxide precipitates at a much lower pH than the corresponding iron (II) compound and preoxidation of the iron salts permits quantitative removal of iron from solutions containing divalent metal ions.

The stability (and solubility) of metal complexes also varies with the oxidation state of the metal ion, and this fact has been used to advantage in the separation of mixtures of metal ions. For example, in the determination of titanium in samples which are primarily composed of iron, such as alloy steels, the iron salts are prereduced to the divalent state. This prevents reaction of the iron with cupferron, an organic precipitant for titanium and iron (III).

Reduction is also used to eliminate the interference of iron in the colorimetric determination of molybdenum in steel. Solution of the steel in fuming perchloric acid converts most of the transition elements present into their higher valency state. To the cold, diluted solution thiocyanate ions are added. This immediately results in the appearance of the intense red, iron (III) thiocyanate complex. On the addition of tin (II) chloride, the iron is reduced; the red color disappears; and the characteristic orange color of the molybdenum thiocyanate complex which is used in the final evaluation step is observed.

In titrimetric methods based on oxidation-reduction, the initial steps in the analytical procedure must ensure that the element of interest is in a definite oxidation state. Accordingly, in the determination of the iron content of a sample by redox titration, the first step is to reduce all the iron to the II state by the addition of tin (II), sulfur dioxide, zinc amalgam, etc. The excess reductant has then to be removed or oxidized prior to titration of the iron. It is obvious that if excess reductant is to be removed by oxidation, the reagent selected must not be capable of accepting electrons from iron (II).

Elements differ in respect to the amount of energy required to change the oxidation state, and by careful selection of reactants it is possible to achieve transformation of one element in the presence of other ions of variable valency. Since energy changes are involved in the oxidation-reduction reactions, selection of the most suitable reagent for a given transformation can be made on the basis of energy calculations. In the same way, it is possible to predict how a group of elements will react towards some particular oxidizing or reducing agent.

It has been noted in the previous chapter that the standard free energy change associated with the reaction

$$a \text{ Ox}_1 + b \text{ Red}_2 \rightleftarrows a \text{ Red}_1 + b \text{ Ox}_2$$

is given by the expression

$$\Delta G^\circ = -abF(E_1^\circ - E_2^\circ).$$

Feasible reactions possess negative ΔG° values; hence any redox reaction which yields a positive value for $(E_1^\circ - E_2^\circ)$ should proceed to an appreciable extent unless kinetically hindered.

·Let us consider what happens to an acid mixture containing iron (III), vanadium (V), and chromium (III) on the addition of (a) zinc dust and (b) potassium iodide.

The half-cell reactions involved are

$VO_2^+ + 2H^+ + e \rightleftarrows VO^{2+} + H_2O$	$E^\circ = 1.00$ V
$Fe^{3+} + e \rightleftarrows Fe^{2+}$	$E^\circ = 0.77$ V
$I_2 + 2e \rightleftarrows 2I^-$	$E^\circ = 0.54$ V
$VO^{2+} + 2H^+ + e \rightleftarrows V^{3+} + H_2O$	$E^\circ = 0.36$ V
$V^{3+} + e \rightleftarrows V^{2+}$	$E^\circ = -0.26$
$Cr^{3+} + e \rightleftarrows Cr^{2+}$	$E^\circ = -0.41$
$Zn^{2+} + 2e \rightleftarrows Zn$	$E^\circ = -0.76$

For the reaction with zinc, $(E_1^\circ - E_2^\circ)$ is positive in all cases. Hence, the addition of zinc dust should reduce the vanadium (V) to vanadium (II), iron (III) to iron (II), and chromium (III) to chromium (II). In other words, zinc dust is a bulk reductant, not a selective reagent.

On the other hand, using iodide as the reductant, $E_2^\circ = 0.54$, and $E_1^\circ - E_2^\circ$ is positive only in respect to the VO_2^+ and Fe^{3+} systems. Thus the addition of iodide ions should cause reduction of the vanadium (V) to vanadium (IV) and iron (III) to iron (II). This is a more selective reaction.

Thus, by judicious use of a table of formal potentials, it is possible to select reagents which react with a minimum number of the species in solution. However, other considerations, such as cost and rates of reaction, have tended to limit the materials used to a few.

B. METHODS OF PREOXIDATION AND REDUCTION

Examples of selective preoxidation are not common. Two which may be quoted are the oxidation of vanadium (IV) salts to vanadium (V) by potassium permanganate in the presence of manganese (II) and chromium (III) compounds; and the oxidation of iron (II) salts by ammonium persulfate (cold solution, no catalyst) in the presence of manganese (II) and chromium (III) compounds.

For group oxidation, the transfer of electrons can be induced during dissolution of the sample or at a later stage.

For highly refractory materials, a method of sample dissolution which simultaneously achieves oxidation is fusion with sodium peroxide. While this procedure has much to recommend it as a method of sample dissolution, acidification of the melt yields hydrogen peroxide, and this compound can partially reduce material fully oxidized in the fusion process.

In acid dissolutions, hot nitric acid has been used to oxidize organic matter, metal carbides, and iron (II) salts. Hot, concentrated perchloric acid is a more vigorous oxidant [e.g., converts Cr(III) to Cr(VI)], but in the absence of mineral acid it reacts explosively with organic matter and this tends to inhibit its use by some chemists. The oxidizing power of perchloric acid can be nullified by merely cooling and diluting the solution.

Ammonium persulfate is another reagent which reacts slowly in the cold, and to ensure a reasonable rate of oxidation, silver (I) ions have to be added as catalyst and the solutions heated. Solutions containing bromine and chlorine are also boiled, but in this case the aim of the heating is to remove excess oxidant.

To simplify the removal of excess oxidant, a number of workers have elected to use solid oxidants. Among the compounds which have been investigated are sodium bismuthate, silver (II) oxide, and lead (IV) oxide.

In chemical analysis most attention has been paid to methods of selective reduction. The reductants used include gases, metals, and ions in solution.

With gaseous reductants some selectivity is achieved, and excess is readily removed by boiling the solution. For example, sulfur dioxide in dilute sulfuric acid solutions reduces iron (III) to iron (II), vanadium (V) to vanadium (IV), and antimony (V) to antimony (III); but compounds of molybdenum, tungsten, and uranium are not reduced. This provides a convenient means of determining vanadium in the presence of molybdenum. Similar reductions are achieved using hydrogen sulfide, with the added advantage that many other metal ions are precipitated as the sulfides.

Examples of ionic reactions are the reduction of iron (III) by tin (II) and permanganate ions by chloride ions.

Reduction by granulated metal has proved very popular, particularly where column operation has been adopted.

In the column technique, an acid solution of metal ions is allowed to flow through a column packed with granulated metal. Metals which have been used include zinc, silver, cadmium, and lead. Amalgamation of the metal surface facilitates reduction. The standard reduction potential for the reaction $Zn^{2+} + 2e \rightleftarrows Zn$ is -0.76 V. The corresponding potentials for the cadmium and lead systems are -0.40 V and -0.13 V respectively; hence a column of these materials provides a more restricted reducing action. The silver column is used in conjunction with hydrochloric acid solutions and the reduction potential of 0.22 V is determined by the redox couple

$$AgCl + e \rightleftarrows Ag + Cl^-$$

One advantage of using columns is that no subsequent separation of excess reductant is required. Separation is a minor problem when liquid amalgams are used instead of amalgamated columns. In the liquid amalgam technique, 3–4% of metal is dissolved in mercury, and the amalgam is shaken with the acid solution to be reduced. Before titration the amalgam has to be run off. The reducing power can be controlled by varying the nature of the metal used (amalgams of zinc, cadmium, lead, bismuth, tin, etc.) and by controlling the acidity and temperature of the solution under study.

The ability of mercury to form amalgams, coupled with the high hydrogen overvoltage on this metal, provides the basis of another method of selective reduction, namely mercury cathode separations. In this technique, a pool of mercury is used as the cathode in an electrical circuit, and by controlling the cathode potential it is possible to subject an acidified test solution to any desired reduction potential. The scope of the technique has been fully outlined in review form [J. A. Page, J. A. Maxwell, and R. P. Graham, *Analyst* **87** (1962) 745].

II. ORGANIC REAGENTS

A. APPLICATIONS

The ability of many organic compounds to react with inorganic species has been utilized in a number of analytical techniques.

Some of these applications are well known and have been discussed in previous sections, e.g., the titration of metal ions with EDTA and the oxidation of oxalic acid by potassium permanganate.

A more extensive application of organic reagents has been in gravimetric analysis. Organic precipitants tend to react selectively with a limited number of elements, and the reaction can often be made specific by the control of pH and by the addition of other complexing ligands to the solution. Because the precipitates are usually not ionized salts, they do not coprecipitate impurities in the same way as most inorganic precipitates. Because of the bulk of the organic component, organic reagents have been used with success for the recovery and concentration of microcomponents of the sample.

Many organic precipitates are soluble in organic solvents; hence the selectivity of precipitation can often be followed by the convenience of solvent extraction. This usually leads to a colored extract which is suitable for determinations based on colorimetry.

Most solid, metal-organic reagent complexes are anhydrous, and are easily dried for weighing. However, in most cases the reagents also have limited solubility in water and have to be added as a solution in another solvent. Too large an excess of reagent leads to contamination of the precipitate, and for gravimetric evaluation the metal chelate then has to be ignited to the metal oxide.

The solubility of a metal chelate can often be varied by altering the structure of the organic reagent. Increasing the molecular weight tends to increase the sensitivity of the reaction. Adding hydrocarbon substituents generally decreases the solubility of the precipitate and reagent; the introduction of hydrophilic groups such as —OH and —SO_3H into the reagent molecule can have the opposite effect.

Many of the metal compounds are highly colored; hence the production of a water-soluble analog is of value for direct colorimetric analysis. The molar absorptivity is usually high and, as in precipitation, with control of pH and use of subsidiary masking reagents a fair degree of selectivity can be achieved.

Organic reagents are themselves used as masking agents. This masking effect can lead to a change in the redox properties of the metal ion system, thus facilitating indicator selection in titrimetry or the separation of species in polarographic studies.

In summary, organic reagents are used extensively in titrimetry, polarography, colorimetry, and gravimetric analysis. In all these applications, the prime attribute is selectivity.

B. FACTORS INFLUENCING SELECTIVITY

The selectivity of organic reagents is associated with definite atomic groupings in the reagent molecule. The organic reagents that react with metal ions all contain groups with replaceable hydrogen atoms (—COOH, —SO₃H, —SH, etc.). Reagents which form insoluble chelate compounds contain, in addition, a functional group of basic character such as —NH₂, =N—, =O, with which the reacting metal is coordinated to form a five- or six-membered ring.

The chelate ring, when formed, must be almost free of strain. To achieve this, the molecule must be of the right size and shape to give a strain-free ring with the metal ion of interest. Conversely, for a metal ion to react with a particular reagent it must possess the right size, oxidation state, and coordination number.

The type of rings formed by a number of common organic precipitants are shown in Table 12.1.

The considerations of size and shape lead to the idea of certain combinations of functional groups being specific in their action. Thus numerous α-dioximes, containing the structural group I react with nickel salts to form colored, usually red, precipitates having a structure similar to II.

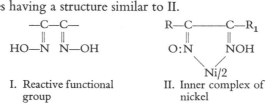

I. Reactive functional   II. Inner complex of
       group   nickel

Actually, the dioxime grouping represents only one particular arrangement of the nickel binding group. Characteristic inner complex nickel salts are also formed by 2-pyridyl ketoximes, e.g.,

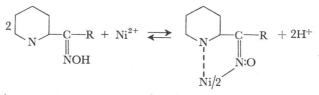

While the most important use of such reagents is for the detection and determination of nickel, other metal ions can react with this grouping. Thus, in acid media, palladium, platinum, and bismuth form yellow precipitates.

The copper specific groups are considered to be

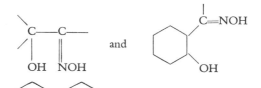

while the group, ⬡⬡, is the determining factor in the formation of red inner complexes of iron (II) and the specificity of this iron reaction.

TABLE 12.1

CHELATES FORMED BY TYPICAL ORGANIC REAGENTS

Organic reagent	Compound formed with metal ion (M^{n+})	Remarks and applications
 8-hydroxyquinoline		Used in the determination of Mg, Zn, Al, and many other metals. Group separations by pH control.
 Nitroso-R-Salt		Colorimetric detn. of Co, indirect detn. of potassium.
 Cupferron		A precipitant for Fe, Ti, Zr, U, Sn, Al, Bi, Cu, Nb, Th, and others.
 o-Phenanthroline		A sensitive reagent for colorimetric determination and detection of iron (II).
 Benzoin α-oxime	 Used in determination of Cu and Mo.	
 Thionalide	 Used for determination of elements of H_2S group.	

The thioamide group reacts preferentially with Bi^{3+}, Fe^{3+}, and Ag^+ ions, while substituted arsonic acids react selectively with zirconium.

The introduction of substituents into the molecule in the vicinity of the reactive functional groups can lead to a loss of analytical value. This can usually be attributed to steric effects, bond formation through orbital overlap being hindered by bulky neighboring groups.

TABLE 12.2

ORGANIC REAGENTS FORMING SALTLIKE PRECIPITATES

Reagent	Species determined
Sodium tetraphenyl boron $NaB(C_6H_5)_4$	K^+
Sodium diethyldithiocarbamate $\quad\quad \overset{C_2H_5}{\underset{C_2H_5}{\diagdown}}NCSSNa$	Cu, Zn
Phenylarsonic acid $C_6H_5\!-\!As\!\!=\!\!O$ (with OH, OH)	Sn, Th, Zr, Bi, Nb, Hf, Ta
Tetraphenylarsonium chloride $(C_6H_5)_4AsCl$	Tl, Sn, Au, Zn, Pt, Hg, Cd, Re, Tl
Benzidine $H_2N\!-\!\langle\,\rangle\!-\!\langle\,\rangle\!-\!NH_2$	SO_4^{2-}, PO_4^{3-}, IO_3^-, Cd, W
Oxalic acid $(COOH)_2$	Ca, Ag, Au, Hg, Pb, Sc, Th, Zn

The introduction of substituents into the reagent molecule beyond the limits of the functional-analytical group can have a favorable effect on the solubility. color, and stability of the reaction product.

Introducing polar groups tends to increase the solubility; extending a conjugated system can increase color intensity. The stability of the product is enhanced by increased basicity in the reactive functional groups, and this can be achieved through the introduction of electron donor groups.

The number of atomic groupings that act as functional-analytical groups is quite limited, but the number of substituent combinations is almost infinite. Thus for the gravimetric or colorimetric determination of any particular ion, it is usually possible to unearth from the literature many different reagents of recommended quality. Conversely, each organic reagent can usually be applied to the determination of several elements. To make a choice between the alternatives, an analyst should first define the problem in terms of concentration range to be handled, solubility required, and interfering ions likely to be encountered. Then reference books devoted to organic reagents should be consulted to find the compound which most closely meets the defined specifications. The pH is always important in precipitations with complex-forming organic reagents, because these are all weak acids, and the higher the pH, the more readily are their hydrogen atoms displaced.

Some organic precipitants form salts rather than chelate complexes with inorganic ions, e.g., oxalic acid is well known for its use in the precipitation of calcium. Other sparingly soluble salt-formers are listed in Table 12.2.

These reagents represent a bridge between the chelate formers and the inorganic precipitate formers discussed in the next section.

III. PRECIPITATE FORMATION

A. KINETICS OF PRECIPITATE FORMATION

A precipitate is suitable for the purposes of gravimetric analysis if,

1. Its solubility is negligibly small;
2. It is readily filtered and washed, i.e., its morphological structure is suitable;
3. Its composition at the time of weighing corresponds to a definite formula. That is, the degree of contamination by foreign ions is negligibly small.

All of these factors are related, to a greater or lesser extent, to the rate of precipitate formation.

Analytical precipitates are in most cases slightly soluble salts. At equilibrium, the number of ions going into solution is equal to the number reaching the surface of the crystal in a given time unit.

$$AB_{(s)} \underset{k_2}{\overset{k_1}{\rightleftharpoons}} A^+ + B^-$$

The rate of dissolution, $v_d = k_1 a_{AB} = k_1$ since the activity of the solid (a_{AB}) may be taken as unity.

The rate of deposition or precipitation, v_p, is described by the expression

$$v_p = k_2 a_{A+} \cdot a_{B-}$$

where a_{A+} and a_{B-} are the activities of the ions A^+ and B^- respectively.

At equilibrium the two rates are equal, i.e., $v_d = v_p$, hence

$$k_1 = k_2 \cdot a_{A+} \cdot a_{B-}$$

$$a_{A+} \cdot a_{B-} = k_1/k_2 = K_{sp}$$

This is the well-known solubility product equation. An alternative form of this equation is:

$$[A^+][B^-] = k_1/k_2 f_A f_B$$

where k_1 is the specific rate constant for dissolution; k_2 is the specific rate constant for deposition; and f_A and f_B are the activity coefficients for the ions A^+ and B^- respectively.

1. Solubility Considerations

If A^+ is the species being determined through the addition of reagent B^-, $[A^+]$ is a measure of the solubility of the precipitate.

For quantitative separations, the aim is to make $[A^+]$ as small as possible. At a given temperature, this requires $[B^-], f_A$ and f_B to be as large as possible; $[B^-]$ can be increased by adding excess reagent, but this effect is countered to some extent by the influence of added electrolyte on f_A and f_B.

The magnitude of the activity coefficients decreases as the ionic strength of the solution increases. Thus the presence of electrolytes of any kind tends to increase solubility. This effect is enhanced if some of the added species tend to interact with the components of the precipitate. For example, if B is the conjugate base of the weak acid HB, solubility becomes pH dependent since the equilibrium

$$H^+ + B^- \rightleftharpoons HB$$

effectively reduces the amount of B^- available for the deposition reaction.

Similarly, the addition of a ligand (L^{m-}) which forms complexes with A^+ can increase the amount of this species present in a soluble form.

$$A^+ + xL^{m-} \rightleftharpoons AL_x^{(mx-1)-}$$

If, as in this example, the ligand is the conjugate base of another weak acid, the amount of ligand present is pH dependent and the solubility of the precipitate is determined by the stability of the complex, the amount of ligand present, and the pH.

The magnitude of the rate constants k_1 and k_2 are temperature variable, and since the dissolution of most precipitates is an endothermic process, k_1 tends to increase more with temperature than k_2. Thus an increase in temperature leads generally to an increase in solubility.

In organic solvents, with small dielectric constants, the activity of water as well as the heat of solvation of the ions is decreased. This slows the dissolution process and the probability of deposition is increased. Accordingly, ionic type precipitates are usually less soluble in organic solvents or in aqueous mixtures of these solvents than in pure water.

Besides varying with temperature and solvent composition, k_1 can sometimes vary with the particle size. As a general rule, the solubility is independent of the particle size if the size of the particles exceeds 1–2 μ. However, with particles smaller than this, increased solubility is often observed.

The attractive force or sphere of influence of a charged particle decreases with distance and one can imagine each particle to have a limited effective radius. If the charged particle being considered is located at the surface of a crystal of large radius of curvature (i.e., almost linear), half of the sphere of influence is exposed to the solvent, and half to the surrounding solid. In the solid phase, attraction to oppositely charged species creates a bonding force which resists the tendency of the particle to solvate and enter the solution. This force can be termed the surface energy of the particle.

As the radius of curvature of the solid particle decreases, more than half of the sphere of influence of the particle extends into the surrounding solvent and less than half of the total capacity is involved in binding the particle to the solid core. The force binding the particle to the solid is therefore very small at edges and spiked peaks, as well as on particles which are smaller than the effective sphere. This increases the probability of dissolution of the particle, and more ions (atoms or molecules) dissolve in a unit of time from the surface of small particles and sharp edges than from large particles and smooth surfaces. That is, k_1 is a function of the surface energy or surface tension of the solid σ.

The increase in solubility can be calculated by means of the thermodynamic expression

$$\ln Sr/S = 2\sigma M/rdRT$$

where Sr is the solubility of a small crystal of radius, r, S is the solubility of a large crystal; d is the density of the solid and M is the molecular weight of the compound.

For small values of Sr/S, $\ln Sr/S \simeq (Sr - S)/S$, and for a given precipitate, the relative solubility increase $\{(Sr - S)/S\}$ is thus approximately inversely proportional to the radius of the particles. In the case of particles of the same size but of different composition, the relative solubility increase is directly related to the surface tension of the precipitate (σ). These two generalizations are of importance when considering procedures for purifying prior to evaluation by weighing.

2. *Formation of Precipitates*

In precipitation reactions a new compound of limited solubility is formed by chemical reaction. During the initial mixing of the reactants the solution becomes supersaturated, and the separation of the new compound starts with the formation of very small particles (termed nuclei). These particles grow more or less rapidly and finally settle out from the solution.

The particle size distribution of the precipitate is determined by the relative rates of the two processes, the formation of nuclei, called nucleation, and the growth of nuclei. Thus, if the rate of nucleation is small compared to the rate of growth, fewer particles are finally produced and these particles are of relatively large particle size. Such a material is generally purer and more easily filtered and washed than materials of smaller particle size.

It has been observed that the rate of nucleation depends on the degree of supersaturation. A solution is said to be supersaturated when it contains a concentration of solute which exceeds that found in a saturated solution.

$$\text{Supersaturation} = Q - S$$

where Q is the concentration of the solute in solution at any instant and S is the equilibrium concentration in a saturated solution of the same solute.

Numerous attempts have been made to ascertain the quantitative relationship between these two factors and the most quoted expression is that of von Weimarn, who proposed that the rate of nucleation (R_n) is directly proportional to the relative supersaturation, i.e.,

$$R_n = K(Q - S)/S$$

The rate of growth of particles is a diffusion controlled process. At the surface of the crystal the concentration of common ions may not be greater than S; in the bulk solution the concentration may be higher (Q). The concentration difference driving the diffusion process is thus $(Q - S)$ and one can propose that the rate of growth (R_g) is proportional to the absolute supersaturation

$$R_g = K'(Q - S)$$

Combining these two expressions yields the equation,

$$\text{Rate of nucleation/rate of growth} = K''/S$$

Thus if the solubility of a compound is extremely small (as in the case of the hydrous oxides of iron, aluminum, chromium, etc.), the rate of nucleation is so great relative to the growth process that all the precipitated material tends to appear as nuclei or extremely small particles.

Conversely if S is large, growth can predominate and the product obtained is composed of a smaller number of large particles.

However, nuclei must be formed before growth can occur; hence the relationship of primary importance is the von Weimarn equation.

On the basis that the number of nuclei formed will be directly related to R_n, it can be observed that fewer particles require Q to be small and S to be large.

By slowly adding very dilute reagent with vigorous stirring Q can be kept small. The magnitude of S can be temporarily increased by increasing the temperature, lowering the pH and/or adding a complexing agent. In some cases, S is so small that variations in conditions have no significant effect.

An alternative method for keeping Q small is to very slowly liberate the precipitation agent by means of a kinetically controlled hydrolysis reaction. This procedure is known as homogeneous precipitation.

In this technique, the substance added does not contain the desired precipitating ion as such and the solution remains homogeneous. On the application of heat, hydrolysis occurs and the required species is formed in situ.

Precipitation from homogeneous solutions yields products of greater purity and larger particle size than those obtained in the usual procedure (due to the low degree of supersaturation and slow rate of growth), but these advantages are offset to a certain extent by the time required for the hydrolytic reactions. The scope and advantages of the technique are fully discussed in the monograph by Gordon, Salutsky, and Willard.

Either an anion or a cation can be generated homogeneously to form a precipitate. For example, the sulfate ion can be produced from sulfamic acid or dimethyl sulfate; oxalate ions can be liberated from dimethyl or diethyl oxalate; hydrogen sulfide can be obtained from hydrolysis of thioacetamide.

The generation of a cation in solution may be effected by releasing the cation from a complex by removal of the ligand. This can be achieved by either changing the pH of the solution, changing the temperature or by destroying the complexing agent.

For the neutralization of an acid, the hydrolysis of urea to form ammonia and carbon dioxide is a convenient process; the addition of an organic acid being often used to buffer the solution to the pH desired for precipitation. Apart from urea, hexamethylene tetramine ($C_6H_{12}N_4$), and acetamide have been used to release hydroxyl ions.

B. Properties of Precipitates

It is obvious from the preceding discussion that one can obtain precipitates whose sizes vary considerably. Ions in true solution have radii of a few angstrom units (10^{-10} meters). In the process of forming nuclei the particles achieve sizes ranging from 0.001 to 1 μ, while after growth the final products preferably possess radii greater than 1 μ (10^{-6} meter).

Due to large values of Q during precipitation, or inherent small values of S, many precipitation reactions do not yield particles with radii larger than a micron. This brings them within the classification of colloidal suspensions. Colloidal particles do not settle under the pull of gravity; they pass through ordinary filters and they bear an electrical charge ($+$ or $-$). Because of their small size the surface area per unit weight is much greater than that observed with larger particles. These differences in properties make it desirable to consider two distinct types of precipitate, namely, crystalline products and colloidal aggregates.

Crystalline particles grow by the addition of positive and negative ions on all available surfaces. The ions are not built into the crystals one by one, but they become arranged in swarms which settle on the surface in the form of small blocks. The process continues until the concentration of one of the ions approaches zero. At this stage, the second ion proceeds to take up its normal lattice position to yield a surface that is electrically charged due to the absence of the second component. The monolayer on the surface is known as the primary adsorbed layer and the electrical charge tends to be neutralized by a diffuse zone of oppositely charged ions from the bulk solution. Thus all precipitates are contaminated by material adsorbed on the outer surface, but with large particles (e.g., $>1 \mu$) the surface area is relatively small and much of the adsorbed matter can be removed by washing.

A more important source of error is the impurities introduced into the crystal itself during the precipitation process.

During the growth stage, all charged ions in solution tend to come in contact with the developing surfaces. With slow conditions of growth, foreign ions have time to diffuse away from the surface and growth takes place in the ideal manner indicated above. On the other hand with rapid growth, many foreign ions are trapped inside the growing crystal. The trapped ions are said to be occluded impurities. Their presence not only reduces the purity of the precipitate but it induces faults in the crystal surface, and growth becomes irregular and uneven in different directions. The degree of contamination increases when the foreign ion forms a strong bond with an ion of the precipitate. Should ionic sizes be comparable ($\pm 10\%$), the foreign ions can become integral parts of the crystal lattice. This process is known as solid solution and it is extremely difficult to free a precipitate from such impurities. Examples of solid solutions are $PbSO_4$—$PbCrO_4$, $PbSO_4$—$BaSO_4$, and $MgNH_4PO_4$—$MgKPO_4$ mixtures.

Most of the occluded impurities can be released by digestion of the precipitate in contact with its mother liquor at an elevated temperature for a reasonable period of time (e.g., several hours). In this period there is continual solution of the solid, particularly at sharp edges and with small particles. The ions so liberated tend to deposit on the more regular surfaces (i.e., larger radii of curvature), and medium sized particles tend to grow together. During the slow process of solution and deposition, occluded impurities have the opportunity to escape into solution. Digested material is thus usually purer, larger, and more regular in shape than the initial product.

Precipitates containing particles of colloidal or near colloidal size rarely show any change in form or purity on digestion. The surface area of a colloidal precipitate can be several million times greater than the area of a corresponding weight of crystalline matter. This means that the amount of adsorbed impurity is now significant and usually it can be related to the concentration of electrolytes in the precipitating medium by an empirical relationship such as the Freundlich isotherm,

$$x/m = kc^{1/n}$$

where x is the weight adsorbed, m is the mass of precipitate, c is the electrolyte concentration, and k and n are constants, n being greater than unity.

With colloidal particles, the charge of the primary adsorbed layer is sufficient to create the equivalent of an electrically charged unit, and since all the particles bear the same charge, they tend to repel each other and remain dispersed. However, the particles can be coagulated into a gellike mass by adding an electrolyte whose ions provide a secondary adsorption layer capable of masking the primary charge. (Typical colloidal precipitates are the hydrous oxides of iron (III) and aluminum.) The coagulated mass can be retained on a filter paper, but if washed with water, the electrolyte is removed and the individual particles again separate and pass through the filter; this process is known as peptization. To avoid this process, coagulated colloidal precipitates must be washed with electrolyte solutions.

Purification of colloidal aggregates is achieved by reprecipitation. The initial precipitate is returned to solution by treatment with a minimum amount of acid or other suitable reagent. After dilution with water, the precipitate is reformed using a minimum excess of the precipitating reagent. Due to a lower electrolyte concentration, the amount adsorbed is greatly decreased. The electrolyte added to ensure coagulation should preferably be a species that can be destroyed by heating at a later stage. For example, hydrous iron (III) oxide is usually reprecipitated in the presence of ammonium nitrate because this salt volatilizes at the temperatures required to remove water from the precipitate.

A number of colloidal and crystalline products are subject to another form of contamination known as post-precipitation. In this case the impurity deposits on the surface of the original precipitate subsequent to quantitative deposition of the latter, and the amount formed increases with time. Common examples are the post-precipitation of zinc on copper sulfide and magnesium on calcium oxalate. The phenomena can be explained in terms of the concentration of anions at the surface of the original precipitate being of such a magnitude that the solubility product of the second compound is exceeded in the region of this surface layer. This effect can only be minimized by reducing the time of contact of the solid with the mother liquor.

There is an intermediate type of precipitate which has a surface tension which precludes growth and purification by digestion, and a limited solubility in common solvents which precludes purification by reprecipitation. A typical example of this is silver chloride. In such cases purification is restricted to efficient washing.

The nature of the wash liquor required varies from precipitate to precipitate but the requirements are uniform. The wash liquor should not dissolve the precipitate but it should displace the impurities attached to the surface of the precipitate and replace them with species that can be completely removed in a subsequent heating process. For example, silver chloride precipitates are washed with dilute nitric acid and magnesium ammonium phosphate crystals are washed with ammonia. Barium sulfate precipitates can be washed with hot water while the more soluble lead sulfate is washed first with dilute sulfuric acid (to remove ionic impurities) and then with alcohol (to remove any adhering acid).

C. GRAVIMETRIC ANALYSIS

In gravimetric analysis, the initial step is to produce a precipitate of limited solubility. This material then has to be purified, dried, and converted to some stable form before being weighed. The product weighed should have a definite chemical composition and should preferably be nonhygroscopic. The amount of heating required to achieve this aim is best determined from thermogravimetric curves as described in Chapter IV, Section II. In some cases heating to 105°C to remove water is adequate; at other times temperatures of up to 500°C are needed to destroy filter paper or other organic matter; and in other situations required chemical transformations necessitate the use of temperatures of up to 1000°C. Where the precipitate has been obtained with the aid of an organic reagent, some additional care is needed. Owing to their nonpolar nature, many of the compounds tend to volatilize, without decomposition, at reasonably low temperatures. Thus ignition losses can be heavy if the aim is to convert the precipitate into the corresponding metal oxide.

Despite the limitations imposed by solubility effects, coprecipitation effects and adsorption effects, gravimetric analysis remains as the most accurate method of chemical analysis for a large number of elements.

The isolation and weighing of a precipitate is an absolute method; hence this technique is widely recommended as a standard procedure when accuracy, and not speed, is the prime requisite.

Gravimetric analysis can be time consuming, it can be tedious, and it demands a high degree of manipulative skill and patience from the chemist who uses such procedures. For these reasons, the technique has been largely supplanted as a means of routine analysis, but it is doubtful if it will ever be fully replaced as a means of obtaining the standard results required in the calibration of modern instrumental procedures.

Precipitation is also a very valuable method for separating a sample into its component parts, and until recent years it was the analyst's most widely used separation technique.

The easiest way of describing the scope of gravimetric analysis is to indicate that it is suitable for everything except trace analysis. For nearly every element in the periodic table there exist several recommended gravimetric procedures, and for separating the groups of elements present in common commercial products, many alternative procedures have been proposed.

The "Handbook of Analytical Chemistry" provides a summary of the methods used for elemental determinations while full details of procedures are quoted in the three volume monograph by Erdey and the major reference books quoted in the bibliography.

For the purpose of assessing the possible suitability of a gravimetric procedure for some calibration process or separation problem, reference to the Handbook may be sufficient.

In Table 3.1 there is a list of the various elements precipitated by thirty general analytical reagents, and this table provides an indication of the interference effects which may be encountered in using a particular reagent.

The next table lists the conditions commonly used for the gravimetric precipitation of the different elements, while Table 3.3 summarizes information on precipitations from homogeneous solution.

Since in gravimetric analysis, interference effects are of prime concern, another table (3.6) lists the species which are most likely to interfere in the methods outlined in Table 3.3, and the special precautions needed to minimize the effect are noted in summary form.

Finally, information is provided on suitable heating temperatures for precipitates, the thermal stabilities of analytical precipitates, and gravimetric factors.

Thus by using these tables, the feasibility of any proposal can be rapidly checked. However, before implementing any proposal, it is advisable to cross–check with the more detailed information available in the major reference works.

IV. ANALYTICAL METHODS BASED ON SOLUTION KINETICS

In previous discussions, reference has been made regularly to the kinetics of processes. In titrimetry it was stated that the basic reaction must proceed at a rapid rate; in the discussion of redox equilibria it was pointed out that many electron transfer processes were too slow to be of analytical value; and in the preceding section the influence of kinetic factors on the nature of precipitates was discussed.

Besides providing a better understanding of how chemical reactions occur, kinetic data have many uses. For example, in chemical analysis, rate data can be used to select optimum conditions; i.e., conditions of maximum rate or conditions in which an interfering reaction is suppressed. The use of a catalyst to speed up a reaction has been noted in several previous sections.

In recent years, analytical chemists have extended their interest in kinetics to include the application of kinetics to the direct quantitative evaluation of the amount of a reactant or catalyst present in the system under study.

Kinetics pervades all chemistry, and a detailed study of this topic usually forms an integral part of physical chemistry courses. However, despite the risk of duplication, it is appropriate to begin this section with a brief reiteration of the general principles of kinetic studies.

A. AN INTRODUCTION TO SOLUTION KINETICS

In kinetic studies, the aim is to study the effect of variables such as reactant concentrations and temperature on the rate of a chemical reaction.

A variety of physical and chemical methods have been used to follow the progress of chemical reactions. These can be divided into two groups.

1. Conventional techniques, such as titrimetry, colorimetry, and electro-analytical procedures which are used when it takes more than a minute. for the reaction to proceed halfway towards equilibrium; and
2. Fast techniques, such as rapid mixing methods, continuous flow, temperature and pressure jump procedures, which are used where the reaction is more than half-complete in periods of time ranging from 30 to 10^{-9} sec.

A discussion of the fast techniques is beyond the scope of this book. Most analytical applications of kinetics are concerned with slower reactions, and these

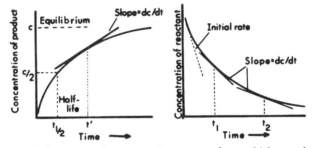

FIGURE 12.1   Typical concentration versus time curves from which rate data can be determined. The slope of the tangents at points 1 and 2 are a measure of the rate of reaction at times t_1 and t_2.

can be followed by the various experimental techniques discussed in other chapters.

The technique selected must be capable of monitoring either the rate of formation of some product with time, or the rate of loss of some reactant with time.

This is achieved by analyzing the reaction system after various periods of time subsequent to the mixing of the reagents. The actual methods of analysis required vary from reaction to reaction. As a general rule, nondestructive physical measurements are preferred, since they do not disturb the course of the reaction and many measurements can be made on a single sample over a period of time. With destructive techniques batch operations are involved. Many identical reaction mixtures are prepared at the same time, and each mixture is analyzed after the lapse of a different time interval.

The results are plotted on a graph to yield rate curves similar to those shown in Fig. 12.1.

Every change in experimental conditions tends to give some change in the shape or position of the experimental curve.

The rate of reaction at any time is ascertained by measuring the slope of the tangent to the curve at this point. Mathematical manipulation is then required to ascertain the relationship between the observed rate and the experimental variables. For the general reaction

$$A + B \rightleftarrows C + D$$

if $d[C]/dt$ and $d[D]/dt$ are the rates of product formation and $-d[A]/dt$ and $-d[B]/dt$ are the rates of consumption of reactants, then, $d[C]/dt = d[D]/dt = -d[A]/dt = -d[B]/dt$.

For the reaction

$$X + 2Y \rightleftarrows Z;$$

$$d[Z]/dt = -d[X]/dt = -1/2 \, d[Y]/dt.$$

Experiments with these systems will probably show that the rates, e.g., $d[C]/dt$, vary with the concentration of reactants, but the nature of the relationship has to be determined for each system.

In order to interpret the experimental data it is necessary to accept two preliminary concepts.

1. The overall stoichiometric equation gives no information as to the reaction sequence or mechanism; e.g., the well-known reaction,

$$Cr_2O_7{}^{2-} + 14H^+ + 6Fe^{2+} \rightleftarrows 2Cr^{3+} + 6Fe^{3+} + 7H_2O$$

does not proceed by the simultaneous collision of 21 reactant ions.

The overall reaction is actually composed of a series of elementary steps, each step involving no more than two chemical species. The sequence of steps is known as the reaction mechanism.

2. Experimental kinetic studies measure the rate of the slowest intermediate step, since this controls the rate of the overall process. Some indication of the nature of this slow step can be deduced from the mathematical relationships derived from the experimental study.

1. *Rate Equations*

The rate equation for the general reaction

$$aA + bB \rightleftarrows cC + dD$$

may be written as

$$-d[A]/dt \propto [A]^n [B]^m$$

or

$$-d[A]/dt = k[A]^n [B]^m$$

so expressing the fact that the rate is proportional to the concentration of A raised to some power n and the concentration of B to some power m; n and m can have integral or fractional values or can equal zero.

It is common for the values n and m to differ significantly from a and b.

The sum $(m + n)$ is called the Overall Order of the Reaction; n and m are called the orders with respect to A and B, respectively; e.g.,

First order in A: $-d[A]/dt = k[A]$ or $k[A][B]^m$

Second order in A: $-d[A]/dt = k[A]^2$ or $k[A]^2[B]^m$

Third order in A: $-d[A]/dt = k[A]^3$ or $k[A]^3[B]^m$

Zero order in A: $-d[A]/dt = k[A]^0$ or $k[B]^m$

If the reaction is first order in A, and first order in B, it is said to be second order overall.

The k's in these equations are referred to as rate constants or velocity constants or as the specific reaction rate.

Not all rate equations have simple forms and more complicated functions of the concentrations are often encountered; e.g., the rate data for the reaction

$$H_2 + Br_2 \rightleftarrows 2HBr$$

corresponds to a rather complicated rate law:

$$d[HBr]/dt = \frac{k[H_2][Br_2]^{\frac{1}{2}}}{1 + k'[HBr][Br_2]}$$

The rate law or equation for any reaction can be determined only by experimental kinetic studies. The order of a reaction may be determined in several ways. One approach involves substituting experimental values into a range of possible general equations. If the postulation, e.g., second order, is correct, a constant value of k is obtained for all stages of the reaction. Instead of using the basic equation, integrated forms are often substituted, since in most cases the integrated equations can be arranged in a form suitable for graphical plotting. The search for the appropriate reaction order then develops into preparing a series of plots in the hope that one will produce a straight line.

For systems with simple, small integral overall orders, there is also a relationship between the half-life of a reaction and the order.

However, for complex systems, the easiest way to gain some indication of the order is to study initial rates. If one considers only the initial reaction rate (e.g., over a period in which there is less than 10% reaction) the rate equation can be expressed in terms of the initial concentration of reactants. If the concentration of one reactant (e.g., A) is held constant while the other (B) is varied, the influence of the variable can be noted. In fact, a logarithmic plot of the initial rate against the concentration of the variable reactant should be a straight line of slope m. Repetition, holding $[B]$ constant and varying $[A]$, provides data on n. The validity of the conclusions derived from initial rate studies should subsequently be checked over the whole course of the reaction by substituting experimental values in the derived equation.

While values of the rate constant, k, are sometimes calculated by measuring the gradient of the concentration-time curves at different intervals of time, and substituting in the general equation, the better method is to use the integrated form of the rate equation.

For a first order reaction

$$-d[A]/dt = k[A]$$

or, rearranging,

$$-d[A]/[A] = k \, dt$$

Integration of this equation yields

$$-\ln [A] = kt + B$$

where B is the integration constant.

If the concentration of A at time t_1 is $[A]_1$, and at time t_2 is $[A]_2$, then

$$-\int_{[A]_1}^{[A]_2} dA/[A] = k \cdot \int_{t_1}^{t_2} dt$$

i.e., $-\ln [A]_2/[A]_1 = k(t_2 - t_1)$.

If t_1 is taken as zero (when the concentration of A is $[A]_o$), and letting $[A]$ represent the concentration at any time t, then,

$$\ln [A]_o/[A] = kt$$
$$[A] = [A]_o e^{-kt}$$
$$\log [A] = \frac{-kt}{2.303} + \log [A]_o$$

The last form indicates that the rate constant k may be evaluated from a plot of $\log [A]$ versus t, the slope of such a plot being $-k/2.303$.

For a second order reaction where

$$-dA/dt = k[A]^2$$
$$-dA/[A]^2 = k \, dt$$

and integrating between the limits $A_o(t_o)$ and $A(t)$ yields

$$\frac{1}{[A]} - \frac{1}{[A]_o} = kt$$

Thus a plot of $1/[A]$ versus t would be linear, with a slope equal to the second order rate constant.

Other integrated equations are somewhat more complex. For example, if

$$A + B \rightarrow \text{products}$$

is found by experiment to follow the rate equation

$$-dA/dt = k[A][B]$$

the integrated form of the equation is

$$kt = \frac{1}{([A]_o - [B]_o)} \ln \frac{[A][B]_o}{[A]_o[B]}$$

which on rearranging gives

$$\log \frac{[A]}{[B]} = \frac{([A]_o - [B]_o)kt}{2.303} + \log \frac{[A]_o}{[B]_o}$$

When one of the reactants is present in large excess (at least tenfold) the concentration of that reactant can be assumed to remain constant throughout the run and the reaction is said to be pseudo first order.

For example, if B is in large excess

$$-d[A]/dt = k'[A] \quad \text{where} \quad k' = k[B]$$

On integration $\log [A] = -k't/2.303 + \log [A]_o$ and k' (hence k from $k'/[B]$) can be obtained from the slope of the linear plot, $\log [A]$ versus t.

Although a few reactions show third order behavior, it is unlikely that these occur in one termolecular step, since such reactions are statistically improbable. The third order usually arises from an equilibrium step which occurs prior to the slowest, i.e., the rate determining step in the mechanism.

Zero order reactions are those in which the rate is unaffected by changes in the concentration of one or more reactants. In these cases the limiting factor is something other than the concentrations, e.g., the amount of light absorbed in a photochemical reaction or the amount of catalyst in a catalytic reaction.

Fractional orders are generally characteristic of chain reactions or heterogeneous reactions, e.g., gas reactions at a solid surface on which one reactant is only moderately absorbed. Alternatively, an initial step may be dissociation of a molecule e.g., $Br_2 \rightleftarrows 2Br$.

2. *Postulation of a Reaction Mechanism*

The elementary reactions which together make up a reaction mechanism are almost invariably bimolecular processes.

After the stoichiometry and products of a reaction have been established, it is generally possible to propose more than one series of intermediary steps, depending, of course, on the complexity of the system.

For a proposed mechanism to be acceptable, the concentration–time data from kinetic experiments must fit the rate equation derived from that mechanism. However, any mechanism is a hypothesis which may be amended or disproved by more erudite experimental studies.

Elucidation of the intermediate steps in a chemical reaction is not easy, since only rarely can one positively identify all the intermediate species. Many of the intermediate reactions occur almost instantaneously.

A discussion on reaction mechanisms and the derivation of rate equations applicable to a series of reactions is beyond the aim of this chapter, but the following three examples should indicate what is meant by a reaction mechanism.

The reaction, $H_2 + Br_2 \rightleftarrows 2HBr$, whose rate equation was quoted earlier, is thought to involve the steps

$$Br_2 + M \rightleftarrows 2Br + M \text{ (M is a surface)}$$

$$Br + H_2 \rightleftarrows HBr + H$$

$$H + Br_2 \rightarrow HBr + Br \text{ (slow step)}$$

Oxidation of substrates (S) by peroxodisulfate ($S_2O_8^{2-}$) is an integral part of many analytical procedures. If the letter R is used to represent a radical derived from the oxidizable substrate, and P represents the product of the reaction, the following reaction scheme can be suggested.

$$S_2O_8^{2-} \xrightarrow{k_1} 2SO_4^-$$

$$SO_4^- + H_2O \xrightarrow{k_2} OH + HSO_4^-$$

$$OH + S \xrightarrow{k_3} OH^- + R$$

$$R + S_2O_8^{2-} \xrightarrow{k_4} P + SO_4^{2-} + SO_4^-$$

$$R + SO_4^- \xrightarrow{k_5} P + SO_4^{2-}$$

The well-known analytical reaction between iodide and iodate ions in strongly acid solution,

$$5I^- + IO_3^- + 6H^+ \rightleftarrows 3I_2 + 3H_2O,$$

is considered to involve the following steps:

$$HIO_3 \rightleftarrows IO_2^+ + OH^-$$

$$IO_2^+ + 2I^- \rightarrow I^+ + 2IO^-$$

$$IO_2^+ + I^- \rightarrow IO^+ + IO^-$$

$$IO^+ + I^- \rightarrow I^+ + IO^-$$

$$IO^- + H^+ \rightarrow HOI$$

$$HOI + H^+ \rightarrow H_2OI^+$$

$$H_2OI^+ + I^- \rightarrow I_2 + H_2O$$

The rate of this reaction is described by the very complex rate law equation

$$\text{Rate} = k'[I_2]^{3/5}[H^+]^{9/5}[IO_3^-]^{9/5}$$

These examples may be considered as extreme cases. A large proportion of the reactions used in analytical kinetic studies conform to far simpler laws, e.g., rate $\propto [A][B]$, with correspondingly simple mechanisms. The most complex systems used involve competing or consecutive reactions, i.e.,

$$\text{Reactants, } A + B \xrightarrow{k_1} \text{Product 1}$$

$$A + B \xrightarrow{k_2} \text{Product 2}$$

$$A + B \xrightarrow{k_3} \text{Product 1, or}$$

$$\text{Product 1} + B \xrightarrow{k_4} \text{Product 2, etc.}$$

3. The Effect of Temperature on Reaction Rates

For most chemical reactions the specific rate constant, k, varies with temperature, T, and a plot of log k against $1/T$ generally yields a straight line.

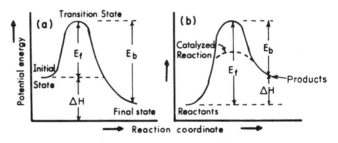

FIGURE 12.2   Potential energy changes along the reaction coordinate: (a) exothermic reaction; (b) endothermic reaction.

The mathematical relationship between k and T is called the Arrhenius Equation and this is written as

$$k = A \exp\left(-E_a/RT\right)$$

where A is termed the preexponential factor or frequency factor or action constant; E_a is the activation energy and R is the gas constant.

It follows from this equation that

$$\ln k = \ln A - E_a/RT$$

and the slope of the plot, log k versus $1/T$, must equal $-E_a/2.303R$. Experiments indicate that E_a is either zero or positive.

Arrhenius postulated that an energy barrier exists between the reactants and products, and only those molecules with sufficient energy are able to react. The extra energy required to pass over the barrier is called the activation energy. This state of affairs is illustrated in Fig. 12.2, in which the potential energy of the system is plotted against the reaction coordinate. It is difficult to define the reaction coordinate in concrete terms; it is any variable which measures the extent of the reaction as it proceeds from left to right.

The state of the system at the top of the barrier is called the transition state.

In this state one does not have normal molecules but rather an unstable complex of high energy. Vibration of this activated complex in a particular mode leads to decomposition and the formation of products.

If E_f represents the activation energy associated with the forward reaction, and E_b represents the activation energy associated with the reverse reaction, then

$$\Delta H = E_f - E_b$$

where ΔH is the enthalpy change associated with the reaction.

It can also be shown that the equilibrium constant, K, for the reaction is equal to k_f/k_b where k_f and k_b are the specific rate constants for the forward and backward reactions, respectively.

In kinetic studies one is concerned primarily with the observation of the rate of progress in one particular direction, i.e., one studies the system prior to equilibrium.

For any given reaction, the value of the preexponential factor may be taken as constant. Hence the magnitude of the rate constant is a function of only two variables, the temperature and the activation energy.

4. Catalysis

A catalyst is a substance which changes the rate of a chemical reaction without appearing in the net reaction and without changing the equilibrium position of the main reaction. This effect is achieved by providing a parallel reaction path or mechanism with a lower activation energy. The catalyst becomes part of the activated complex and is regenerated during the decomposition process which yields the final products.

Nearly all chemical reactions are subject to catalysis or inhibition (negative catalysis), and the following sequence can be proposed for a simple system.

$$S + C \underset{k_{-1}}{\overset{k_1}{\rightleftharpoons}} SC$$

$$SC + R \underset{k_2}{\longrightarrow} P + C \text{ slow}$$

where S is one reactant, R is a second reactant, C is the catalyst and P represents the products.

If the second step is the rate determining step

$$d[P]/dt = k_2[SC][R]$$

But from the first equation

$$k_1/k_{-1} = K = [SC]/[S][C]$$
$$[SC] = K[S][C]$$

Incorporation of this value in the rate equation yields

$$\text{Rate} = k_2 \cdot K[S][C][R]$$

It can be observed from this equation that in reactions of this type, the rate of reaction is directly related to the amount of catalyst present. The amount of catalyst required is usually very small; hence the kinetic approach can be used to determine the presence of trace amounts of catalytic material in a standard reaction system.

B. KINETIC METHODS FOR CATALYZED REACTIONS

In some particular cases, e.g., enzyme-catalyzed reactions, the action of the catalyst is selective in regard to the type of reagent with which it reacts. Accordingly, catalytic reactions are potentially of great value in chemical analysis. The amount of catalyst that can be detected is several orders of magnitude smaller than can be found by most methods which employ equilibrium conditions. In addition, where interaction is selective, analyses can be performed in situ in the presence of large excesses of other species.

In the example quoted above, the values of $[S]$, $[C]$, and $[R]$ do not correspond to the original concentrations of these species.

If $[C]_o$ and $[S]_o$ are the original concentrations of these species, then due to complex formation

$$[C] = [C]_o - [CS]$$
$$[S] = [S]_o - [CS]$$

In most catalysis reactions, $[S]_o \gg [C]_o$, hence $[S]_o - [CS] \simeq [S]_o$. Therefore, considering only the formation of the catalyst complex

$$[SC] = K[S]_o([C]_o - [CS])$$
$$[SC] = K[S]_o[C]_o/(1 + K[S]_o)$$
$$\text{Rate} = k_2 K[S]_o[C]_o[R]/(1 + K[S]_o)$$

There are many systems in which the intermediate complex is very unstable, and the rate of decomposition is large compared to the rate of reaction. In this case the concentration of $[SC]$ is very small but one can calculate the dynamic steady state concentration.

In the steady state treatment, the net rate of formation of the intermediate is considered to equal zero.

$$d[SC]/dt = 0 = k_1[S][C] - k_{-1}[CS] - k_2[CS][R]$$

or substituting initial concentrations

$$k_1[S]_o([C]_o - [SC]) - k_{-1}[CS] - k_2[CS][R] = 0$$

Rearrangement and simplification yields

$$[CS] = k_1[S]_o[C]_o \,/\, k_1[S]_o + k_{-1} + k_2[R]$$
$$\therefore \text{ Rate of Reaction} = k_1 k_2[S]_o[C]_o[R] \,/\, k_1[S]_o + k_{-1} + k_2[R]$$

These two equations can be simplified to a more general form by choosing conditions in which S is present in large excess, so that $[S]$ can be considered to remain virtually constant during the period of measurement.

The rate constants and this term can then be combined to give a new pseudo constant K'

$$K' = k_2 K[S]_o/1 + K[S]_o \text{ or } k_1 k_2[S]_o/k_1[S]_o + k_{-1} + k_2[R]$$

Thus where $[R]$ is $\ll [S]_o$,

$$\text{Rate (cat)} = K'[C]_o[R]$$

where $[R]$ is the observed concentration of R at any time t. If x is the amount of product formed, $[R] = [R]_o - x$ and

$$dx(\text{cat})/dt = K'[C]_o([R]_o - x)$$

The rate of reaction does not necessarily become zero in the absence of catalyst, i.e.,

$$S + R \xrightarrow{k_3} P$$

and the rate of this reaction, which may proceed simultaneously with the catalyzed reaction, is described by the equation:

$$dx(\text{uncat})/dt = k_3[S]_o([R]_o - x)$$

The total reaction rate then becomes

$$dx(\text{total})/dt = dx(\text{cat})/dt + dx(\text{uncat})/dt$$
$$= K'[C]_o([R]_o - x) + k_3[S]_o([R]_o - x)$$

If very small amounts of x can be measured accurately, the initial rate of reaction can be determined. In the initial stages, the amount of reagent reacting, x, is very small compared to $[R]_o$. The term $([R]_o - x)$ therefore approximately equals $[R]_o$.

Thus if the concentrations $[S]_o$ and $[R]_o$ are maintained constant in a series of studies,

$$\text{Initial } dx(\text{total})/dt = A[C]_o + B$$

where $A = K'[R]_o$ and B is the rate of the uncatalyzed reaction.

A plot of the observed initial rate against catalyst concentration should therefore yield a straight line of slope A and intercept B, as shown in Fig. 12.3.

There are many advantages in using the initial rate approach, and two general procedures have been employed to estimate the initial rate without drawing the entire rate curve. In the variable time procedure, the time required to produce a fixed change in composition is measured. In the fixed time procedure the change in composition, Δx, produced over a fixed time interval is determined.

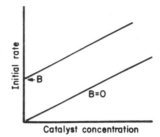

FIGURE 12.3   Calibration curves for systems obeying the general relationship of initial rate $= A[C]_o + B$, where $[C]_o$ is the initial concentration of catalyst and B is the rate of the uncatalyzed reaction.

In experimental studies, calibration curves are prepared by adding varying amounts of catalyst to a standard reaction mixture containing known amounts of S and R.

If $[S]_o$ is at least fifty times greater than $[R]_o$, the $dx(\text{total})/dt$ equation can be integrated to yield the relationship

$$\ln\left([R]_1/[R]_2\right) = (A[C]_o + B)(t_2 - t_1)$$

where $[R]_1$ and $[R]_2$ are the observed concentrations of R at times t_1 and t_2, respectively.

The time for the initial measurement thus need not be zero, and if an induction period (delay in appearance of desired reaction) is characteristic of a particular catalyzed reaction, one can wait till this period is completed before making the first measurement of $[R]$.

Some workers recommend that the calibration graph should consist of a K^* versus catalyst concentration plot, K^* being the calculated value of $\ln([R]_1/[R]_2)/(t_2 - t_1)$.

The greater the difference in rate between the catalyzed and uncatalyzed reaction, the greater is the sensitivity attainable.

1.   *Applications*

The above general discussion is best amplified by considering two actual examples.

The reaction between arsenic (III) and cerium (IV) ions is exceedingly slow unless a catalyst (such as iodide ions, osmium, or ruthenium salts) is present. The effect of these species on the reaction rate has been used as a means of determining microgram liter^{-1} concentrations of the catalyst. For example, minute concentrations of iodide ion have been evaluated by adding the sample to an excess of arsenious acid. A small amount of cerium (IV) sulfate is then added and the cerium (IV) concentration determined at various times. The determinations are calibrated in terms of K^* values because the initial rate method is not applicable (pseudo first order behavior is only observed after thirty seconds). One application of this method has been the determination of the iodine content of common salt. In this application, the extent of cerium (IV) reduction was measured after the expiration of a fixed time interval, and it was observed that very low concentrations of iodide ion acted catalytically only in the presence of chloride ions.

A probable sequence of the steps in the overall reaction is:

$$Ce^{4+} + I^- \rightarrow I^0 + Ce^{3+}$$
$$2I^0 \rightarrow I_2$$
$$I_2 + H_2O \rightarrow HOI + H^+ + I^-$$
$$H_3AsO_3 + HOI \rightarrow H_3AsO_4 + H^+ + I^-$$

For the determination of osmium and ruthenium catalyst concentrations, a variable time procedure is usually adopted. One approach involves measuring the time required to decrease the absorption of the yellow solution to a specified value. The alternative approach measures the time required for the solution to reach a specified oxidation potential.

There are many redox reactions whose rate can be increased by the addition of a catalyst, but in a large number of cases, the analytical determinations based on the kinetic studies lack specificity. For example, the cerium (IV)–arsenic (III) reaction is catalyzed by three species, and when they are present together in a sample, it becomes necessary to attempt a preliminary separation and isolation of the microgram amounts of each catalyst.

The elements which have been determined through their catalytic behavior include Fe, Cu, Co, Cr, Mn, V, Mo, Ti, Zr, Th, W, Re, Ta, Pd, Pt, Ag, Pb, and Au together with a few anions such as S^{2-}, Se^{2-}, $S_2O_3^{2-}$, and CNS^-.

Enzyme action can be highly specific and enzymes are frequently used in clinical laboratories for chemical analysis. As catalysts they are very potent, being very much more effective than nonbiological catalysts.

The general mechanism for a simple enzyme-catalyzed reaction follows the pattern outlined in the previous section. However, the catalyzed reactions are used to determine substrate, activator (substance which causes certain enzymes to act catalytically) and inhibitor (causes a decrease in rate) concentrations as well as enzyme concentrations.

An example of a substrate analysis is the determination of glucose using the enzyme, glucose oxidase. This enzyme catalyzes the oxidation of glucose by oxygen.

$$\text{Glucose} + O_2 \xrightarrow[\text{glucose oxidase}]{} \text{Gluconic acid} + H_2O_2$$

The amount of hydrogen peroxide formed can be determined by reducing it with a colorless dye that has a colored oxidation product.

By maintaining the glucose oxidase activity constant, and ensuring the presence of excess oxygen, reduced dye and peroxidase, the glucose concentration becomes directly responsible for the change in absorbance of the solution.

Other substrates which have been determined kinetically are amino acids, alcohols, and thioesters.

The determination of enzyme concentrations follows the pattern outlined for metal catalysts.

Besides catalyzing redox reactions, a number of metal ions have been observed to influence the rate of ligand exchange in metal complexes. Some of these exchange reactions may prove quite useful for the determination of catalyst concentrations.

C.   Kinetic Methods Involving Uncatalyzed Reactions

The concentration of a single reacting species can be determined by measuring the rate of an uncatalyzed reaction. However, the species of interest, e.g., A, should be present in reasonably high concentration to ensure that $[A]$ does not vary significantly during reaction with an added reagent, R.

$$A + R \xrightarrow{k} \text{Products}$$

If a simple bimolecular reaction is involved and A is present in excess, the kinetics may be treated as a pseudo first order system and

$$dx/dt = k[A]_o([R]_o - x)$$

If initial rates are used for analysis purposes,

$$[A]_o = dx/dt \text{ (initial)}/k[R]_o$$

Either fixed time or variable time methods may be employed to calculate dx/dt (initial), and in this case, it is not essential for $[A]_o \gg [R]_o$, as the measurement is made after a small fraction of the reaction has taken place.

On the other hand, if induction periods or other factors make measurements of the initial rates undesirable, the integrated method has to be employed. Under these circumstances the concentration of the added reagent, R, should be at least one-fiftieth of the concentration of A. This places a limit on the sensitivity of the determination, since small amounts of R lead to slow rates and problems in accurately evaluating concentrations. The appropriate equation, using the integrated method, is

$$[A]_o = \ln ([R]_1/[R]_2)/k(t_2 - t_1)$$

In general, the accuracy of kinetic methods is lower than corresponding equilibrium or static methods. However, there are some cases where the kinetic approach is more straightforward and convenient. Examples of this are the determination of acetylacetone (with hydroxylamine hydrochloride) and phenolic substances (by bromination).

When the solution contains two or more species (X, Y) which can react with the selected reagent, evaluation of one of the species is fairly simple, provided that there is a marked difference in the rates at which the components react. For example, if one of the species (Y) reacts very slowly, its contribution to the overall rate in a given time interval may be neglected and the kinetic system treated as a solution containing component X only. If the difference in rates is great enough, the X reaction may proceed to near completion before appreciable amounts of Y have been consumed. The second part of the kinetic study can then be used to ascertain $[Y]$.

Consider a mixture of X and Y reacting with reagent R to give products Z_1 and Z_2.

$$X + R \xrightarrow{k_x} Z_1$$

$$Y + R \xrightarrow{k_y} Z_2$$

If R is present in large excess, the rate of disappearance of X and Y can be treated as pseudo first order reactions, i.e.,

$$-d[X]/dt = k_x[R][X] \text{ and}$$

$$-d[Y]/dt = k_y[R][Y]$$

and integrating with respect to time gives:

$$\ln\{[X]_t/[X]_o\} = (k_x[R])(t - t_o)$$

$$\ln\{[Y]_t/[Y]_o\} = (k_y[R])(t - t_o)$$

Division of these two equations and rearrangement of the terms gives the relationship

$$\log\{[X]_o/[X_t]\} = \log\{[Y]_o/[Y]_t\} \cdot k_x/k_y$$

This expression can be used to evaluate the extent to which reaction Y will have proceeded when different fractions of X have been consumed.

It has been proposed that the ratio k_x/k_y must be greater than 500 if the contribution of Y is to be neglected. If a percentage error greater than 1% can be tolerated, lower k_x/k_y ratios are satisfactory. The limit on k_x/k_y is an arbitrary one, since one has to consider the relative concentrations of X and Y, and the fraction of X which will react during the period of measurement.

With large values of k_x/k_y, e.g., >500, over 98% of Y remains when 99.9% of X has been consumed. This means that Y can now be readily determined as a separate entity, provided that the reaction can be speeded up to give results in a reasonable time. The rate of reaction with Y can be increased by increasing the temperature, by adding a catalyst, or by increasing the concentration of reagent R.

For mixtures in which the ratio k_x/k_y is not so large, differential reaction rate methods have been developed.

The mathematical approach involved varies with the relative concentrations of the reagent, R, and the mixture ($X + Y$).

The simplest cases are those in which $[R] \gg ([X] + [Y])$ or $[R] \ll ([X] + [Y])$. These can be treated as pseudo first order reactions, and the rate may be measured either in terms of product formation or reactant consumption.

For the pseudo first order reaction in which $[R] \gg ([X]_o + [Y]_o)$ and the two components of the mixture react to give a common product, P, the system may be represented by the equations

$$X \xrightarrow{k_x'} P$$

$$Y \xrightarrow{k_y'} P$$

The concentration of product at any time t is given by the expression

$$[P]_\infty - [P]_t = [X]_t + [Y]_t = [X]_o e^{-k_x't} + [Y]_o e^{-k_y't}$$

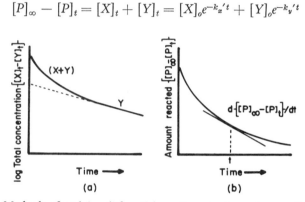

FIGURE 12.4   Methods of studying differential reaction rates in a mixture of species (X and Y) following (pseudo) first order kinetics: (a) logarithmic extrapolation method; (b) overall rate of change of concentration method.

In the situation where $k_x' > k_y'$, the term $[X]_o e^{-k_x't}$ becomes very small compared to $[Y]_o e^{-k_y't}$ at all times after virtually all X has reacted.

Thus when $[X]_t \simeq 0$, the logarithmic form of the above expression becomes

$$\ln \{[X]_t + [Y]_t\} = \ln \{[P]_\infty - [P]_t\} = -k_y't + \ln [Y]_o$$

This predicts that a plot of the logarithm of the concentrations, $\{[X]_t + [Y]_t\}$ or $\{[P]_\infty - [P]_t\}$, against time t, should yield a straight line with a slope $-k_y'$ and an intercept (at $t = 0$) equal to $\ln [Y]_o$. The value of $[X]_o$ may then be obtained by subtracting $[Y]_o$ from the total initial concentration of the mixture which can be determined either from P_∞ or by independent means. The type of plot obtained is shown in Fig. 12.4(a).

Because of its simplicity, this method is one of the most widely used differential methods.

Another approach involves plotting the rate of change of total concentration of the mixture, i.e., $([X]_t + [Y]_t)$ or $([P]_\infty - [P]_t)$, against time [cf. Fig. 12.4(b)].

By differentiating the basic equation, one obtains the expression,

$$-d([P]_\infty - [P]_t)/dt = k_x'[X]_t + k_y'[Y]_t$$

As the concentration of the faster reacting species $[X]_t$ approaches zero, $-d([P]_\infty - [P]_t)/dt$ approaches $k_y'[Y]_t$.

Thus, if the tangent to the curved plot is measured at different intervals, a point is passed beyond which the ratio of the slope to $([P]_\infty - [P]_t)$ has a constant value (equal to k_y').

As $([P]_\infty - [P]_t) = [Y]_t$ at this point, $[Y]_o$ can be calculated from the equation

$$\ln [Y]_o = \ln [Y]_t + k_y't$$

Other workers have proposed an alternative graphical procedure, known as the single-point method. In this procedure, the initial concentrations are determined from a knowledge of the total concentration of the mixture and the extent of the reaction at a single selected time.

By dividing the first basic equation by the total concentration $([X]_o + [Y]_o)$ and rearranging the terms, one obtains the expression

$$\frac{[X]_t + [Y]_t}{[X]_o + [Y]_o} = \frac{[P]_\infty - [P]_t}{[P]_\infty}$$

$$= (e^{-k_x't} - e^{-k_y't}) \cdot \frac{[X]_o}{[X]_o + [Y]_o} + e^{-k_y't}$$

A plot of $([X]_t + [Y]_t)/([X]_o + [Y]_o)$ at any time, t, against the initial mole fraction of X in the mixture yields a straight line of slope $(e^{-k_x't} - e^{-k_y't})$ and intercepts $e^{-k_y't}$ and $e^{-k_x't}$ at mole fractions of X equal to zero and unity, respectively.

A calibration graph can thus be constructed by measuring the extent of reaction of pure X and pure Y, at a particular time, t, and drawing a straight line between these points.

The extent of the reaction after time t, in a mixture of $(X + Y)$, is then measured, and the graph used to determine the mole fraction of Y present.

Another alternative mathematical approach is the method of proportional equations, which is based on the principle of constant fractional life.

Where reaction conditions are such that the initial concentration of added reagent, $[R]_o$, is not very much greater or less than the combined concentration of reactants, $([X]_o + [Y]_o)$, the system has to be considered as a second order reaction.

The same general approaches can be applied, but in these cases the mathematics are a little more complex and will not be outlined here.

The differential rate methods have been used to determine the composition of binary mixtures of many closely related species. The systems examined have included compounds such as alcohols, amines, carbonyl homologs, polymer functional groups, and isomeric sugars. The procedures are of particular interest in studies of organic compounds, since modern physical-organic studies have demonstrated a clear relationship between reaction rate and structure.

Kinetic methods represent a completely new area of research in analytical chemistry.

A brief account, such as given in this chapter, can merely outline the broad principles and indicate the potential scope of the technique. For fuller details, the interested reader should study the monographs written by Yatsimerskii and by Mark and Rechnitz.

The procedures are in many cases prone to interference and care is required to ensure that conditions are adequately stabilized in a series of kinetic runs. On the other hand, the technique is extremely sensitive and could develop into a method of major importance in trace metal analysis. Its position in clinical analysis is reasonably well established already.

The advent of automatic sampling units and recording instruments has eliminated much of the tedium of kinetic studies and has facilitated control of measurement procedures. However, to any chemist, the new technique offers an interesting challenge. To be completely successful one needs to be able to combine the patience required for elucidating the rate equations of complex systems; the imagination needed to postulate probable reaction mechanisms; and the manipulative skill demanded by microanalysis of any form.

V.　LITERATURE ASSIGNMENTS

1. Most general textbooks on quantitative chemical analysis provide full descriptions of the precipitation and purification of barium sulfate, lead sulfate, hydrous oxide of iron (III), and silver chloride.

   Prepare a summary of the methods advocated for each type of material and discuss the relevance of each step in terms of the von Weimarn equation and the kinetics of precipitate formation.

2. A. Using books devoted to titrimetric oxidation-reduction procedures or the summary of redox titration methods listed in the "Handbook of Analytical Chemistry," prepare a series of lists showing:

   a. The chemical species most widely used as oxidants—these should be subdivided in terms of (i) titrants and (ii) general reagents added prior to the titration step.

   b. Recommended methods of reduction. This list may be subdivided in terms of (i) solid reductants; (ii) ionic reductants; and (iii) titration with a reducing agent. Note the formal or standard reduction potential associated with each reagent.

   B. In view of the range of potentials covered in these lists, do you consider that new means of oxidation or more selective modes of reduction need to be discovered? Why?

C. Aliquots of an acid solution containing the salts of titanium (IV), vanadium (V), chromium (VI), tungsten (VI), and iron (III) are to be shaken with amalgams of different reducing power.

If the amalgams selected are those prepared from bismuth, lead, and zinc (the formal reduction potentials of these three systems being approximately equal to 0.2, −0.1, and −0.7 V respectively), use a table of standard reduction potentials to predict the valency state to which each element will be reduced by each reductant.

If the original aliquots contained one millimole of each metal salt, how many milliliters of 0.2 *M* potassium permanganate would be required to titrate the reduced solutions?

3. a. Why is the selectivity of organic reagents usually associated with a particular functional-analytical group?

    b. The properties of an organic reagent can be modified through substitution in the structural skeleton. Indicate the type of substituents required to change the solubility of a metal chelate system. What other effects can substitution have on the behavior of an organic molecule?

    c. Quote some examples of organic reagents which form metal salts of limited solubility.

        Would one expect these compounds to be contaminated with foreign ions, excess organic reagent, or both? Give reasons for your answer.

    d. Four well-known organic reagents are 2,2′-dipyridyl, dithizone (*sym*-diphenyl-thiocarbazone), dimethylglyoxime, and alizarin red S. These represent four different types of organic reagents. For each compound list the structure, possible mode of bonding to the functional-analytical group, and most common applications. Discuss the role that structure might play in determining the observed behavior of these reagents.

4. In classic rock analysis, the general procedure is to separate the components of the sample by selective group precipitation using general reagents. For example, silica is first removed as the dehydrated oxide; Ag, Pb, Hg are precipitated as chlorides; Cu, Cd, Bi are precipitated from acid media as sulfides; and Fe, Al, Ti, etc., are removed as hydrous oxides.

Using the tables in the "Handbook of Analytical Chemistry" or an appropriate reference book, prepare a list of the elements which might accompany each of the species listed in the above group.

Assuming that only the above elements are present, how might the group precipitates be purified to release trapped foreign ions?

Would pH control permit separations to be made within the general group separation, e.g., could iron be separated from aluminum?

5. "The evolution of instrumental techniques has made it unnecessary for chemists to persist in synthesizing more selective organic reagents." Discuss this statement.

6. a. Outline the principles of precipitation from homogeneous solution.
   b. Critically evaluate the advantages claimed for this technique.
   c. Produce brief summaries of two determinations based on homogeneous solution precipitation.

7. Thermodynamic data are so readily available and well organized that an astute chemist can predict the feasibility of most projected reactions.

   Accordingly, most undergraduate courses in analytical chemistry devote a significant amount of time to equilibrium prediction studies, e.g., calculation of pH changes, competing equilibria, solubility products, etc.

   However, many predicted reactions are too slow for practical use. On this basis, it is proposed that future analytical chemists should develop a marked interest in solution kinetics and reaction mechanisms.

   Prepare an argument in favor of this proposal.

8. Write brief notes on the following:
   a. Overall order of a chemical reaction.
   b. Pseudo first order kinetics.
   c. Arrhenius equation.
   d. Role of a catalyst.
   e. Determination of rate constants.
   f. Complex reaction mechanisms.

9. Write an essay on the determination of traces of metal ions by kinetic methods.

   Indicate the possible advantages and disadvantages of the technique.

10. Many organic compounds exhibit similar chemical behavior but different kinetic behavior. This property has been used as a means of analyzing mixtures of isomers and other chemically related compounds.

    Outline the basic principles associated with the alternative kinetic proposals, and use reference books to provide specific examples of each particular approach.

References

A. I. Busev and N. G. Polianskii, *The Use of Organic Reagents in Inorganic Analysis*, Pergamon, Oxford, 1960.

L. Erdey, *Gravimetric Analysis*, Vols. I, II, III, Pergamon, Oxford, 1963.

J. F. Flagg, *Organic Reagents*, Wiley (Interscience), New York, 1958.

N. H. Furman and F. J. Welcher (eds.), *Standard Methods of Chemical Analysis*, 6th Ed., Vols. I and II, Van Nostrand, London, 1962.

L. Gordon, M. L. Salutsky, and H. H. Willard, *Precipitation from Homogeneous Solution*, Wiley, New York, 1959.

I. M. Kolthoff and P. J. Elving (eds.), *Treatise on Analytical Chemistry*, Wiley (Interscience), New York, 1968.

H. B. Mark and G. A. Rechnitz, *Kinetics in Analytical Chemistry*, Wiley (Interscience), New York, 1968.

A. E. Nielsen, *Kinetics of Precipitation*, Pergamon, Oxford, 1964.

A. G. Walton, *The Formation and Properties of Precipitates*, Wiley (Interscience), New York, 1967.

F. J. Welcher (ed.), *Organic Analytical Reagents*, Vols. I to IV, Van Nostrand, New York, 1947.

C. L. Wilson and D. W. Wilson (eds.), *Comprehensive Analytical Chemistry*, Elsevier, Amsterdam, 1959.

K. B. Yatsimerskii, *Kinetic Methods of Analysis*, Pergamon, Oxford, 1966.

13

ELECTRICAL TRANSFORMATIONS

I. Equilibrium and Kinetic Considerations. 430
   A. Nernst Equation 430
   B. Kinetics of Electron Transfer Reactions. 432
II. Conductivity of Electrolyte Solutions 435
   A. Conductance 435
   B. Measurement of Conductance 437
   C. Conductometric Titrations 439
III. Potentiometric Methods 440
   A. Electrochemical Cells 440
   B. Types of Electrodes 442
   C. Direct Potentiometric Measurements 448
   D. Potentiometric Titrations 450
   E. Potentiometry with Polarized Indicator Electrodes 452
IV. Electrodeposition 453
   A. Electrolysis 453
   B. Electrogravimetry 456
   C. Electroseparations 458
   D. Mercury Cathode Separations 461
V. Coulometric Analysis 462
   A. Faraday's Laws. 462
   B. Coulometry at Constant Electrode Potential 462
   C. Coulometry at Controlled Current 463
   D. Chronopotentiometry 466
VI. Polarography 467
   A. Current-Voltage Relationships 467
   B. Dropping Mercury Electrode. 468
   C. Polarographic Circuit 471
   D. Equation of the Polarographic Wave 473
   E. Irreversible Waves and Departures from Pure Diffusion . . . 476
   F. Instrumental Techniques 479
   G. Applications 481
VII. Amperometric Titrations 482
VIII. Revision and Review Questions 482
   References 486

I. EQUILIBRIUM AND KINETIC CONSIDERATIONS

A. NERNST EQUATION

In any electron-transfer process, the thermodynamics and kinetics of the reaction have to be considered simultaneously, since many "energetically feasible" reactions proceed at too slow a rate to be of practical value.

Electrolytic processes may be classified as heterogeneous electron-transfer reactions. In such heterogeneous systems, the rate controlling step can be one of many processes, e.g.,

> Diffusion of reactant to the solid surface, or,
> Adsorption of reactant on the electrode, or,
> Electron transfer at the surface, or,
> Migration of species on the surface or within the solid phase, or,
> Diffusion of products from the surface.

These kinetic aspects are discussed in Section I,B, and the analytical techniques based on electrolytic processes form the subject matter of this chapter.

For the prediction of equilibrium conditions in redox systems, the basic equation is that attributed to Nernst, and in previous chapters brief mention has been made of this equation.

The standard free energy change (ΔG_r°) associated with the electron transfer reaction,

$$a \, Ox_1 + b \, Red_2 \rightleftarrows a \, Red_1 + b \, Ox_2$$

is equal to the difference in free energy content of the component half-cells.

$$Ox_1 + be \rightleftarrows Red_1; \quad G_1^\circ$$

$$Ox_2 + ae \rightleftarrows Red_2; \quad G_2^\circ$$

$$\Delta G_r^\circ = aG_1^\circ - bG_2^\circ$$

The free energy associated with the half-cell reactions is related to the position of equilibrium and the absolute temperature (T).

$$G_1^\circ = -kT \ln [Red_1]/[Ox_1][e]^b$$

$$G_2^\circ = -kT \ln [Red_2]/[Ox_2][e]^a$$

Therefore, $\Delta G_r^\circ = -akT \ln [Red_1]/[Ox_1][e]^b + bkT \ln[Red_2]/[Ox_2][e]^a$.

If a and b are taken into the logarithmic terms, and the latter are combined, one obtains the expression

$$\Delta G_r^\circ = -kT \ln \frac{[Red_1]^a[Ox_2]^b}{[Ox_1]^a[Red_2]^b}$$

$$= -kT \ln K$$

where K corresponds to the equilibrium constant for the electron transfer reaction, ΔG_r° is in units of electron volts per electron, and k is equal to 8.615×10^{-5} eV per degree electron.

A scale of relative electron free energies (at a given temperature) has been established by considering reactions in which $Red_2 = H_2$, and then arbitrarily assigning G_H° a value of zero.

Thus, for the system

$$Ox_1 + n/2\ H_2 \rightleftarrows Red_1 + nH^+$$

$$\Delta G^\circ = -kT \ln K = G_1^\circ - G_H^\circ = G_1^\circ - 0$$

The relative electron free energy of system one is, therefore,

$$G_1^\circ = -kT \ln K = -kT \ln [Red_1]/[Ox_1][e]^n$$

Rearrangement yields the relationship

$$-kT \ln [e] = \frac{-G_1^\circ}{n} - \frac{kT}{n} \ln \frac{[Red_1]}{[Ox_1]}$$

The oxidized form of a species can be considered to possess vacant electron levels while in the reduced form these electron levels are occupied. The potential, E, constitutes a measure of the average free energy per electron in an assemblage of occupied and unoccupied electron free-energy levels. Numerically, $E = -kT \ln [e]$.

When the number of occupied and unoccupied levels of a given system are equal (i.e., $[Red_1] = [Ox_1]$), the potential is known as the standard potential, E°.

In the above equations, activity coefficients have been omitted for the sake of simplicity. Thermodynamic equations relate to active masses, and not concentrations; hence it is more correct to state that a system exhibits its standard potential when all components are in their standard state. For a gas the standard state is a pressure of one atmosphere; for a solute the standard state is unit activity, and by definition solids are considered to possess unit activity.

If $[Red_1] = [Ox_1]$ in the above equation, $E_1^\circ = -G_1^\circ/n$, and the equation may be rewritten as

$$E = E_1^\circ - \frac{kT}{n} \ln \frac{[Red_1]}{[Ox_1]}$$

To convert the units of free energy from electron volts per electron to the heat unit of joules mole^{-1}, the potential has to be multiplied by the Faraday, F, which is the quantity of electricity equivalent to one mole of electrons, i.e., 96,491 coulombs.

$$\therefore \Delta G_1^\circ = -nE_1^\circ \text{ eV per electron}$$

$$= -nFE_1^\circ \text{ joules or } -nFE_1^\circ/4.184 \text{ cal}$$

. *Electrical Transformations*

Expressed in heat units, the equation takes the form

$$EF = E°F - \frac{RT}{n} \ln \frac{[\text{Red}]}{[\text{Ox}]}$$

In the change of units, Boltzmann's constant, k, has to be multiplied by N (Avogadro's number) and the product Nk is the gas constant R (8.31 joules $\deg^{-1}$ mole^{-1}).

Dividing this expression by F, and taking the negative sign into the logarithmic term yields the better known form of the Nernst equation

$$E = E° + \frac{2.303RT}{nF} \log \frac{[\text{Ox}]}{[\text{Red}]}$$

At 25°C, 2.303 RT/F equals 0.059.

When electrons are added to a system they tend to populate the lowest free energy state first. As soon as this level approaches saturation, continued addition of electrons leads to population of the next highest level, and a change in potential, E, occurs.

When a solute species possessing a populated high energy level is mixed with a species with an unpopulated lower energy level, electrons tend to be transferred and a conducting electrode placed in such a system acquires the equilibrium potential of the system.

On the other hand, if an electrical potential is applied to the electrode from an external source, electrons are transferred until such time as the equilibrium potential of the redox system equals the superimposed potentials.

When any system capable of donating electrons is brought in contact with a system capable of accepting electrons, the reaction proceeds in the direction which results in a decrease of free energy. The position at equilibrium is determined primarily by the difference in energy of the two systems involved, for as discussed in Chapter 11 (Section VI,C), the equilibrium state for the general reaction,

$$a\text{Ox}_1 + b\text{Red}_2 \rightleftarrows a\text{Red}_1 + b\text{Ox}_2$$

is described by the expression

$$\ln K = \ln [\text{Red}_1]^a[\text{Ox}_2]^b/[\text{Ox}_1]^a[\text{Red}_2]^b = (E_1° - E_2°)abF/RT$$

B. KINETICS OF ELECTRON TRANSFER REACTIONS

In Chapter 12, it was pointed out that the rate of a chemical reaction is proportional to the concentration of the species involved in the slowest stage of the reaction. These species are often intermediates formed during the reaction and their concentrations can be complex functions of the reagent concentrations initially brought in contact. If the initial reagents are represented by A, B, and

C, the general form of the rate equation is thus

$$\text{Rate} = k[A]^p[B]^q[C]^r$$

where k is the rate constant and p, q, r are integers having values of 0, 1, 2, etc., or fractional values.

With heterogeneous reactions the rate controlling step often involves a species present at the phase interface. The concentration of this species may be controlled by a diffusion process (in either phase), by the number of active sites occupied at the interface (adsorption phenomena) or by surface reactions (chemisorption, electron transfer, etc). Thus in heterogeneous systems the surface area and nature of the more rigid phase can be of prime importance, but in most cases it is difficult to define either of these factors with accuracy. For example, the active surface area of a solid can be a function of the particle size, purity and porosity of the material.

In electron transfer reactions, as in all other forms of chemical reaction, there is an energy barrier which has to be surmounted before reaction takes place. Where this activation energy is large, the electron transfer process can become the rate controlling step.

The free energy barrier to be surmounted during electron transfer is equal to $\Delta G^{\ddagger}$, where $\Delta G^{\ddagger}$ is the free energy required to convert a mole of reactant to the activated state. The rate of the reaction is equal to the number of activated complexes passing over the energy barrier per unit of time, and by transition state theory the rate constant for the process equals $(kT/h) \exp(-\Delta G^{\ddagger}/RT)$, k being Boltzmann's constant and h being Planck's constant.

In redox reactions the movement of activated species is associated with a flow of electrons; hence the rate of the process can be expressed in terms of a current.

If a reduction process is represented by

$$\text{Ox} + ne \xrightarrow{k_f} \text{Red}$$

the conjugate oxidation process is

$$\text{Red} \xrightarrow{k_r} \text{Ox} + ne$$

and the net current, i, is the sum of the cathodic current (reduction), i_c, and the anodic current (oxidation), i_a.

$$i = i_c + i_a$$

For an electrode of area A, the current flowing at any potential, E, may be written as,

$$i_c = nFAk_f a_{\text{ox}} = nFAk_f' [\text{Ox}]$$
$$i_a = nFAk_r a_{\text{red}} = -nFAk_r' [\text{Red}]$$

At equilibrium the net current flow is zero. Hence at the standard potential, E°,

$$k_f = k_r = k^\circ$$

while at the formal potential, E'

$$k_f' = k_r' = k'$$

When the electrode potential is changed by an amount ΔE, the free energy of the system is changed by an amount equal to $-nF\,\Delta E$ and this alters the magnitude of the activation energies of the forward and reverse processes. Thus, if the potential change favors the cathodic reaction, the new cathodic free energy of activation becomes $\Delta G_c^{\ddagger}$ which equals $(\Delta G^{\ddagger} - \alpha nF\,\Delta E)$. The symbol, α, is called the transfer coefficient and represents the fraction of the total energy which acts to decrease the height of the energy barrier for the cathodic reaction.

A change in activation energy is reflected in a change in the magnitude of the rate constant. Thus at the new potential, E,

$$k_f = k_f^{\circ} \exp\{-\alpha nF(E - E'')/RT\}$$

The term k_f° is defined as the rate constant for the reaction when $E = E''$, the reference potential of the electrode couple. If the reversible formal potential of the couple is known, it is convenient to define E'' as this formal potential. Alternatively, the reference potential may be taken as the potential of the standard hydrogen electrode.

Accordingly at potential E the cathodic current is described by the equation

$$i_c = nFAk_f^{\circ}[\text{Ox}] \exp\{-\alpha nF(E - E'')/RT\}$$

Changing the potential of an equilibrium system by ΔE increases the activation energy for the back reaction, and it can be shown that the anodic current is described by the relationship

$$i_a = -nFAk_r^{\circ}[\text{Red}] \exp\{(1 - \alpha)nF(E - E'')/RT\}$$

At the equilibrium potential (E_{eq}) the magnitudes of the anodic and cathodic currents are equal, and equation of the two current equations yields the expression

$$[\text{Ox}]/[\text{Red}] = (k_r^{\circ}/k_f^{\circ}) \exp\{nF(E_{eq} - E'')/RT\}$$

When $[\text{Ox}] = [\text{Red}]$, $E_{eq} = E^{\circ}$. Accordingly,

$$k_r^{\circ}/k_f^{\circ} = 1/\exp\{nF(E^{\circ} - E'')/RT\}$$

Substituting for k_r°/k_f°

$$[\text{Ox}]/[\text{Red}] = \exp\{nF(E_{eq} - E^{\circ})/RT\}$$

This is the Nernst equation in an alternative form. It must be emphasized that the activities and concentrations appearing in the above equations refer to the actual interface, and there are many examples in later sections in which these concentrations are controlled by mass transport processes.

II. CONDUCTIVITY OF ELECTROLYTE SOLUTIONS

A. CONDUCTANCE

In studies of electrical behavior, one of the most basic equations is Ohm's law,

$$I = E/R$$

where I is the current (in amps), which flows on imposition of a potential difference, E (volts), across a resistance, R (ohms).

Electroanalytical techniques are concerned, for the most part, with the electrical behavior of electrolyte solutions and the usual aim is to measure I, or E, or R (or in some cases a combination of two of these), and to relate the measurement to the concentration of some species in solution.

The simplest, and least specific, technique is that based on the measurement of the conductance of a solution.

The conduction of an electrical current through an electrolyte solution involves a migration of positively charged species towards the cathode (negative electrode of the circuit) and movement of negatively charged species towards the anode (positive electrode). Thus all charged particles contribute to the conduction process.

The conductance of a solution, L, is defined as the inverse of its resistance, R, However, a solution is a three dimensional conductor and the exact resistance of any given solution depends on the geometry of the cell used in the measurement. Accordingly, a more useful unit is the specific conductance, k, which is defined by the equation

$$L = 1/R = kA/b$$

where A is the cross-sectional area (cm²) of two similar electrodes, b cm apart.

The ratio b/A is referred to as the cell constant θ. Practical cells are never of simple geometry and calculation of b/A can be difficult. However, this problem can be overcome by measuring the conductance of a solution of known specific conductance. The standard usually used is potassium chloride since tabulations of the specific conductance of KCl solutions of different concentrations at different temperatures are readily available. The conductance of a solution varies with temperature (approximately 2% per degree); hence this parameter should be controlled in a series of experiments.

The equivalent conductance, Λ, equals $1000\ k/C$ where C is the concentration of solution in the cell in gram equivalents per liter. Equivalent conductance increases with dilution due to a reduction in interionic forces and Λ_∞ is known as the equivalent conductance at infinite dilution.

Both cations and anions contribute to the total conductance; hence for a simple salt solution $\Lambda_\infty = \lambda_+^\circ + \lambda_-^\circ$ where λ_+° and λ_-° are the equivalent ionic conductances of the cation and anion at infinite dilution.

Where more than one electrolyte is present, the conductance of very dilute solutions is described by the expression,

$$L = \sum_j C_j Z_j \lambda_j / \theta$$

The summation is made over all the ions in the solution; C_j is the concentration of the jth ion in moles liter^{-1}, Z_j is the absolute value of its charge, and λ_j is its equivalent ionic conductance expressed in mho-cm^2 equivalent^{-1}.

TABLE 13.1

EQUIVALENT IONIC CONDUCTANCES AT INFINITE DILUTION

Cations	λ_+° 25°C mho-cm^2 equiv^{-1}	Anions	λ_-° 25°C mho-cm^2 equiv^{-1}
H$^+$	349.8	OH$^-$	198.0
Co(NH$_3$)$_6$$^{3+}$	102.3	Fe(CN)$_6$$^{3-}$	100.0
Cs$^+$	77.0	SO$_4$$^{2-}$	80.0
K$^+$, NH$_4$$^+$	73.5	Br$^-$	78.5
Pb^{2+}	73.0	Cl$^-$	76.3
Ag$^+$	61.9	NO$_3$$^-$	71.4
Ca^{2+}, Sr^{2+}	59.5	CO$_3$$^{2-}$	69.3
		ClO$_3$$^-$	64.6
Co^{2+}, Zn^{2+}	52.8	F$^-$	55.0
Na$^+$	50.1	HCO$_3$$^-$	44.5
(CH$_3$)$_4$N$^+$	44.9	CH$_3$COO$^-$	40.9
Li$^+$	38.7	HC$_2$O$_4$$^-$	40.2
(C$_2$H$_5$)$_4$N$^+$	32.7	C$_6$H$_5$COO$^-$	32.3
(n-C$_4$H$_9$)$_4$N$^+$	19.5	B(C$_6$H$_5$)$_4$$^-$	18.0
(n-C$_5$H$_{11}$)$_4$N$^+$	17.5		

This relationship becomes more approximate as the ionic strength of the solution increases, due to changes in the activity coefficients of the various species.

The magnitude of equivalent ionic conductances at infinite dilution depends principally on the hydrated ionic size. A selected list is given in Table 13.1 and it can be observed that there are appreciable differences in the equivalent conductances of different ions. Two notable cases are λ° H$_3$O$^+$ (349.8) and λ° OH$^-$ (198.0) which are several times larger than most ions (λ° generally 40 to 80). A more extensive list is provided in the "Handbook of Analytical Chemistry"—Table 5–11—and other electrochemical reference books. The tables can be used, in conjunction with the summation rule quoted above, to calculate the conductance of any solution. In most cases the accuracy is sufficient to define the experimental conditions required for measurement. Such calculations are also useful for predicting the changes in conductance which may be observed during the course of a titration. The conductance of a solution can change markedly during titration due to the addition or removal of ions having high equivalent ionic conductances. Alternatively a product of the reaction may be un-ionized and hence a poor conductor.

B. Measurement of Conductance

All measurements of conductance require that some current pass through the solution. In the most direct method, a potential is applied between two inert metal electrodes placed directly into the solution under test. The application of a dc potential source tends to give false readings because polarization potentials appear at the electrode—solution interface and large capacitance effects can arise from electrical double layers.

One means of overcoming polarization effects is to apply a dc potential across the electrodes, E_{app}, which is very large compared with any polarization potential, E_p.

Since E_p can be of the order of one or two volts, E_{app} should be greater than 100 V if E_p is to be neglected in the basic equation.

$$\text{Resistance } (R) = (E_{app} - E_p)/\text{Current}$$

Unless the current is small (i.e., resistance high) and the duration of measurement is short, electrolysis can seriously change the composition of the solution under test. On the other hand, the technique is particularly useful for conductance measurements in nonaqueous solvents.

The more popular method for reducing polarization effects to negligible values is to use alternating applied voltage and platinum electrodes coated with platinum black. If the applied voltage is sufficiently large, e.g., >5 V, the principal reaction occurring at the electrode surfaces is the oxidation and reduction of hydrogen produced from the decomposition of water. Since the current reverses with each half-cycle there is no net change in solution composition.

The resistance may be determined by measuring the current flowing between the electrodes, but the classic circuit makes the cell containing the two electrodes one arm of a Wheatstone Bridge circuit.

A schematic representation of the circuit is shown in Fig. 13.1. The applied voltage E is impressed across two resistance voltage dividers. An ac null detector (e.g., an ac galvanometer or "magic eye" unit) is connected between the midpoints of the two divider arms. The voltage appearing at the midpoint of one arm is $ER_S/(R_S + R_1)$ where R_S is the resistance of the conductance cell and R_1 is a standard resistor. The voltage at the midpoint of the other arm is $ER_V/(R_V + R_2)$ where R_V is a variable resistor and R_2 is a standard resistor. If R_V is adjusted until no voltage appears across the midpoints, as indicated by the null meter, then

$$R_S/(R_S + R_1) = R_V/(R_V + R_2)$$

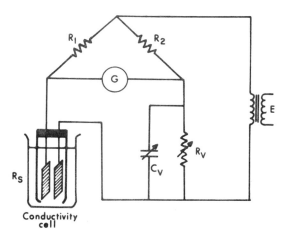

FIGURE 13.1 Wheatstone Bridge circuit for measuring the conductance of solutions; E is a source of alternating current; G is a galvanometer for detecting the position of zero current flow; R_1 and R_2 are fixed resistors; R_V is a variable resistor; C_V a variable capacitor; and R_S is the measuring cell.

Knowing R_1, R_2 and R_V, R_S can be calculated. With high resistance solutions the capacity associated with the cell can prevent an accurate balance, and compensation for this effect is achieved by introducing a variable capacitor, C_V, across R_V.

Two nonreactive electrodes immersed in a liquid medium are electrically equivalent to a parallel resistance, R, and capacitance, C. On the application of an ac potential the current which flows through R is in phase with the applied potential, while the current through C is 90° ahead of the potential. The total current flow is the vectorial sum of the two and the capacitance in the circuit serves to oppose the flow of current. The magnitude of this effect, designated the capacitive reactance, X_C, depends on the ac frequency, f, as well as the capacitance C.

$$X_C = 1/2\pi fC$$

For a liquid, the magnitude of C is governed by its dielectric constant.

If either R or X_C is considerably larger than the other, essentially all of the current passes through the path of minimum opposition and the branch of higher energy may be ignored in interpreting the electrical behavior.

Thus, with the frequencies used in most conductivity bridges (60 to 1000 cps), the contribution of X_C is much larger than R, and reasonably accurate values of conductance are obtained.

Conversely, using a high frequency signal (one to 100 megacycles), the contribution of the capacitive reactance becomes very small, and the circuit can be used to measure the dielectric of the medium.

C.   CONDUCTOMETRIC TITRATIONS

The electrical conductance of a solution is a nonspecific property; hence direct analytical applications are limited to the analysis of binary water-electrolyte mixtures and the determination of the total electrolyte concentration of a solution.

However, changes in the number and type of ions present in a solution occur during a titration, and in many cases, the course of the titration can be followed

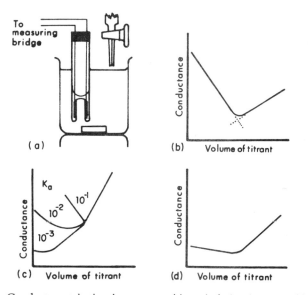

FIGURE 13.2   Conductometric titration curves: (a) typical titration assembly; (b) strong acid–strong base titration; (c) titration of weak acids (K_a 10^{-1} to 10^{-4}) with strong base; and (d) precipitation reaction, e.g., titration of sodium chloride with silver nitrate.

by measuring the conductance of the solution at regular intervals. Consider the titration of the acid HCl by NaOH solution.

$$H^+ + Cl^- + Na^+ + OH^- \rightleftarrows Na^+ + Cl^- + H_2O$$

Initially the conductivity of the solution can be attributed to the H^+ and Cl^- present. During titration, the highly mobile H^+ are removed as water and are replaced by Na^+. Since the equivalent ionic mobility of the sodium ion is only one-seventh that of the hydroxonium ion, the conductance of the solution decreases markedly. After equivalence point, continued addition of titrant adds both Na^+ and OH^- and conductance increases. The titration curve shown in Fig. 13.2(b) indicates the type of variation in conductance observed during the titration.

During titration the volume of the solution changes and the measured conductance must be corrected for this factor. This effect can be minimized by using a titrant which is more concentrated than the assay solution.

Conductometric titration curves always depart from linearity in the vicinity of equivalence point, and the linear portions of the curves are best defined by points removed from the actual end point. Accordingly, in a titration it is common practice to take a few conductance readings before and after equivalence point and then join these points by straight lines. The end point of the titration is marked by the point of intersection of the two lines.

The apparatus required is fairly simple. As indicated in Fig. 13.2(a), the titration assembly generally consists of a dipping electrode probe (connected to a measuring bridge) immersed in a solution which is adequately stirred. A normal burette is used for adding the titrant.

The main advantage of the conductometric end-point detection method is its applicability to very dilute solutions. The titration technique is subject to many interference effects and becomes less accurate and satisfactory as the electrolyte concentration of the solution increases. The complete lack of specificity of conductance measurements severely restricts the number of occasions when this technique can be utilized, but the Handbook (Table 5.14) lists nearly fifty methods.

The applications may be subdivided into groups such as acid-base titrations (aqueous and nonaqueous), precipitation reactions, complexometric titrations and oxidation–reduction studies. The concentration ranges studied range from 10^{-1} to $10^{-6} M$ and in most methods a precision better than $\pm 0.5\%$ is claimed.

The type of titration curves obtained in acid-base and precipitation studies are shown in Fig. 13.2, segments c and d.

III.   POTENTIOMETRIC METHODS

A.   ELECTROCHEMICAL CELLS

A system consisting of two single electrodes in contact with their ions (half-cells), such as that shown in Fig. 11.4, is called an electrochemical cell.

If the electrode on the left hand side is considered to be made of metal A (the solution containing the ions A^{m+}) and the half-cell on the right hand side consists of metal B in contact with its ions B^{n+}, then the cell can be represented as follows:

$$A \mid A^{m+}(xM) \parallel B^{n+}(yM) \mid B$$
$$E_1 \qquad\quad E_3 \qquad\quad E_2$$

The left-hand electrode is first specified, followed by the ion or ions in solution in equilibrium with the electrode. The concentrations of the ions are given in parentheses. The right hand electrode is specified in the reverse direction, as indicated. A phase boundary across which a potential exists (E_1

and E_2 above) is indicated by a single vertical line (some authors use a slant line or a semicolon). If no difference in potential exists between two phases, the two are separated by a comma; e.g., the hydrogen electrode is written

$$\text{Pt, H}_2 \text{ (1 atm)} \mid \text{H}^+ \text{ (1 } M \text{)}.$$

A double line indicates a junction of two liquid phases across which a potential (E_3) exists. In most cases the potential is small enough to be disregarded in calculations.

Liquid junction potentials arise because the various ionic species diffuse or migrate across the interface at characteristically different rates. The rate of diffusion of a particular ion across the interface depends on the ionic mobility of the ion and the concentration gradient (in respect to this ion) which exists across the interface. The migration of anions and cations across the barrier need not occur in equal numbers so that one side of the liquid junction may become slightly negative in respect to the other side of the barrier.

Liquid junction potentials usually cannot be evaluated accurately, but if the solutions in the two half-cells are similar, and if a salt bridge is employed, its magnitude is made negligibly small (e.g., 2 mV). Because of the similarity in the ionic conductances of potassium, ammonium, chloride, and nitrate ions, salt bridges usually contain a concentrated solution of potassium chloride or ammonium nitrate.

Any variations or uncertainties in the liquid junction potentials are reflected in the observed potential of the galvanic cell, and these uncertainties constitute one of the drawbacks to the widespread analytical use of direct potentiometry. The most successful analytical applications of direct potentiometry, notably pH measurements, rely upon an empirical calibration technique to minimize the effect of errors arising from liquid junction potentials.

The following examples illustrate the conventional method of recording the composition of some galvanic cells of analytical interest.

$$\text{Cu} \mid \text{Cu}^{2+} \text{ (1 } M \text{)} \parallel \text{Zn}^{2+} \text{ (1 } M \text{)} \mid \text{Zn}$$

$$\text{Pt, H}_2 \text{ (1 atm)} \mid \text{H}^+ \text{ (0.1 } M \text{)} \parallel \text{KCl (0.1 } M \text{)}, \text{AgCl} \mid \text{Ag}$$

$$\text{Ag} \mid \text{AgCl, Cl}^- \parallel \text{Cl}^-, \text{Hg}_2\text{Cl}_2 \mid \text{Hg}$$

$$\text{Pt} \mid \text{Fe}^{2+}, \text{Fe}^{3+} \parallel \text{Cr}^{3+}, \text{Cr}_2\text{O}_7{}^{2-}, \text{H}^+ \mid \text{Pt}$$

$$\text{Cu} \mid \text{CuY}^{2-}, \text{H}_2\text{Y}^{2-}, \text{H}^+ \parallel \text{Cu}^{2+} \mid \text{Cu}$$

The cell voltage, E_{cell}, is the algebraic difference between the two electrode potentials. The potential of an electrode has a sign, but the cell voltage is given an absolute value.

Consider the cell,

$$\text{Cu} \mid \text{Cu}^{2+} \text{ (1 } M \text{)} \parallel \text{Zn}^{2+} \text{ (1 } M \text{)} \mid \text{Zn}$$

For the half-cells,

$$Cu^{2+} + 2e \rightleftarrows Cu, \ E^{\circ} = 0.34 \ V$$

$$Zn^{2+} + 2e \rightleftarrows Zn, \ E^{\circ} = -0.76 \ V$$

Since the concentration of the metal ions is one molar, the potential of each half-cell equals E°. If the concentrations had differed from unity, the Nernst equation would have been used to calculate the appropriate values of E_1 and E_2.

$$E_{cell} = E_1 - E_2$$

$$= 0.34 - (-0.76) = 1.10 \ V$$

The cell potential is a measure of the chemical force tending to drive the cell reaction towards equilibrium.

The spontaneous cell reaction is the reaction which occurs when the electrodes are connected with a wire and the cell is allowed to discharge. For a reaction to be spontaneous, ΔG° must be negative, and since $\Delta G^{\circ} = -nFE$, this means that the calculated value of E_{cell} must be positive.

The spontaneous reaction in this example is

$$Cu^{2+} + Zn \rightleftarrows Cu + Zn^{2+}$$

To achieve the reverse of this reaction, that is, oxidation of Cu to Cu^{2+} and reduction of Zn^{2+} to Zn, energy must be supplied to the system from an external source, e.g., a battery. This is the principle of electrolysis.

In calculating cell potentials it is usually tacitly assumed that the electrode system behaves reversibly. By this we mean that the components within the half-cell system are always in equilibrium with the electrode. That is, a very small change in the charge of an electrode should result in an immediate change in concentrations to maintain the equilibrium, or vice versa. Expressed in other words, the rate of the half-cell reaction must be extremely fast.

A few metals (Ag, Hg, Cu, Pb, Zn) and their respective ions meet this criterion very well. Others behave reversibly if the metal is amalgamated with mercury (Na, Cd, Bi, Tl).

On the other hand, the mechanisms of most electrode reactions are very complicated and the electrode process may take place very slowly and irreversibly or not at all. Under these conditions the Nernst equation provides a poor approximation of the observed behavior.

For absolute accuracy, the Nernst equation also requires the use of activities rather than concentrations.

B.   Types of Electrodes

The electrodes used in analytical, potentiometric procedures can be divided into two broad groups: reference and indicator electrodes.

1. *Reference Electrode*

A reference electrode must maintain a constant, reproducible potential, even if small currents are passed. For convenience, they are constructed as compact units that can be immersed in the test solution, a salt bridge between the two solutions being achieved by means of a slight leak in the outer sheath.

The most commonly used reference electrode is the calomel electrode which is composed of metallic mercury and solid mercury (I) chloride, in equilibrium

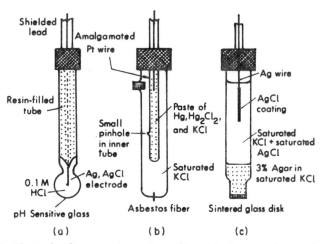

FIGURE 13.3   Electrodes for potentiometric studies: (a) glass membrane electrode; (b) calomel reference electrode; and (c) silver–silver chloride reference electrode.

with an aqueous solution of potassium chloride. A common form of commercial electrode is illustrated in Fig. 13.3(b). Contact with the solution under study is made by means of a salt bridge which can consist of a small asbestos fiber or a controlled leak through a ground glass sleeve mounted on the base of the electrode.

The electrochemical equilibrium which characterizes the behavior of the electrode is represented by the half-reaction

$$Hg_2Cl_{2(s)} + 2e \rightleftarrows 2Hg_{(s)} + 2Cl^-$$

and the corresponding Nernst equation can be written as

$$E_{Hg_2Cl_2, Hg} = E^\circ_{Hg_2Cl_2, Hg} - 2.303RT \log{[Cl^-]}/F$$

An inspection of this equation reveals that the potential of a calomel reference electrode depends only on the chloride ion concentration and the temperature. The amount of chloride ion is governed almost entirely by the amount of potassium chloride in the aqueous solution, and three different calomel electrodes are referred to in chemical literature. The more common is the saturated calomel electrode (SCE) in which the aqueous solution is saturated with potassium chloride; the normal calomel electrode is filled with a 1 M potassium chloride solution; and the decinormal calomel electrode uses 0.1 M potassium chloride solution. At 20°, the potentials of these three electrodes versus the normal hydrogen electrode (NHE) are $+0.2444$ V, $+0.2815$ V, and 0.3340 V, respectively.

Another reference electrode which is widely used is the silver-silver chloride electrode. It is analogous to the calomel electrode except that silver metal and silver chloride replace the mercury and mercury (I) chloride, respectively. The pertinent half-reaction is

$$AgCl_{(s)} + e \rightleftarrows Ag_{(s)} + Cl^-$$

Using a saturated potassium chloride aqueous solution, the potential of the electrode at 25° is $+0.1988$ V versus the NHE [cf. Fig. 13.3(c)].

In other electrodes 0.1 M KCl or saturated NaCl is substituted for the saturated KCl solution.

Reference electrodes have also been prepared using the redox couples Hg/HgO (in presence of 1 or 0.1 M NaOH) and Hg/Hg_2SO_4 (in presence of 0.5 M H_2SO_4 or saturated K_2SO_4).

2. *Indicator Electrode*

An indicator electrode responds to changes in the concentration of the species being measured. They vary in nature with the type of species being examined.

For example, in the titration of halides with silver salts, a silver wire is an ideal indicator electrode, the potential of the wire varying with the concentration of silver ions in the solution. In titrations involving the removal of copper ions from a solution, a suitable indicating electrode is a piece of copper rod.

With systems containing the two components of a redox couple, the only requirements of the indicating electrode are that it be inert in respect to the reactants and a good electrical conductor. A platinum electrode meets these requirements admirably.

By analogy with the above examples, a hydrogen electrode would represent the most logical means of measuring hydroxonium ion concentrations. While such electrodes have been used, it has been found more convenient to use another type of electrode, known as the glass electrode, to detect changes in the pH of solutions.

The glass electrode is an example of a membrane electrode, and in recent years similar types of electrodes have been developed for studying ions other than H^+.

3. *Glass Electrode*

A glass electrode responds rapidly and accurately to sudden changes in pH and is not subject to interference by oxidizing and reducing agents in the sample solution. However, the mechanism responsible for the response of the electrode to hydrogen ions is different from that of the hydrogen electrode, since it involves an ion exchange reaction rather than an electron transfer process.

The most familiar form of the glass membrane electrode is depicted in Fig. 13.3(a). A thin-walled bulb, fabricated from a special glass which is highly sensitive to the hydrogen ion activity of a solution, is sealed to the bottom of a glass tube. Inside the glass bulb is placed a dilute (usually 0.1 M) solution of hydrochloric acid. Immersed into the hydrochloric acid is a silver wire coated with a layer of silver chloride. The silver wire is extended upward through a resin-filled tube to provide electrical contact with the external circuit.

A complete theory for the mechanism of glass electrode operation remains to be formulated, but the presence of water in the outer and inner layers has been proved to be essential.

When the glass electrode is placed in an aqueous solution, water molecules are considered to penetrate the silicate lattice forming an inner and outer hydrated layer.

unknown solution	outer hydrated layer	dry glass layer	inner hydrated layer	0.1 M HCl

glass membrane

In the hydration process, sodium ions, loosely associated with the silicate lattice in interstitial holes, are thought to exchange with hydrogen ions,

$$Na^+_{glass} + H^+_{(aq)} \rightleftarrows Na^+_{(aq)} + H^+_{glass}$$

If an ion exchange process takes place within the inner and outer layers, the quantity of hydrogen ion occupying the exchange sites on either side of the membrane becomes proportional to the activity of the hydrogen ions in the adjacent solution. On this basis, it can be proposed that a phase-boundary potential is associated with the two hydrated layers, and what one ultimately measures is the difference between the two phase boundary potentials. In other words, the electrode may be considered to be equivalent to the concentration cell,

reference electrode	$H^+_{(a_1)}$	glass membrane	$H^+_{(a_2)}$	reference electrode.

In this model, the cell consists of a reference electrode immersed in each of two solutions having different hydrogen ion activities, a_1 and a_2 respectively, and separated by a glass membrane capable of acting as a "dry" salt bridge.

The emf of the cell at 25° is found to be governed by the relationship

$$E_{cell} = k - 0.059 \log (a_2)/(a_1)$$

The constant k accounts for any inequality in the potentials of the reference electrodes and/or liquid junction potentials together with a parameter called the asymmetry potential of the glass membrane. (The asymmetry potential is caused by strains and imperfections in the glass membrane.)

The activity (a_1) of the hydrogen ion in the inner compartment remains constant; hence the equation can be rewritten as

$$E_{cell} = k' - 0.059 \log (a_2) = k' + 0.059 \text{ pH}$$

It is this equation which serves as the basis of all practical pH determinations. Since k' cannot be determined accurately, the electrodes are calibrated by using solutions of known pH.

The composition of the glass membrane has an important bearing on the characteristics of the electrode. At high pH values (>10), many electrodes exhibit an alkaline error, i.e., the observed pH is lower than the true value. The size of the alkaline error varies with the nature and concentration of the extraneous alkali ions as well as with the temperature. At the other end of the pH scale, the glass electrode exhibits an acid error. In very strong acid solutions, the observed pH is higher than the true value.

4. *Ion-Selective Electrodes*

A major development in recent years has been the introduction of other ion-selective membranes.

By varying the composition of the glass used in making the lower bulb, it is possible to prepare electrodes which exhibit a preferential response to potassium ions, or sodium ions, or silver ions, or lithium ions.

Anion-selective indicator electrodes have been prepared by impregnating a solid matrix with an insoluble precipitate containing the desired ion. For example, an electrode for bromide ions can be prepared by causing monomeric silicone rubber to polymerize in the presence of an equal weight of silver bromide particles. After the mixture has solidified, the uniform matrix is sealed to the bottom of a glass tube. Potassium bromide and a silver electrode are placed in the tube. A solid tends to adsorb preferentially the ions of which it is composed; hence when the silver–bromide–silicone rubber matrix is immersed in a test solution, it tends to adsorb bromide ions in proportion to the activity of the ions in the solution. In turn, this adsorption of bromide ion produces a phase boundary potential which can be measured. In a similar manner, electrodes have been prepared for the determination of chloride, iodide, fluoride, sulfide and cyanide ion concentrations. The properties of a group of such electrodes are summarized in Table 13.2. Solid state electrodes have also been used for the determination of silver and copper.

Liquid ion exchangers are used in another type of highly specific indicator electrode. In these electrodes, an inner silver-silver chloride reference electrode dips into a liquid ion exchanger charged with the ion of interest. For example,

TABLE 13.2

THE PROPERTIES OF ION SELECTIVE MEMBRANES[a]

Type	Ion (X)	Range pX	Range pH	Interfering ions
Glass membrane	H^+ Na^+ K^+, Ag^+ Li^+	0–10		Na^+, K^+, Li^+
Solid state	F^-	0–6	0–8	OH^-
	Cl^-	0–4	0–14	S^{2-} must be absent, I^-, Br^-, CN^-
	Br^-	0–7	0–14	S^{2-} must be absent
	I^-	0–7	0–14	S^{2-} must be absent
	CN^-	2–6	0–14	S^{2-} must be absent, I^-
	S^{2-}	0–17	0–14	
				Selectivity Constants[b]
	Cl^-	1–5	2–10	ClO_4^- 32, I^- 17, NO_3^- 4.2, Br^- 1.6, OH^- 1.0, OAc^- 0.32, HCO_3^- 0.19, SO_4^{2-} 0.14, F^- 0.10.
Liquid ion exchange	NO_3^-	1–5	2–12	ClO_4^- 10^3, I^- 20, ClO_3^- 2, Br^- 0.9, S^{2-} 0.57, CN^-, HCO_3^- 0.02, Cl^-, OAc^-, CO_3^{2-}, $S_2O_3^{2-}$, SO_3^{2-}, 0.006.
	ClO_3^-	1–5	4–11	I^- 0.012, NO_3^- 0.0015, Cl^-, F^- 0.0002, Br^- 0.0006.
	BF_4^-	1–5	2–12	
	Ca^{2+}	0–5	5.5–11	Zn^{2+} 3.2, Fe^{2+} 0.8, Pb^{2+} 0.63 Cu^{2+} 0.27, Ni^{2+} 0.08, Mg^{2+}, Ba^{2+} 0.01
	Cu^{2+}	1–5	4–7	Fe^{2+} 1, Ni^{2+} 0.005, Zn^{2+} 0.001, $Na^+, K^+ < 0.001$, Ca^{2+} 0.0005, Ba^{2+}, Sr^{2+} 0.0002

[a] Based on data published by Orion Research Incorporated, Cambridge, Massachusetts.
[b] % Error caused by interferants = (Selectivity ratio)(Conc. interferant ion) · 100/(Conc. of measured ion).

a calcium electrode is filled with a calcium organophosphorus compound, the cell being sealed at the lower end with a sintered glass disk or a plastic membrane. The primary purpose of the disk or membrane is to prevent the liquid exchanger from dissolving in the unknown samples. Such an electrode has been shown to follow the Nernst equation for calcium concentrations as low as 10^{-5} M and to exhibit marked specificity for calcium. Electrodes of this type have been developed for the determination of chloride, nitrate, perchlorate, fluoborate, calcium, copper, and water hardness (expressed as concentration of divalent cations).

Since several ions may compete for positions on the exchanger, these electrodes are subject to some interference effects. The magnitude of the interference effect is usually reported in terms of a selectivity constant. The values quoted for one brand of electrodes are listed in Table 13.2.

Where contact between the inner chamber solution and the sample solution is highly undesirable, double-junction electrodes are recommended. In these electrodes there is an outer chamber on the cell which is filled with a solution

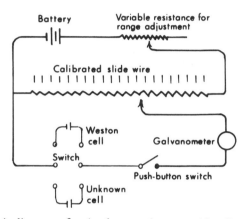

FIGURE 13.4   Circuit diagram of a simple potentiometer with a linear voltage divider.

which provides the required contact between the liquid exchanger retaining membrane and the outer seal of the cell.

C.   DIRECT POTENTIOMETRIC MEASUREMENTS

A meaningful value of the emf of an electrochemical cell can be obtained only if virtually no current flows through the cell during the period of measurement. The most reliable values are derived from measurements made with a potentiometer. In a potentiometer circuit the cell voltage is opposed by an auxiliary voltage source, and when the two voltages are identical, no current flows. The null point is detected by a simple galvanometer. The details of a potentiometer circuit are outlined in Fig. 13.4.

To standardize potential measuring instruments, one requires a galvanic cell whose emf is known very accurately. The internationally accepted standard is the Weston cell, which can be represented as

$$-Cd(Hg) \,|\, CdSO_4 \cdot 8/3 \, H_2O_{(s)}, \, Hg_2SO_{4(s)} \,|\, Hg+$$

The cathode consists of a pool of mercury in equilibrium with the aqueous phase which is saturated in respect to both cadmium sulfate and mercury (I) sulfate. The anode in equilibrium with this aqueous phase is a saturated cadmium amalgam. At 20°, the emf of the Weston cell is 1.086 V.

Alternatively an electronic voltmeter can be used to measure the cell voltages. The vacuum tube voltmeter incorporates an electronic amplifier so that only minute currents are required. This type of circuit is particularly useful where high resistance electrodes (e.g., the glass electrode) are being used as the indicating electrode in the measuring circuit.

Under ideal conditions, the potentiometric measurement of the electromotive force of an electrochemical cell should be directly related, through the Nernst equation, to the activity of a single desired species (provided that one half-cell is a standard reference electrode).

For example, in order to analyze a dilute aqueous solution of silver nitrate one could construct a cell in which the indicating electrode is a piece of silver wire and the reference half-cell is a saturated calomel electrode. To prevent precipitation of silver chloride the reference electrode would need to be connected to the test solution via a salt bridge (e.g., 3% agar in $1M$ KNO_3). The galvanic cell being measured may then be represented as

$$Hg \mid Hg_2Cl_{2(s)}, KCl_{(s)} \mid KNO_3(1\ M) \mid AgNO_3 \mid Ag$$
$$E_{Hg,Hg_2Cl_2} \qquad E_j \qquad E_j{}' \qquad E_{Ag^+,Ag}$$

The cell potential measured by a potentiometer would be the sum of all the components; i.e.,

$$E_{cell} = E_{Hg,Hg_2Cl_2} + E_j + E_j{}' + E_{Ag^+,Ag}$$

From the Nernst equation, at 25°

$$E_{Ag^+,Ag} = E^0_{Ag^+,Ag} + 0.059 \log (Ag^+),$$

hence if E_{cell}, E_{Hg,Hg_2Cl_2}, E_j and $E_j{}'$ are known, the activity of the silver ion in solution can be calculated.

Of these terms, the liquid-junction potentials are of uncertain magnitude, and the only way of accurately compensating for them is to measure the potential of electrochemical cells containing known concentrations of silver ion, and having an ionic strength comparable to the unknown samples to be studied. The results so obtained are then used to prepare a calibration graph of E_{cell} versus concentration of silver ion.

Another practical problem in direct potentiometry is the tendency of some indicator electrodes to respond to several species. This effect is of particular importance when using the selective ion membranes discussed in the preceding section.

On the other hand, direct potentiometry possesses some important virtues. The technique is nondestructive, it requires only small samples, it is suitable for studies of relatively dilute solutions, and it is amenable to automatic recording.

The latter property means that the technique can be employed to monitor industrial processes and biochemical systems.

The advent of commercially available ion-selective membranes has greatly expanded the scope of the procedure, and the number of applications can be predicted to increase manyfold in the next decade.

Until recently, however, the most important application of this principle was the determination of the pH of solutions.

As in the measurement of other ion activities by direct potentiometry, it is necessary to compensate for liquid-junction potentials and electrode effects by calibrating the equipment against standards. In all practical pH determinations, the pH of the unknown solution (pH_x) is compared to the pH of a standard buffer (pH_s).

The two are related by the equation

$$pH_x = pH_s + \{(E_{cell})_x - (E_{cell})_s/0.059\}$$

For greatest accuracy the standard buffer should be of similar pH to the unknown.

The pH values of a selected group of standard buffers have been determined by the National Bureau of Standards from electromotive force measurements using galvanic cells without liquid-junction potentials. The pH values of these standard buffers are known with an accuracy of ± 0.005 pH units at 25°.

D. POTENTIOMETRIC TITRATIONS

Although the number of applications of direct potentiometry are increasing, most potentiometric methods of analysis involve titrimetry.

In a potentiometric titration the change in concentration of one or more of the species present is measured potentiometrically. The titration beaker becomes an electrochemical cell, the indicating electrode and the titrated solution acting as one half-cell, a reference electrode the other half-cell.

The change in electromotive force of the cell during the titration is recorded as a function of the added titrant. The major goal of this procedure is to locate precisely the equivalence point or points in the titration, but other information such as the dissociation constants of acids and complex ions, solubility products, and oxidation–reduction potentials can be deduced from the curves.

The experimental setup used for potentiometric titrations is shown diagrammatically in Fig. 13.5(a). The plot of the cell potential against volume yields an S-type curve, the point of inflection of the curve occurring at the equivalence point of the titration.

The slope of the curve in the vicinity of the equivalence point depends on a number of factors including reaction of the titration products with the solvent; e.g., the slope is influenced by the hydrolysis of salts or the solubility of precipitates. To facilitate the detection of the equivalence point, it is sometimes

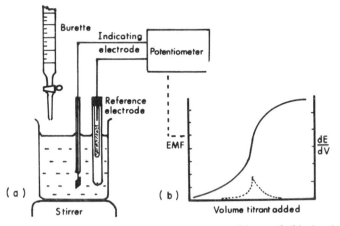

FIGURE 13.5 Potentiometric titrations: (a) titration assembly; and (b) titration curves. The full line represents the normal potentiometric curve while the dotted line is the differential curve obtained by plotting dE/dV versus volume of titrant added.

preferable to prepare a differential curve (dE/dV versus V) as shown in the dotted line superimposed on Fig. 13.5(b).

Potentiometric titrations are especially useful in studies of chemical systems which are too brightly colored for ordinary visual indicators to be used, and in titrations in nonaqueous media. The approach provides a valuable means of studying systems where there is marked interaction of the products with the solvent, or where no internal indicator is available. The technique also provides a means of assessing the suitability of visual indicators for new applications.

Micro and macro amounts of material can be titrated, and in most cases the accuracy is equal to a fraction of a percent of content. For routine purposes, the technique can be simplified by merely titrating the assay solution until the potential of the solution equals the potential at equivalence point as determined from an initial complete titration curve. This process can be made automatic by using the critical equivalence potential to switch off the device that supplies the titrant. For the repetitive titration of a single type of product, automatic titrators are of great value.

Other titration units eliminate the tedium of plotting titration curves by incorporating a chart recorder in the system.

Potentiometric titrations may be used for acid-base, complexometric, oxidation–reduction and precipitation reactions. With appropriate modification of the electrode and/or salt bridge systems, titrations are readily performed in nonaqueous media.

Tables outlining the procedures for the execution of a number of potentiometric titrations form part of the "Handbook of Analytical Chemistry." More complete information may be found in the reference works listed in the bibliography.

E. POTENTIOMETRY WITH POLARIZED INDICATOR ELECTRODES

A number of commonly used redox couples, such as $S_2O_3^{2-}/S_4O_6^{2-}$ and $Cr_2O_7^{2-}/Cr^{3+}$, behave irreversibly at a platinum electrode, and when measurements are made by ordinary (zero current) potentiometry, constant potentials are established very slowly.

With such systems, a sharp jump in potential can sometimes be achieved by the use of polarized indicator electrodes. The polarization is achieved by forcing a slight amount of electrolysis to occur, the current flowing in the circuit being maintained at a small constant value (a few microamps).

Potentiometric titrations at constant current may be divided into two classes, depending on whether one or two indicator electrodes are used. In the former case the potential of a single polarized platinum indicating electrode is measured against a reference electrode. Alternatively, two platinum electrodes may be used, in which case one acts as the anode and the other as cathode.

The potential between the electrodes is measured during the course of the titration. If both the titrant couple and the reactant couple are reversible, the use of one polarized indicator electrode results in a curve similar to the ordinary (zero current) case; but two polarized indicator electrodes yield a "peak" type curve.

The main advantage of the technique arises when one of the reactant couples behaves irreversibly. In such cases the resultant curve is usually characterized by a sharp increase (or decrease) in potential at or near the equivalence point.

A particularly useful modification of this procedure is the "dead-stop" technique in which a small potential (order of millivolts) is applied to two platinum electrodes. Such an applied potential is too small to exceed the decomposition potential of most of the ionic species normally present in a titration system. Under these circumstances, no current flows in the circuit until the solution gains both a species which is readily reduced at the cathode and a species which is readily oxidized at the anode. For example, in the titration of sodium thiosulfate with iodine, the solution before equivalence point contains $S_2O_3^{2-}$, $S_4O_6^{2-}$, and I^-. The iodide ion depolarizes the anode but no current is observed since the cathode remains polarized. The first excess of titrant, however, yields iodine which is readily reduced at the cathode and a current flows in the circuit. The sharp change from zero to maximum current marks the end point of the titration. The reverse effect is observed when iodine solutions are titrated with sodium thiosulfate.

The dead-stop technique can be used whenever the equivalence point in a titration corresponds to a transition in conditions such that the electrodes go from the state of at least one being polarized to a state of both being depolarized, or vice versa. An important application of this method is the titration of water using the Karl Fischer reagent.

The concept of polarization is considered in more detail in the following section.

IV. ELECTRODEPOSITION

A. ELECTROLYSIS

In oxidation-reduction titrations, ionic species are used either as sources of electrons or as electron acceptors.

An alternative method of providing the electrons required for electrical transformations is to apply an external voltage to chemically inert electrodes

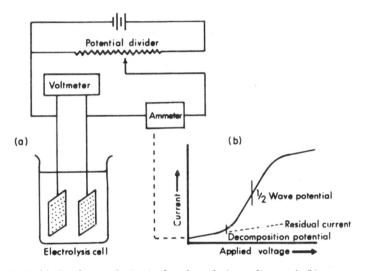

FIGURE 13.6 (a) Fundamental circuit for electrolysis studies, and (b) current-voltage relationship.

(e.g., platinum) immersed in the test solution. In order to achieve electron transfer, the applied potential must be great enough to reverse the direction of the reactions responsible for the development of the normal galvanic cell potential.

The transformations induced in this way are said to be caused by electrolysis, and a simple circuit for electrolytic studies is shown in Fig. 13.6. Charged and uncharged species are brought to the electrodes by convection and diffusion processes, and in addition, on the imposition of an applied potential, charged ions migrate to the electrode of opposite sign. Thus as the potential is gradually increased from zero, an electrical double layer of anions and cations is set up in the vicinity of the electrodes. At the same time as the double layer is built up, small concentrations of impurities (e.g., oxygen) are oxidized or reduced at the electrodes. In maintaining the double layer and the transformation of the highly reactive trace impurities, a small current flows in the circuit. This current is known as the residual current [cf. Fig. 13.6(b)].

Beyond a point known as the decomposition potential, continuous electrolysis occurs and the current increases rapidly with increasing applied voltage. At potentials just a little greater than the decomposition potential, more charged species reach the electrodes than can be discharged. With higher applied voltages, the migration of the reacting species becomes the rate controlling step and the current flowing in the circuit tends towards a maximum value.

The decomposition potential may be defined as the minimum external voltage required to cause electrolysis which proceeds at a measurable rate.

During electrolysis, oxidation occurs at the anode and reduction occurs at the cathode. The species oxidized or reduced at the electrodes are those species in solution which require the least amount of energy for the transformation. The energy requirements of a given species depend on its concentration, its standard potential, and a variable known as overvoltage.

If, in the course of electrolysis, the battery is replaced by a conducting wire, a current may be observed flowing for a short time in the opposite direction. In other words, the electrode system can act as a cell with a potential equal to $E_c - E_a$, E_c and E_a being the potential of the cathode half-cell and anode half-cell, respectively.

The temporary cell potential is known as the polarization, counter or back emf. Values for E_c and E_a can be calculated for reversible systems by substitution in the appropriate Nernst equation. For example, during the electrolysis of an 0.1 M copper nitrate solution, copper is deposited at the cathode in accordance with the equation

$$Cu^{2+} + 2e \rightleftarrows Cu$$

The potential of the cathode, E_c, is therefore described by the relationship

$$E_c = E^{\circ}_{Cu^{2+}, \, Cu} + \ln [Cu^{2+}] \cdot RT/nF$$
$$= 0.34 + 0.0295 \log [Cu^{2+}]$$

A similar equation, based on the oxidation of water, can be written for the anode reaction.

$$4H^+ + O_2 + 4e \rightleftarrows 2H_2O, \qquad E^{\circ} = 1.229$$
$$E_a = E^{\circ} + 0.059 \log [H^+]$$

The minimum potential required to cause electrolysis is one which is opposite in direction but greater in magnitude than the electrochemical cell.
That is,

$$E_R \geq -(\text{polarization emf}) \geq E_a - E_c$$

Comparison of calculated E_R values with the decomposition potentials, E_D, observed in experiments often shows a marked discrepancy. Part of the difference can be attributed to the potential between the electrodes arising from the resistance of the solution ($E_S = IR$) but most of the effect has to be attributed

to the influence of irreversible processes occurring at the electrodes. The combined effect of the irreversible processes is known as the overvoltage:

$$E_D = E_a - E_c + IR + \text{overvoltage}$$

Overvoltage is the excess of voltage, over the polarization emf, needed to
maintain a certain rate of deposition on an electrode. There are several types of

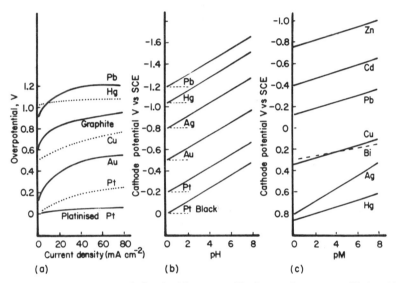

(a)   (b)   (c)

FIGURE 13.7   (a) Overpotentials for the liberation of hydrogen from 1 M sulfuric acid on
solid electrodes; and (b) variation of cathode potential required for the liberation of
hydrogen on different metals at different pH values; and (c) calculated values of the cathode
potential required for the deposition of metal from solutions of differing metal ion concentration. Based on data taken from the *Handbook of Analytical Chemistry*, edited by L.
Meites. Copyright © 1963, McGraw-Hill Book Company. Used with permission of
McGraw-Hill Book Company.

overvoltage, and it can be attributed to processes occurring at the anode and/or
cathode.

One type of overvoltage which reaches fairly high values is gas overvoltage.
It varies markedly with the metal surface condition and current density of the
electrode associated with the liberation of the gaseous product. The influence
of these factors on the overvoltage observed with hydrogen liberation is shown
in Fig. 13.7(a). Where the product of electrolysis is a metal, overvoltage at the
cathode is usually very small.

Another source of overvoltage is the concentration gradient which develops
in the immediate vicinity of the electrode. During electrolysis the zone near
the electrodes has a lower concentration of the reacting species than the bulk
solution. In the absence of stirring, diffusion of ions across the concentration

gradient becomes the rate controlling process. In order to maintain a reasonable rate of electrolysis, the rate of diffusion has to be increased either by heating the solution, by vigorous stirring, or by the addition of more charge on the electrode. Overvoltage of this type is minimized by using electrodes of large surface area, by limiting the current to a small value, by electrolyzing a hot solution, and by efficient stirring.

B. Electrogravimetry

In analysis by electrogravimetry, deposition of an element is carried to completion by passage of a large excess of current, the amount of material deposited being determined by weighing the cathode before and after electrolysis. A few determinations are based on the measurement of the weight of a metal oxide deposit formed at the anode.

The aim in electrogravimetry is to obtain a pure, coherent, dense, and smooth deposit which can be washed, dried, and weighed. Adherence to the electrode is a most important property. Thus hydrogen evolution at the cathode at the same time as metal deposition is objectionable, since it produces spongy deposits. Moreover, the discharge of hydrogen tends to leave an alkaline film close to the electrode, and this can cause oxides or basic salts to form.

Factors which help to produce smooth, adherent deposits are efficient stirring, low current densities, and the proper selection of anions. The role of the anions is varied. As a general rule, smoother deposits are produced from solutions of complex ions than from simple salt solutions. Halide ions facilitate the deposition of many metals, probably because the overpotential is lower for ions of the type MCl_x^{m-x} than for the corresponding aquocomplex. Other anions act as depolarizers. For example, the liberation of hydrogen can be avoided by having nitrate ions in the solution, for these are reduced in preference to hydrogen ions.

$$NO_3^- + 10H^+ + 8e \rightleftarrows NH_4^+ + 3H_2O$$

In the most widely adopted mode of electrodeposition, the current is held more or less constant by periodic or continuous adjustment of the voltage applied to the cell. Under this system the reaction with the most positive reduction potential is the first to occur at the cathode. As the concentration of the cation falls through reduction, the rate at which it reaches the electrode decreases, and to maintain a constant current the applied voltage has to be increased. In this process a point can be reached where a second species begins to deposit.

To minimize this effect, it is necessary to calculate the minimum applied emf required to quantitatively deposit the material of interest and to calculate the maximum voltage which can be used without introducing some codeposited material.

In most analytical separations, the product liberated at the anode is oxygen.

For this reaction

$$E_a = 1.23 - 0.059 \text{ pH} + \text{overvoltage}$$

During electrolysis, neither the pH nor the overvoltage varies very significantly, and in molar acid solution using platinum electrodes, E_a remains at about 1.7 V.

The decomposition potential required to cause continuous electrolysis is controlled in such cases by the variable term E_c + overvoltage (cathode).

Consider a solution which is 0.1 M in respect to lead ions and molar in respect to hydrogen ions,

For hydrogen evolution on platinum electrodes

$$\begin{aligned} E_D &= E_a - (E° + 0.059 \log [\text{H}^+]) + \text{overvoltage} \\ &= 1.70 - (0.00 + 0) + 0.2 \\ &= 1.90 \text{ V} \end{aligned}$$

For lead deposition on platinum electrodes

$$\begin{aligned} E_D &= E_a - (E° + 0.0295 \log [\text{Pb}^{2+}]) + \text{overvoltage} \\ &= 1.70 - (-0.126 - 0.0295) + 0 \\ &= 1.85 \text{ V} \end{aligned}$$

Thus, because of gas overvoltage, lead would be preferentially deposited from the acid solution at an applied voltage between 1.85 and 1.90 V, even though the standard potential for hydrogen is more positive than that of lead.

By using copper electrodes, the hydrogen overvoltage is increased to 0.7 V, and the applied voltage could then be increased to nearly 2.4 V before hydrogen would tend to be liberated simultaneously with the lead. However, at this point, the electrode would be coated with lead, on which the overvoltage for hydrogen is nearly 1.2 V; hence gas evolution may not occur unless the applied voltage exceeds 2.8 V.

The recommended experimental procedure deposits lead onto a copper-coated platinum electrode, from a phosphoric acid solution of the lead salt.

While the deposition of lead would begin at an applied potential of 1.85 V, quantitative removal of the lead ions from solution needs a higher value, since E_c varies with the concentration of reducible ions remaining.

The value of E_c for different concentrations of ions is most conveniently recorded in a graphical form.

For the reversible reaction

$$M^{n+} + ne \rightleftarrows M$$

$$E_c = E° + (\log M)(0.059/n) + \text{overvoltage}$$

or

$$E_c = E° - pM(0.059/n) + \text{overvoltage.}$$

A plot of E_c against pM should give a straight line of slope $0.059/n$ and an intercept at $pM = 0$ of $E°$ + overvoltage.

The effect of the overvoltage term on the intercept is shown in Fig. 13.7(b) which records the cathode potential required for the liberation of hydrogen from solutions of different pH when using electrodes made of different metals. The influence of pM on the cathode potential required for the deposition of metals is illustrated in Fig. 13.7(c).

For quantitative deposition of a metal, the applied decomposition potential should exceed the value required when the concentration of metal ion has been reduced to 10^{-4} of its original value. This corresponds to a change of four pM units on the graph.

Returning to the example of the deposition of lead, for quantitative removal of this element pM needs to change from 1 to 5. At $pPb = 5$, $E_c = -0.274$ V, and if $E_a = 1.70$ V the minimum voltage for quantitative deposition is 1.97 V.

The minimum equipment required for constant current electrolysis is a dc power supply capable of providing a potential variable between 0 and 3 V, a pair of electrodes, and a stirrer. The anode and cathode are usually cylindrical in shape and made of platinum gauze or perforated sheet, the cathode being larger in size. The electrodes are cleaned and weighed before deposition. After continuous electrolysis has quantitatively removed the desired ion, the deposit is washed with water and alcohol before being dried and the electrode reweighed.

Metals such as zinc, cadium, lead, and mercury, alloy with platinum and for the deposition of these metals it is preferable to use electrodes coated with copper or silver.

Most of the deposits that are collected for analytical purposes are metals formed at the cathode, and the Handbook summarizes procedures for the determination of Ag, Au, Bi, Cd, Co, Cu, Fe, Hg, Ni, Pb, Pd, Pt, Rh, Sb, Sn, U, and Zn.

Commercially, the elements most regularly determined are Cu, Pb, Co, and Ni.

Elements such as As, Ga, and Te are determined through coprecipitation with a known added amount of copper.

In the presence of nitric acid, lead forms a weighable deposit of PbO_2 at the anode. Manganese also deposits as a hydrated oxide on the anode. By using a silver anode, chlorine and bromine can be determined through the weight of silver halide deposited on this electrode.

In all the Handbook summaries, up to eight elements are listed in the column headed "codeposit or interfere." Thus, while electrogravimetric procedures can be extremely accurate, great care is required if the final deposit is to possess maximum purity.

C. ELECTROSEPARATIONS

The problem of coprecipitation arises when the applied potential is sufficient to exceed the deposition potential of more than one species. If the second species produced is hydrogen there is no real problem in regard to codeposition, although the gas liberation process can influence the physical properties of the deposit. In fact, hydrogen liberation can aid the separation procedure. For example, in the determination of copper in a brass, copper is completely separated from zinc provided that the pH is less than 5. Reference to Fig. 13.7 shows that a 0.1 M solution of copper should begin to deposit when E_c is more negative than $+0.31$ V. If the acid is 0.1 M also, hydrogen evolution would begin at $E_c = -0.66$, a value which is more positive than the zinc line. However, between the copper line and the hydrogen on copper line (Fig. 13.7) there are lines due to many elements, e.g., Bi, Pb, Sn, etc. and below the copper line, Ag, Au, etc. All these elements are potential interferants.

Some are removed chemically, e.g., in the presence of nitric acid, lead deposits as PbO_2 on the anode; others are eliminated by electrochemical separation procedures.

Consider a copper alloy which contains 1% bismuth. A solution of the alloy which was approximately 0.1 M in respect to copper ion would therefore be about 0.001 M in respect to bismuth.

Use of the chart (Fig. 13.7) indicates that the copper would begin to deposit at a potential of $(E_a - 0.31)$ V and would be quantitatively deposited at potentials greater than $(E_a - 0.19)$ V. The bismuth would begin to codeposit when the applied potential reached $(E_a - 0.26)$ and would be virtually completely deposited at $(E_a - 0.19)$ V.

The problem of the bismuth contamination could be solved in several ways.

One way requires the addition of a chemical to selectively precipitate bismuth before electrolysis. For example, the addition of hydrogen sulfide to an ammoniacal solution containing cyanide ions would produce a precipitate of bismuth sulfide which could be removed by filtration.

Alternatively the cyanide solution could be electrolyzed directly. Due to the formation of the stable cyanide complex, pCu becomes $\gg 3$ (e.g., >20) and bismuth becomes the species preferentially deposited. After determination of the weight of bismuth deposited, the potential could be increased to the higher value required to deposit the copper from the cyanide solution.

A third approach is to control the cathode potential at a value below that required for bismuth deposition. That is, in this case, $E_D \leq E_a - 0.26$ V. At this voltage, pCu in equilibrium with the deposit is 2.7 or $[Cu^{2+}] = 2 \times 10^{-3}$ M which corresponds to 98% removal of the original copper. Thus by sacrificing accuracy, a separation can be achieved. This method is known as the controlled cathode potential technique and the basic equipment required is shown in Fig. 13.8. The potential of the cathode in respect to the solution is measured against a reference half-cell using a standard potentiometer circuit. The voltage applied to the primary electrolysis circuit is then varied to ensure that the cathode potential does not exceed the calculated maximum permissable value. The polarization emf eventually equals the applied potential and current ceases to flow. This marks the end of the determination. The tedium of manual adjustment has been overcome by the development of automatically controlled potentiostats.

Controlled potential electrolysis has been used for the determination of Cu, Pb, and Sn in brasses and bronzes and for the analysis of nickel bronzes.

An alternative method of preventing the cathode potential exceeding a selected value is to add a depolarizer. A depolarizer is a species in solution

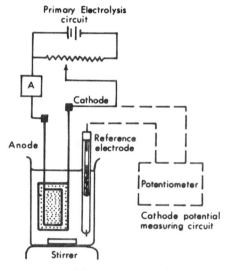

FIGURE 13.8 Basic components of the apparatus required for controlled cathode potential electrolysis. The system consists of a primary electrolysis circuit and a potentiometer circuit for measuring the potential difference between the cathode of the primary circuit and standard reference half-cell.

which becomes oxidized or reduced in preference to other species which contaminate the deposit or reduce the efficiency of deposition.

The hydrogen ions which are reduced in preference to the zinc ions in the determination of copper in brass may be regarded as depolarizing the cathode. A better depolarizer for this reaction is the nitrate ion, since it is reduced to ammonia at a much lower potential.

If iron is present in copper alloys it becomes necessary to add hydroxylamine as an anodic depolarizer. In the absence of hydroxylamine, the electrode processes of minimum energy are oxidation of iron (II) at the anode and reduction of iron (III) at the cathode. The addition of hydroxylamine reduces the iron (III) chemically to iron (II) and excess reagent is preferentially oxidized at the anode. This eliminates the processes which tend to prevent copper deposition. A similar effect can be achieved by adding fluoride or phosphate ions to form a stable complex with iron (III).

D. MERCURY CATHODE SEPARATIONS

It can be observed from the above discussion that some selectivity in electro-gravimetric procedures can be achieved by judicious use of complexing agents, control of cathode potentials, and addition of depolarizers.

The one interferant which has been mentioned only in passing is the hydrogen ion. Reference to Fig. 13.7(b) indicates that the cathodic potential for hydrogen liberation is a function of both the pH and the nature of the metal cathode.

A comparison of the applied emf required to liberate hydrogen on platinum with the emf required for the reduction of cobalt and nickel ions shows that at pH values <7, hydrogen is the preferred reduction product. At higher pH values, however, the reduction of the metal ions requires the lower potential, and cobalt and nickel can be quantitatively deposited from ammoniacal solutions.

The deposition of zinc at a cathode has been achieved by combining the effect of increased gas overvoltage (e.g., a copper electrode is used) with pH effects (the electrolyte contains free ammonia).

The hydrogen overvoltage has a maximum value on mercury and the use of a mercury cathode allows one to deposit a large number of metals from acid solutions. Thus, by means of a mercury cathode, a large group of elements can be removed from solution without resorce to chemical additives. The circuit generally consists of a platinum anode and a mercury pool cathode. Four or more volts are applied across the electrodes, while the solution is being stirred. This removes as an amalgam or deposit all those elements which can be reduced to the metallic state at potentials lower than that required to liberate hydrogen from a mercury surface. From a 0.1 to 0.2 M sulfuric acid solution the elements quantitatively deposited include Ag, Au, Bi, Cd, Co, Cr, Cu, Hg, Fe, Ni, Mo, Pd, Pt, Sn, Tl, and Zn. At the end of the electrolysis period the mercury is separated from the aqueous solution. The potential is maintained across the system during the phase separation to prevent dissolution of the deposit in the acid solution which may still contain a number of elements such as Al, Be, Mg, Ti, V, W, alkaline earths, and rare earths. The common analytical applications are based on the complete removal of the first group of elements in order to simplify the determination of a member of the group which remains in solution. For example, a preliminary mercury cathode separation has been recommended in the determination of Al and Mg in zinc-base alloys, and in the determination of Al, V, Zr, Ce, or La in steels.

The usefulness of this technique is extended considerably by control of the cathode potential. This aspect is considered in some detail in the review by Page, Maxwell, and Graham [*Analyst*, **87** (1962) 745].

V. COULOMETRIC ANALYSIS

A. FARADAY'S LAWS

In analysis by electrodeposition, the redox reaction is carried to completion by passage of a large excess of current and the amount of material deposited is determined by weighing.

Instead of using excess quantities of electricity, it is sometimes possible to determine directly the exact quantity of electricity required to quantitatively reduce or oxidize a given species. Procedures which measure the quantity of electricity consumed are termed coulometric methods of analysis.

Such methods are founded upon Faraday's Laws of Electrolysis which state that the amount of chemical change that occurs as a result of electrolysis is directly proportional to the quantity of electricity which passes.

For the general reaction

$$A^{a+} + ne \rightleftarrows A^{(a-n)+}$$

the quantity of electricity required for reduction of one mole of species A is nF coulombs where n is the number of electrons transferred and F is the Faraday ($96,487 \pm 1.6$ coulombs).

In any electrical circuit, the quantity of electricity (Q) involved is given by the integral of the current flow (i amperes) over the time interval (t seconds). Thus

$$Q = \int_0^t i\, dt = nwF/M$$

where w is the weight in grams of the species that is consumed or produced during electrolysis and M is its gram molecular weight.

This is the basic equation of coulometry. It indicates that w can be determined directly if Q can be accurately evaluated and if the experimental conditions are so arranged that a single reaction of known stoichiometry occurs during electrolysis.

Coulometric procedures may be divided into two groups: coulometry at controlled electrode potential and coulometry at constant current.

B. COULOMETRY AT CONSTANT ELECTRODE POTENTIAL

In coulometry at constant potential, the substance being determined reacts at an electrode whose potential is maintained at a value that precludes other unwanted electrode reactions (cf. controlled cathode potential technique, Section IV,C). The current decreases exponentially as the electrolysis proceeds; hence Q has to be evaluated by an integration technique. The simplest method of determining Q is to place a coulometer in series with the reaction cell. A coulometer is an electrolysis cell which produces a product (with 100%

efficiency) that can be weighed or measured accurately. Three common types are the silver, copper, and gas coulometers. The passage of one coulomb of electricity causes the deposition of 1.118 mg of silver in a silver coulometer, 0.659 mg of copper in a copper coulometer, and the liberation of 0.1739 ml of gas (STP) in a hydrogen-oxygen coulometer. Since weighings can be performed to a fraction of a milligram, Q can be determined to a fraction of a coulomb. The technique is therefore very sensitive and accurate. However, for some time the number of applications was limited by the difficulties associated with maintaining a constant cathode potential. Precise potentiostats are now readily available and advances in analytical instrumentation have made potentiostatic coulometry a simple and rapid technique, suitable for routine use. By varying the cathode potential, successive determinations are possible. Modern instruments electronically monitor the potential and integrate the current flow. The titration cells are designed to give optimum conditions for the electron transfer process. The electrode area is made large, the solution volume, small, and a high rate of stirring is maintained. Three electrodes are necessary: a working electrode at which the desired reaction takes place, an auxiliary electrode to complete the electrolysis circuit, and a reference electrode for measurement of the potential of the working electrode.

The experimental conditions chosen are those which give the best compromise between desired accuracy, selectivity and speed of analysis. For reversible redox systems, selection of electrode potentials, and solution conditions, follow the procedures outlined for controlled cathode potential electrodepositions. Many irreversible reactions are also amenable to potentiostatic coulometric analysis, but for these the correct electrode potential cannot be calculated; it must be chosen empirically.

Tables 5.56 and 5.57 of the "Handbook of Analytical Chemistry" summarize the conditions that have been used in determining various substances by controlled-potential coulometry. Methods for twenty-eight different elements and a similar number of organic compounds are tabulated, and due to the increased interest in recent years, these listings can only be regarded as a sample group.

C. COULOMETRY AT CONTROLLED CURRENT

In the technique known as coulometry at controlled current, the electrolysis is performed with a current which is maintained at a constant value. Evaluation of Q thus requires merely the measurement of the current and the time of electrolysis.

However, with this technique it is more difficult to establish experimental conditions which ensure that a single reaction occurs during electrolysis. As the concentration of the substance being determined falls, a point is reached where migration of this species is insufficient to support all of the current. Some other electrolytic processes must then occur to maintain the current at the selected constant value.

The electrolysis can only be effectively attributed to a single stoichiometric reaction, if the products of the secondary process react rapidly with the substance being determined. The overall result is the same as if a titrant were being generated electrolytically; hence this technique is frequently referred to as coulometric titration.

Consider as an example the reduction of Ce(IV) in 1 M H_2SO_4 solution. On the application of a constant current Ce(IV) is reduced to Ce(III). However, as

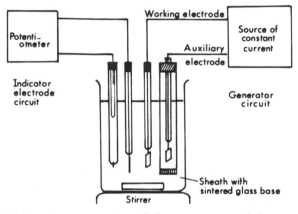

FIGURE 13.9 Schematic representation of the components of the reaction cell used for constant current coulometry.

electrolysis proceeds, the limiting current for cerium reduction gradually drops below the applied current and hydrogen ion reduction takes place to maintain the current. Hydrogen gas does not react instantaneously with cerium (IV); hence some of the electricity is consumed in the secondary process and the Faraday relationship between Q and amount of cerium fails. On the other hand, if some iron (III) is placed in the acid solution, the secondary process of minimum energy is the reduction of iron (III) to iron (II). The iron (II) is immediately reoxidized by cerium (IV) and the efficiency of the electrical reduction process remains 100%.

$$Ce^{4+} + Fe^{2+} \rightleftarrows Ce^{3+} + Fe^{3+}$$

As in potentiometric titrations, the detection system commonly used to observe the end of the reaction is a pair of electrodes, one of which is a reference half-cell, the other an indicating electrode. The experimental setup is shown schematically in Fig. 13.9. The design of the unit must ensure that the contents of the cell are mixed rapidly, and that the indicator system responds quickly. In some arrangements, the titrant is produced in an external generation cell prior to flowing into the test assay.

For small constant currents (1–2 mA), the source of current can be as simple as a few 45 V dry cells in series with a large variable resistor and a smaller standard resistor.   For larger currents, or greater control, more complex electronic amperostats are used.

More advanced instruments have been designed which stop the current and a timer once the indicator system shows that the end point of the reaction has been reached.   For this purpose, electrometric techniques of detection (e.g., potentiometry, conductimetry) are preferable to optical methods which include visual indicators and photometric end point techniques.

The generator electrode is usually made of platinum and has an area of 2 to 5 cm². The precursor concentration of reagents generally lies between 0.05 and 1 M and generating currents up to 50 mA have been used.   A number of reagents are generated by ion exchange, i.e., an ion exchange membrane charged in the appropriate ionic form is placed in the cell.   The reactants (e.g., Cl^-, Br^-, I^-, H_2, $EDTA^{2-}$, and Ca^{2+}) are subsequently displaced by competing species liberated by electrolysis, for example H^+ and OH^-. These ions are formed by electrolysis of sodium sulfate or other salt solutions.   The halogens, Cl_2, Br_2, and I_2, are produced by electrolyzing the appropriate halide salt. Metal ions, such as iron (II), tin (II), and vanadium (IV), are prepared by reducing a compound of higher valency.   The ions silver (I), mercury (I), and mercury (II) are generated by using the parent metals as components of the cell anode.

Every well-known type of titration, e.g., acid–base, precipitation, complexo-metric and redox, has been successfully performed coulometrically.   The "Handbook of Analytical Chemistry" presents summaries of procedures for the determination of about fifty inorganic substances and a larger number of organic compounds (Table 5.5).

The titrations are most conveniently applied to samples containing 10^{-7} to 10^{-5} mole (0.01 to 1 mg) of reactive material.   Errors are characteristically of the order of 0.1 to 0.3%.   Coulometric methods are ideally suited for both routine and remote analysis, and accordingly they have been applied to some unusual analytical problems, such as titrations of highly hazardous materials and titrations in molten salt media.

By coulometric methods it is possible to perform many titrations which cannot be done by classical techniques.   Examples are titrations which utilize unstable or difficult to prepare titrants such as bromine, chlorine, chromium (II), copper (I), silver (II), titanium (III), and uranium (IV) or (V).

Besides the two procedures briefly outlined in this section, there are several other analytical techniques which may be classified as "coulometric procedures." These include measurement of the quantity of electricity required to completely remove a coating of metal off a sample or an electrode, and coulometric internal electrolysis.   Full details of all the techniques are provided in the reference books listed in the bibliography section.

D. CHRONOPOTENTIOMETRY

When a constant current is impressed across an electrolysis cell, the potentials of the electrodes vary with time. If several conditions are fulfilled, the potential–time curve is characteristic of the nature and concentration of the species reacting at the electrode, and may be used for analytical purposes.

Consider the reduction of a metal ion, M^{n+}, on a mercury cathode, in an unstirred solution. When the constant current is first applied an equilibrium potential is rapidly established. The magnitude of the potential depends on the concentration of M^{n+} at the surface of the electrode and can be calculated from the appropriate Nernst equation.

As the current continues to pass, the surface concentration of M^{n+} decreases and the electrode potential shifts to more negative values [cf. Fig. 13.7(c)]. Since stirring is absent, an extreme case of concentration polarization is eventually attained, and the M^{n+} is reduced as rapidly as it diffuses to the electrode. When the M^{n+} concentration at the electrode surface becomes very small, the potential of the electrode shifts rapidly until some other species starts to be reduced.

The elapsed time between the start of the electrolysis and the rapid potential shift is called the "transition time," τ.

$$\tau^{1/2} = \pi^{1/2} \cdot n \cdot F \cdot A \cdot D^{1/2} C / 2i$$

where n is the number of electrons involved in the reduction, F is the Faraday, A is the area of the electrode, D is the diffusion coefficient and C the concentration of the species of interest, and i is the constant current.

For this equation to be applicable, diffusion must be the only method by which the ions M^{n+} reach the electrode. Thus electrical migration effects have to be minimized by having a large excess of some "nonreducible" salt in the test solution.

Several inflection points may be observed in the curve if the system contains two or more electroactive species whose redox potentials differ by more than 0.1 V. The current efficiency for the second process is less than 100% (since some of the first species will still be diffusing to the electrode) and the basic equation for the second transition time τ_2 is

$$(\tau_1 - \tau_2)^{1/2} - \tau_1^{1/2} = \pi^{1/2} F A n_2 D_2^{1/2} C_2 / 2i$$

where τ_2 is measured from the first to the second transition points, and n_2, D_2, and C_2 pertain to the second substance.

For a single electrode reaction producing soluble products, it has been shown that

$$E = E_{\tau/4} - (RT/nF) \ln t^{1/2} / (\tau^{1/2} - t^{1/2})$$

where E is the observed potential, $E_{\tau/4}$ is the quarter-transition-time potential ($E = E_{\tau/4}$ when $t = \tau/4$) and t is the time.

The basic instrumentation required is a source of constant current, a timer, a working electrode and an auxiliary electrode. Often the auxiliary electrode is isolated in a separate compartment, a sintered glass plug being used to provide electrical contact. A potentiometer or recorder to measure the potential of the working electrode against some reference electrode completes the unit.

The optimum concentration ranges for study seem to be 0.001 to about 0.05 M and transition times between 10 and 100 sec are best for analytical purposes.

VI. POLAROGRAPHY

A. CURRENT-VOLTAGE RELATIONSHIPS

In polarography, electrolysis which occurs in a cell containing a microelectrode is observed by simultaneously measuring the current flowing in the circuit and the applied potential. The current-voltage relationship is then used to determine the identity and/or concentration of the species reacting at the microelectrode.

For valid interpretation of the current-voltage curve, diffusion of the electroactive species to the microelectrode has to be the rate controlling step.

As mentioned in Section IV,A, there are three mass transport processes which can convey solute species to the surface of an electrode, namely, migration of charged ions under the influence of a potential gradient, convection caused by stirring, and diffusion.

In polarography the migration of the species of interest is minimized by introducing into the sample solution a fifty to one hundredfold excess of some inert "supporting electrolyte." The ions of this salt tend to migrate under the influence of the field in preference to the small concentration of electroactive species. Convection processes are minimized by maintaining the solution at a constant temperature, and by eliminating stirring.

During electrolysis, ions are removed from the region of the microelectrode surface and this makes the concentration of the electroactive species smaller in this region than in the bulk solution. Accordingly, a concentration gradient develops across a zone of solution adjacent to the electrode. Nernst proposed that this zone could be assumed to have a definite thickness, δ, and to possess a linear concentration gradient, i.e.,

$$dc/dx = (C - C_o)/\delta$$

where C is the concentration of the ion in the bulk solution and C_o is the concentration at the phase interface.

Fick's Law of Diffusion states that the number of moles, N, of diffusing substance which diffuses across a cross-sectional plane of unit area in the small time interval, dt, is proportional to the concentration gradient at the plane in question:

$$\text{Flux} = (dN/dt)/A = D(\delta c/\delta x)$$

where D is the diffusion coefficient, c refers to the concentration at distance x from the surface at time t, and A is the area of the plane.

Combining these two equations gives the relationship

$$dN/dt = AD(C - C_o)/\delta$$

Where diffusion is the rate controlling process in an electrolysis cell, the current flowing is, in theory, directly proportional to dN/dt.

Thus,

$$\text{Diffusion current } i_d = k\, dN/dt = kAD(C - C_o)/\delta$$

At the decomposition potential, only some of the ions reaching the electrode surface are discharged. At the half-wave potential half the ions reaching the electrode are discharged (i.e., $C_o = C/2$). At higher potentials, the ions are reduced as fast as they arrive. In this case $C_o \to 0$, and by the above relationship the current should reach a limiting value and be proportional to C, the concentration of ion in the bulk of solution. The limiting value of the current is called the limiting current, $i_{\lim}$, where $i_{\lim} = nFA\, DC/\delta$. The limiting current recorded during electrolysis includes the nonfaradaic contribution known as the residual current. Correction for this can be made by observing the current which flows in a solution that contains all the original chemical compounds except the electroactive species. For most practical purposes, a reasonable correction is obtained by extrapolating the first leg of the current–voltage curve, as shown in Fig. 13.10.

The difference between the limiting current and the residual current is the diffusion current, i_d. The potential corresponding to a current equal to half the diffusion current is characteristic of the species being reduced, and is known as the half-wave potential.

B. Dropping Mercury Electrode

The electrode at which the electrical transformation of interest occurs should be fully polarized. To achieve this aim, the electrode needs to be small in area. A small stationary wire, however, is unsatisfactory because the thickness of the diffusion layer, δ, tends to increase with time and causes the limiting current to decrease with time.

This effect can be overcome by using a rapidly rotating platinum microelectrode since in stirred solutions, diffusion layers at phase interfaces tend to remain constant and are small in magnitude. However, stirring also aids convection; hence a larger limiting current is observed, the size varying with the rate of stirring. In addition, a platinum microelectrode suffers from the disadvantage that deposition of metal alters the nature of the electrode.

The most useful electrode has proved to be a dropping mercury electrode, DME. This is composed of a fine capillary tube through which mercury drops slowly from an elevated reservoir. With the dropping mercury electrode a constant limiting current is generally observed which indicates that the growth of the drops offsets the effect of a widening diffusion layer. In addition, each successive mercury drop presents a new clean surface for the electron transfer process. Many metals form amalgams, and so there is no accumulation of products to cause variable behavior during the life of a drop. Finally, the high overpotential for the discharge of hydrogen on mercury allows studies to be made in acid solution.

The magnitude of the current flowing in the electrical circuit is a function of the area of the electrode, and with a DME this varies during the lifetime of a drop from near zero to a maximum value it achieves just before falling. The size of a drop can be characterized in terms of the rate of flow of mercury through the capillary, m (mg sec^{-1}) and the length of time the drop grows before falling, τ (sec). As the electrode varies in size in a regular cycle, the observed current fluctuates in a reproducible manner from a minimum to a maximum value. The degree of fluctuation is usually minimized in the electrical circuit by introducing capacitors and by using galvanometers with a long natural period. The average current (averaged over the life of the drop) which is measured in this way has been shown to be equal to six-sevenths of the maximum current flowing just before the drop falls.

The average polarographic diffusion current is described mathematically by the Ilkovic equation which in its simplest form is

$$i_{ave} - 607 \ nCD^{1/2}m^{2/3}\tau^{1/6}$$

This equation indicates the experimental factors which need to be controlled in order to obtain average diffusion currents which retain a direct relationship to the concentration C.

The diffusion coefficient, D, varies with the size and charge of the diffusing species and with the viscosity of the medium. These properties, in turn, vary with the nature of the solvent, the composition of the solution, and with the temperature.

The addition of electrolyte to minimize migration currents alters the viscosity of the medium, particularly where the concentration exceeds 0.1 M. Interaction of the electroactive species with added ions or molecules can result in the formation of complex ions which possess a different size and shape, and accordingly, diffusion coefficient. The relationship between i_d and C can thus only be considered constant for a particular solution composition.

The diffusion coefficient, D, has a temperature coefficient of about 2% deg^{-1} for most ions, so in order that the temperature factor may not cause an error of more than 1%, it is necessary to control the temperature to at least ± 0.5 degree.

The mass of mercury flowing through the capillary per second, m, and the drop time, τ, are related to the height of the mercury reservoir above the capillary tip.

A plot of the drop time of a DME against the applied potential yields a curve that is parabolic in shape. To allow for this factor, a term $[3.1 \ (m\tau)^{-1/3}]$ is usually subtracted from the measured height of the mercury level. This gives the "effective" height of mercury h_{cor}, and to a first approximation, m is directly proportional to h_{cor} and τ is inversely proportional to h_{cor}.

The average current, by the Ilkovic equation is proportional to $m^{2/3}\tau^{1/6}$, and by combining constants, this yields the expression

$$i_{ave} = km^{2/3}\tau^{1/6} = k(k'h_{cor})^{2/3}(k''/h_{cor})^{1/6}$$
$$= k'''(h_{cor})^{1/2}$$

The variation of the drop time with potential can be explained in terms of the charged double layer formed at the surface of the mercury. The ions or dipoles in the solution which become oriented in respect to the mercury-solution interface induce a charge on the mercury surface. For example, adsorption of negative ions induces positive "image" charges at the inside surface of the drop by repelling electrons away from the drop and toward the mercury in the reservoir. This process is repeated as each drop grows, and if the potential of the DME is maintained constant, an anodic current flows indefinitely. This is the source of the "charging" or residual current observed on polarograms.

To balance the positive charge on the drop, the electrode has to be made more negative in respect to the solution by the imposition of a larger applied potential. A point is eventually reached where the anion adsorption is just counteracted by the negative charge built up on the mercury. The potential at this point corresponds to the electrocapillary maximum for that solution, and at this stage the charging current decreases to zero, the surface tension of the mercury is at its maximum value, and the observed drop time is a maximum. The interfacial tension at the mercury-solution interface reaches a maximum value at the electrocapillary zero because less work is required to increase the area of an interface if the surface carries a charge.

With potentials more negative than the electrocapillary maximum, the mercury surface is negatively charged in respect to the solution and the direction of the residual or charging current is reversed.

The stronger the forces of attraction between the dipole or anion and mercury, the more negative is the potential of the electrocapillary maximum. For example, in 0.1 M solutions of chloride, bromide, iodide, and sulfide the electrocapillary maximums are reported to be respectively -0.46, -0.53, -0.69, and -0.88 V versus a saturated calomel electrode.

A serious disadvantage of the dropping mercury electrode is the relative ease with which mercury is oxidized. Hence it is not suitable for use as an anode; in nitrate media oxidation occurs at potentials greater than about $+0.3$ V versus the SCE. Anodic dissolution occurs at less positive potentials in the presence of anions that form insoluble mercury salts or stable complexes, e.g., in the presence of molar sodium cyanide, the oxidation of mercury proceeds at potentials more positive than -0.7 V versus the SCE.

C. Polarographic Circuit

The major components of the apparatus required for polarographic measurements are pictured in Fig. 13.10.

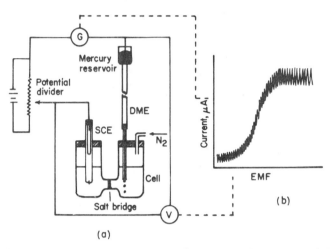

FIGURE 13.10   (a) Schematic representation of a polarographic circuit; and (b) a typical diffusion current-voltage curve.

In the electrolysis circuit, the auxiliary electrode should have a large surface area and a more or less standard potential. In most instances a saturated calomel electrode is used. The currents resulting from the electrode processes at the surface of the microelectrode are so small (order of several microamperes) that the potential of the calomel electrode remains virtually unchanged by the passage of this tiny current. In the H-cell design shown in Fig. 13.10, the reference cell is separated from the dropping mercury electrode by a sintered glass disk and an agar-saturated potassium chloride plug. Other cell designs place the reference electrode directly in the test solution, and for routine analytical determinations a pool of mercury has been used as the circuit anode.

The drop time of the DME is usually adjusted to between three and six seconds by varying the height of the mercury reservoir. The bore of the capillary should be such that a head of mercury of about 30 cm is required to give this drop rate, the drops having diameters of approximately 0.05 mm. To ensure uniform and reproducible capillary characteristics, the tube must have a perfectly flat end and should be mounted below the surface of the solution in a truly vertical position.

A potential of 0 to 2.5 V is applied across the electrodes, and the current flowing in the circuit is either measured with a sensitive galvanometer, or is amplified and fed to a recorder. Since the current flowing is of the order of microamperes, few ions are removed and the bulk concentration of the electroactive species in solution remains virtually unchanged.

Since oxygen undergoes stepwise reduction at the mercury cathode,

$$O_2 + 2H^+ + 2e \rightleftarrows H_2O_2$$
$$H_2O_2 + 2H^+ + 2e \rightleftarrows 2H_2O,$$

nitrogen or some other inert gas is bubbled through the sample solution to remove oxygen prior to the recording of a polarogram and is passed over the solution during the actual measurements.

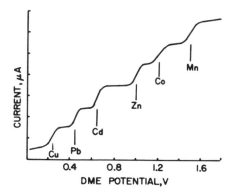

FIGURE 13.11. Polarogram of a solution containing several reducible ions. Base electrolyte 0.1 M KCl.

The capacity of the test solution cell is usually 15 to 25 ml, but units catering to volumes of less than 1 ml have been operated successfully.

A complete plot of the current as a function of the dropping electrode potential is termed a polarogram, and a representative polarogram is shown in Fig. 13.10. Where more than one reducible species is present in solution, multiple curves such as shown in Fig. 13.11 can be obtained. The position of the half-wave potentials and the magnitude of the current produced are similar to the values obtained in simple solutions of equal metal ion concentrations. The current fluctuation due to the growth and fall of the mercury drops is usually minimized by means of a condenser damping circuit and the average current is noted from the polarogram. Under ideal conditions, the current-voltage curves are S-shaped with near horizontal lower and upper arms. In practice, the elevation of these two limbs can vary from the horizontal, and graphical procedures are required to locate the position of the half-wave potential and to estimate the magnitude of the diffusion current.

To ensure that the current being measured can be attributed to diffusion alone, the test solution contains an excess of supporting electrolyte, and in most studies a small amount of surface active agent is added to eliminate "maxima" in the curves. Where it is necessary to vary the half-wave potentials of some components, a complexing agent may be included in the base solution. The role of the two latter additions is explained in succeeding sections.

D. EQUATION OF THE POLAROGRAPHIC WAVE

A polarographic wave may be considered to be reversible if the concentrations (or rather the activities) of the reactants and products at the electrode-solution interface are related to the potential of the electrode through the Nernst equation (within experimental error). That is, for the reaction

$$Ox + ne \rightleftarrows Red;$$

$$E_{Ox, \, Red} = E^{\circ}_{Ox, \, Red} - \frac{RT}{nF} \ln \frac{[Red]_{surface}}{[Ox]_{surface}}$$

In this expression, $E_{Ox, \, Red}$ and $E^{\circ}_{Ox, \, Red}$ represent, respectively, the potential of the dropping mercury electrode and the standard potential for the pertinent half-reaction, each measured with respect to the same reference electrode. The observed potential of the dropping electrode, and the surface concentrations of the oxidized and reduced species refer to the average value during the lifetime of a mercury drop.

In terms of kinetics, reversible behavior is observed when the rates of both the forward and reverse reactions are rapid enough to ensure that equilibrium is maintained at the surface of the electrode, in spite of changes in concentration produced by the flow of current or diffusion. An irreversible reaction is one in which the concentrations of the species at the electrode surface are determined by the rate of some reaction or process other than electrolysis and diffusion.

For a reversible reaction, in which a soluble oxidized species yields a reduction product which is soluble in either the solution or in the mercury as an amalgam, the flow of current in the electrolysis circuit establishes a concentration gradient in the vicinity of the interface.

With diffusion control of the current,

$$i = k_{Ox}(C_{Ox} - C^{\circ}_{Ox}) \quad \text{or} \quad i = k_{Red}(C^{\circ}_{Red} - C_{Red})$$

where $C - C^{\circ}$ (or $C^{\circ} - C$) represents the concentration gradient at the interface; i is the current; and k_{Ox} and k_{Red} are proportionality constants which include the diffusion constants of the respective species. From the Ilkovic equation $k_{Ox} = 607 n D_{Ox}^{1/2} m^{2/3} \tau^{1/6}$.

When the electrolysis system is passing the limiting current, i_d, $C^{\circ} = 0$ and the limiting cathodic current is given by the expression

$$(i_d)_c = k_{Ox} C_{Ox}$$

and the limiting anodic current is described by

$$(i_d)_a = -k_{Red}C_{Red}$$

Substituting these values in the general equation for the current i, one obtains

$$i = (i_d)_c - k_{Ox}C_{Ox}^\circ \quad \text{or} \quad i = k_{Red}C_{Red}^\circ + (i_d)_a$$

Rearranging these terms

$$C_{Ox}^\circ = (i_d)_c - i/k_{Ox}; \qquad C_{Red}^\circ = i - (i_d)_a/k_{Red}$$

The values may now be substituted in the Nernst equation since C_{Ox}° and C_{Red}° represent the concentration of species at the interface.

$$E_{DME} = E_{Ox, Red}^\circ - \frac{RT}{nF} \ln \frac{i - (i_d)_a}{(i_d)_c - i} \cdot \frac{k_{Ox}}{k_{Red}}$$

Usually, a cathodic or an anodic wave is studied separately. For a reversible cathodic wave, $C_{Red} = 0$ and accordingly $(i_d)_a = 0$; and the equation becomes

$$E_{DME} = E_{Ox, Red}^\circ - \frac{RT}{nF} \ln \frac{i}{(i_d - i)} \cdot \frac{k_{Ox}}{k_{Red}}$$

The values of k_{Ox} and k_{Red} differ only in respect to the magnitude of the diffusion coefficients of the two species;

$$k_{Ox}/k_{Red} = D_{Ox}^{1/2}/D_{Red}^{1/2}$$

$$\therefore E_{DME} = E_{Ox, Red}^\circ - \frac{RT}{2nF} \ln \frac{D_{Ox}}{D_{Red}} - \frac{RT}{nF} \ln \frac{i}{i_d - i}$$

$$= E_{Ox, Red}^\circ - k_o - \frac{RT}{nF} \ln \frac{i}{i_d - i}$$

In the derivation of this equation, concentrations rather than activities have been considered, so the logarithmic term represented by k_o should also include the ratio of the activity coefficients of the two species, f_{Red}/f_{Ox}.

At the point in the wave where $i = i_d/2$, the last term in the equation equals zero and

$$E_{DME} = E_{Ox, Red}^\circ - k_o = E_{1/2}.$$

The values of k_o are usually quite small; hence the half-wave potential, $E_{1/2}$, is generally similar in magnitude to the standard reduction potential.

The basic equation for a reversible cathodic wave can, therefore, be written in the form

$$E_{DME} = E_{1/2} - \frac{0.059}{n} \log \frac{i}{i_d - i}$$

where $0.059 = 2.303\,RT/F$ at 25°C.

Manipulation of the preceding equations for an anodic wave yields the same basic wave equation.

This equation can be used to test the reversibility of a reaction. If the system is reversible, a plot of E_{DME} versus $\log i/(i_d - i)$ should yield a straight line with a slope of $0.059/n$. The slope of the line also permits evaluation of n, the number of electrons involved in the reaction, and, the value of the potential when the log term equals zero represents the half-wave potential for the system.

An alternative means of evaluating the reversibility of a system is to measure the potential at points which correspond to one-fourth and three-fourths the wave height. For reversible waves the difference, $(E_{1/4} - E_{3/4})$ should be $56.4/n$ mV at 25°.

Provided that the reactant and product species are completely soluble, the half-wave potential is independent of the concentration of the reactant and may serve to identify an unknown polarographic wave. Extensive lists of half-wave potentials are presented in the monographs devoted to polarography and there is a comprehensive tabulation in the "Handbook of Analytical Chemistry."

When the reduction process leads to the formation of an amalgam

$$M^{n+} + x\,Hg + ne \rightleftarrows M(Hg)_x$$

the basic equation for the polarographic wave is the same, but the value of $E^{\circ'}$ refers to the standard potential of the amalgam electrode reaction and the values of $E_{1/2}$ can be much more positive than tabulated values of E° for aqueous reactions.

If the product formed on reduction is insoluble at the DME its activity becomes constant and independent of the current. The equation for a cathodic wave then becomes

$$E_{DME} = \dot{E}^{\circ} - \frac{0.059}{n} \log{(D_{ox}^{1/2}/f_{ox})} + \frac{0.059}{n} \log{(i_d - i)}$$

and the half-wave potential is no longer independent of concentration, since at this potential $i = i_d/2$ and the last term of the above equation becomes $0.059\,(\log i_d/2)/n$ instead of the zero observed in the case of soluble products.

The addition of a complexing agent to the test solution alters the location of the half-wave potential. Consider the case of a simple metal ion and a co-ordination compound of that metal ion (in the same oxidation state) which both undergo reversible electrode processes.

For the simple hydrated metal ion, the reduction may be represented as

$$M^{n+} + ne + Hg \rightleftarrows M(Hg)$$

and the potential of the electrode in a solution of the ions may be described by the equation

$$E_{DME} = (E_{1/2})_m - \frac{0.059}{n} \log{\frac{[M(Hg)]}{[M^{n+}]}}$$

where $M(Hg)$ is the concentration of metal, M, in the amalgam.

On the addition of a ligand, X^{a-}, a complex may be formed,

$$M^{n+} + pX^{a-} \rightleftarrows MX_p^{(n-pa)+}$$

and the concentration of free metal ions is then determined by the equilibrium expression

$$K = [M^{n+}][X^{a-}]^p / [MX_p^{(n-pa)+}]$$

Substituting for $[M^{n+}]$ in the polarographic wave equation yields

$$E_{\mathrm{DME}} = (E_{1/2})_m - \frac{0.059}{n} \log \frac{[M(Hg)] \cdot [X^{a-}]^p}{[MX_p^{(n-pa)+}] \cdot K}$$

Reduction of the complex at the cathode may be described as

$$MX_p^{(n-pa)+} + ne + Hg \rightleftarrows M(Hg) + pX^{a-}$$

and the half-wave potential for the complex ion, $(E_{1/2})_c$, is the value of E_{DME} when $[MX_p^{(n-pa)+}] = [M(Hg)]$.

The wave equation may now be written as

$$(E_{1/2})_c = (E_{1/2})_m - \frac{0.059}{n} \log \frac{[X^{a-}]^p}{K}$$

$$= (E_{1/2})_m + \frac{0.059}{n} \log K - \frac{0.059 \cdot p}{n} \log [X^{a-}]$$

From this equation, it can be observed that a plot of $(E_{1/2})_c - (E_{1/2})_m$ against $\log [X^{a-}]$ should yield a straight line of slope $-0.059p/n$ and an intercept when $\log [X^{a-}] = 0$ equal to $0.059 \log K/n$.

Thus, by determining the half-wave potential in the presence of varying amounts of ligand, it is possible to evaluate both p and K.

It may also be deduced that the greater the stability of the complex, the more negative will be its $(E_{1/2})_c$ value.

Solutions containing more than one reducible species sometimes yield a single polarographic wave because the reduction potentials of the hydrated ions are similar. This undesired effect can be overcome by adding a complexing agent which reacts preferentially with one species to form a coordination compound of such stability that the half-wave potential is altered by a significant amount.

In nonaqueous solvents, ion pair formation is important and the half-wave potential depends on both the identity and concentration of the supporting electrolyte.

E. IRREVERSIBLE WAVES AND DEPARTURES FROM PURE DIFFUSION

Irreversible waves do not satisfy the criteria of reversibility outlined above because the current flowing during the rising portion of the wave is governed by the rate of reaction, by diffusion, or both.

In a totally irreversible wave, the rate constant for the forward reaction is much greater than the rate constant for the backward reaction, i.e., $k_f \gg k_r$.

The current flowing at any potential for a cathodic reaction may then be written as

$$i = nFAk_f[\text{Ox}] \text{ (cf. Section A)}$$

$$= nFA[\text{Ox}]k_f^{\circ}e^{-\alpha nF(E-E')/RT}$$

Because the transport coefficient, α, is $\ll 1$ in a totally irreversible system, only a fraction of a given change in electrode potential serves to increase the rate of the cathodic reaction. The current-voltage waves therefore tend to be drawn out and distorted. At sufficiently large negative potentials, [Ox] at the electrode surface decreases to negligibly small values, and at these potentials the current can be diffusion limited and vary in direct proportion to the concentration in the bulk solution.

Thus, irreversible waves can be used for quantitative studies provided the measurement of the diffusion current is made at potentials where the current is diffusion controlled.

An equation for an irreversible wave can be written in terms of an observed half-wave potential and the limiting average diffusion current, i_d.

$$E \simeq E_{1/2} - \frac{0.059}{\alpha n} \log i/(i_d - i)$$

where $E_{1/2} = 0.059 \log [1.35k_f^{\circ}\tau^{\frac{1}{2}}/D_{\text{Ox}}^{\frac{1}{2}}]/\alpha n$.

The half-wave potential becomes increasingly negative as the rate constant k_f° becomes smaller; i.e., the overvoltage increases as the reaction rate decreases. Conversely, overpotential is caused by a slow kinetic step in the electron transfer process.

At equilibrium in any redox system the net current flow is zero, $i_c = i_a = i_o$, where i_o is known as the exchange current.

The net current $(i_c + i_a)$ at any other potential may be described in terms of the exchange current.

$$i = i_o \exp \{-\alpha nF(E - E_{eq})/RT\} - i_o \exp \{(1 - \alpha)nF(E - E_{eq})/RT\}$$

where E is the applied potential and E_{eq} is the equilibrium potential.

In order to have a flow of current, the potential of the system has to deviate from its equilibrium state by an amount $(E - E_{eq})$ which is called the activation overpotential, symbolized by η.

When η is small the exponential terms in the current equation can be expanded in the form of $e^{-x} \simeq 1 - x$ and the equation becomes

$$i = -i_o nF\eta/RT$$

For large values of η, the back reaction can be neglected. Thus with cathodic polarization

$$i_c = i_o \exp\left(-\alpha n F \eta / RT\right)$$

which in logarithmic form is

$$\eta = 2.303RT \log i_o/\alpha nF - 2.303RT \log i_c/\alpha nF$$

This is of the same form as the Tafel equation

$$\eta = a + b \log i_c$$

which describes the experimental results obtained in many studies of activation overpotential.

There are chemical systems in which the limiting current is governed by the rate of production of the electroactive species, rather than by the rate of diffusion to the electrode surface, e.g.,

$$A \rightleftarrows B \text{ (slow)}$$

$$B + ne \rightarrow \text{Reduction product}$$

In such cases the current does not follow the Ilkovic equation and is termed a kinetic current.

A kinetic current has two distinguishing characteristics. First, it is independent of the values of the drop time and the mass of the mercury drop, i.e., altering the height of the mercury reservoir does not alter the magnitude of the current.

Second, the current is smaller than would be predicted for direct conversion of A to products but larger than that predicted from the equilibrium concentration of B.

Kinetic currents are not particularly useful for practical quantitative analyses.

Current enhancement occurs in systems containing two oxidants, only one of which is reduced at the mercury electrode. The reduction product of one component (e.g., A) may then be reoxidized by component two (B) at the surface of the drop. If the quantity of A regenerated chemically at the surface is large compared to the amount of A reaching the surface by diffusion, the observed current is much larger than that predicted for a pure diffusion process. Such currents are termed catalytic currents, and they have been found useful for the determination of trace concentrations of certain substances.

Examples are the reduction of iron (III), molybdates, vanadates, or tungstates in the presence of hydrogen peroxide.

Hydrogen peroxide is not reduced at the DME at the potential required for the reduction of the other species because it is associated with a significant activation overpotential effect. The diffusion current from the electrode process, e.g.,

$$Fe^{3+} + e \rightleftarrows Fe^{2+}$$

is greatly enhanced by the cyclic process in which higher valency material is produced at the electrode surface by oxidation with H_2O_2 prior to being again reduced at the electrode.

$$2Fe^{2+} + H_2O_2 \rightleftarrows 2Fe^{3+} + 2OH^-$$

Frequently, the current-voltage curves obtained with a dropping mercury electrode show a distinct "maxima." The shapes of maxima vary from rounded humps to sharp peaks, the height and shape depending on the concentration of the reducible species, the drop time of the capillary, the concentration and charge of the supporting electrolyte.

For a maximum to occur, more ions or molecules must be reacting at the electrode than reach the electrode surface by diffusion through an unstirred solution. There is evidence that growth of the drop results in "streaming" of the solution past the electrode, and it is considered by some that the motion of the solution brings the additional reducible ions in contact with the electrode surface. Another theory attributes the higher concentration to an adsorption process.

Whatever the real explanation, maxima can usually be suppressed by adding to the solution a surface active agent which is adsorbed on the mercury surface at the potential of the maximum. Recommended maximum suppressors include gelatin ($<0.005\%$), methyl red ($<0.0004\%$) and Triton X-100, a commercial surface-active agent ($<0.002\%$).

Where more than one electrode reaction occurs at the cathode, there is a possibility that the products of one of the reactions may interfere with the behavior of another; this leads to nonadditive and mixed currents.

For example, it has been shown that the production of OH^-, S^{2-}, Se^{2-}, and Te^{2-} as a result of an electrode reaction can interfere with the magnitude of the diffusion current observed for metal ions which form insoluble precipitates with these anions.

F.   Instrumental Techniques

The essential components of a conventional dc polarograph have been described in Section C and in Fig. 13.10. This type of circuit is still widely used but the significant advances in instrumentation which have occurred in recent years have led to modified techniques capable of greater resolution and sensitivity.

For example, derivative dc polarography has been shown to have several advantages over conventional dc polarography.

First, the derivative of a linearly increasing residual current is a constant; hence in derivative curves this factor is automatically eliminated. This allows concentrations of electroactive species as low as 10^{-6} M to be detected.

Second, derivative polarography yields a peak at $E_{1/2}$ whose height is concentration dependent.

$$(di/dt)_{max} = -kn^2C(dE/dt)$$

k is a proportionality constant; di/dt is the derivative of the average current recorded with respect to time; dE/dt is the voltage scanning rate.

Because of the sharpness of the peaks, the derivative technique can yield quantitative data for species whose conventional waves are incompletely resolved because they overlap slightly.

Various techniques for recording the first derivatives of polarographic waves have been proposed. They include electronic differentiating circuits, ac signals superimposed on a dc scan, and measurement of the difference in current between two dropping electrodes maintained at slightly different potentials throughout the scan.

The application of a linear voltage scan to stationary electrodes of constant area in unstirred solutions gives rise to peak-shaped polarograms, the peak height varying with the scanning rate and with the concentration of electro-active material.

. If the electrode is rotated, or if the solution is stirred rapidly, the current-voltage curve tends to become independent of the scanning rate, and a level limiting current results.

Some stirring action also occurs in rapid dropping mercury electrode polarography. In this technique the drop time is shorter and the scanning rate is faster than that used in a conventional circuit. These modifications do not yield greater sensitivity, but they do decrease the time required to run a polarogram.

The ultimate in rapid scanning is achieved in cathode ray polarography. Here the entire voltage sweep (0.5 V in 2 sec) is made toward the end of the lifetime of a single drop.

The current voltage curve is displayed on an oscilloscope or cathode-ray tube. The rapid scan gives rise to a peaked polarogram, the height of the peak being directly proportional to concentration. The sensitivity is claimed to be about $4\sqrt{n}$ times greater than the usual dc procedure.

If a small alternating voltage signal of relatively low frequency is super-imposed on a slow, linear dc potential scan, an alternating current signal which is enhanced in the region of a reversible polarographic wave is observed. The ac current attains its maximum value at the half-wave potential, and if rectified and damped, the signal corresponds to the first derivative of the polarographic wave. The optimum frequency for the alternating potential appears to lie in the range 10 to 60 cps and the optimum amplitude between 1 and 35 mV. The sensitivity of this approach is similar to conventional dc polarography.

However, the sensitivity is enhanced in modifications which aim to separate the faradaic component of the ac current from the capacitive component.

In the square wave polarograph and the pulse polarograph, a short square wave voltage signal or pulse is imposed at regular intervals onto a linearly increasing dc voltage scan. The current is measured during the last few milli-seconds of the square wave. As the capacitance current decays much more rapidly than the faradaic current, the measured signal records essentially only faradaic current. Instruments of this type are said to possess increased resolution, and a limit of detectability approaching 10^{-8} M in some special cases.

G. Applications

Polarography is an accepted method for trace analysis, the concentration range recommended for standard instruments being 10^{-3} to 10^{-4} M. Most applications involve reduction at the dropping mercury electrode but oxidation processes are also regularly studied. In fact, it has been claimed that polarography can be used to study any reaction involving the transfer of electrons.

Many inorganic cations, anions (e.g., iodate, bromate, nitrate, and permanganate) and molecules (e.g., oxygen, sulfur dioxide, nitric oxide, and hydrogen peroxide) undergo reduction at a dropping mercury electrode to yield cathodic waves.

Organic compounds which are reducible at the DME include those which contain carbonyl groups, carbon-carbon double bonds, carbon-halogen bonds, nitrogen-oxygen bonds, sulfur-sulfur bonds, diazo groups, epoxide and peroxide groups. The reductions are generally irreversible but the diffusion current is usually proportional to concentration. Since many organic compounds have very limited solubility in water, organic liquids are often used as solvents.

Other inorganic and organic substances can be directly oxidized at the DME to give anodic waves.

The popularity of the technique and the wide range of applications is reflected in the number and magnitude of the monographs on this topic.

In these monographs and in the "Handbook of Analytical Chemistry," extensive tabulations have been prepared which summarize the experimental conditions which are recommended for a wide range of polarographic studies. The tables usually include a series of alternative base solutions (carrier electrolyte plus complexing agents) and indicate the observed position of half-wave potentials.

The tabulations can be used to select the conditions required for clearly resolved waves of all trace components of a complex mixture. Two or more base solutions may be required in the overall study, but more than one determination may be achieved on a single polarogram.

The relationship between diffusion current and concentration has to be established by examining the polarograms of standard samples present in identical base solutions.

With strict adherence to established technique, fairly rigid temperature control and careful calibration, the reproducibility of duplicate analyses may be as good as $\pm 2\%$.

VII. AMPEROMETRIC TITRATIONS

The end point of a titration can be located by plotting current against volume of titrant, and the technique has proved particularly useful in determining the end points of titrations involving precipitation and complex formation.

The amperometric titration technique can be employed if either reactant gives a diffusion current. The electrical circuit required is similar to that used in polarography except that the potential applied to the circuit is set at a value large enough to cause reduction of either the species in solution or the titrant.

During the titration process, the concentration of reducible ions in solution varies and causes a change in the limiting current. To minimize the effect of dilution by the titrant, the latter is usually more concentrated than the assay solution and is delivered from a microburette. Several readings of the diffusion current are made at points before and after equivalence point. The plots of current (at constant potential) versus volume of titrant added resemble the curves obtained in conductometric titrations (cf. Fig. 13.2) and the end point is detected by extrapolation of the linear segments.

The titration media employed are the same as those required for polarography and the concentration of the substance titrated is usually in the range of 10^{-1} to 10^{-4} M. The precision and accuracy attainable vary with the systems but most procedures appear to have a standard deviation of less than 1%. Unlike conductometric titrations, the procedure can be made fairly specific for a given species through careful selection of titrant and operating potential.

A rotating platinum electrode has proved to be a satisfactory substitute for the DME in many applications. The polarographic monographs and the Handbook list many of the hundreds of titrations which have been satisfactorily followed by this technique.

VIII. REVISION AND REVIEW QUESTIONS

Question 1
a. Describe what you understand by the terms
   i. Specific conductance
   ii. Equivalent conductance at infinite dilution
   iii. Decomposition potential
   iv. Overvoltage
   v. Diffusion current
b. Using the equivalent ionic conductances listed in the "Handbook of Analytical Chemistry" or other references, prepare sketches of the conductometric titration curves which might be observed in each of the following titrations:
   i. Hydrochloric acid with sodium hydroxide

    ii. Hydrochloric acid with aqueous ammonia
   iii. Mixture of equal amounts of hydrochloric acid and acetic acid with aqueous ammonia
   iv. Acetic acid with sodium hydroxide
    v. Silver nitrate with potassium chloride
   vi. Iron (III) nitrate with sodium fluoride
  vii. Calcium chloride with ammonium oxalate
 viii. Barium nitrate with sulfuric acid
   ix. Copper (II) nitrate with disodium ethylenediaminetetraacetate
    x. Mixture of ammonium chloride and aluminum chloride with sodium hydroxide

c. Conductivity meters are regularly used to meter the saline content of tidal rivers.

Discuss the possible accuracy or significance of the readings obtained (i) well upstream and (ii) near the estuary.

If the river flowed through an area in which the soil was rich in mineral salts, would you recommend this analytical technique?

Question 2

Prepare a seminar on the theory and applications of ion selective membrane electrodes (one source of information and references is *Anal. Chem.* **39** (1967) 29A).

Question 3

a. Discuss the theoretical and practical limitations of direct potentiometry for analytical purposes.
b. Describe how the iodine–iodide couple, in conjunction with a pair of polarized platinum indicator electrodes, might be used to follow the progress of the titration of
    i. Sodium thiosulfate with potassium triiodide
   ii. Silver ion with standard potassium iodide solution
  iii. Moisture in transformer oil with Karl Fischer reagent
c. Prepare a list of the indicator–reference electrode combinations which have proved useful for potentiometric titration in nonaqueous media.
d. Using the appropriate tables in the "Handbook of Analytical Chemistry," extract the relevant experimental conditions required for the potentiometric titration of
    i. Mercaptans with alcoholic silver nitrate
   ii. Barium ions with potassium chromate
  iii. Carbohydrates with iron (II) solutions
  iv. Potassium dichromate with arsenic (III) solutions
   v. Manganese (II) salts with potassium permanganate
  vi. Aluminum ions with EDTA

Question 4

a. "Electrodeposition is the most selective and least troublesome of the gravimetric procedures used in chemical analysis, because the addition of chemical reagents is kept to a minimum." Critically discuss the validity of this statement.

b. Discuss the feasibility of using an electrolytic method to quantitatively separate zinc and cadmium ions from a solution which is approximately 10^{-1} M in respect to zinc ions and 10^{-2} M in respect to cadmium ions. Comment on the possible effect of having iron (III) impurity in the sample.

   Relevant reduction potentials, overpotentials for hydrogen, etc., should be extracted from reference tables.

c. Use the review by Page, Maxwell, and Graham [*Analyst* **87** (1962) 745] to prepare an essay on "The Analytical Uses of a Mercury Cathode."

Question 5

a. You are provided with a source of variable dc voltage (0 to 10 V), a potentiometer, a sensitive galvanometer, a calomel reference electrode, a pair of platinum gauze electrodes, a pound of mercury, and a short length of fine bore capillary tube; together with standard laboratory gear such as burettes, stirrer, beakers, balance, tubing, etc.

   How many electroanalytical techniques could you undertake?

   Draw a circuit diagram for each technique and tabulate a series of possible applications for each technique.

b. Prepare a statement confirming, or denying, the statement, "Modern electroanalytical techniques are of more interest to the electronic gadgeteer than to a practicing industrial chemist."

Question 6

a. Distinguish between activation and concentration overpotential. Under what conditions are these terms negligible?

b. What is the role of a depolarizer?

c. i. Calculate the applied potential required to quantitatively deposit 0.635 gm copper from 200 ml of solution which is 1 M in respect to nitric acid

   ii. Assuming that the cathode reaction is 100% efficient, what quantity of electricity would be required for the electrodeposition process?

   iii. If the sample solution contained lead ions, would this affect the copper determination?

d. The iodine required for a titration is generated electrically, and the gas coulometer in series with the generator liberates 4.35 ml gas (STP). What volume 0.01 M sodium thiosulfate would be required to react with the iodine liberated?

e. If the overpotential for hydrogen is 0.20 V, what solution pH (theoretical) would allow the concentration of nickel to be reduced by electrolysis to 10^{-6} M before evolution of hydrogen?

Repeat the calculation for zinc.

Calculate the pH required (for nickel and zinc) if a mercury cathode were used (overpotential 1.0 V, activity of metal ion in the amalgam approximately 10^{-6}).

f. What should be the ratio of iron (III) to iron (II) ions in a potential buffer in order to limit the cathode potential to $+0.71$ V?

Question 7

a. Define clearly each of the following terms: working electrode, auxiliary electrode, limiting current, electrocapillary maximum, maximum suppressor, and controlled cathode potential.

b. Carefully distinguish between coulometry at constant potential and coulometry at constant current.

c. From tabulations in the Handbook or from reference books, select two inorganic and two organic determinations which may be performed with relative ease using coulometry but which could be troublesome using alternative procedures.

d. Why is it sometimes advantageous to generate the reagent external to the titration cell?

e. List some of the advantages of coulometry.

f. Outline the principles of chronopotentiometry.

Question 8

a. What are the advantages of the dropping mercury electrode over a solid microelectrode? Compare the usefulness of the DME as a cathode and as an anode.

b. Prepare a precis of the types of solid microelectrodes which have been used in analysis (cf. Table 5.45 in Handbook) and a brief summary of typical applications.

c. Explain the source of the residual current.

d. In a polarographic cell containing a suitable supporting electrolyte, the observed current is sometimes larger or smaller than the diffusion current. Explain why?

e. What is the role of a complexing agent in the supporting electrolyte?

f. Briefly explain why, in tne absence of stirring, the diffusion current observed in polarography is proportional to the concentration of electro-reducible species.

g. A lead solution of unknown concentration gives a diffusion current of 4.8 μA. To 100 ml of this solution are added 5 ml of a 0.004 M solution of lead, and the polarogram is run again giving a diffusion current of 15 μA. Calculate the concentration of the unknown lead solution.

h. Indicate how the many tabulations of half-wave potentials of different elements in different supporting electrolyte compositions may be used to select the best experimental conditions for the polarographic determination of trace quantities of Pb, Cu, Cd, Bi, Fe, and Ag in pure zinc.

Question 9

a. Compare and contrast the various modified forms of polarography such as derivative polarography, ac polarography, etc., with the more conventional dc polarographic technique.

b. Describe how polarography can be used to determine the composition and stability of a metal complex.

c. Prepare a short review article on organic polarography.

Question 10

a. Outline the principles of amperometric titrations.

b. Sketch the various types of amperometric titration curves which may be observed (readily obtained from reference books).

c. Explain why so many applications of this technique are listed in the literature.

d. Would you predict that amperometric titrations in nonaqueous solvents could develop rapidly and become a popular means of organic analysis?

References

G. Charlot, J. Badoz-Lambling, and R. Tremillon, *Electrochemical Reactions*, Elsevier, Amsterdam, 1962.

P. Delahay, *New Instrumental Methods in Electrochemistry*, Wiley (Interscience), New York, 1954.

I. M. Kolthoff and P. J. Elving (eds.), *Treatise on Analytical Chemistry*, Part I, Vol. 4, Wiley (Interscience), New York, 1959.

J. J. Lingane, *Electroanalytical Chemistry*, 2nd Ed., Wiley (Interscience), New York, 1958.

R. W. Murray and C. N. Reilley, *Electroanalytical Principles*, Wiley (Interscience), New York, 1963.

W. C. Purdy, *Electroanalytical Methods in Biochemistry*, McGraw-Hill, New York, 1965.

F. J. Welcher, (ed.), *Standard Methods of Chemical Analysis*, sixth Ed., Vol. 3-A, Van Nostrand Co., Inc., New York, 1966.

C. L. Wilson and D. W. Wilson, (eds.), *Comprehensive Analytical Chemistry*, Part A, Vol. 2, Elsevier, 1964.

SPECIALIZED MONOGRAPHS

K. Abresch and I. Claassen, *Coulometric Analysis*, Chapman and Hall, London, 1965.

R. G. Bates, *Determination of pH*, Wiley, New York, 1964.

H. T. S. Britton, *Conductometric Analysis*, Chapman and Hall, London, 1934.

G. Eisenman, G. Mattock, R. Bates, and S. M. Friedman, *The Glass Electrode*, Wiley (Interscience), New York, 1965.

I. M. Kolthoff and N. H. Furman, *Potentiometric Titrations*, 2nd Ed., Wiley, New York, 1931.
G. W. C. Milner and G. Phillips, *Coulometry in Analytical Chemistry*, Pergamon, Oxford, 1968.
G. A. Rechnitz, *Controlled-Potential Analysis*, Macmillan, New York, 1963.
H. J. S. Sand, *Electrochemistry and Electrochemical Analysis*, Blackie, 1939.
J. T. Stock, *Amperometric Titrations*, Wiley (Interscience), New York, 1965.

POLAROGRAPHY

B. Breyer and H. Bauer, *Alternating Current Polarography and Tensammetry*, Wiley (Interscience), New York, 1963.
M. Brezina and P. Zuman, *Polarography in Medicine, Biochemistry and Pharmacy*, Wiley (Interscience), New York, 1958.
D. R. Crow, *Polarography of Metal Complexes*, Academic Press, New York, 1969.
D. R. Crow and J. V. Westwood, *Polarography*, Methuen, London, 1968.
J. Heyrovsky and J. Kuta, *Principles of Polarography*, Academic Press, New York, 1966.
T. Kambara, *Modern Aspects of Polarography*, Plenum, New York, 1966.
R. Kalvoda, *Techniques of Oscillographic Polarography*, Elsevier, Amsterdam, 1965.
I. M. Kolthoff and J. J. Lingane, *Polarography*, Vols. I and 2, 2nd Ed., Wiley (Interscience), New York, 1952.
L. Meites, *Polarographic Techniques*, 2nd Ed., Wiley (Interscience), New York, 1965.
G. W. C. Milner, *The Principles and Applications of Polarography*, Longmans, Green, London, 1957.
P. Zuman, *Organic Polarography*, Macmillan, New York, 1964.

14

ADSORPTION, DIFFUSION, AND ION EXCHANGE

I.	Separation Techniques	489
II.	Adsorption	491
	A. Equilibrium Relationships	491
	B. Analytical Procedures Involving Adsorption	492
	C. Adsorption Chromatography	493
III.	Separation Based on Electromigration	502
	A. Electrography	502
	B. Electrophoresis	504
	C. Electrochromatography	507
IV.	Diffusion	508
	A. The Permeability of Polymers	508
	B. Types of Polymers Used in Analysis	512
	C. Gel Filtration	516
V.	Ion Exchange	517
	A. Exchange Materials	517
	B. The Ion Exchange Process	521
	C. Selectivity	525
	D. Analytical Applications of Ion Exchange	526
VI.	Literature Exercises	531
	References	533

I. SEPARATION TECHNIQUES

In the development of a comprehensive method of chemical analysis five basic factors have to be considered. These are sampling problems, selection of the most appropriate analytical technique, sample preparation, procedures for the elimination of interfering species or effects, and the assessed validity of the experimental result.

When mixtures and complex commercial samples are involved, several of the constituents may contribute to the observed response of the selected measuring technique. In such cases, elimination of the influence of the interfering species becomes the most critical part of the analytical operation.

With instrumental techniques, increased selectivity is sometimes sought through closer control of the instrumental variables. In solution reactions, a reduction in the reactivity of the interfering species is often achieved through the addition of masking agents. In this approach the interfering substance usually remains in the solution phase, but in a chemical form which is analytically inert.

The more general approach to the problem, however, involves the actual physical segregation of the desired constituent from interfering substances. This procedure is normally described by the term separation.

TABLE 14.1

SEPARATION TECHNIQUES

Phase 1	Phase 2	Number of equilibrations	Common name of separation procedure
Gas	Liquid	$\simeq 1$	Gas analysis
		1	Volatilization
		$\gg 1$	Fractionation
		$\gg 1$	Gas–liquid chromatography
Gas	Solid	1	Thermal reactions
		$\simeq 1$	Gas analysis
		$\gg 1$	Gas–solid chromatography
Liquid	Liquid	$\simeq 1$	Solvent extraction
		$\simeq 1$	Liquid ion exchange
		$\gg 1$	Partition chromatography
		$\gg 1$	Counter current extraction
Liquid	Solid	1	Precipitation
		1	Electrodeposition
		> 1	Zone refining
		$\gg 1$	Ion exchange
		$\gg 1$	Adsorption chromatography
		$\gg 1$	Thin–layer chromatography

All separation procedures share a common principle; they are based on the selective distribution of sample components between two distinct phases.

A phase is a physically discernible portion of matter separated by definite boundaries from other such portions; it is a homogeneous region separated from other phases by surfaces of discontinuity. The phase material may be gaseous, liquid, or solid and separation procedures can be broadly classified in terms of the physical nature of the phases involved in the distribution process. Table 14.1 summarizes some of the techniques based on different phase combinations. The procedures can be further subdivided on the basis of whether separation is achieved in a single equilibrium process or whether multiple distributions are required to separate the components.

With single equilibrium techniques, the separation process generally culminates in a mechanical separation of the two phases, e.g., by filtration or a "cut" with a separatory funnel.

Multiple distributions are sometimes achieved by repetition of the single equilibrium procedure, but the more usual approach involves movement of one phase relative to the other. In many of these procedures the separated components appear as "blocks" or "zones" of material in the mobile phase and instrumental procedures are required to detect their presence.

The general principles of several of the techniques listed in Table 14.1 have been discussed in preceding chapters. For example, precipitation is discussed in Chapter 12, Section III; Chapter 10, Section IV,B; and Chapter 13, Section IV,C. Thermogravimetry (Chapter 4, Section III) often involves separating a gas from a solid, and fractionation is mentioned briefly in the same chapter (Chapter 4, Section V).

In the remaining techniques a number of processes are responsible for the selective distribution of chemical species between phases. Three of these processes are adsorption, diffusion, and ion exchange.

II. ADSORPTION

A. EQUILIBRIUM RELATIONSHIPS

The equilibrium relationships associated with the adsorption of solutes at a solid–liquid interface have been discussed briefly in Chapter 10, Sections III and IV,D.

The equations quoted are essentially empirical, but they can be used to predict the influence of solution variables on the extent of adsorption. Using general terms, it can be stated that the amount of material adsorbed on a surface is proportional to the surface area exposed and the concentration of adsorbate raised to a fractional power. The magnitude of the power term is determined by the nature of the surface, the nature of the adsorbate and the degree of competition for the active sites; i.e., Amount adsorbed $\propto$ Surface area $\times$ [Conc. Adsorbate]$^{1/n}$ where n is an integer.

The adsorption of gases on solids is more amenable to theoretical treatment than the liquid–solid system and several equations have been derived from fundamental principles. The simplest and best known is the Langmuir isotherm. For localized adsorption with one molecule of gas per adsorption site

$$\theta = Bp/(1 + Bp)$$

where θ is the fraction of the available sites covered; p is the gas pressure; and B is a constant whose magnitude is a function of the nature of the solid–gas interaction.

The mode of bonding at the interface can vary from weak van der Waals forces and weak dipole–dipole interactions to strong electrostatic attraction and orbital overlap. Weakly bonded adsorbates are said to be physically adsorbed and the process is characterized by the relative ease with which the adsorbed material can be displaced and the low activation energy (few kilocalories per mole) involved. A physically adsorbed species may be displaced from the surface by a reduction in the pressure (or concentration of solutes in solution) or by the addition of species which compete for the surface sites.

Strong bonding at the surface resembles the formation of a chemical compound, the activation energy of the process lying usually between 10 and 30 kcal per mole. The process may be irreversible, in the sense that the desorbed species has a different chemical composition to the adsorbate. Adsorption of this type is known as chemisorption. `

The line of demarcation between physisorption and chemisorption is very indistinct, some systems exhibiting the characteristics of both. Adsorption in all its forms is of interest to analytical chemists.

B.   Analytical Procedures Involving Adsorption

A simple but important application of gas adsorption is the selective removal of vapors from a gas stream. For example, the carbon and hydrogen content of organic materials is determined by heating the combustible sample in a stream of oxygen. The resultant gaseous products are fed into tared adsorption bulbs containing specific reagents. The moisture content is first removed by a column of desiccant (phosphorus pentoxide or magnesium perchlorate) and then all the carbon dioxide formed is retained on a trade preparation (usually potassium hydroxide mounted on an inert support). The increase in weight of the bulbs is used to calculate the amount of hydrogen and carbon in the original sample.

In other applications, adsorption of a vapor by a solid results in the formation of a colored or volatile product.

Many of the standard procedures for detecting traces of toxic gases in the atmosphere are based on drawing a known volume of gas through a tube of suitable adsorbent. The extent of any color change provides a semiquantitative measure of the concentration of impurity present. This principle is also used in the "breathalyzer" tests given to motorists suspected of driving under the influence of alcohol. A sample of the breath is fed through a tube containing finely divided, acidified potassium dichromate mixed with silica gel. Any alcohol vapor present is adsorbed and reduces the yellow solid to the green chromium (III) state. As an example of the formation of a volatile product one may quote the determination of minute amounts of carbon monoxide. Adsorption of carbon monoxide by a heated column of iodine pentoxide leads to the formation of carbon dioxide and an equivalent amount of iodine. The iodine vapor in the effluent gas stream is trapped in a potassium iodide solution and subsequently determined by titration with standard sodium thiosulfate solution.

The repeated distribution of gases and vapors between a mobile gas phase and a suitable solid adsorbent is the principle process involved in the separation technique known as gas–solid chromatography. This technique is discussed in Section II,D.

The adsorption of solutes on to a solid surface is an integral component of many analytical procedures. For example, the contamination of colloidal precipitates is attributed to the adsorption of foreign electrolytes, and the dissolution of solid samples in acid is preceded by the adsorption of protons at the surface.

Multiple distribution of different solutes between a liquid and a solid phase can lead to separation, and the techniques known as adsorption chromatography and thin-layer chromatography are based on this principle.

C.   Adsorption Chromatography

The name, chromatography, is applied to processes in which species are separated by differential migration in a porous medium, the migration being caused by a flow of some mobile phase.

In chromatography a small volume of sample is placed at one end of a column packed with some powdered solid, and a gas or liquid is allowed to flow at a controlled rate through the column. As the mobile phase passes over the sample and moves down the column, the components of the sample mixture are subjected to a series of distributions between the phases. Individual components tend to migrate down the column at different rates and leave the column of solid at different times.

If the composition of the mobile phase is monitored continually as it leaves the column, the individual species may be detected as they appear. Suitable detectors for monitoring gas flows, e.g., thermal conductivity cells, were discussed in Chapter 4. With liquid mobile phases the absorption spectra of the eluted species is sometimes sufficiently characteristic to permit continuous observation but in a large proportion of studies it is necessary to add selective, color-forming reagents to successive aliquots of the solution leaving the column. To achieve efficient separations, very small samples (microgram quantities) are required and this restricts the number of analytical techniques that can be used to study the eluted components.

Figure 14.1 is a diagrammatic representation of column chromatography.

1.   *Column Development*

The separation of the sample components into independent waves, zones, or bands under the influence of the moving phase is known as "development" of the chromatogram.

In adsorption chromatography there are three conventional development procedures.

The procedure outlined in preceding paragraphs is known as elution analysis. The flow of the mobile phase through the column is continued until all the zones

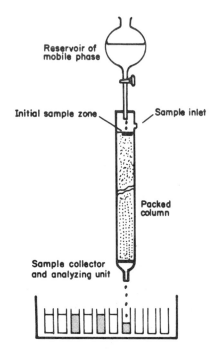

Reservoir of
mobile phase

Initial sample zone

Sample inlet

Packed
column

Sample collector
and analyzing unit

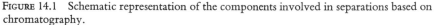

FIGURE 14.1   Schematic representation of the components involved in separations based on chromatography.

due to the sample components have left the column, and if the separation process is efficient the zones of significant detector response are separated by volumes of pure mobile phase as indicated in Fig. 14.2(a). The volume of gas or solvent required to remove a component from the column (the retention volume) may be used for identification purposes.

In stepwise elution, several solvents are used in succession, each one being a more effective eluting agent than the one preceding. This procedure can speed up the elution of strongly adsorbed species.

In adsorption chromatography the major process responsible for the separation is selective adsorption, and for undistorted elution zones to be observed a number of conditions must be satisfied.

1. The column must not be overloaded with solutes; this imposes a limit on the sample size.
2. Diffusion effects must be negligible; diffusion is minimized by efficient packing of the column and by increasing the flow rate of the mobile phase.
3. Adsorption and desorption from the fixed phase must be almost instantaneous so that equilibrium is rapidly attained.

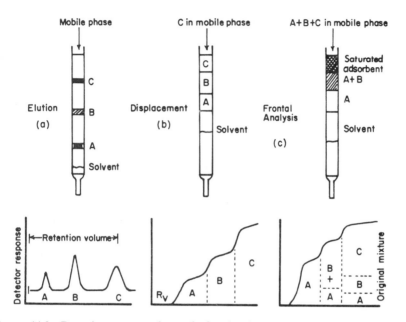

FIGURE 14.2   Detection patterns observed after development of an adsorption chromatogram using (a) elution analysis, (b) displacement analysis, and (c) frontal analysis.

4. The distribution of the solute between the two phases must vary linearly with concentration.

While conditions 1 to 3 apply to many adsorption systems, condition 4 is rarely satisfied. The amount of solute adsorbed during each equilibrium step usually varies as a fractional power of the solute concentration, and the distribution curve is curved instead of linear. As a result, the elution zones are asymmetrical in shape and the tailing or leading of the zones tends to cause overlap of the components. The influence of the shape of the distribution curve on the nature of the elution zone is shown diagrammatically in Fig. 14.3.

The degree of tailing can be reduced by using a modified procedure known as gradient elution analysis. An eluting agent more powerful than the one already in use is added to the column in gradually increasing concentration. This produces a concentration gradient down the column which results in the rear portion of any chromatographic band being always in contact with a stronger eluting solution than the front portion.

Because of the distorted shape of most elution bands and the limited number of suitable adsorbents available, most modern chemists elect to use alternative forms of chromatography. On the other hand, adsorption chromatography is the only form of this technique which can be subjected to frontal analysis or developed by a displacement procedure.

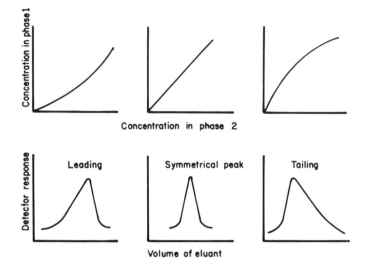

FIGURE 14.3 Diagram showing the relationship between the form of the distribution isotherm and the shape of the eluted zone obtained from a chromatogram. From "Fundamental Principles of Chemical Analysis," W. F. Pickering. Used with permission of Elsevier Publishing Company, Amsterdam, copyright © 1966.

For displacement development, the mobile phase is a solution of some substance which is more strongly held by the adsorbent than any component of the sample mixture. The solute in the eluting solution [i.e., C in Fig. 14.2(b)] displaces sample components (e.g., A and B) from the zone at the top of the column. If B is more strongly adsorbed than A, B successively displaces A. Consequently the zones should pass down the column in the order A, B, C [cf. Fig. 14.2(b)]. However, one of the disadvantages of this procedure is that the zones are not separated by regions of pure solvent and there can be extensive overlapping. In some instances this problem has been overcome by introducing another substance into the eluting medium which has adsorption properties intermediate between those of A and B.

In frontal analysis, a small column is packed with a known weight of adsorbent, and the sample solution is added continuously, i.e., it becomes the mobile phase. Initially all the solutes in the mixture are adsorbed, but as further sample solution is added the adsorbent becomes saturated and the more weakly bound species (e.g., A) is displaced and eventually passes out of the column. Analysis of the mobile phase leaving the column yields a pattern of the type shown in Fig. 14.2(c). The first step consists of pure A, and the second step consists mainly of species B, although it is contaminated with A which is continually supplied with the main body of the mixture. The third step consists of the original mixture A + B + C, at first slightly depleted in B. The number of steps equals the number of components in the mixture. The amount of substance A in the mixture can be calculated from the volume of solvent which passes through the column before substance A appears. Other components are estimated from volume-concentration relationships.

2   Adsorbents and Solvents

Apart from the mode of developments the two major variables in adsorption chromatography are the nature of the adsorbent and the composition of the mobile phase.

The tendency of species to be adsorbed from solution can be varied by selecting adsorbents and/or solvents of differing polarity. For example, alumina is a polar compound and polar organic substances are strongly adsorbed from nonpolar solvents. These adsorbed species are readily displaced by a flow of polar solvent. The development of an adsorption chromatogram may thus be considered to involve competition between the solvent and solute for the adsorbent; the amount of adsorption that occurs depends on the relative polarities of the three components.

Adsorbents may be classified in terms of relative polarity (i.e. strength of adsorption of a selected species) and capacity (i.e. grams solute adsorbed per gram of solid).

The adsorption capacity is often reported in terms of the amount of organic dye adsorbed using standard conditions, but since this value varies with the surface area (particle size) of the adsorbent, it is difficult to arrange materials in any particular order. However, the experimental capacity does influence the size of column required to achieve a particular separation.

The bonding strength of adsorbents is more specific and in Table 14.2 a number of materials are arranged in an order which approximates to an increase in capacity, and binding strength, as one progresses down the left hand column. The right hand column of this table lists the type of compounds commonly separated on the various adsorbents.

With a particular polar adsorbent, the strength of adsorption of polar groups decreases in the order:

$$-COOH > -OH > -NH_2 > -SH > -CHO$$

$$> =C=O > -COOR > -OCH_3 > -CH=CH-$$

For adsorption on charcoal, the order is almost the direct reverse.

In a similar manner solvents can be arranged in an order based on their eluting power. The order of the eluotropic series varies a little with the nature of the adsorbent and the species being eluted, but in general it corresponds to the sequence obtained by arranging the solvents in terms of their dielectric constants. Hence, for polar adsorbents, the eluting power decreases in the order:

Mixtures of acids and bases > organic acids > pyridine > pure water > methanol > ethanol > propanol > acetone > dichloroethane > ethyl acetate > chloroform and diethyl ether > dichloromethane > benzene and

TABLE 14.2

ADSORBENTS FOR CHROMATOGRAPHY

Solid adsorbent	Types of compounds separated
Sugar	Chlorophyll, xanthophyll
Starch	Enzymes
Calcium carbonate	Carotenoids, xanthophylls
Calcium phosphate	Enzymes, proteins, polynucleotides
Magnesium carbonate	Porphyrins
Slaked lime	Carotenoids
Magnesia	Sterols, dyestuffs, vitamins, esters, alkaloids
Aluminum silicate	Sterols
Activated magnesium silicate	Sterols, esters, glycerides, alkaloids
Activated alumina	Sterols, dyestuffs, vitamins, esters, alkaloids, inorganic compounds
Silica gel	Sterols, amino acids
Activated carbon	Peptides, carbohydrates, amino acids

toluene > triochloroethylene > carbon tetrachloride > cyclohexane > hexane > light petroleum fractions. For the elution of substances adsorbed on charcoal, this order is reversed.

3. R_f Values and Applications

Elution development of chromatograms is usually continued until all the different zones of solute have left the column. However, there is an alternative approach in which the solvent flow is stopped before the solvent front reaches the end of the column. The complete column is then extruded and the individual zones, located along its length, are visualized by treatment of the surface with a suitable color-forming reagent. After location of the zones a figure known as the R_f value is determined.

The term R_f can be defined by the expression: R_f = Distance traveled by zone/Distance traveled by solvent front.

The determination of an R_f value thus involves measuring the distance from the point of application of the sample to the center of an individual zone and to the edge of the solvent front (cf. Fig. 14.4).

The R_f values facilitate identification of the sample components, since they can be compared with the values obtained with standard substances subjected to the same chromatographic treatment. Reference books dealing with adsorption and partition chromatography usually contain extensive compilations of R_f data. Similarity of R_f values does not positively identify a species but it provides circumstantial evidence which can be confirmed by the application of some specific reagent.

In the continuous flow methods the confirmatory tests or physical measurements (e.g., infrared spectra) are made directly using the collected aliquots of eluted material.

The applications of adsorption chromatography include the detection and characterization of the components of mixtures, the resolution of mixtures of substances, and the purification of chemical preparations. The technique has been used in studies of organic, inorganic and biological substances.

A summary of the experimental conditions which have proved satisfactory for the separation of a wide range of compound types is provided in Table 10.13 of the "Handbook of Analytical Chemistry." Reference to this table indicates, e.g., that alcohols may be separated on an alumina column using petroleum ether as eluant. The same solvent can be used to selectively elute carotenoids from adsorption columns packed with finely ground lime, alumina, magnesia, or sugar. For the separation of alkaloids on a kaolin column, water is shown as the recommended eluant.

Not listed in such compilations are the disadvantages of the technique, which include the time required to achieve satisfactory separations and the distortion of the zones which arises from diffusion effects in the adsorbent mass. Both these effects can be minimized by reducing the size of the column. A limit in this direction was reached when sheets of filter paper proved to be highly satisfactory substitutes for cellulose columns. The advantages of thin films, faster development, and less diffusion were immediately apparent and a new technique was evolved.

4.   Thin-Layer Chromatography

In thin-layer chromatography, a thin (200–300 mμ) layer of fine adsorbent (a particle size of 1–25 mμ is recommended) is evenly spread over the surface of a glass plate. A binder such as starch, plaster of paris, collodion, or a plastic dispersion is mixed with the adsorbent (e.g., alumina, silica gel, etc.) to ensure adhesion to the glass. The plates are dried in an oven and are used in a similar manner to paper strips.

The separation of complex mixtures is achieved by placing small spots of the sample near one end of the prepared plate. This end is then placed in a shallow solvent reservoir and the solvent moves up the adsorbent film by capillary flow.

After the solvent has traveled some distance along the sorbent layer, the plate is removed and the components of the mixture are visualized by spraying the adsorbent with a suitable reagent. Partial identification of the components can then be based on R_f measurements, the R_f values of the separated zones being compared with the movement observed with samples of pure compounds treated in an identical manner; the ideal result is shown in Fig. 14.4. In this diagram, the three components of a mixture are shown to have similar R_f values to three standard compounds.

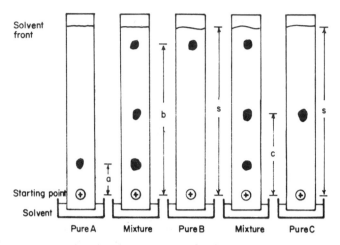

FIGURE 14.4   Diagram showing the separation of a three component mixture on a thin-layer chromatogram. The R_f values of suspected components, A, B, and C have been identified from the comparison chromatograms of pure compounds; the R_f of A is a/s, the R_f of B is b/s, etc.

For chemical confirmation of the nature of the constituents, the individual zones of isolated material may be scraped off the plate and subjected to extraction with a suitable solvent. After removal of the adsorbent by filtration, the extracted material is examined by an appropriate microanalytical technique.

This technique separates and isolates molecular species; hence it is an important supplement to gas-liquid chromatography in studies of natural products and related problems.

As in column chromatography, the type of adsorbent required is determined somewhat by the type of compounds being separated, and the solvent chosen should satisfy the same conditions as solvents used in column studies. The solvent has to facilitate reversible dynamic sorption of the components of a mixture without altering the components or the adsorbent; it must be less strongly adsorbed than the substances being examined; and it should not interfere with the detection of the separated substances.

5.   Gas-Solid Chromatography

An extremely efficient means of separating gases and vapors of low boiling point (e.g., O_2, CO, CO_2, NO, rare gases, gaseous hydrocarbons up to C_5, etc.) is provided by gas-solid chromatography.

The components of the equipment required for this technique are shown in Fig. 14.5.

The gases and/or vapors are separated on a packed column which is commonly about one-fourth inch in diameter and can vary in length from two to twenty feet or more. The tube may be made of copper, aluminum, stainless steel, or

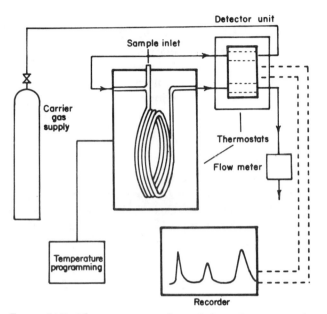

FIGURE 14.5   The components of a typical gas chromatograph.

glass; and to accommodate long lengths in a small oven space, it is often formed into a coil or arranged as a series of U tubes.

The packing which fills the column is a granulated adsorbent such as activated carbon, silica gel, alumina, or molecular sieve. (With molecular sieves differences in the adsorption characteristics of the gaseous components are enhanced by diffusion effects.)

The temperature of the column is an important operational variable; hence both the column and the detector unit are mounted in thermostats. The optimum operating temperature is controlled by the nature of the sample and for complex samples it is often necessary to vary the temperature linearly with time. Such temperature programming has enhanced many gas chromatographic procedures.

The gaseous sample, which is introduced at the top end of the column, is eluted through the column by a controlled flow of carrier gas. The samples are admitted by means of fine bore syringes (the syringe needle passing through a serum cap mounted on a side arm) or by means of micropipettes, bypass sample collectors, or mechanical sampling valves.

The nature of the carrier gas used (e.g., nitrogen, argon, carbon dioxide, hydrogen, or helium) depends partly on the composition of the sample and partly on the type of detector unit used to monitor the composition of the eluant gas. For every separation procedure there is an optimum carrier gas flow rate which has to be determined experimentally.

Gas–solid chromatography usually leads to asymmetric elution peaks which have a sharp front flank and a relatively long, drawnout tail. The peaks tend to overlap and with some gas or vapor mixtures resolution is poor.

Qualitative identification of the eluted species is usually based on retention volumes. The retention volume is the volume of gas which passes through the column in the interval between the addition of the sample and the appearance of the detector peak for the component of interest. The retention volumes of the sample constituents are compared with the volumes required by standard compounds, the experimental conditions (temperature, type of column, etc.) being retained constant.

The technique is somewhat limited by the small range of adsorbents available and the need to replace or degas the adsorption column repeatedly. For this reason, gas–liquid chromatography (Chapter 15, Section V) is preferred for the separation of species other than the low–boiling point materials mentioned at the beginning of this topic.

III. SEPARATION BASED ON ELECTROMIGRATION

A. ELECTROGRAPHY

In the adsorption techniques discussed in the previous section, differential migration of solute species is obtained by a flow of mobile phase. The movement of charged solutes can also be induced by the imposition of an electrical potential across the body of a solid–liquid phase system.

The resulting current flow causes electrolytic reactions at the electrode interfaces and the products of electrolysis can include ionic species derived from metallic electrodes. The newly formed ions then migrate towards the electrode of opposite charge sign.

This electrolytic dissolution and subsequent migration process is utilized in the technique known as electrography which is used to study the composition of, and any nonhomogeneity in surface coatings. Another application is the detection of pores and discontinuities in protective surface coatings.

The basic assembly for electrography is shown in Fig. 14.6(a). A pad of paper or other adsorbent material, soaked with an electrolyte solution, is squeezed between two flat metal electrodes. One of the flat polished surfaces in contact with the moist adsorbent is made of an inert (e.g., gold) or noninterfering metal (e.g., aluminum). The second electrode is the metal specimen to be examined. The contact surface of the sample is initially polished with fine abrasive and is then washed to remove adhering abrasive and any oily films.

Efficient contact between the electrodes and the sandwiched material is maintained by the continual application of some form of pressure. The sample plate is made the anode of the electrolysis circuit if cations are to be removed from the specimen surface. Conversely, if anionic derivatives of the surface are of interest, the sample becomes the cathode.

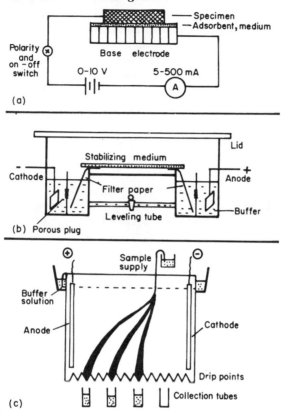

FIGURE 14.6   (a) Basic unit for electrography studies, (b) apparatus for zone electrophoresis, and (c) continuous paper electrophoresis.

If the two electrode surfaces are parallel, a uniform electrical field drives the ions released from the sample surface towards the opposite electrode. The transfer medium has to receive the ions and retain them for subsequent visualization. If there is minimal lateral diffusion in this medium, the observed distribution of the ions reproduces the spatial distribution of the source metals on the specimen surface. In this manner contaminants or segregations on a surface, or pinholes and scratches in a surface coating, are detected.

The color-forming reagent(s) required to identify the products of electrolysis may be added to the initial electrolyte solution or sprayed on at a later stage. With simple systems, e.g., the detection of iron impurities on an aluminum surface, the color reagent is added to the electrolyte. With more complex mixtures of surface components, preliminary washes of the adsorbent layer with selected reagents may be required to remove interfering species. The preliminary treatment is kept to a minimum by using spot test reagents which possess some degree of selectivity but the material in the adsorbent medium must be strong enough to withstand a series of chemical treatments.

Hardened filter papers have proved quite satisfactory; other workers use gelatin-coated paper or unplasticized cellophane sheets. These thin sheets are usually mounted on the top of a pad of thick soft paper soaked with an electrolyte solution having a salt concentration of 0.1 to 0.5 M.

Besides providing electrical contact between the electrodes, the electrolyte has to fulfil several secondary roles. Smooth, continuous solution has to be maintained at the specimen surface; hence, the electrolyte must not introduce species which cause surfaces to become coated with insoluble products. On the other hand, lateral diffusion of the migrating species should be minimal and this can be achieved by buffering the electrolyte with sodium carbonate because the carbonate causes practically all metal ions to form precipitates.

The voltage applied across the electrodes usually does not greatly exceed the solution potential of the metal to be detected, since in this way all the electrical energy is used to dissolve the metal. The actual quantity of electricity required is determined by the quantity of ion transfer needed to give a satisfactory color density in the final "print." Generally, several seconds flow of a current measured in milliamperes is sufficient.

B.   ELECTROPHORESIS

In electrophoresis the item of interest is not the nature and disposition of any ions produced by electrolysis, but rather the rate of migration of added charged species towards the electrodes. Provided the migration path is of reasonable length, differences in individual rates can be used to separate mixtures. The experimental procedures can be divided into two types: free electrophoresis and zone electrophoresis.

In the technique known as free electrophoresis a small volume of solution containing the species to be separated is added to a column of suitable electrolyte solution. Under the influence of an applied field, the various components migrate at different rates towards the electrodes and tend to separate into zones. However, once the power is switched off, all the ions are free to diffuse in any direction. The position of the separating molecules has therefore to be assessed while migration is actually in progress. The various zones are generally located by means of a complex optical system that detects the changes in refractive index which occurs at the boundary between solutions of different composition (the Schlieren effect). The equipment required for this type of measurement is expensive and close control of experimental conditions is required; hence, the majority of electrophoresis studies are now made using the procedure known as zone electrophoresis.

In zone electrophoresis the charged molecules or ions which constitute the sample are caused to migrate along a stabilized medium in which they are fixed when the current is switched off. The separated species can then be detected by treating the medium with suitable color-forming reagents.

The rate of movement of the charged species is a function of the net charge, size and shape of the ion, together with any selective retardation introduced by the stabilizing medium. Thus in theory, zone electrophoresis may be applied to any type of mixture of ions or polar molecules, and the technique has been used to separate simple and complex inorganic ions; organic molecules such as amino acids and nucleotides; and macromolecules, e.g., proteins in human serum.

The basic components of the apparatus required are shown in Fig. 14.6(b). The area upon which the electrophoretic separations occur is called the "bed" and it can be composed of a number of materials including paper, polymer films, gels, and powders. The bed is initially moistened with electrolyte (usually a buffer solution) and the ends are maintained in contact with more electrolyte solution contained in two chambers which also hold the electrodes that are connected to the dc power supply. To minimize diffusion of electrolysis products (e.g., H^+ and OH^-), the electrodes are mounted in compartments which are separated from the electrolyte reservoirs by porous plugs. To prevent siphoning of electrolyte solution through the bed, provision is made for adjusting the level of liquid in all chambers to equal values.

In addition to providing the stationary support for the separation process, the stabilizing medium (or bed) acts as a conducting bridge for the applied potential. As such it becomes heated and evaporation losses have to be minimized by enclosing the separation unit in an airtight chamber. The bed may be mounted horizontally [as in Fig. 14.6(b)] or vertically (if it is in sheet form). Then 1–100 μl of sample is applied to the center of the bed, usually in the form of a spot or streak, and the current is switched on. After sufficient time has elapsed for good separation, the current is switched off and the stabilizing medium is removed from the tank. The components are then fixed and identified by treating the bed with specific reagents. The medium containing the separated zones is referred to as an electrophoretogram.

The rate of movement of a particle under controlled conditions is reproducible, but the mobility of the species has to be determined experimentally for each specific set of conditions because so many variables are involved. Mobility may be defined as the distance (in cm) a particle travels per second per unit strength of electrical field. (The field strength is normally quoted as volts per centimeter along the length of the electrophoretic bed.)

The two features of an ion which are most pertinent to its electrophoretic behavior are its size and charge; the larger the size, the slower the ion travels; the bigger the charge, the faster it goes. Substances carrying equal numbers of positive and negative charge groups, and uncharged species, should remain at their point of origin, but during electrophoresis there is often a flow of water under the influence of the voltage gradient which can cause some movement (this phenomenon is called electroosmosis).

Electromigration is slower in stabilizing media than in free solution because collision of the moving particles with the support can cause them to change direction. In addition the migrating species can be adsorbed by the solid phase. The magnitude of the medium effect depends on the type of substances being separated, the chemical composition of the bed material, and the concentration and pH of the buffer electrolyte solution.

Increasing the buffer concentration generally leads to sharper component zones but it also causes a decrease in mobility and an increase in heat output. Heating causes evaporation of the solvent which is followed by a flow of buffer on to the bed. To minimize this effect the applied voltage is made as large as practicable in order to achieve maximum separation in the minimum time. With many stabilizing media, field strengths of 5–10 V cm^{-1} have proved quite satisfactory.

The pH of the buffer solution can affect the charge of the species being separated. For example, ampholytes such as amino acids or proteins can exist in the following forms:

$$H_3N^+—CH—COOH \underset{H^+}{\overset{OH^-}{\rightleftarrows}} H_3N^+—CH—COO^- \underset{H^+}{\overset{OH^-}{\rightleftarrows}} H_2N—CH—COO^-$$
$$\qquad\quad \underset{R}{|} \qquad\qquad\qquad \underset{R}{|} \qquad\qquad\qquad \underset{R}{|}$$

It follows that by suitable choice of pH, a substance of this type may be made to migrate to the anode or the cathode or to remain at the origin. Since the pH required to effect the transition between different forms varies from amino acid to amino acid, effective separation of groups of these compounds can be achieved by suitable choice of the buffer pH, e.g., for separating proteins the optimum pH range appears to be between 8.6 and 9.2.

With any weak acid or base the pH of the buffer establishes its degree of ionization, and there are pH ranges in which significant amounts of two ionization states coexist. Within these ranges the two ionization states migrate at different rates and the zones become quite diffuse.

The sign and net charge of inorganic ions may be altered by coordination, and the addition of a suitable ligand to the buffer solution is often used to facilitate the separation of groups of metal ions. The rate (and direction) of movement of the individual species is related to the effective stability of any complexes formed, and this factor can be varied by altering the ligand concentration or the pH of the buffer.

Organic substances which do not possess a characteristic charge at any pH can sometimes be separated by causing them to react with other charged species. For example, sugars can be separated by electrophoresis using borate buffer solutions, since they form stable borate anion complexes. Alcohols in the steroid and triterpenoid series have been separated as their acid sulfate and acid succinate esters.

A number of materials have been used as the stabilizing medium in zone electrophoresis; each has its particular merits and each requires a different set of experimental conditions.

Filter paper possessing a high wet strength is a popular support although for some purposes strips of cellulose acetate are preferred because there are fewer adsorption effects and the separated bands are sharper. For clinical investigations gels are regularly used as the support media. The gels (which may contain starch, agar, or synthetic polymers such as polyacrylamide) are prepared by using a buffer solution as the liquid phase and the bed is prepared by casting the gel in the trough (or tube) that joins the electrode chambers.

A number of studies have also been made in which the solid support consists of a block of starch or a packed tube of some other powdery adsorbent material.

The electrophoretic method is extremely versatile since all soluble substances which possess charge; or can be given charges, are suitable subjects for separation operations. The method is widely used by biochemists and in clinical diagnosis. The range of species studied includes proteins, hormones, vitamins, nucleoproteins, toxins, porphyrins, carcinogens, and antibiotics.

A procedure has been designed for continuous paper electrophoresis. As shown in Fig. 14.6(c), a large sheet of filter paper is allowed to hang vertically, the upper edge dipping into a trough containing buffer solution. The solute mixture to be separated is applied continuously at a point near the top of the paper. The buffer solution, which advances down the paper by capillary flow, carries the mixture with it. By means of electrodes mounted on the sides of the paper a potential difference is applied at right angles to the direction of the solvent flow. The various ions in the mixture migrate towards the cathode or anode, according to their charge. This causes the various components to reach the bottom of the sheet at different places, and by serrating the lower edge of the paper, the separated components can be collected as they drip off the appropriate points.

C. ELECTROCHROMATOGRAPHY

Electrochromatography is a separation procedure which simultaneously utilizes both electrophoresis and chromatography in such a manner as to combine the best features of both processes.

The experimental setup is similar to that shown in Fig. 14.6(c), the significant difference being that the upper trough contains an eluant which is capable of achieving chromatographic separation of the sample components in the absence of an applied electrical field. The sample may be applied continuously [as in Fig. 14.6(c)] for preparative purposes or as a single aliquot or spot. The discontinuous sample application procedure is known as analytical electrochromatography. When the adsorption process is reversible, this technique separates the initial spot of sample mixture into a series of round or elliptical spots which are separated both vertically (by chromatography) and horizontally (by electrical migration).

For successful operation all the factors contributing to migration have to be in dynamic equilibrium. Thus one needs to establish the optimum combination of eluant flow, electrical power, temperature and, for continuous separations, sample flow.

The eluant mixture has many roles; it has to ensure that the species being separated are in appropriately charged forms; it must act as the conducting medium for the electrical current; and it must selectively desorb the components of sample mixture from the solid support. Finally, to ensure optimum efficiency in the separation process it must flow freely down the support curtain.

Unless the heat generated by the flow of current is dissipated effectively, there can be excessive evaporation leading to unstable steady state conditions. In addition, excessive heating effects can lead to decomposition of unstable sample components.

The amount of current flowing is determined by the conductivity of the background fluid and the magnitude of the applied potential. The best separation conditions are obtained by keeping the conductivity to a minimum. This allows large applied voltages to be used without creating excessive heating effects.

The nature of the sample can influence the experimental conditions required. For example, in continuous electrochromatography the flow of sample solution can make a significant contribution to the overall conductivity of the liquid phase, and the potential used may have to be decreased to limit heating. Alternatively, if the sample contains suspended material or gellike components, these may clog the pores of the support medium and so reduce the overall efficiency of the process.

The desirable concentration of sample components in the solution fed to the separation unit varies with the molecular weight and the nature of the substances. For inorganic ions and small molecules the concentrations used range from 0.1 to 0.5%, while for large colloidal particles with molecular weights of 10^6 to 10^9 concentrations of 0.0001 to 0.1% are appropriate.

Despite the empirical nature of the technique it has proved very satisfactory for the preparative fractionation of substances of biological origin, and the list of typical applications recorded in the "Handbook of Analytical Chemistry" (Table 10.81) contains entries under many group classifications, e.g., antibiotics, bacteria, bile, blood sera, botanicals, cells, enzymes, hormones, immunizing agents, inorganics, plant juices, protein hydrolyzates, spinal fluid, tissue homogenates and extracts, urine and viruses.

IV. DIFFUSION

A. THE PERMEABILITY OF POLYMERS

The term diffusion has been introduced in several preceding sections. For example, in electromigration techniques it was recommended that lateral diffusion should be kept to a minimum, and diffusion was named as one of the causes of distorted elution zones in chromatographic separations.

In chromatographic separations the diffusion process may involve mass transport within the pores of the solid support and/or within the surrounding mobile phase.

The diffusion of ions and other species in an external liquid phase may be considered to resemble the diffusion effects discussed in some detail in the section dealing with polarography (Chapter 13, Section VI,A).

The diffusion of gases or vapors within a gaseous phase is probably most simply described in terms of Graham's law, which states that, "at a given temperature and pressure, the rate of streaming through an orifice (R_T) varies inversely as the square root of the molecular mass (m)."

$$R_T = PA/\sqrt{2\pi k Tm}$$

where P is driving pressure, A is the area of the aperture (must be very small), k is Boltzmanns constant and T is the absolute temperature.

If the rates of effusion of two different gases at standard conditions are compared

$$R_T^{I}/R_T^{II} = (m^{II}/m^{I})^{1/2}$$

In a packed column the interstices between particles and any pores approximate to narrow orifices and a number of factors combine to create pressure gradients within a column.

This type of diffusion contributes to the distortion of chromatographic elution zones, but the same tendency of gases to diffuse across an air space can also be used to separate volatile reaction products (e.g., water, ammonia, carbon dioxide, halogens) from a given sample. Any volatile compound, retained in a closed system, sets up an equilibrium vapor pressure. If the gaseous vapor is continually trapped by absorption in a second phase, the equilibrium is displaced and eventually all the vapor is removed from the sample through being fully absorbed in the second phase. A simple example of this process is the drying of precipitates over concentrated sulfuric acid or other desiccants. With a small apparatus containing a large surface of absorbent, the rate of diffusion of the vapor from the initial source to the absorbent can be quite rapid.

In the microdiffusion analysis technique developed by Conway, the absorbent solution is placed in the inner well of a small Pyrex glass unit of similar size to a small Petri dish. The circular glass wall that forms the inner well is usually only half the height of the outer wall of the vessel, and during analysis the unit is covered by a ground glass lid sealed by grease to the outer rim. The volatile compound is generated in the outer chamber, the volume of solution in this zone being kept to a minimum (e.g., 2–3 ml) to keep diffusion times to a reasonable value (i.e., <3 hr). The species which diffuses to, and is absorbed in, the inner solution phase is subsequently determined by titration or colorimetric procedures. This technique eliminates distillation and allows very small amounts of material to be determined with fairly high precision.

The mass transport of material within the pores of solid materials involves a number of additional factors, and because this process is important in a number of separation techniques (e.g., gel filtration, ion exchange, and gas–solid chromatography) it needs to be discussed as a separate entity.

The solid support or phase used in separation procedures is usually classifiable as a form of polymer, and a large number of studies have been made of the permeability of polymers to gases and vapors. Somewhat less attention has been paid to the diffusion of solutes.

The transport of small molecules through polymers is considered to involve at least one of the following processes:

1. *Flow of Small Molecules through Preformed Holes or Capillaries*

The flow may consist of molecular streaming or viscous flow, and the rate of transport is influenced by the vapor pressure differential across the holes, and the size and shape of the diffusing molecules and passage ways. The process is unactivated in the sense that the only effect of temperature is to increase the vapor pressure differential. The synthetic zeolites represent a group of polymers in which this process predominates.

2. *Molecular Diffusion through the Mass of Polymer by Way of Temporary Holes Created by the Thermal Motion of Polymer Chains*

This process is definitely temperature dependent and is said to be "activated." The diffusing molecule may be considered as being dissolved in a hole until another hole opens nearby into which it may move. The interchain bonding forces and general chain mobility of the polymer are important because they determine the ease of hole formation.

3. *Activated Diffusion Involving the Interaction of the Diffusing Molecule with Active Spots on the Internal Surfaces of the Polymer*

The diffusing molecule is thought to be adsorbed on one of these active spots and to vibrate in this position until it acquires sufficient energy to evaporate from this point. Once free it moves through the polymer until readsorbed at another active spot.

Active spots on polymer chains tend to interact with other parts of the same chain or with a segment of an adjacent chain. Adsorption on these spots weakens such bonds and increases the probability of hole formation through thermal motion. Thus chances for diffusion by processes 1 and 2 are enhanced. The tendency for species to be adsorbed (or dissolved in the polymer) increases when the structure and polarity of the diffusing material resembles the structure and polarity of the polymer.

An increase in the degree of cross-linking in a polymer tends to decrease the rate of permeation through stabilization of segmental motion. Amorphous polymers have higher permeabilities than crystalline polymers; hence one may conclude that amorphous regions contain a higher proportion of holes and capillaries.

The permeation of gases and vapors through polymer films appears to be primarily diffusion controlled. Thus when a stationary state of flow is obtained, the flux, q (amount of gas passing through the polymer film per unit area and time), satisfies Fick's first law, in that

$$q = DS(p_1 - p_2)/l = P(p_1 - p_2)/l$$

where D is the diffusion constant, S is the solubility coefficient and P is the permeability constant which is equal to DS; p_1 and p_2 are the pressures of the gas on both sides of the barrier and l is the thickness of the film.

For the permanent gases the values of D and S do not alter significantly with changes in pressure, but with organic vapors D and S often increase with increasing pressure. The temperature dependence of these constants follows the usual exponential function; hence the temperature dependence of the permeability constant can be described by an Arrhenius equation of the form: $P = P_o \exp(-E_p/RT)$.

The order of the solubility coefficient (S) may be established by swelling measurements or weight increase measurements, and it has been observed that solution follows the general rule of "like dissolves like." With vapors and organic liquids the contribution of solution (S) to permeability is usually more important than diffusion (D); hence for these materials one can make the generalization, "like permeates through like." This statement must be taken into consideration before adopting plastic containers for storage of solvents or chemical solutions. For example, an aqueous solution of bromine stored in a polythene bottle rapidly loses its halogen content through diffusion of the bromine vapor into the atmosphere.

The same general considerations apply to systems involving contact between solutions and a polymer except that interaction with the polymer chains now involves both solvent and solutes, e.g., the adsorption of solvent causes swelling of the polymer. The degree of swelling usually increases in the presence of solutes because adsorption of these species further weakens inter- and intrachain bonding effects.

The driving force in the diffusion of solute species is a concentration gradient, and with polymer particles surrounded by solution, a gradient initially exists between the surface openings of all pores and the interior of the pores. The diffusion process is generally activated, because polymers can contain active sites which range from aliphatic or aromatic segments to polar components or substituent groups. The degree of interaction at the active sites can vary from dipole–dipole attraction (adsorption) to electrostatic attractions between ions (ion exchange). The number of active sites can vary with the composition of the polymer, the pH of the surrounding liquid (dissociation of functional groups), and the degree of cross-linking, since this can influence the extent of diffusion between holes.

In separation procedures which utilize molecular sieves and gels, the efficiency of the process is associated with two factors: relative size and diffusion rates. The mixture to be separated is usually passed through a column of polymer particles in some diluent gas or solvent. Molecules which are too large to enter the holes and capillaries of the polymer move with the mobile phase, being retarded only by the external surfaces of individual particles. Smaller ions and molecules are free to diffuse into the polymer as well as around the particles. The extent of internal diffusion varies with the size, shape, and nature of the small species; this causes variations in retardation effects, and individual components are eluted as separate zones.

Molecular sieves adsorb species with greater tenacity than conventional adsorbents. The small molecules readily retained include water, ammonia, methanol, carbon dioxide, and sulfur dioxide. Sieves with larger pores also retain aromatic compounds, higher paraffins, amines, and cyclo-paraffins.

One important use of these materials is as a drying agent for gases and solvents, and in desiccators. The powerful affinity for water makes them particularly efficient, e.g., ethanol, which is extremely difficult to dry, can be dried to a product containing less than 10 ppm of water by passage through a column of zeolite type molecular sieve.

The separations achieved with columns of molecular sieves include the separation of straight-chain from branch-chain molecules, of compounds of varying degree of unsaturation (e.g., ethylene and acetylene), and of hydrocarbons of different molecular weights.

Dialysis is another separation technique in which diffusion through a polymer is important.

The apparatus required for simple dialysis is a bag-shaped membrane containing the aqueous solution to be dialyzed. The bag is suspended in water and while the small molecules in the solution succeed in diffusing through the membrane, the larger molecules are held back. (Cellophane is most commonly used as the membrane; other materials tried include collodion, parchment paper, and synthetic polymers.)

One of the most important applications of this technique is the removal of salts and other low molecular weight impurities from proteins. In many cases the process can be speeded up by applying an electrical potential between the two aqueous phases separated by the membrane.

B.   Types of Polymers Used in Analysis

A widely used group of inorganic polymers are the alumino silicate compounds known as zeolites. The zeolite lattice consists of SiO_4 and AlO_4 tetrahedra which share oxygen atoms. This leads to a relatively open three-dimensional framework with channels and interconnecting cavities. In recent years, zeolites with completely regular crystal structure have been prepared through crystallization of the material at an elevated temperature from solutions containing appropriate amounts of silica, alumina, and alkali. Because of their narrow, rigid, and strictly uniform pore structure they act as molecular sieves. Several types are synthesized by the Linde Corporation, the pore size being adjusted by converting the materials into different chemical forms. Thus Linde type A has pore diameters of 3 to 5 Å while type X has pore diameters of 10 to 13 Å. A typical member of the type A group is the sodium alumino silicate, $Na_{12}[(AlO_2)_{12}(SiO_2)_{12}]$; a larger molecule, $Na_{86}[(AlO_2)_{86}(SiO_2)_{106}]$, belongs to the X group.

Other water-insoluble inorganic materials have proved useful for separation purposes, e.g., zirconium phosphates. The compounds, with variable ZrO_2: P_2O_5 ratios, are prepared by precipitating zirconyl ions with alkali phosphates or phosphoric acid. Other compounds of similar type can be prepared by using arsenic, molybdic and tungstic acids instead of phosphoric acid and titanium, tin and thorium instead of zirconium.

These nonstoichiometric species, like natural zeolites, have lattices which carry a residual negative electric charge. The charge is balanced by alkali or alkaline earth cations which do not occupy fixed positions, but are free to move in the channels of the lattice framework. These ions act as counter ions and can be replaced by other cations. Such materials are used in ion exchange studies.

The natural polymer most widely used in analytical chemistry is cellulose, generally in the form of filter paper. The cellulose molecule is a linear chain polymer composed of anhydroglucose units linked at the 1 and 4 position through β glycosidic bonds.

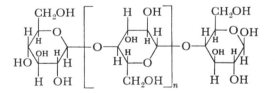

Oxidation of this material gives a product which contains carboxylic acid groups while sulfonation or treatment with phosphoric acid yields a cellulose matrix possessing the active groups $—O—C_2H_4 \cdot SO_3H$ or $—O—PO—(OH)_2$. The introduction of these functional groups confers cation exchange properties on the cellulose. By introducing diethylamine groups ($—O—C_2H_4—N-(C_2H_5)_3Cl$, $—O—C_2H_4—N(C_2H_5)_2$, etc.) into the polymer structure, a material capable of exchanging anions is obtained.

For electrochromatography and several other applications sheets of cellulose acetate are preferred to natural cellulose because there is a smaller degree of swelling and few adsorption effects.

A number of products are based on cross-linked polysaccharides. The material known as Sephadex is a modified dextran made by cross-linking dextran macromolecules into a three-dimensional network by means of a reagent such as epichlorohydrin. Because of the high content of hydroxy groups, Sephadex is strongly hydrophilic and beads of this material swell strongly in water and electrolyte solutions, the degree of swelling varying with the extent of cross-linking. The resultant gels differ in pore size and gels prepared from materials possessing a low degree of cross-linking are used for the fractionation of high molecular weight substances. More compact gels are used in studies of low molecular weight compounds.

If the fractionation process requires the use of nonaqueous solvents, a material capable of swelling in such solvents is needed. In Sephadex LH-20 this desirable property is achieved by alkylating some of the hydroxy groups of the dextran polymer. The alkylated product swells to a gel in the presence of many polar organic solvents, e.g., methanol, ethanol, *n*-butanol, chloroform, tetrahydro-furan, acetone, ethyl acetate, toluene, or aqueous mixtures containing these solvents.

A series of gels which are chemically more inert have been prepared by copolymerizing acrylamide with N,N'-1–methylene-bis-acrylamide. These gels, trade name Bio-Gels, can be used successfully in the pH range of 2–11 and the inert polyacrylamide matrix reduces the adsorption of polar materials to a minimum.

Styragel is a rigid cross-linked polystyrene gel which can be used at tempera-tures up to 150°C with a range of organic solvents such as tetrahydrofuran, benzene, trichlorobenzene, perchloroethylene, cresol, dimethyl sulfoxide, chloroform, carbon tetrachloride, aromatics, etc.

Of major interest to analytical chemists are the synthetic organic polymers used as the matrix material for ion exchange resins.

One type of synthetic resin ion exchanger is prepared through a condensation reaction. For a cation exchanger, the base material is either phenol, resorcinol, hydroquinone, *p*- and *m*-phenol sulfonic acid, α-resorcylic acid or a silicone derivative, and this is condensed with formaldehyde or some other aldehyde in the presence of an acid or base as a catalyst. A typical example is the product formed by condensation of phenol sulfonic acid with formaldehyde. At some stage of the reaction the structure may be represented as:

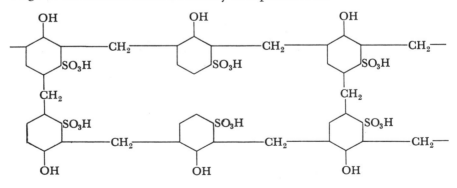

Anion exchangers are made by condensation of different organic bases (aromatic amines, polyamines, carbamide, guanidine, melamine, etc.) with formaldehyde.

The condensed product, after washing and drying, is broken up into grains of desired size.

By proper choice of base materials, the degree of cross-linking of the product can be controlled to a certain extent. However, one disadvantage of the resins is that they contain phenolic groups as well as the functional groups purposely incorporated to give the polymer specific ion exchange properties. For this and other reasons, condensation type resins are being superseded by resins produced by polymerization reactions.

The new resins are produced in the form of beads by polymerizing an aqueous solution of base (e.g., styrene monomer) and a cross-linking reagent (e.g., divinyl benzene) in the presence of a catalyst (e.g., benzoyl peroxide). The degree of agitation of the suspension during polymerization determines the size of the hard, transparent beads of solid material produced in the reaction. Sulfonic acid groups can be introduced by treating the polymer with sulfuric acid or oleum. The structure of sulfonated, cross-linked polystyrene may be represented as

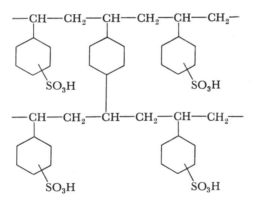

By using other bases (e.g., divinyl benzene, acrylic acid, methacrylic acid, vinyl alcohol, etc.) and other chemical treatments, a range of ion exchange materials can be prepared.

The degree of cross-linking can be controlled by varying the amount of bifunctional monomer (divinyl benzene) present in the original mixture. The composition of the product is fairly uniform and ions diffuse rapidly in the flexible network.

The pore size of any ion exchanger varies over a range of values and one can speak, therefore, only of mean pore size. The range of pore sizes in resins extends from about 4 to 20 Å. This can be compared with the 4–5 Å diameter of the pores in inorganic molecular sieves, the range of 10^4 to 10^5 Å quoted for filter papers, and the pore diameters of 10^3 to 10^5 Å observed in samples of active coal.

C. GEL FILTRATION

It has been noted previously that diffusion within the pores is a significant process in the separation techniques known as gel filtration, ion exchange and chromatography.

The use of molecular sieves in gas-solid chromatography was mentioned in Section II,C and the role of diffusion in other forms of chromatography is discussed in some detail in Chapter 15. Ion exchange is the subject of Section V of this chapter. Hence, the only analytical application of diffusion in polymers which need be discussed in this segment is gel filtration.

Gel filtration fractionates and separates molecules according to size. The technique is simple and rapid and may be applied to both analytical determinations and preparative studies. It is particularly useful for separating substances of biological origin.

In a gel filtration experiment the gel, for example, swollen Sephadex, is formed in a column and the sample is eluted through the gel by a flow of solvent. Molecules larger than the largest pores of the swollen polymer are unable to penetrate the particles and therefore pass through the bed via the solution filling the external voids. These species are the first eluted. Smaller molecules, however, penetrate the gel particles to varying extents depending on their size and shape. These molecules tend to be eluted from the bed in the order of decreasing molecular size.

The choice of the gel polymer depends on the solvent to be used, and the molecular size and chemical properties of the substances to be separated.

Certain molecules deviate strongly from the elution behavior predicted on the basis of molecular size. For example, aromatic substances (homocyclic or heterocyclic) show a greater affinity for Sephadex gels than nonaromatic substances of similar size, and are thus usually retarded to a greater extent. Using chloroform as solvent, solutes containing hydroxyl and carboxyl groups are retarded more than would be expected from their molecular size.

The distribution of a substance between the two phases (i.e., the inner and outer solvent) can be defined by a distribution coefficient, K_d. The magnitude of K_d is a function of the molecular weight of the substance and the type of gel involved. For large molecules that are excluded from the pores $K_d = 0$, and for small molecules or ions where accessibility is complete $K_d = 1$. Between these limits K_d varies in proportion to the molecular weight.

The elution volume (V_e) of a substance depends on the volume external to the gel particles (V_o) and on $K_d V_i$ where V_i is the volume of solvent inside the gel grains.

$$V_e = V_o + K_d V_i$$

For a series of chemically similar compounds the elution volume varies more or less linearly with the logarithm of the molecular weight.

If two substances, on a certain type of gel, have distribution coefficients K_d' and K_d'', the elution volumes differ by the separation volume (V_s) which is defined by the equation

$$V_s = (K_d' - K_d'')V_i$$

For complete separation of the two materials the sample volume must be less than V_s.

The concentration of the test solution does not play an important part in gel filtration except where it increases the viscosity of the solution. At high viscosities diffusion is so reduced that exchange between the mobile and stationary phases is prevented. Since the distribution process occurs at a finite rate, sharper elution peaks are obtained using small gel particles and slow flow rates.

Gel filtration is widely used in clinical analysis and as a separation procedure in biochemical studies.

For desalting operations gel filtration has many advantages over dialysis. Other low molecular weight substances, which are not salts, can also be separated easily from macromolecules. In this way nucleic acids have been separated from nucleotides and phenols; polysaccharides separated from sugars; and proteins isolated from amino acids and other low molecular weight substances.

The selective properties of the various gel filtration materials have led to their introduction in other separation techniques. Thus Sephadex gels are used in adsorption, thin-layer and partition chromatography procedures and as a support in zone electrophoresis studies. Sephadex ion exchangers, which are the diethyl-, aminoethyl-, carboxymethyl-, or sulfoethyl- derivatives of the dextran polymer, are claimed to combine the advantages of both resin-based and cellulose-based exchangers.

V.   ION EXCHANGE

A.   EXCHANGE MATERIALS

An ion exchange material is an insoluble substance containing labile ions which exchanges these ions with other ions in a surrounding solution without any apparent physical change taking place in the material.

Materials in which the labile ions are cations are known as cation exchange resins, e.g.,

$$2\,Ex{\cdot}Na^+ + Ca^{2+} \rightleftarrows Ex_2{\cdot}Ca^{2+} + 2Na^+$$

while materials which exchange anions are known as anion exchangers, e.g.,

$$2\,Ex{\cdot}Cl^- + SO_4^{2-} \rightleftarrows Ex_2{\cdot}SO_4^{2-} + 2Cl^-$$

In these equations, *Ex* represents the solid insoluble matrix to which are attached the monovalent functional groups responsible for the exchange.

A wide variety of materials exhibit ion exchange properties. The phenomena were observed in soils and clays late in the nineteenth century and zeolites were used for water purification by early in the twentieth century. Zeolites are unstable in the presence of acids and alkalis; hence many attempts were made to find substitutes, e.g., by chemical treatment of other natural materials. Sulfonated coals (which contain sulfonic and carboxylic acid functional groups) are still in limited use today.

Interest in ion exchange heightened after the development of the synthetic resin exchangers (discussed in the preceding section), and the introduction of exchange materials based on cellulose and dextran. The latter have proved very popular for separating large molecules such as proteins, nucleic acids, hormones, viruses, and germicides.

For inorganic analysis, an important new class of exchanger is that based on the oxides of the elements of the fourth, fifth, and sixth groups of the periodic table, e.g., zirconium phosphates, tungstates and molybdates. The latter compounds are cation exchangers whose capacity depends strongly on the pH of the solution. The hydrous oxides of zirconium, thorium, and tin (IV) change behavior with pH, being anion exchangers in acid media and cation exchangers in alkaline conditions. This class of exchanger is remarkably selective, particularly in respect to the alkali and alkaline earth metal series. Because of the density of the material, the capacity (i.e., exchangeable ions per unit volume) is high and exchange is rapid.

The only requirements for an ion exchanger are that it be of high molecular weight, insoluble in water, and capable of exchanging mobile ions with an aqueous phase. It can therefore be a liquid.

Typical liquid cation exchangers are the bivalent alkyl phosphoric acids and phosphonic acids $[PO(OH)_2]$ (e.g., mono-dodecylphosphoric acid or mono-*n*-butyl phosphoric acid); univalent dialkylphosphoric and phosphinic acids [POOH] (e.g., diphenylphosphoric acid, dicyclohexylphosphoric acid), and alkyl sulfonic acids (e.g., dinonylnaphthalene sulfonic acid) or carboxylic acids (e.g., salicylic acid, perfluorobutyric acid).

Liquid anion exchangers are usually primary (I), secondary (II) or tertiary (III) amines or are quaternary ammonium ions (IV). Typical examples are 1-(3-ethylpentyl)-4-ethyloctylamine (I), trialkylmethylamine (I), *N*-dodecenyl (trialkylmethyl)amine (II), dilaurylamine (II), *N*-benzyl-1-(3-ethylpentyl)-4-ethyloctylamine (II), methyloctylamine (III), tribenzylamine (III), didodecenyl-*n*-butylamine (III), dodecenyl (trialkyl) ammonium (IV), tetrapropylammonium (IV), and tetra-*n*-heptylammonium (IV).

For analytical purposes these liquid exchangers are diluted with water-immiscible solvents such as benzene, toluene, kerosine, carbon tetrachloride, petroleum ether, octane or cyclohexane, and they are normally used as 5–10%

solutions. The general behavior of these materials is somewhat similar to that observed with solid counterparts which contain the same functional groups.

Ion exchange materials are commonly divided into four main classes: strongly acidic or weakly acidic cation exchangers, strongly basic or weakly basic anion exchangers. In addition there are some special types known as chelating resins, and redox exchangers.

A strongly acidic cation exchanger resembles a strong acid, i.e., it is completely ionized over a wide pH range. Such a material exchanges ions rapidly and the capacity is not a function of pH. The functional group usually associated with this type of exchanger is the sulfonic group, $-SO_3H$.

$$2\,Ex \cdot SO_3^-H^+ + CaCl_2 \rightleftarrows (Ex \cdot SO_3^-)_2 Ca^{2+} + 2HCl$$

A weakly acid exchanger contains substituents such as carboxylic acid groups, $-COOH$, or phenolic hydroxyl groups, $-OH$. The materials are ionized completely only under alkaline conditions and are thus used with solutions of pH >7. Since solutions containing the salts of weak acids have a pH >7, weak acid exchangers are able to split such salts. The exchange rate and capacity both increase with increasing pH.

$$2\,Ex \cdot COOH + CaCl_2 \rightarrow \text{no reaction}$$
$$2\,Ex \cdot COOH + Ca(OAc)_2 \rightarrow (Ex \cdot COO^-)_2 \cdot Ca^{2+} + 2HOAc$$

The introduction of phosphonic groups, $-PO(OH)_2$, into an exchange medium yields a product whose behavior is intermediate between the two extremes just discussed.

Cation exchangers can contain two types of ionic groups (i.e., bifunctional) and exchangers with a specific preference for particular metal ions have been prepared by incorporating groups which form strong chelates with the selected cation (e.g., iminodiacetic acid groups).

Information on the type of active group present in an exchanger is usually obtained from potentiometric titration of the material in the presence of a displacing salt, e.g., NaCl.

The presence of primary, secondary and tertiary amino groups usually confers weak-base properties on an anion exchange material. Substitution of quaternary ammonium groups or quaternary phosphonium or tertiary sulfonium groups yields exchangers with strong base properties.

Strongly basic anion exchange resins can be used over the whole pH range without change of capacity. Exchange is rapid and gives rise to stable salts which require an excess of strong base for conversion to the hydroxide form.

$$Ex \cdot NH_3^+OH^- + HCl \rightleftarrows Ex \cdot NH_3^+Cl^- + H_2O$$
$$Ex \cdot NH_3^+OH^- + NaCl \rightleftarrows Ex \cdot NH_3^+Cl^- + NaOH$$
$$2Ex \cdot NH_3^+OH^- + CO_2 \rightleftarrows (Ex \cdot NH_3^+)_2 CO_3^= + H_2O$$

Weakly basic resins are ionized only under acidic conditions, with both capacity and exchange rates decreasing with increasing pH. The salt forms are not stable (e.g., they tend to hydrolyze on washing), and the exchanger can be readily converted to free-base form by treatment with mild alkalis.

$$Ex{\cdot}NH_2 + HCl \rightleftarrows Ex{\cdot}NH_2{\cdot}HCl \rightleftarrows Ex{\cdot}NH_3{}^+Cl^-$$

$$Ex{\cdot}NH_2 + NaCl \rightleftarrows \text{no reaction}$$

$$Ex{\cdot}NH_2 + CO_2 \rightleftarrows \text{no reaction}$$

Weakly basic materials do react, however, with the salts of very weak bases, e.g., aniline hydrochloride:

$$Ex{\cdot}NH_2 + C_6H_5NH_2{\cdot}HCl \rightleftarrows Ex{\cdot}NH_3{}^+Cl^- + C_6H_5NH_2$$

Amphoteric ion exchangers contain both acidic and basic groups.

Ion exchange materials are usually used in the form of small beads or granules, but ion exchange membranes have been prepared by several methods.

"Homogeneous" membranes are formed by reinforcing coherent gels with wide mesh plastic tissues. "Heterogeneous" membranes are made by incorporating colloidal ion exchanger particles into an inert polymer binder or matrix.

Redox ion exchangers are conventional ion exchanger materials into which reversible oxidation-reduction couples such as Cu^{2+}/Cu, Fe^{3+}/Fe^{2+}, methylene blue/leukomethylene blue, etc. have been introduced. The redox potential of the couple is not significantly changed by incorporation into the resin and the exchangers are used for oxidizing or reducing substrates in solutions. Cross-linked polymers which carry built-in reversible redox couples (e.g., quinone/hydroquinone) are known as electron exchangers. Electron exchangers lack the functional groups of redox ion exchangers but behave in a similar manner and both are characterized by their redox capacity, their standard reduction potential, and their reaction rate. The most serious limitations of these materials in practical applications are slow reaction rates and a lack of chemical stability.

The capacity of an ion exchanger is normally defined in terms of the number of ionogenic groups in the material, and is usually quoted in milliequivalents per gram of dry H^+ form (cation exchangers) or Cl^- form (anion exchangers). However, a majority of ion exchange studies utilize resin-based materials which are elastic gels that swell considerably in water and other polar solvents. The capacity of an exchanger is thus often quoted in terms of the number of ionogenic groups per milliliter of swollen resin bed.

Most common ion exchangers have weight capacities of between two and ten milliequivalents per gram; hence the interior of a swollen resin bed ordinarily resembles a drop of electrolyte solution whose concentration is of the order of 2 to 10 molal.

B. The Ion Exchange Process

An exchange reaction involving ions of the same charge may be represented by an equation of the type:

$$Ex{\cdot}A + B \rightleftarrows Ex{\cdot}B + A$$

where A and B represent ions in solution and $Ex{\cdot}A$ and $Ex{\cdot}B$ represent the corresponding resin forms.

Ion exchange is usually reversible and stoichiometric, but both the position of equilibrium and the rate of reaction depend on many factors.

Since the interior of an exchange resin is like a drop of concentrated electrolyte solution, the movement of ions in and out of the bead involves a diffusion process, modified by electrical potential gradients.

The ion exchange process can be divided into several steps.

1. Diffusion of the ion in solution across the liquid film or Nernst layer immediately adjacent to, and surrounding, the exchanger particle.
2. Diffusion of the ion within the pores of the bead or particle.
3. Exchange reaction at the functional site.
4. Exit of the exchanged ion by diffusion through the pores and across the Nernst layer.

The actual exchange reaction is ionic and fast; hence the rate of the overall process is controlled by one of the diffusion steps.

By Fick's law the rate of diffusion is proportional to the existing concentration gradient. The proportionality constant is known as the diffusion coefficient, and it has been observed that the diffusion coefficients of counter-ions are many times smaller in the resin phase than in ordinary aqueous solutions. The higher the charge, the more slowly the ions diffuse. Large ions whose size is commensurate with that of the resin "pores" exchange very slowly and incompletely.

With strongly acidic or basic exchangers, the rate controlling step in the exchange of simple inorganic ions using dilute solutions is film diffusion (i.e., step 1). The rate of exchange can be increased by increasing the concentration of the ion B in solution, by agitation (reduces film thickness) and by increasing the temperature.

With high solute concentrations or with very large ions, diffusion within the particles tends to become rate controlling and in this case the rate can be increased by decreasing the bead size, raising the temperature or by using a material with a lower degree of cross-linking.

The position of the exchange equilibrium depends on the nature of the exchanger and the ions being exchanged, and to a first approximation, can be described in terms of the Law of Mass Action.

Consider the general case where ions A (valence x) are exchanged with ions B (valence y).

$$y\,Ex{\cdot}A + x\,B^{y+} \rightleftarrows x\,Ex{\cdot}B + y A^{x+}$$

Applying the mass action law

$$K = (a_{r_B})^x \cdot (a_A)^y / (a_{r_A})^y \cdot (a_B)^x$$

where a_A and a_B are the activities of the ions A^{x+} and B^{y+} in solution, and a_{r_A} and a_{r_B} are the mean activities of the species A and B in the solid phase; K is the thermodynamic equilibrium constant.

The activity terms can be replaced by analytical concentrations and activity coefficients. If (A) and (B) are taken to represent the equilibrium concentrations in the exchanger phase and $[A]$ and $[B]$ represent solution concentrations, then

$$K = \frac{(B)^x \cdot \gamma_B{}^x \cdot [A]^y \cdot f_A{}^y}{(A)^y \cdot \gamma_A{}^y \cdot [B]^x \cdot f_B{}^x}$$

γ_A and γ_B are the activity coefficients of the ions in the exchanger phase, f_A and f_B are the activity coefficients of the ions in solution.

The activity of ions in the exchanger is not easy to define since the swelling of the polymer matrix can lead to a swelling pressure within the pores of the order of several hundred atmospheres and electrolyte concentrations within these pores are greater than one molar.

Accordingly, it is easier to express experimental data in terms of an apparent equilibrium constant K' where

$$K' = K \cdot \gamma_A{}^y f_B{}^x / \gamma_B{}^x \cdot f_A{}^y = [A]^y (B)^x / [B]^x (A)^y$$

The magnitude of the apparent equilibrium constant depends on the properties of the ion exchanger, on the chemical composition of the electrolytes, on the concentration of the solution, and on the nature of the ions involved in the exchange reaction.

Another term used to describe the equilibrium state is the selectivity coefficient K_d, which for the reaction under discussion is defined by:

$$K_d = C_A \cdot g_B / C_B \cdot g_A$$

g_A and g_B are the concentrations of ions A and B in the exchanger and C_A and C_B are the concentrations in the solution (expressed in mg equiv/ml).

In the case of the exchange of univalent ions, $K_d = K'$. In the case of multivalent ions, if $y > x$

$$K_d{}^x = K' g_A^{(y-x)} / C_A^{(y-x)}$$

It can be seen from this equation that the magnitude of the selectivity coefficient can be influenced by the concentration and charge of the ion initially on the exchanger.

On the basis of selectivity coefficients, the relative affinities of ions for an ion exchanger can be quantitatively evaluated.

Besides exchanging ions, ion exchange materials can take up nonelectrolytes in much the same way as nonionic sorbents. The isotherms for nonelectrolyte sorption are of the same general type as the Langmuir and Freundlich relationships, i.e., the distribution coefficient gradually decreases with increasing concentration of the external solution.

The early workers who studied natural exchange materials and polyfunctional materials often described ion exchange equilibrium in terms of the adsorption isotherms and empirical relationships of Langmuir, Freundlich, etc.; but a characteristic of the adsorption of electrolytes is an increase in the distribution coefficient with increasing solution concentration.

The tendency of electrolytes to be sorbed into an exchanger material is opposed by a potential difference at the interface known as the Donnan Potential.

Electrolyte-sorption equilibria have a formal resemblance to equilibria in systems such as

$$A^+, Y^-, R^-, H_2O \quad \begin{vmatrix} \text{membrane} \\ \text{impermeable for} \\ R^- \end{vmatrix} \quad A^+, Y^-, H_2O$$

in which a membrane permeable to all species except R^- is situated between two solutions containing the ions A^+ and Y^-.

In a cation exchange resin, the nondiffusible ion R^- is represented by the charged functional groups and the phase interface can be regarded as the semipermeable membrane.

When an exchanger charged in the A^+ form is placed in a dilute solution of the electrolyte AY, cations tend to diffuse from the exchanger into the solution and anions tend to diffuse into the exchanger. This migration is induced by the difference in concentrations in the two phases. The movement of the ions results in an accumulation of positive charge in the solution and of negative charge in the ion exchanger. In this way an electric potential difference (the Donnan potential) builds up at the interface. At equilibrium the tendency of ions to migrate under the influence of the concentration gradient is balanced by the action of the electric field, and the chemical potential in the two phases becomes equal, i.e.,

$$\mu_A^\circ + RT \ln a_{rA} + \mu_Y^\circ + RT \ln a_{rY} = \mu_A^\circ + RT \ln a_A + \mu_Y^\circ + RT \ln a_Y$$

In this equation, μ° is the standard chemical potential; a_A and a_Y are the activities of the mobile ions in solution; and a_{rA} and a_{rY} are the activities of the ions in the resin phase.

Simplification of the equation gives the relationship:

$$(a_{rA})(a_{rY}) = (a_A)(a_Y)$$

For electrical neutrality $C_A = C_Y$.

For the purpose of illustrating the significance of these equations let us assume that the activity coefficients cancel (i.e., $g_A \cdot g_Y = C_A \cdot C_Y$) and let us consider the situations where the electrolyte concentration in the external solutions is either $0.1 M$ or $0.001 M$ (i.e., $C_A = C_Y = 10^{-1}$ or 10^{-3}). The capacity of cation exchangers is usually about two milliequivalents per milliliter of swollen resin; hence $g_A \simeq 2$ mEq ml^{-1}.

In the two hypothetical situations $g_A \cdot g_Y$ equals 10^{-2} or 10^{-6} respectively, and substituting for g_A, g_Y equals 5×10^{-3} or 5×10^{-7} mEq ml^{-1} respectively. Expressed in words, this result indicates that the uptake of electrolyte (expressed as g_Y) decreases with decreasing concentration in the external solution and the amount of anion taken into the exchanger is extremely small when the capacity of the resin is high.

Conversely, the uptake of electrolyte is relatively high when the capacity of the exchanger is low. For natural cellulose, the capacity, $C_r \simeq 0.1$ mEq ml^{-1} and in the above situations g_Y would equal 10^{-1} or 10^{-5} respectively. Thus in $0.1\ M$ electrolyte solution, the concentration of the salt AY in the polymer phase would be similar to the concentration in the external solution.

The degree of absorption also depends on the valence of the ions. The Donnan equilibrium relationship for the salt BX_2 may be simplified to

$$(a_{rB})(a_{rX})^2 = (a_B)(a_X)^2$$

Should the external solution contain another salt, AX, containing the same anion

$$(A_{rA})(a_{rX}) = (a_A)(a_X)$$

By squaring both sides of this equation and dividing by the equation for BX_2, one obtains the relationship,

$$(a_{rA})^2/(a_{rB}) = (a_A)^2/a_B.$$

If, perchance, $a_A = a_B = 0.1$, then $(a_{rA})^2/(a_{rB}) = 0.1$. Now if the solution activities are reduced to 0.001 (e.g., by dilution of the external phase with water) a new equilibrium results and $(a_{rA})^2/(a_{rB})$ becomes equal to 0.001. The large change in magnitude of this ratio suggests that dilution favors the uptake by the exchanger of ions of higher valency. This generalization can be confirmed by taking the mathematics one stage further. Thus, if in this hypothetical situation $C_r = 1$ mEq ml^{-1}, $a_{rA} + a_{rB} \simeq 1$, and when $a_A = a_B = 0.1$, the values calculated for a_{rA} and a_{rB} become 0.27 and 0.73 respectively. Repeat calculations with $a_A = a_B = 0.001$ give values of 0.03 and 0.97 for a_{rA} and a_{rB}.

A similar type of conclusion can be derived from the Law of Mass Action equation. It is an important result, because of the practical implications. From dilute aqueous solutions, multivalent cations are preferentially taken up by the exchanger, hence may be removed from test solutions. The process can then be reversed by treating the exchanger with a concentrated solution of monovalent counter ion. This regenerates the resin and releases the multivalent cations into solution for further study if required.

While the above argument has been based on cation exchangers, an analogous relationship can be evolved for anion exchange materials.

C. SELECTIVITY

It is difficult to define the order of relative affinity for a series of ions, because the order can vary with the type of exchanger involved. As the degree of cross-linking is increased, resins become more selective and different functional groups show different affinities for a given ion.

In the absence of rigorous quantitative relationships, relative exchange affinities are best predicted from empirical rules. For cation exchanges at room temperature from dilute aqueous solutions, the following conditions appear to apply:

1. *Ionic Size*

For ions of the same valence, the absorption strength increases with decreasing diameter of the hydrated ion. $Li^+ < H^+ < Na^+ < K^+ = NH_4^+ < Rb^+ < Cs^+ < Ag^+ < Tl^+$; $Hg^{2+} < Cd^{2+} < Mn^{2+} < Mg^{2+} < Zn^{2+} < Cu^{2+} < Ni^{2+} < Co^{2+} < Ca^{2+} < Sr^{2+} < Pb^{2+} < Ba^{2+}$; $Al^{3+} < Sc^{3+} < Y^{3+} < Eu^{3+} < Sm^{3+} < Nd^{3+} < Pr^{3+} < Ce^{3+} < La^{3+}$.

2. *Cationic Charge*

The affinity generally increases with increasing charge, but, as explained in the preceding section, in the exchange of ions of different valency, the relative affinity for the higher valency ion decreases as the concentration increases.

3. *Dissociation*

Incomplete dissociation of a salt in solution decreases its apparent affinity.

Incomplete dissociation of the functional groups can result in reduced capacity and varied selectivity effects. For example, weakly acidic ion exchangers absorb hydrogen ions strongly and a slight excess of acid in solution tends to displace all other cations from the matrix. Such materials are particularly selective for multivalent metal ions, although they also strongly absorb those bivalent ions which form sparingly soluble carbonate salts.

Weakly acidic exchangers in the hydrogen form do not take up any appreciable quantity of cation from a salt solution unless the cation is associated with the anion of a weak acid.

Organic ions are bound by van der Waals forces as well as by Coulomb forces and their affinities in consequence increase with increasing ionic size. Even when the ionic sieve effect is operating, the small amount of organic ion that is taken up by the resin is bound very tightly.

An order of affinity for anions can also be prepared for strongly basic anion exchange resins. The valence of the ion and the volume of the hydrated ion are again important factors in relation to the order. The position of the hydroxyl ion depends on the base strength of the anion exchanger but the apparent order for a number of resin exchangers is: $OH^- < F^- <$ propionate $<$ acetate $<$ formate $< IO_3^- < HCO_3^- < Cl^- < NO_2^- < BrO_3^- = HSO_{3_9}^- < CN^- < Br^- < NO_3^- < ClO_3^- < HSO_4^- <$ phenolate $< I^-$.

The order of affinity of ions on medium strong and weakly basic materials do not appear to follow any well–defined order because the order is determined by four factors: diameter of the hydrated ion, valence, strength of the acid corresponding to the anion, and the structure of the ion. Weakly basic resins in the free base form do not take up anions to any appreciable extent except from free acids.

The order of affinity for a weakly basic anion exchanger is approximately: $HCO_3^- <$ acetate $< F^- < Cl^- < SCN^- < Br^- < I^- < NO_3^- < H_2AsO_3^- < H_2PO_4^- <$ tartrate $<$ citrate $< HCrO_4^- < HSO_4^- < OH^-$.

Small variations in affinity can be enhanced by adding a complexing agent to the aqueous phase. Combination of the equilibrium equations for the exchange reaction and complex formation permits prediction of the degree of enhancement.

D.   Analytical Applications of Ion Exchange

Ion exchange can be achieved by adding solid exchanger to volumes of the test solution, but as ion exchange is an equilibrium process a large excess of exchanger is required to ensure complete replacement of any particular species. For this reason, columns of material are usually used.

In a single batch operation (equivalent to one segment of the column) not all of the ions in the solution are exchanged. A repeat of the process, using fresh exchanger and the residual solution, removes a further fraction of the remaining ions from the aqueous phase. Slow passage of a solution through a column is equivalent to multiple batch treatments and quantitative exchange is eventually achieved.

Ion exchange columns can be of any size, but for analytical work the columns are generally 8–15 mm in diameter and are packed with swollen resin to give a column length which is 10–20 times the diameter. The exchanger column is supported on a coarse sintered glass disc. In operating the exchanger column, the bed must be covered with liquid at all times and the packed material must be free of entrapped air.

The exchanger is charged in the desired form by running several bed volumes of concentrated electrolyte slowly through the column. For example, a cation exchanger may be converted to the hydrogen form by initially removing water in the void space by a rapid flow of 3–4 M hydrochloric acid. The flow through the column is then reduced to 0.5 to 2 ml per square centimeter of bed per minute. (This is the recommended flow rate for most exchange operations.)

After rinsing with distilled water the column is ready for use. The same procedure is required to regenerate the exchanger materials and release entrapped ions.

The exchange of ions on a column is used in analytical procedures for concentrating dilute solutions, determining the total salt content of aqueous solutions, and for removing interfering ions.

Ion exchange columns are able to filter out or retain strongly absorbed ions from large volumes of solutions containing trace amounts of these ions. For example, the mineral content of the water in a river or stream may be determined by passing many gallons of the water through a cation exchange column. The cations collected are then displaced from the column by the addition of a small volume of concentrated acid. The concentration of displaced metal ions in the acid solution can be great enough to permit analysis by standard techniques.

If the affinity of a given species for the exchanger is much greater than the affinity of the other ions present, retention and concentration may be achieved in the presence of significant amounts of the other ions.

Increased selectivity in the exchange process can be brought about in several ways. For example, one can exploit the preference of an exchanger for highly charged ions in dilute solutions or one can choose a chelating resin. The most effective way, however, is to make use of complex formation. Thus species which form stable anionic complexes can be separated from a wide range of other cations through absorption on an anion exchanger.

Well-known applications of the concentration procedure include the determination of traces of lead, copper, and iron in wine; sodium, potassium, calcium, magnesium, iron, aluminum, zinc, gold, etc. in natural waters; calcium, magnesium, and manganese in salt brines; calcium ions in urine and other biological liquids; strontium, calcium and copper ions in milk; etc.

Expressed in general terms, ion exchange is a highly convenient means of separating electrolytes from nonelectrolytes.

The total electrolyte concentration of a dilute solution can be determined by causing all the cations or anions of the sample solution to be quantitatively exchanged for another ion, (e.g., hydrogen or hydroxyl ions) which can be easily determined. The sample is allowed to flow through a column of exchanger at flow rates of up to 5 ml/cm^2/min, after which the column is rinsed with 4 or 5 bed volumes of distilled water. The effluent solution from a cation exchanger originally in the hydrogen form contains all of the anions of the original solution in the form of their free acids. Titration of the liberated acid provides a simple means of determining the total salt content, since the number of protons displaced equals the number of gram equivalents of the cations present in the original mixture.

Salts of acids which readily decompose (e.g., $NaHCO_3$, Na_2SO_3, and $NaNO_2$) are best studied by using an anion exchanger, followed by titration of the liberated base. The application of anion exchangers to this purpose is limited by the fact that they can only be used if the liberated base is soluble, as in the case of alkali salts.

By a slight modification of the procedure, an accurately known amount of acid (using a cation exchanger in the H^+ form) or base (strong base anion exchanger in the OH^- form) can be produced from a pure salt (such as sodium chloride) and used for standardizing bases or acids in volumetric analysis.

Demineralized water, suitable for use in chemical analysis, is prepared by passing natural water through a bed containing a cation exchanger in the hydrogen form and an anion exchanger in the hydroxyl form. The protons liberated by the cations are neutralized by the hydroxyl ions liberated by the equivalent concentration of anions in the natural water.

There are many experimental situations where the accurate analytical determination of a species is hindered by the presence of certain other ions of opposite charge. For example, sulfate ion concentrations are regularly determined by weighing the amount of precipitate formed on the addition of barium ions, but this method is unsatisfactory in the presence of iron (III), chromium (III), and aluminum ions because these cations tend to coprecipitate.

This, and many other interference problems, can be solved by first passing the sample through a cation exchanger charged in the hydrogen form.

The removal of interfering cations by a cation exchanger facilitates the titrimetric and photometric determination of many anionic species. For example, the determination of the phosphate content of rocks, superphosphate, steel, etc. is achieved by passing the phosphate solutions through a cation exchanger (H^+ form) after some prereduction step to reduce the valency of iron, vanadium, etc. (this prevents retention of phosphate as a metal complex). The eluted phosphoric acid is then titrated with a base or with a specific titrant for phosphate ions; small amounts are determined colorimetrically.

An anion exchanger charged in an appropriate form provides a convenient means of replacing anionic species which interfere in analytical procedures. For example, the selective precipitation of elements of the hydrous oxide group (Fe, Al, Cr, Mn, Zr, etc.) is hindered by the presence of phosphate ions because many of these metal ions form insoluble phosphate compounds. This interference problem can be eliminated by passing the test sample through an anion exchanger charged in the chloride form. The alternative solution involves trapping the cations of interest on a cation exchanger, followed by elution with an acid whose anions do not interfere in any subsequent analytical operations.

If the eluting acid forms complexes of different stabilities with the metal ions on the exchanger, individual elements may be selectively displaced and the isolated components used for quantitative analysis.

The separation of a mixture of ionic species on an ion exchange column is known as ion exchange chromatography. This technique closely resembles adsorption chromatography.

As discussed in a previous section, ions differ in their affinity for the exchanger and in some cases this difference can be exploited. Thus elution with a dilute acid may bring about the separation of ions of lower valency from ions of higher valency. With ions of similar charge, ionic size can determine the order of elution. Sharper elution peaks can often be achieved by elution with a solution of continually increasing ionic strength.

The addition of complexing agents to the mobile phase often shifts the ion exchange equilibrium in a direction which favors separation. For example, ions of marked similarity such as the rare earths have been separated using citrate buffers as the displacing medium.

The effective stability of most complexes is a function of the concentration of ligand and the pH of the solution.

A very effective tool for separating metal ions is the anion exchange of chloride complexes. Most metal ions are taken up by anion exchangers from hydrochloric acid solutions, the strength of acid required varying with the metal ion involved (cf. Table 10.84 of the "Handbook of Analytical Chemistry"). As a general rule, the adsorption of metal ion increases to a maximum and then decreases as the concentration of acid is increased. At the higher concentrations of acid the additional chloride ions compete successfully with the stabilized metal complex anion for exchange sites. The only metals that are not appreciably absorbed in one or other of their common oxidation states are the alkali metals, alkaline earths, rare earths, and nickel.

The distribution coefficients of the other metal ions vary greatly and many separations can be performed efficiently. For example, if a mixture of iron (III), cobalt, nickel, and zinc ions in 9 M HCl is passed slowly through a column of strong-base type anion exchanger, all four metal species are initially retained. Washing the column with approximately 1.5 bed volumes of 9 M HCl usually elutes all the nickel in the original sample. The cobalt can be subsequently removed by passing 1.5 bed volumes of 4 M HCl while the iron (III) is retained until the column is treated with about two bed volumes of 0.5 M HCl. To displace the stable zinc chloride complex, the column has to be eluted with a couple of bed volumes of 3 M HNO$_3$.

Separations based on differences in the stabilities of fluoride and sulfate complexes have also proved successful.

Cations retained on a strong-acid type cation exchanger can be selectively displaced by adding to the eluting solution a ligand which forms negatively charged complexes. The most quoted example in this field is the separation of the rare earth metals using citrate solutions for elution. The smallest rare earth cation Lu^{3+} forms the most stable complex and is eluted first; the largest cation La^{3+} is eluted last. Gradient elution effects are achieved by systematically increasing the pH of the solution, since this increases the proportion of the complexing species in the desired ligand form.

Other complexing agents which have been used include EDTA, ammonia, lactate ions, cyanide ions, oxalate ions, and phosphate ions.

A summary of the eluants and conditions which have proved suitable for the separation of inorganic species is provided in Table 10.85 of the "Handbook of Analytical Chemistry," and more extensive compilations appear in ion exchange monographs.

Ion exchange separations are also used very widely in organic chemistry and biochemistry. The applications are varied, but of major importance are the studies which involve amino acids, nucleotides, antibiotics, vitamins, hormones, organic acids, and bases in general.

The size of organic ions can differ by factors of 100 or 1000; hence the size of the pores in the ion exchange material becomes very important. This factor is determined by the degree of cross-linking and the degree of swelling, and the latter can be small if solubility considerations dictate that an organic solvent be substituted for water. In addition, organic compounds tend to molecular adsorption, the extent depending on structure. For these reasons the order of elution is not always predictable.

Nonionic substances can be separated solely on the basis of selective molecular adsorption. Alternatively, the separation can be achieved by making a derivative which is ionic in character. Derivatives can be formed by reaction with anions (e.g., aldehydes or ketones plus bisulfite ions; sugars or polyhydroxy compounds plus borate ions; etc.) or cations (e.g., formation of the amine complexes of Ag^+, Ni^{2+}, and Cu^{2+}).

The ion exchange procedures used in organic analysis are somewhat similar to those applied to inorganic analysis. Besides separating mixtures by ion exchange chromatography, there are procedures for removing interfering ions, methods for determining the total concentration of ionic forms in a mixture, and applications based on the concentration of trace amounts of material.

For example, organic acids (citric, malic, succinic, ascorbic, galacturonic, etc.) can be isolated from plant liquors by first passing the sample solution through a cation exchanger in the H^+ form to obtain a solution of the free acids. This solution is then passed through an anion exchange column. The degree of retention varies with the acid strength and the degree of dissociation, and chromatographic separation of the components is achieved through elution with water, dilute formic acid, acetic acid, hydrochloric acid, or a buffer solution.

Amino acids contain basic and acidic functional groups, i.e., can exist as anions or cations, depending on the pH. By eluting a cation exchange column (Na^+ form) with citrate and phosphate buffers whose pH increases stepwise from 3.4 to 11, and by progressively raising the temperature from 30 to 75°, very complex mixtures of amino acids can be separated.

Other methods of separation are based on ion exclusion and selective adsorption. The ion exclusion method allows electrolytes and nonelectrolytes soluble in water to be separated from each other (e.g., glycerin and sodium chloride); the principle of this technique has been outlined in the section on gel filtration (Section IV,C).

The degree of adsorption of nonionic (or weakly dissociating) water-soluble organic substances can be affected by the salt concentration of the solution in contact with the resin.   In salting out chromatography, organic compounds of polar character (e.g., aldehydes, ketones, amines, and carboxylic acids) are placed on top of a column of open-structured exchanger material. The column is then eluted with a succession of electrolyte (e.g., ammonium sulfate) solutions ranging in concentration from saturated to more dilute. The ammonium sulfate salts out the various compounds, i.e., they go from the aqueous solution into the resin. The more hydrophilic the solute, the greater the ammonium sulfate concentration needed to keep it on the resin. Thus the constituents of a mixture are eluted roughly in the order of their molecular weight, those with highest molecular weight coming out last.

The degree of adsorption of nonionic organic substances which are only slightly soluble in water can also be affected by the composition of the solvent. In solubilization chromatography the solutes are eluted from the column first with water, and then with aqueous solutions of organic solvents (e.g. acetic acid, methanol) of progressively increasing concentration. Again, the compounds of highest molecular weight come out last.

Other laboratory applications of ion exchange materials include the determination of the charge and stability of metal complexes, and the purification of solvents.

Liquid ion exchangers are used primarily in separation procedures which resemble the technique of solvent extraction and are discussed in Chapter 15.

For the separation of biological compounds, ion exchange materials based on polymeric gels have proved extremely useful and many inorganic separations have been conveniently achieved using sheets of filter paper treated to increase the ion exchange capacity of the cellulose. Ion exchange columns handle macro amounts of material (e.g., gram quantities) while thin film approaches (including ion exchange papers) are ideally suited for microscale studies.

VI.   LITERATURE EXERCISES

Question 1
   Write brief notes on each of the following topics:

   a. Difference between physical adsorption and chemisorption.
   b. Inorganic ion exchange materials.
   c. Importance of separation procedures in chemical analysis.
   d. Measurement and significance of R_f values.
   e. Gradient elution.
   f. Free electrophoresis.
   g. Liquid ion exchangers.
   h. Salting out chromatography.

   i. Conway microdiffusion technique.

   j. Water softening and demineralization by ion exchange.

Question 2

a. Discuss the role of polarity in adsorption column chromatography.

b. Outline the principles of displacement, elution, and frontal analysis development procedures.

c. Prepare a summary of two chemical separations (one inorganic, one organic) satisfactorily achieved by adsorption chromatography.

Question 3

Prepare a seminar on the principles, practice, advantages, and applications of thin-layer chromatography.

Question 4

a. Sketch the components of a gas chromatograph.

b. Explain why the determination of elution patterns using a gaseous mobile phase is experimentally easier than the procedure required with liquid eluants.

c. Indicate why the elution peaks in gas–solid chromatography are usually distorted in shape.

d. Discuss the types of applications in which GSC is particularly useful.

Question 5

"The migration of ions under the influence of an applied electric field is an important component of the techniques known as electrography and electrochromatography."

Expand this statement by comparing and contrasting the principles of the two techniques mentioned.

Question 6

In gel filtration and ion exchange separations, an important consideration is pore size.

a. Explain how the pore size in synthetic polymers is influenced by the degree of cross-linking and the degree of swelling.

b. Outline the role of diffusion in these two techniques.

c. Indicate the significance of molecular adsorption in nonionic retention processes.

d. Explain why it is difficult to assess the activity of an ionic species held in the pores of an ion exchanger.

e. Suggest reasons for the observation that species are eluted in the order of decreasing molecular size.

f. What do you understand by the term "solubilization chromatography"?

Question 7

It has been stated that the three outstanding procedures for separating compounds of biological interest are zone electrophoresis, gel filtration, and ion exchange on polymers with a low degree of cross-linking.

Confirm the validity of this contention by outlining some of the biochemical separations which have been achieved by these techniques.

Question 8
a. "Ion exchange materials may be classified in terms of the acidic or basic strength of the functional groups attached to the polymer matrix." Explain this statement.
b. Define the term "selectivity coefficient" and indicate its relationship to the equilibrium constant for an exchange reaction.
c. Explain what you understand by the term "Donnan potential" and indicate how the Donnan theory can be used to explain the significant absorption of electrolytes by ion exchangers of low capacity.

Question 9
   "The efficient separation of inorganic species by zone electrophoresis and ion exchange chromatography depends greatly on the pH of the buffer solution used and the judicious selection of a suitable complexing agent. For the separation of organic acids, amino acids and many other species of biological interest, complex formation plays a secondary role to pH effects."
   Explain this statement.

Question 10
a. Write an essay on the analytical applications of ion exchangers, including discussion of the use of these materials to remove interferants, purify chemicals, and concentrate trace components in solution.
b. Briefly outline the advantages which may be gained by preparing specific ion exchange resins which contain functional groups known to react selectively with particular groups of cations.
c. Outline some uses for ion exchange membranes.
d. Ion exchange resins charged in particular forms have been proposed as catalysts for a range of reactions. Suggest some advantages and disadvantages which might arise from this approach.

References

E. W. Berg, *Physical and Chemical Methods of Separation*, McGraw-Hill, New York, 1963.
J. Bobbitt, *Introduction to Chromatography*, Reinhold, New York, 1968.
H. G. Cassidy, *Fundamentals of Chromatography*, Wiley (Interscience), New York, 1957.
E. J. Conway, *Microdiffusion Analysis and Volumetric Error*, 4th Ed., Crosby Lockwood, London, 1957.
E. Heftmann (ed.), *Chromatography*, 2nd Ed., Reinhold, New York, 1966.
E. Lederer and M. Lederer, *Chromatography: a Review of Principles and Applications*, 2nd Ed., Elsevier, Amsterdam, 1957.
C. J. O. R. Morris and P. Morris, *Separation Methods in Biochemistry*, Wiley (Interscience), New York, 1964.
I. Smith, (ed.), *Chromatographic and Electrophoretic Techniques*, Heinemann, London, 1960.
O. C. Smith, *Inorganic Chromatography*, Van Nostrand, Princeton, New Jersey, 1953.

ADSORPTION AND THIN-LAYER CHROMATOGRAPHY

J. M. Bobbitt, *Thin Layer Chromatography*, Reinhold, New York, 1963.
J. G. Kirchner, *Thin Layer Chromatography*, Wiley (Interscience) New York 1967.
G. B. Marini-Bettolo (ed.), *Thin Layer Chromatography*, Elsevier, Amsterdam, 1964.
L. R. Snyder, *Principles of Adsorption Chromatography*, Marcel Dekker, New York, 1968.
E. Stahl (ed.), *Thin Layer Chromatography*, Academic, New York, 1965.
H. Strain, *Chromatographic Adsorption Analysis*, Wiley (Interscience), New York, 1942.
E. V. Truter, *Thin Film Chromatography*, Cleaver-Hume, London, 1963.

ELECTROPHORETIC METHODS

H. Bloemendal, *Zone Electrophoresis in Blocks and Columns*, Elsevier, Amsterdam, 1963.
H. W. Hermance and H. V. Wadlow, *Electrography and Electrospot Testing*, in F. J. Welcher (ed.), *Standard Methods of Chemical Analysis*, 6th Ed., Volume IIIA, Chapter 25, Van Nostrand, New York, 1966.
L. P. Ribeiro, E. Mitidieri, and O. R. Affonso, *Paper Electrophoresis*, Elsevier, Amsterdam, 1961.
Ch. Wunderly, *Principles and Applications of Paper Electrophoresis*, Elsevier, Amsterdam, 1961.

GEL CHROMATOGRAPHY

H. Determann, *Gel Chromatography*, Springer, New York, 1967.
J. F. Johnson and R. S. Porter, *Analytical Gel Permeation Chromatography*, Wiley (Interscience), New York, 1968.

ION EXCHANGE

C. B. Amphlett, *Inorganic Ion Exchangers*, Elsevier, Amsterdam, 1964.
F. Helfferich, *Ion Exchange*, McGraw-Hill, New York, 1962.
J. Inczedy, *Analytical Applications of Ion Exchangers*, Pergamon, Oxford, 1966.
R. Kunin, *Ion Exchange Resins*, 2nd Ed., Wiley, New York, 1958.
Y. Marcus and A. S. Kertes, *Ion Exchange and Solvent Extraction of Metal Complexes*, Wiley (Interscience), New York, 1968.
G. H. Osborn, *Synthetic Ion Exchangers*, 2nd Ed., Chapman & Hall, London, 1961.
O. Samuelson, *Ion Exchange Separations in Analytical Chemistry*, 2nd Ed., Wiley, New York, 1963.

GAS CHROMATOGRAPHY

E. Bayer, *Gas Chromatography*, Elsevier, Amsterdam, 1961.
L. S. Ettre and A. Zlatkis, *Practice of Gas Chromatography*, Wiley (Interscience), New York, 1967.
J. C. Giddings, *Dynamics of Chromatography; Principles and Theory*, Marcel Dekker, New York, 1965.
R. Kaiser and R. A. Keller, *Chromatography in the Gas Phase: Quantitative Techniques*, Marcel Dekker, New York, 1969.
A. I. M. Keulemanns, *Gas Chromatography*, Reinhold, New York, 1960.
J. H. Purnell, *Gas Chromatography*, Wiley, New York, 1962.

15

HETEROGENEOUS EQUILIBRIA

I. Two Phase Systems	535
A. The Phase Rule	535
B. Zone Refining and Fractional Crystallization	538
II. Solvent Extraction	542
A. Simple Liquid–Liquid Extraction	542
B. Multiple and Continuous Extraction	549
C. Countercurrent Extraction	553
III. Factors Influencing the Separation of Mixtures by Multiple Distribution Processes	556
A. Separation Factor	556
B. Multiple Distributions	558
C. Theoretical Treatments of the Chromatographic Process	560
IV. Partition Chromatography	563
A. Column Operation	563
B. Paper Chromatography	565
V. Gas-Liquid Chromatography	569
A. Column Packings	569
B. Capillary Columns	570
C. Sample Introduction and Detection	572
D. Retention Volumes	573
E. The Effect of Temperature	576
F. Applications	578
VI. Assignments and Questions	580
References	584

I. TWO PHASE SYSTEMS

A. THE PHASE RULE

While the treatment of chemical equilibrium in terms of the Law of Mass Action is primarily concerned with homogeneous equilibria, it was shown in the last chapter that it could also be applied to heterogeneous systems (cf. Ion Exchange, Chapter 14, Section V,A).

Another method of treating heterogeneous equilibria is in terms of Gibbs' Phase Rule, which may be expressed mathematically by the equation

$$P + F = C + 2$$

P is the number of phases present, C represents the number of components in the system, and F is the number of degrees of freedom.

In order to apply the phase rule the meaning of the terms involved must be fully understood.

As indicated in the previous chapter, a phase is any part of a system which is homogeneous throughout, and is separated by a bounding surface from other homogeneous parts of a system.

Consider ice, liquid water, and water vapor. These are three phases of the one substance. Under a specific set of conditions all three can exist together, but it is more usual for only two to exist side by side. A beaker of water placed in a sealed space gives a two phase system, vapor in equilibrium with liquid. Saturation of the liquid phase with sodium chloride converts this into a three phase system, the excess solid constituting the third phase.

Each phase must be in one of the states: solid, liquid, or gas. In the gaseous state there can be only one phase since gases are always completely miscible. Solids are invariably regarded as separate phases, except in the instance of a solid solution. Liquids may, or may not, form a single phase. If one liquid is immiscible with another, there are two phases; if it is partly miscible, there may be one phase, or two, depending on the concentration.

The number of components in a system is the smallest number of substances required to describe separately the compositions of each of the phases in the system. This is not necessarily the same as the number of chemical substances present. For example, consider a system in which heated ammonium chloride is in equilibrium with the vapor arising from decomposition of the solid:

$$NH_4Cl_{(s)} \rightleftharpoons NH_{3(g)} + HCl_{(g)}$$

This system contains three chemical species, but since all are obtainable from NH_4Cl, it is a one component system.

For the system,

$$CaCO_{3(s)} \rightleftharpoons CaO_{(s)} + CO_{2(g)}$$

three chemical species are in equilibrium but the composition of each phase can be expressed in terms of any two of the molecular compounds. For example, let $CaCO_{3(s)}$ and $CO_{2(g)}$ be chosen as the components of the system. Then

$$\text{Phase 1, } CaCO_3, = CaCO_3 + 0\ CO_2$$

$$\text{Phase 2, } CaO, = CaCO_3 - CO_2$$

$$\text{Phase 3, } CO_2, = 0\ CaCO_3 + CO_2$$

The total system may thus be described as a three phase, two component system.

The system in which ice, water, and water vapor are in equilibrium is a three phase, one component system.

The degrees of freedom or variance of a system is the smallest number of independent variables (pressure, temperature, and concentrations of the various phases) that must be specified to describe completely the state of the system. The degrees of freedom are the number of properties which can be altered (within limits) without causing the disappearance of a phase or the appearance of a new one.

A pure gas has two degrees of freedom, since to describe the state of a pure gas it is necessary to specify only two variables. For a real gas, $PV = RT$; hence if P and V are fixed, the value of T is fixed also. Similarly, if P and T are held constant, then the value of V is fixed. Alternatively, P is fixed if V and T have some specified values.

In the case of a liquid in equilibrium with its vapor, the vapor pressure depends on the temperature but not on the amount of liquid present. Fixing the temperature fixes the vapor pressure; hence this is a system with one degree of freedom.

Let us now consider heterogeneous equilibria. In the systems of interest to analytical chemists, there are always at least two phases and usually more than two components. For example, in the technique known as solvent extraction, a solute is distributed between two immiscible liquid phases. Here $P = 2$ and $C = 3$, hence

$$F = 3 + 2 - 2 = 3$$

If the extraction is performed at room temperature and pressure, two degrees of freedom are taken up in defining these conditions. This leaves one degree of freedom, e.g., the concentration of solute in one of the liquid phases. If this concentration is defined, then the concentration of solute in the other solvent phase is fixed by the system. In other words, at a given temperature and pressure there is a definite fixed relationship between the solute concentrations in the two liquid phases, i.e.,

$$[\text{Solute}]_1 / [\text{Solute}]_2 = K(T, P \text{ constant})$$

Provided the composition of the two solvents remains unchanged, the concentration ratio is independent of phase volumes. However, in practical situations, the transfer of solutes and attendant solvent sheaths can change the composition of the phases. For example, the introduction of hydrogen chloride into a water-ether system substantially alters the percentage of ether (and water) in each phase, and the addition of sufficient acid transforms the system of two immiscible liquid phases into a monophase. Accordingly, the value of the distribution constant, K, tends to vary with the amount of solute

being distributed, trace amounts of substances not necessarily following the
same pattern as large quantities.

In gas-solid chromatography, a vapor or gas is in contact with a solid adsorbent and is dispersed in a carrier gas. Again $P = 2$ and $C = 3$; hence in such
systems there are three degrees of freedom. With the columns operating at a
fixed temperature and constant pressure, one degree of freedom remains. If
this is used to fix the concentration of vapor in one phase, then the concentration
in the other phase is fixed, provided the nature of the phases remains unaltered.
However, it has been noted that the adsorption of species on solids does not
usually yield a linear distribution (i.e., $C_1/C_2 \neq K$), and it may be suggested
that after the formation of the first monolayer on the surface the process also
involves solution of vapor in a newly formed liquid phase.

In the simple case in which a gas dissolves in a liquid, there are two phases
(gas and solution) and two components (gas and solvent); hence, according to
the phase rule there must be two degrees of freedom. Thus if the composition
is fixed, pressure and temperature are interrelated; if the temperature is fixed,
the composition of the solution is fixed by the pressure. This relationship is
generally quoted in the form of Henry's law which states that the amount of
gas (m) dissolved by a given volume of liquid at constant temperature is proportional to the pressure of the gas (p) (T constant).

$$m/p = K'$$

Now the mass of gas dissolved in a given volume can be expressed as a
concentration, (e.g., C_1) and the pressure of a gas is another way of expressing
its concentration (e.g., C_2 instead of p), which leads to the generalized expression
of Henry's law.

$$C_1/C_2 = K''$$

Again, it must be borne in mind that Henry's law only holds when the same
molecular species is present in the two phases. In a gas-liquid system this condition is not always satisfied. For example, on solution the gas may associate
into more complex molecules; there may be compound formation between the
solvent and the gas (e.g., $NH_3 + H_2O \rightleftarrows NH_3 \cdot H_2O$); or there may be
electrolytic or other dissociation of the substance (e.g., $HCl + H_2O \rightleftarrows$
$H_3O^+ + Cl^-$).

Henry's law only applies to ideal systems in which there is ordinary physical
solution; hence in practice the behavior of most systems deviates from that
predicted on the basis of this law alone.

Accordingly, while the phase rule is useful for predicting the number of
experimental variables that need to be controlled in a distribution process, the
actual nature of the distribution relationship has to be determined by experiment.

B. ZONE REFINING AND FRACTIONAL CRYSTALLIZATION

With solid two phase systems, such as metal alloys and impure organic
compounds, the vapor pressures of the solids are so small that virtually no

gaseous phase exists. Any experiments are normally conducted at atmospheric pressure; hence it has become accepted practice to call such a system a "condensed system" and to describe it in terms of a "reduced" phase rule, viz.,

$$P + F' = C + 1$$

where F' is the number of degrees of freedom which the system can possess in addition to the pressure. If the number of components is two, it is possible to have three phases in equilibrium, but the system would be invariant. With one degree of freedom, two phases are in equilibrium.

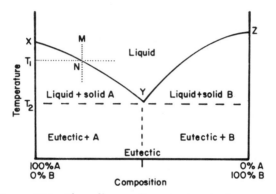

FIGURE 15.1   Phase diagram for a simple eutectic system.

Figure 15.1 represents a phase diagram for a two component, simple eutectic system. The curve XY represents the freezing point curve of a compound A to which successive small quantities of compound B have been added. The curve ZY is the freezing point curve of B to which small amounts of A have been added; Y is the point at which solid A and solid B are in equilibrium with the fused mass. At this point there are three phases and therefore no degrees of freedom. The point Y is called the eutectic point.

If a fused mass, rich in B, is cooled, B separates out and the composition of the residue moves along the line ZY until the eutectic point is reached. At this stage the whole mass crystallizes. Similar remarks apply when a molten mixture rich in A is cooled.

This principle is used in the technique known as zone refining which is a popular procedure for purifying organic chemicals.

The crude material, which must melt without decomposition, is packed into a vertical cylindrical tube. Around the tube is placed a small annular heating coil which fits loosely and can be moved up and down mechanically (cf. Fig. 15.2).

The coil is initially located at the top of the column and sufficient heat is applied to produce a molten zone inside the column. The coil is then moved slowly downwards, causing the molten zone to gradually descend, and to carry

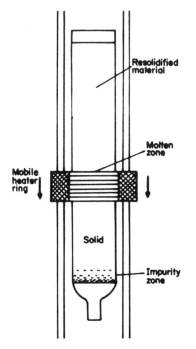

FIGURE 15.2 Principle of zone refining.

impurities with it. Left in the wake of the heater is purified, resolidified material. The purification can be explained as follows: Suppose that in Fig. 15.1, A represents the material being purified and B is the impurity to be removed. Let M be the composition of the crude material. At the top of the column the liquid phase has composition M when the heater is located at this zone. As the heater moves down, the top layer begins to cool, and at temperature T_1 (corresponding to the point N) pure component A will crystallize out from the melt, leaving the small amount of B initially present in the liquid phase. In the zone refining column the impure liquid successively passes into the next lower zone and the impurity passes gradually down to the bottom, from which position it may be removed. By repeating the passage of the coil down the column several times products of very high purity can be obtained.

When the two components form a solid solution, only two phases can be present (i.e., liquid and solid solution) and the reduced phase rule indicates that F' must always be at least one. Accordingly, there is no eutectic point and the phase diagram for a system in which no compounds are formed resembles that shown in Fig. 15.3.

Since there are two solutions, the liquid and the solid, and since the concentration of the components in these two phases is not, in general, the same, two curves are required. The first, ALB, is the melting point curve, and is known as

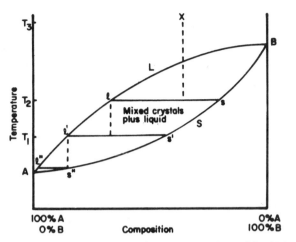

FIGURE 15.3   Equilibrium curve for mixture giving solid solutions.

the liquidus curve. It indicates the equilibrium between a single liquid phase and mixed crystals plus liquid. The second curve, *ASB* (the solidus curve), is the freezing point curve and represents the equilibrium between mixed crystals plus liquid and solid.

Substances which form solid solutions can be separated by fractional crystallization.

Consider the case of a fused mixture of composition X at temperature T_3 (cf. Fig. 15.3), which is allowed to cool. When the temperature falls to the liquidus curve, a solid solution will deposit, and if the temperature is held at some value T_2, the equilibrium mixture will consist of a liquid phase of composition l and a solid phase of composition s. The liquid phase is richer in A, and the solid phase richer in B, than the original mixture. In fractional crystallization these two phases are separated at this temperature. If the isolated liquid phase is now cooled to temperature T_1, the system consists of a mixture of liquid (composition l') and solid (composition s'). Separation of these two phases gives a liquid which is very rich in A. Repetition of the cooling and separation process soon leads to almost pure A (e.g., point l'' in Fig. 15.3). Almost pure B is obtained by taking the isolated solid phases and allowing them to come to equilibrium at higher temperatures than that used in the initial phase separation.

Similar results are obtained if the two components are dissolved in a solvent that does not form solid solutions with the solutes. The solvent may be regarded as an inert medium which dilutes the system and makes it easier to approach equilibrium.

All separation procedures involving fractionation, e.g., distillation, liquid–liquid extraction, fractional crystallization, or fractional precipitation, are much the same in general principle.

II. SOLVENT EXTRACTION

A. SIMPLE LIQUID–LIQUID EXTRACTION

1. *Types of Extraction Systems*

As discussed in Chapter 10 (Section IV,C), the distribution of a solute between two immiscible liquid phases is best described in terms of the distribution ratio, D, where

$$D = \frac{\text{Total concentration of solute in phase 1}}{\text{Total concentration of solute in phase 2}}$$

The concentrations refer to actual experimental values and include all chemical forms of the species of interest. By convention, phase 1 is usually an organic phase.

The laboratory procedure for liquid–liquid extraction is exceedingly simple. An aqueous solution (phase 2), containing the species to be extracted, is shaken in a separating funnel with a volume of water-immiscible organic solvent. After standing, the two phases are separated. If the distribution ratio is very large (e.g., >100) a single extraction is sufficient to transfer the solute quantitatively from the aqueous to the organic phase.

However, with smaller values of D it is better to extract the aqueous solution with several small volumes of organic solvent, than to use a single large volume. For example, consider the extraction of a fatty acid into ether, where the distribution ratio is about 5. Let one gram of acid, RCOOH, be present in the initial 20 ml of solution, and let x gram be transferred to the organic layer on shaking with 20 ml of ether,

$$D = C_{(\text{ether})}/C_{(\text{water})} = 5 = \frac{x \cdot 1000}{20} \times \frac{20}{(1 - x) \cdot 1000}$$

or $x = 5/6$ gram.

In other words, a single extraction with 20 ml of ether removes 83% of the fatty acid from the aqueous phase.

If the ether is added in two, 10 ml portions, the degree of extraction with the first portion can be calculated from the equation

$$5 = \frac{y}{10} \times \frac{20}{1 - y}$$

from which $y = 5/7$ or 0.714 gm.

On separation of the two phases, 0.286 gm of RCOOH remain in the aqueous phase. On shaking this solution with the second 10 ml portion of ether, z gm of fatty acid transfer to the organic layer.

$$5 = \frac{z}{10} \times \frac{20}{(0.286 - z)} \quad \text{or } z = 0.2 \text{ gm}$$

The total amount extracted in the two, 10 ml portions $(y + z)$ equals 0.914 gm or 91.4% recovery.

Treatment of the same initial aqueous solution with four 5 ml aliquots of ether theoretically gives a recovery of 0.961 gm or 96.1%.

Successive extractions are tedious; hence the general aim is to find conditions where the percentage extracted in a single operation is greater than 99%. The percentage extracted (E) is determined by the relative volumes of the two phases (V_w/V_o) and the distribution ratio.

$$\%E = 100\ D/D + (V_w/V_o).$$

In most systems the magnitude of D can be varied by changing the solvent or by changing the chemical composition of the species to be extracted.

A number of covalent compounds, such as the oxides of ruthenium and osmium, sulfur dioxide, 8-hydroxyquinoline, the halogens, etc., are naturally highly soluble in a number of organic solvents, and extraction is based on simple physical solution of these compounds in the organic phase.

The distribution of covalent material between two solvents possessing a low degree of miscibility approximates to a simple partition process, that is, $C_1/C_2 = p$.

However, the overall distribution can be distorted by dissociation, association or chemical reaction within one or both of the phases.

For example, the amount of iodine (I_2) available for partition into a carbon tetrachloride phase is significantly reduced in the presence of iodide ions due to the formation of I_3^-.

The extraction of osmium tetroxide is pH dependent because in alkaline solution this compound tends to hydrolyze to OsO_5^{2-} and OsO_5H^-. There is also a tendency for the osmium tetroxide to form a tetramer in the organic solvent.

Accordingly, $[OsO_4]_0/[OsO_4]_w = p$, but

$$D = \{[(OsO_4)_0 + 4[(OsO_4)_4]_0\}/\{[OsO_4]_w + [OsO_5H^-]_w + [OsO_5^{2-}]_w\}$$

From a knowledge of the equilibrium constants for the hydrolysis reactions and tetramer formation, it is possible to calculate the effect of pH on the magnitude of D (cf. example of 8-hydroxyquinoline quoted in Chapter 10, Section IV,C).

Most organic compounds are highly soluble in organic reagents, but many are subject to ionization in polar solvents and/or polymerization in nonpolar solvents. However, with this type of compound the effect of aqueous variables on the extent of extraction is usually predictable.

In contrast, few simple inorganic compounds are naturally soluble in non-aqueous solvents, and for the solvent extraction of metal ions, the polar nature of the metal species has to be masked.

One means of achieving this aim is to convert the metal ion into a neutral inner complex by the addition of a chelating agent. Some typical examples of the chelating agents used for this purpose are shown in Table 15.1. As a class, the neutral chelates are most readily extracted into such organic solvents as benzene, carbon tetrachloride and chloroform. This type of extraction system is amenable to prediction of the effect of solution variables on the overall equilibrium (cf. Chapter 10).

A second means of masking metal ion polarity is through the formation of ion pairs. In ion-association extraction systems, the metal ion is either incorporated in a large ion containing organic groups, or it becomes associated with an organic ion of great size.

$$M^{n+} + \gamma(\text{Org}) \rightleftharpoons M(\text{Org})_y{}^{n+}$$

$$M(\text{Org})_y{}^{n+} + nA^- \rightleftharpoons (M(\text{Org})_y{}^{n+}, nA^-)$$

$$M^{n+} + \gamma L^- \rightleftharpoons ML_y^{(y-n)-}$$

$$ML_y^{(y-n)-} + (\gamma - n)\text{Org}^+ \rightleftharpoons (ML_y^{(y-n)-}, (\gamma - n)\text{Org}^+)$$

In the scheme shown, A^- usually represents a simple inorganic anion such as perchlorate, nitrate, etc., and dipyridyl or 1,10-phenanthroline are typical of the reagents which form cationic metal complexes.

The ligands (Y) used to form anionic metal complexes are normally the halide, nitrate, or thiocyanate ions. The organic cation associated with metal containing anions is usually of an "onium" type. For example, oxyanions such as MnO_4^-, ReO_4^-, IO_4^- associate readily with the tetraphenyl arsonium ion. Other useful onium reagents include substituted ammonium compounds, $RNH_3^+ \ldots R_4N^+$; oxonium species, ROH_2^+, R_2OH^+, R_2COH^+; sulfonium anions, R_3S; stibonium anions, R_4Sb^+; and phosphonium ions, R_4P^+.

As may be predicted from the general rule, like dissolves like, oxonium type solvents such as alcohols, ethers, ketones, etc. are most effective with ion-association extraction systems.

The solvent can play an important secondary role in a number of systems. For example, the water initially coordinated to the metal ion may be displaced by the combined action of ligands, such as the halides, and oxonium type solvents. Thus, in the extraction of iron (III) into diethyl ether from hydrochloric acid solution, it is suggested that the species extracted is

$$[(C_2H_5)_2OH^+, FeCl_4\{(C_2H_5)_2O\}_2{}^-]$$

In other cases, the extracted species is a neutral molecule-solvent adduct. For example, in the extraction of uranium from nitric acid solutions into an organic phase composed of tributyl phosphate and ether, the species extracted is $[UO_2(NO_3)_2 2Bu_3PO_4]$.

This idea has been extended by using reagents in the extractive phase which behave somewhat like liquid ion exchangers. The reactive components are

TABLE 15.1

CLASSIFICATION OF METAL EXTRACTION SYSTEMS

Type	Classification	Typical reagents	
		Structure	Name
Chelate systems	4-Membered rings	$(C_2H_5)_2NC$ with $=S$ and S^-	Diethyl dithiocarbamate ion
	5-Membered rings	$CH_3-C=NOH$ $CH_3-C=NOH$	Dimethyl glyoxime
		8-Hydroxyquinoline structure (OH, N)	8-Hydroxyquinoline
		Diphenylthiocarbazone structure (H H, –N–N, C=S, –N=N)	Diphenylthiocarbazone (dithizone)
	6-Membered rings	$CH_3\,C\,CH_2\,C\,CH_3$ with O, O	Acetyl acetone
		$-CCH_2CCF_3$ with O O (thiophene ring)	Thenoyltrifluoroacetone (TTA)
		$CH=NOH$, OH (benzene ring)	Salicylaldoxime
		$N=O$, OH (naphthol ring)	1-Nitroso-2-napthol
Ion association systems	Metal in cationic member of ion pair	Dipyridyl structure ($=N$, $N=$)	Dipyridyl
		$(C_8H_{17})_3P\rightarrow O$	Tri-n-octylphosphine oxide (TOPO)
		$(C_4H_9O)_2P$ with O and OH	Di-n-butylphosphoric acid (DBP)
	Metal in anionic member of ion pair	HF, HCl, HBr, HI NH_4SCN	

usually derivatives of phosphoric acid or substituted ammonium ions. For simplicity they are known by initials, typical examples being heptadecyldihydrogen phosphate (HDPA), di-(2-ethylhexyl)phosphoric acid (D2EHPA), and monododecylphosphoric acid (DDPA).

These organophosphorus reagents are usually used as 5–10% solutions in inert solvents such as benzene, toluene, kerosine, carbon tetrachloride, petroleum ether, octane, cyclohexane, etc. In the inert solvent, many of the reagents exist as dimers and the reaction responsible for the formation of an extractable species is typified by the case of the uranyl ion.

$$UO_{2(aq)}^{2+} + 2(HX)_2 \rightleftarrows UO_2 \cdot 2HX_{(org)} + 2H_{(aq)}^+$$

where $(HX)_2$ represents the dimerized reagent.

Basic organonitrogen compounds act as liquid anion exchange materials and can be used to extract simple and complex anions from highly acid solutions. Common reagents and their abbreviations are methyl octylamine (MDOA), tri-iso-octylamine (IOA), dodecenyl(trialkyl)ammonium (DTA), tetrabutylammonium (TBuA), and tetra-*n*-heptylammonium (THA).

A synergistic effect has been observed with mixtures of organic phosphorus compounds, i.e., the total extracting power of a mixture of two of these compounds is often much greater than the combined extracting power of the individual components. For example, the distribution ratio for the extraction of uranium decreases by a factor of about four if tributyl phosphate is used instead of diethylhexyl hydrogen phosphate (D2EHPA). However, the ratio increases by a factor of fifty if some tributyl phosphine oxide (e.g., 0.05 M) is added to the D2EHPA phase.

2. *Extraction Curves*

In most extraction procedures, the efficiency of the extraction varies with the composition of the aqueous phase. Thus, in summaries of extraction procedures (e.g., Table 10.2 in the "Handbook of Analytical Chemistry") there are tabulations of recommended organic phases and recommended aqueous solution conditions.

The aqueous phase column lists the reagent required to convert the selected metal ion into an extractable form, the optimum pH for extraction, and, in some cases, the masking agent needed to prevent extraction of other metal species.

For each element several alternative procedures are usually listed. These can differ in selectivity, and the table column headed "separated from" indicates the type of separation which may be satisfactorily achieved using any particular combination of aqueous and organic phases.

For example, gold may be separated from bismuth, indium, molybdenum, tellurium, and zinc by extracting a solution which is 6.9 M in respect to hydriodic acid with diethyl ether.

Aluminum can be separated from iron, nickel, titanium and vanadium by adding 8-hydroxyquinoline to an aqueous phase adjusted to pH 5–9, followed by extraction of the precipitate into benzene or chloroform.

An organic phase consisting of dithizone dissolved in carbon tetrachloride or chloroform has proved useful in a number of extraction studies. However, because so many metal ions react with this organic reagent, selectivity is achieved only by efficient masking (see Chapter 10, Section IV,B) procedures. Thus, for the extraction of zinc the aqueous phase is treated with potassium cyanide, tartrate ions, and acetate ions and the pH is adjusted to lie between 5.7 and 7.0.

Irving, Rossotti, and Williams [*J. Chem. Soc.*, (1955), 1906] developed a quantitative generalized treatment of solvent extraction equilibria which provides a basis for predicting the behavior of all types of inorganic extraction systems. However, the data required are available for only a few systems and in most extraction studies it is necessary to plot experimental extraction curves in order to ascertain the optimum conditions for metal transfer.

The experimental curve is prepared by determining the percentage of the original metal ion transferred when one of the solution variables (e.g., pH or ligand concentration) is altered systematically.

The effect of pH on the extraction of 8-hydroxyquinoline is indicated in Fig. 10.4. Similarly, with metal chelates, the pH influences the "apparent" stability of the complex, that is, the fraction of the metal ion present in the extractable complex form. This significantly influences the distribution ratio, and even if the complex has a partition coefficient greater than a hundred, the extraction is pH dependent and the experimental curve resembles Fig. 15.4(a). The curve for a given metal ion shifts along the abscissa with changes in the total concentration of ligand added, and metals possessing the same valency, but different stability constants, yield a series of parallel curves.

Thus curves 1 and 2 on Fig. 15.4(a) could represent the extraction curves obtained using the chelates of two different metal ions and standard experimental conditions. It can be observed that at pH >3.4 extraction of species 1 is 100% complete in a single distribution. However, once the pH exceeds 3.8, some of species 2 is also extracted. To quantitatively separate species 1 from species 2, the pH would therefore need to be controlled between 3.4 and 3.8. For removal of both complexes, a pH >6 in the aqueous phase is a necessary preliminary requirement.

The shape and position of the extraction curves observed with ion–association systems vary with the metal ion involved and the concentration of acid, etc. Figure 15.4(b) shows the distribution of various metallic species into diethyl ether from hydrochloric acid solutions. It can be observed that some metals are only partially extracted using optimum conditions (that is, $D \ll 100$) while others are quantitatively distributed into the organic phase within defined solution conditions. It should be emphasized that the position of the curves

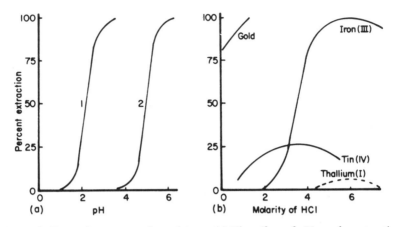

FIGURE 15.4   Extraction curves of metal ions. (a) The effect of pH on the extraction of metal chelate complexes; (b) the effect of acid concentration on the extraction of ion association systems. Data taken from R. Bock and E. Bock, *Z. Anorg. Chem.* **263** (1950) 146 with permission of the publishers, Johann Ambrosius Barth.

can vary with the concentration of metal ion and with the concentration of other salts present.

Many added salts tend to reduce dissociation of the extractable species and thus enhance the percentage removed. On the other hand, if the salt (or acid) is itself extractable into the organic phase, increasing the concentration leads to competition for the organic solvent molecules and this can result in a drop in the percentage extracted.

Solvent extraction applied as a batch procedure is used mainly as a means of isolating one particular compound or group of compounds, and the conditions for effective separation are best determined by examination of a series of extraction curves.

The extracted materials may be discarded or used as the basis of quantitative analysis. For example, the absorbance of UV-visible light by the organic layer can be related to the concentration of metal ion in the aqueous phase; or the organic layer can be aspirated directly into the burner system of an atomic absorption unit. For other analytical procedures it is sometimes necessary to strip the organic phase by shaking the organic extract with a fresh aqueous solution, adjusted chemically to favor distribution back into the water.

The separation of organic species by selective solvent extraction is a well-established procedure, particularly in natural product studies, and is described in detail in monographs dealing with the techniques of organic chemistry.

For many separations, simple inversion of the separating funnel containing the two phases (when repeated many tens of times in a period of minutes) is sufficient to achieve equilibrium in the extraction process. On the other hand, the extraction of metal chelates can proceed at a slow rate, especially if low concentrations are involved. (Among the reagents which tend to be slow in forming extractable chelates are dithizone, TTA and acetylacetone.) The rate of extraction can also be reduced by the presence of masking agents in the aqueous phase. Thus, in developing a new procedure, it is essential to consider both the kinetics of the extraction and the position of equilibrium.

B. MULTIPLE AND CONTINUOUS EXTRACTION

Where extraction efficiencies are less than 100%, complete recovery of the species of interest requires a series of extractions with fresh organic phase.

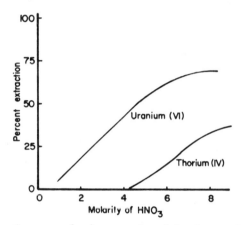

FIGURE 15.5 Extraction curves for the extraction of the nitrates of uranium and thorium from nitric acid solutions into diethyl ether. Based on data of R. Bock and E. Bock, *Z. Anorg. Chem.* **263** (1950) 146.

Consider the extraction of uranyl nitrate into diethyl ether from a nitric acid solution. The extraction curves for uranyl and thorium salts using these two phases are shown in Fig. 15.5.

Reference to this figure shows that from 8 M nitric acid, 67% of the uranium is extracted, leaving 33% in the aqueous phase. Shaking this residual aqueous layer with an equal volume of fresh solvent should remove two-thirds of the remaining metal ion. Thus, after two extractions the aqueous layer still retains 11% of the initial uranium content. Three further extractions with fresh solvent would reduce this figure to 0.4%. Or, expressed in another way, combination of the five ether layers would give a solution containing 99.6% of the uranium.

Such a procedure can become tedious if carried out manually, particularly if the percent extraction is numerically less than 50%. For example, if the uranium were extracted from a 2 M HNO_3 solution, where the percent extracted is about 18%, thirty extractions with fresh solvent would be required to remove 99.6% of the initial uranyl salt from the aqueous solution.

Some of the tedium of multiple extractions can be avoided by using the process of continuous extraction. In this process the organic solvent is distilled in a side tube and is condensed above the aqueous layer. The droplets of organic solvent so formed pass through the aqueous layer before excess overflows into the heating vessel. As the process continues, the concentration of extracted species builds up in the heated side vessel. The design of continuous extractors

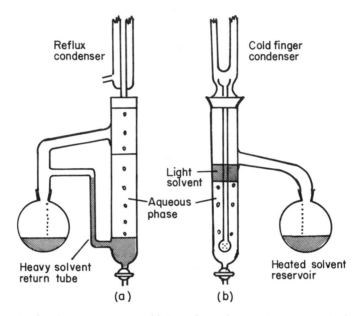

FIGURE 15.6   Continuous extractors. (a) For solvents heavier than water; (b) for solvents lighter than water.

suitable for use with solvents heavier or lighter than water is indicated by the sketches shown in Fig. 15.6.

The extraction of the soluble components of a solid sample may be achieved by replacing the aqueous phase in Fig. 15.6(a) with a porous (e.g., filter paper) cup containing the solid.

Continuous extraction and simple multiple extractions are useful procedures if only one species is extractable, but they are not suitable for separating mixtures of species which differ only in the degree of extraction.

For example, consider an 8 M nitric acid solution containing 100 mg of both uranyl and thorium nitrates. A single extraction of this solution with an equal volume of diethyl ether would transfer 67 mg of the uranyl salt and 33 mg of the thorium salt to the organic phase (cf. Fig. 15.5). In this organic layer, the ratio of uranium to thorium is greater than in the initial aqueous solution, but the two species are far from being separated.

An increase in individual purity can be achieved by taking the organic layer and extracting it with several volumes of fresh aqueous phase (8 *M* HNO$_3$), or by taking the initial aqueous layer and extracting it with successive portions of fresh organic solvent. The distribution of the original material after several

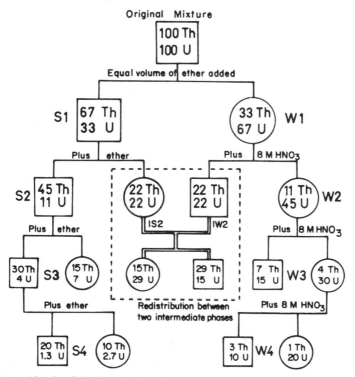

FIGURE 15.7   Sketch of the distribution of uranyl and thorium nitrates in 8 *M* nitric acid after multiple extractions with diethyl ether. The percentage extraction of the metal ions is 67% and 33% respectively under the selected conditions. The organic phase is represented by the symbol S and a circular enclosure, the aqueous phase by the symbol W and square enclosures. From "Fundamental Principles of Chemical Analysis," W. F. Pickering. Used with permission of Elsevier Publishing Company, Amsterdam, copyright © 1966.

extraction processes is best illustrated in a box diagram such as that drawn in Fig. 15.7.

It can be observed from this figure that treatment of the original solution with four successive volumes of ether yields an aqueous phase which contains about 20 mg of thorium salt and 1.3 mg uranyl nitrate (pathway S1 to S4). It can be claimed that purification has occurred since the percentage of thorium in the mixture is over 94%, but the amount of the initial material which is present in this purified form is only one-fifth of the original amount. Similarly, back-washing of the first organic layer with three successive volumes of 8 *M*

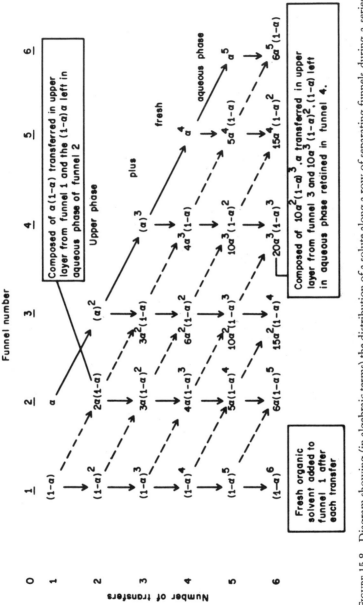

FIGURE 15.8   Diagram showing (in algebraic terms) the distribution of a solute along a row of separating funnels during a series of transfer operations in which the upper layer is successively transferred into the funnel of next highest number; α represents the fraction of the original material extracted into the organic layer at each phase distribution.   The terms in each row of the table represent the binomial expansion of $[(\alpha) + (1 - \alpha)]^n$ where n is the number of transfers.   From "Fundamental Principles of Chemical Analysis," W. F. Pickering.   Used with permission of Elsevier Publishing Company, Amsterdam, copyright © 1966.

HNO_3 gives an ether layer containing 20 mg U and 1 mg Th (i.e., 95% U with a recovery of about a fifth of the original material).

These procedures are satisfactory if one merely desires to obtain a small amount of pure component. However, if quantitative recovery is required, it is obvious that the large bulk of the material present in the separated intermediate phases must be recovered and distributed preferentially into one or other of the phases. Mixing of any two of these intermediates produces a new distribution, as shown in Fig. 15.7, phases IS2 and IW2. This type of redistribution can be achieved by transferring the lower layer present after one separation (e.g., W2) into the funnel retaining the ether layer of another separation (e.g., S2). The process could be continued along a series of separating funnels.

C.   COUNTERCURRENT EXTRACTION

Let us consider a row of separating funnels numbered 1, 2, 3, 4 ... n. In the first funnel is placed an aqueous solution containing some species A, together with an equal volume of immiscible organic solvent. After shaking, let the upper layer be transferred to funnel 2, fraction α of the original material having been extracted into this phase. To funnel 1 another volume of organic solvent is added; to funnel 2 is added an equal volume of fresh aqueous phase (e.g., 8 M HNO_3 in the above example).

In the next transfer, let the upper phase in each separating funnel be transferred to the funnel of the next highest number, and let the transferred organic layer be equilibrated with the lower (aqueous) phase present in the funnel. For each operation, fresh organic solvent is needed in funnel 1 and a fresh aliquot of aqueous phase has to be present in the funnel of highest number. This process of upper layer transfer can be continued over a large series of funnels and the distribution of A after six transfer processes is described algebraically in Fig. 15.8.

This procedure may be described as discontinuous countercurrent extraction and the analytical significance becomes more apparent if absolute values are assigned to α and the results are plotted in the form of a histogram.

Figure 15.9 shows the distribution in successive funnels after six transfer and equilibration processes for systems in which $\alpha = 0.2$, 0.4, 0.6, and 0.8 (i.e., systems having distribution ratios, D, of 0.25, 0.67, 1.5, and 4 respectively).

This figure demonstrates that the result of successive transfers is to cause the solute to move through the system of separating funnels as a sort of concentration wave of steadily diminishing amplitude. As the process proceeds the position of maximum (total) concentration occurs in separating funnels of successively higher serial numbers. The rate of travel from left to right increases as the value of D increases. The curves shown apply to six transfers only. When the number is much larger the distribution of concentrations between the various funnels is less distorted and closely approximates the normal Gaussian distribution curve.

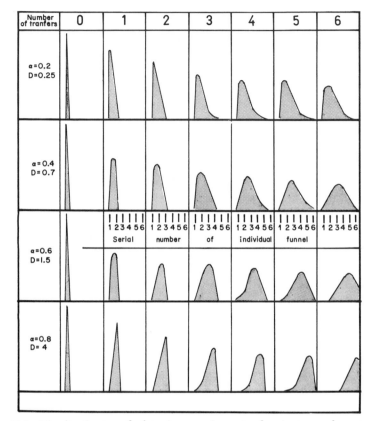

FIGURE 15.9   The distribution of solutes in successive units after six stages of countercurrent distribution.

It can be observed from Fig. 15.9 that if two substances have different distribution ratios, separation may be achieved through the two waves moving at different rates. For example, after six transfers, a substance having a D value of 0.25 is present mainly in funnels 1, 2, and 3 while another substance with a distribution ratio of 4.0 is present mainly in funnels 5 and 6 at this stage. A few more transfer steps and these two compounds would be quantitatively isolated from each other. The greater the similarity in D values, the larger the number of discontinuous counter current extractions required to give a satisfactory separation.

The theory of this type of countercurrent extraction was first postulated by Craig, who also developed an all glass multiple-stage apparatus capable of handling many tens of equilibrations and transfers. The apparatus designed by Craig is particularly valuable for the separation of complex mixtures of organic substances and for purity studies. It consists of a large number of connected glass tubes of the type shown in Fig. 15.10.

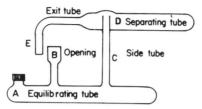

FIGURE 15.10   A unit of a Craig countercurrent extractor.

Each of these glass units consists of an equilibrating tube (A) (about twelve inches long and half inch diameter), which is fitted with an opening (B) through which equal volumes of the two phases are introduced, and a side tube (C) connected to a separation cell (D). In an assembly of units the offset exit tube (E) of one extractor cell fits into the opening (B) of the next unit on its left. The full set of units (e.g., 50 to 100) is held in a rigid framework which can be rotated about an axis at right angles to the plane of AB.

Phase equilibrium is brought about by tilting the equilibrium tube backward and forward through an angle of about 35°. The phases are then allowed to separate, and by tilting the equilibrium tube (A) at an angle to the horizontal (about 100°), the upper layer is caused to decant through the side tube (C) into the separation tube (D), while the lower phase remains in (A). When the apparatus is returned to its original position, the contents of (D) (the upper phase) pass through the exit tube (E) into the equilibration tube of the adjacent unit. The whole sequence comprises a single transfer.

The sample containing the solutes to be separated is placed in unit 1, and an appropriate amount of solvent 1 is added to bring the volume to the desired mark. To each of the other Craig tubes is added pure solvent 1. The volume added to each tube is adjusted to ensure that solvent 1 does not pass through the side arm (C) when the tube is tilted to its maximum angle. An equal volume of pure extracting solvent 2 is then introduced into the Craig unit containing the sample (i.e., unit 1).

The equilibration and transfer sequence is then invoked, and after each transfer step a fresh portion of solvent 2 is added to the Craig unit No. 1. The process is continued until all the units in the apparatus have been involved in the separation process.

In a system operating with equal volumes of solvents 1 and 2, the number of the tube, n_{max}, containing the maximum concentration of solute can be established from the equation

$$n_{max} = n \cdot D/D + 1$$

where n is the total number of tubes used and D is the distribution ratio.

The fraction of the solute, f, present in this tube is given by the expression:

$$f = (D + 1)/\sqrt{2\pi n D}$$

Thus, if one subjected a mixture of the four species discussed in Fig. 15.9 ($D = 0.25$, 0.67, 1.5, and 4, respectively) to separation in a 100 tube Craig apparatus, the maximum concentration of the four components should occur in tubes 20, 40, 60, and 80, respectively.

While it is quite efficient, the Craig apparatus is somewhat large and cumbersome and it was soon realized that similar types of equilibria were involved in chromatographic processes.

For example, in partition chromatography a column packed with some inert solid is moistened with water. In this state the column may be considered to consist of several hundred separating funnels (or Craig units) stacked on top of each other, each being filled with a particular volume of water. The top segment becomes equivalent to filter funnel 1 when the mixture of solutes to be separated is placed on top of the packed column. For elution development, an organic solvent is added at a constant rate to the top of the column. On the first addition of organic solvent to segment 1, the solutes are distributed between the small volume of organic phase and the volume of water present in this segment.

Movement of the solvent down the column to segment 2 is then equivalent to transferring the upper layer of one funnel (or Craig unit) to the funnel of next highest number where it meets fresh aqueous phase. The continual addition of organic solvent to the top of the column is equivalent to adding fresh solvent to funnel 1 (cf. Fig. 15.8).

As the organic solvent front moves down the column, equilibrium processes similar to those summarized in Figs. 15.8 and 15.9 occur and the components of the initial mixture travel down the column as waves or zones of material, the rate of movement varying with the magnitude of the distribution ratio for each component.

The analogy is somewhat approximate because the system is dynamic and true equilibrium may not be achieved at any given point. However, the distribution observed within a small segment or volume of column can, for the purposes of discussion, be considered to be equivalent to one equilibration between an organic phase and an aqueous phase. This segment (by analogy with distillation processes) is usually described as a theoretical plate.

Under ideal conditions, migration of the solute through the column during elution development can correspond to thousands of transfers and equilibrations. This large number of transfers allows the band or wave of one species to be separated from that of a second species whose distribution ratio is only slightly different from that of the first species.

III.   FACTORS INFLUENCING THE SEPARATION OF MIXTURES BY MULTIPLE DISTRIBUTION PROCESSES

A.   SEPARATION FACTOR

In the preceding discussion on solvent extraction, the volumes of the two immiscible phases have been considered to be equal. However, in separations based on partition chromatography the volumes of the two phases can differ by a factor as great as one hundred. It is therefore desirable to consider the situation where the volumes of the two phases are not equal.

If x is the fraction of solute extracted into the organic phase and V_o and V_w represent the volume of the organic and aqueous phases respectively, then the distribution ratio, D, is described by the expression

$$D = [A]_o/[A]_w = \{x/V_o\}/\{(1 - x)/V_w\}$$
$$= x/(1 - x) \cdot V_w/V_o$$

If the phase volume ratio, V_o/V_w, is represented by the symbol R,

$$x/(1 - x) = RD$$

or by rearrangement

$$x = RD/(1 + RD) \quad \text{and} \quad (1 - x) = 1/(RD + 1)$$

It has been noted in Fig. 15.8 that the distribution of solute between the various funnels can be represented by the expansion of the binomial expression $[\alpha + (1 - \alpha)]^n$ where α is the fraction of solute in the organic phase, $(1 - \alpha)$ is the fraction remaining in the aqueous phase, and n is the number of transfers.

This expression refers to the particular case where the phase volume ratio is unity (i.e., $V_o = V_w$), and to allow for other phase ratios, the terms α and $(1 - \alpha)$ have to be replaced by the more general terms x and $(1 - x)$ derived above.

Accordingly, after n distributions or transfer processes, the solute is distributed between n funnels, or segments, in accordance with the expansion of the expression:

$$\left[\frac{RD}{RD + 1} + \frac{1}{RD + 1} \right]^n$$

While this expression correctly indicates that some solute is present in all n segments, the amount present in a large number of these is usually so small that it may be neglected. However, one advantage of writing the equation in this form is that it emphasizes the importance of the distribution ratio in the distribution of the material after n transfers.

When a mixture of compounds is allowed to come to equilibrium with a two-phase system, every component tends to distribute itself between the phases, and the distribution of each species can be described in terms of a distribution ratio; e.g.,

$$D_A = [A]_1/[A]_2; \quad D_B = [B]_1/[B]_2$$

where D_A or D_B are the distribution ratios for species A and B respectively, between the phases 1 and 2.

For any two species, one can define a factor, known as the separation factor, α, which relates the two distribution ratios.

$$\alpha = D_A/D_B$$

When α is very large (e.g., >100), or very small (e.g., $<10^{-2}$), a single distribution between the two phases results in almost complete separation of the species A and B. When $10^2 > \alpha > 10^{-2}$, separation is possible, but only through multiple distributions. No separation is possible in the special case where α equals unity.

Since optimum separations are achieved when the separation factor differs significantly from zero, it is desirable to select experimental conditions which maximize this factor; α can be varied in a number of ways:

1. By changing the nature of one or both of the phases involved. For example, in liquid–liquid distributions a different organic solvent may be selected. In a liquid–solid distribution, either the solid phase or liquid phase may be replaced systematically until a suitable combination is found.

2. By changing the chemical composition. Thus the salts of organic acids may be protonated by adjustment of pH or converted to organic esters. The form of inorganic compounds is altered by complex or chelate formation.

   Where chemical reaction is involved, α can often be varied through a range of values merely by adjusting the pH of the aqueous phase and the concentration of reagent present.

   To calculate (from published data) the approximate value of the separation factor it is necessary to consider:

   a. The appropriate equilibrium equation for the heterogeneous phase distribution. The form of the equation usually varies with the physical nature of the phases involved.

   b. The nature of the chemical equilibria associated with both phases 1 and 2, and the magnitude of the appropriate equilibrium constants.

   c. The magnitude of the activity coefficients for each of the species in each phase.

Since all of this information is rarely available, all calculations are approximate and have to be checked by experiment.

B. MULTIPLE DISTRIBUTIONS

As a general statement, it may be said that the efficiency of any multi-distribution process is controlled by

   a. The magnitude of the separation factor, α, that is, the degree of separation per individual equilibriation.

   b. The number of distribution processes.

   c. The nature of the equation describing the distribution between phases.

The significance of the shape of the distribution isotherm in regard to the shape of elution peaks was discussed in respect to adsorption chromatography in Chapter 14, Section II,C. Fortunately, the majority of liquid–liquid extractions, partition chromatographic studies, and gas–liquid distributions possess a linear isotherm (i.e., the distribution ratio does not vary with concentration) and yield peaks which are reasonably symmetrical.

The sharpness of elution peaks is influenced by the number of distributions and the magnitude of the separation factor. This is illustrated diagrammatically in Fig. 15.11.

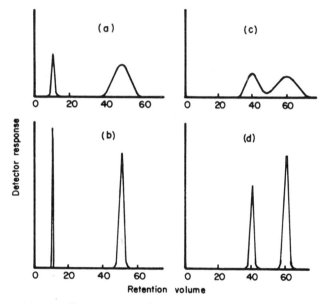

FIGURE 15.11   Diagram illustrating the effect of separation factor and number of theoretical plates on the degree of separation of two substances having linear distribution isotherms.

(a) $\alpha = 5$,   n = 100;   (b) $\alpha = 5$,   n = 1000
(c) $\alpha = 2$,   n = 100;   and (d) $\alpha = 2$,   n = 1000

For any given solute, the sharpness of the peak is a function of the number of transfer processes.

With countercurrent apparatus of the Craig type, the number of transfers is a known factor, but with chromatographic columns it is impossible to determine this value directly. In fact, the usual practice is to estimate the number of distributions by measuring the sharpness of an elution peak.

A number of mathematical expressions have been proposed to relate peak parameters with the number of distributions (n), and one of the simpler ones is

$$n = 16(R_v/\Delta V)^2$$

where R_v is the observed retention volume (or time) and ΔV is the breadth of the peak at the base line (as illustrated in Fig. 15.14). [With a constant rate of flow of mobile solvent, retention volume = (retention time) × (flow rate).]

C. THEORETICAL TREATMENTS OF THE CHROMATOGRAPHIC PROCESS

A number of theoretical treatments have been proposed to explain the behavior of solutes on a chromatographic column. Those of particular interest
are the plate theory of Martin and Synge, subsequently modified by Glueckauf;
the rate theory of van Deemter et al.; and the theory developed by Golay for
capillary columns.

In the plate concept of column operation (introduced in Section II,C), the
chromatographic column is thought of as a series of plates containing a stationary
phase through which a mobile fluid is allowed to percolate and run into the next
plate. It is assumed that the volume of stationary phase is the same in each
plate; that the volume of mobile phase is constant from plate to plate; that in
any plate the two phases are in equilibrium; and that the magnitude of the
distribution coefficient is independent of solute concentration.

Under these conditions, the mathematics is similar to that discussed in regard
to discontinuous counter extraction, and the relative distribution of any solute
among n plates is obtained by expanding the expression $(a + b)^n$ where $a =
RD/(RD + 1)$ and $b = 1/(RD + 1)$, D being the appropriate distribution ratio
(or partition coefficient); and R the phase volume ratio (volume of mobile
phase/volume of stationary phase).

The simple plate theory illustrates several important features of chromatographic development but it does not allow for the continuous flow of solvent.
To allow for this factor Glueckauf has derived an appropriate set of equations
and has shown that with a small number of plates the distribution of solute
approximates to a Poisson type of distribution, while the distribution pattern
approximates a normal or Gaussian curve when the plate number exceeds about
one hundred.

The expressions derived by Glueckauf to relate the number of theoretical
plates and the separation factor with the fractional band impurity (degree of
overlap of bands) are most conveniently presented as a family of curves. The
mathematical treatment clearly shows that when the separation factor is small,
say 1.05, a large number of plates are required to separate the two substances so
that the fractional band impurity is less than one percent. The treatment also
demonstrates that for a given number of plates, the band impurity is greatest
when the mole fractions of the two substances to be separated are equal.

Although the plate concept is the simplest to grasp, it involves a model based
on several assumptions.

Development of a chromatogram is actually a continuous countercurrent
process in which no segment is in a true equilibrium state. The nonideal
behavior can be attributed to nonuniformity in the column packing, diffusion
of solute in the mobile phase, and finite rates of transfer of the solute between
the mobile and nonmobile phases.

The rate theory considers the chromatographic process in terms of the kinetic
factors involved. Most investigators have based their arguments on gas-liquid

chromatography, but the principles of the treatment are identical for all chromatographic systems.

Because time is required for a system to reach equilibrium, the kinetic approach predicts that the effective number of transfers is inversely proportional to the velocity of the mobile phase. Expressed in another way, the slower the rate of elution, the sharper the emergent peaks. This effect is opposed by the other two factors not considered in the plate theory, namely, packing effects and solute diffusion.

Irregular packing leads to variations in flow rate across and down the column. The variable flow rates cause distortion of the zone fronts and result in a broadening of the eluted peak. The velocities in different zones may be assumed to be distributed statistically about some average velocity, which means that the width of the velocity distribution function depends only on the packing and not on the average flow velocity.

Diffusion of the solute in the mobile phase also produces band broadening. As the diffusion of the solute is proportional to the time the band spends in the column, it is inversely proportional to the rate of flow of the mobile phase.

With a large number of plates or transfers the kinetic theory and plate theory give almost identical results. However, the kinetic theory possesses the advantage that it can explain the influence of the rate of mobile phase flow and the observed broadening of bands.

For comparing the efficiencies of different columns, a convenient quantity is the Height Equivalent to a Theoretical Plate (HETP) where HETP = Column length/n, and the two theories can be combined in an equation of the form:

$$\text{HETP} = A + B/u + Cu$$

where A is a measure of column inhomogeneity; B is the diffusion coefficient of the solute in the mobile phase; C is related to the observed nonequilibrium or stochastic broadening effect; and u is the velocity of the mobile phase.

The equation as written is better known as the simplified form of the van Deemter equation.

The full equation derived by van Deemter, using a mass transport approach, is

$$\text{HETP} = 2\lambda\, d_p + \frac{2\gamma D_g}{u} + \frac{8k\, d_f^2 u}{\pi^2(1+k)^2 D_L}$$

$$\underbrace{\text{eddy}\quad\text{molecular}}_{\text{axial diffusion}}\quad \begin{array}{c}\text{nonequilibrium}\\ \text{effect}\\ \\ \text{or resistance to}\\ \text{mass transfer}\end{array}$$

where the symbols represent the following terms:

λ measure of the packing irregularities
d_p average particle diameter
γ correction factor for the tortuosity of the interparticle spaces

D_g diffusion coefficient of species in gas phase
D_L diffusion coefficient of species in liquid phase
u average linear gas velocity
k partition coefficient, i.e., fraction of sample in the liquid phase divided
          by the fraction in the vapor phase
d_f average thickness of the liquid film

If HETP is plotted against the velocity of the mobile phase, a hyperbola is obtained similar to that shown in Fig. 15.12. It can be seen that for maximum efficiency, (minimum HETP), the constants A, B, and C should be as small as

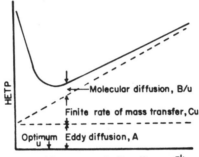

FIGURE 15.12   A plot of the van Deemter equation.

possible, and the velocity u should be adjusted to the optimum value indicated by the graph.

The equation indicates several ways in which the performance of a column may be improved.

For example, the magnitude of A should be reduced by using particles of smaller size, i.e., by varying d_p. In practice, this effect is countered by packing difficulties (larger λ) and the best particle size seems to be within the range of 50–80 BSS.

The value of the molecular diffusion coefficient B increases with decreasing molecular weight of the carrier gas; hence at low flow rates (where the contribution of B/u is significant) the best performance is obtained with the heavier gases such as carbon dioxide and nitrogen. On the other hand, the sensitivity of some gas detectors is greatly enhanced by using the lighter gases, hydrogen and helium, and these gases are better carriers at high flow rates, because B/u is negligible and resistance to mass transfer in the gas phase is much less with these gases.

The term designated resistance to mass transfer also includes the factors of partition coefficient and volume ratio considered in preceding paragraphs. Decreasing the film thickness should improve efficiency significantly (since d_f

is squared) but such benefits can be reduced by the change in phase volume and hence change in k'.

These theories predict that very high efficiencies should be achieved in columns which are simply long capillary tubes whose inner wall is coated with a thin layer of stationary phase. The equation developed by Golay for calculating the HETP in a round capillary column of radius r is:

$$\text{HETP} = \frac{2D_g}{u} + \left[\frac{1 + 6k + 11(k)^2}{24(1 + k)^2} \cdot \frac{r^2}{D_g} + \frac{2}{3}\frac{k}{(1 + k)^2}\frac{d_f^2}{D_L}\right]u$$

which can be simplified to

$$\text{HETP} = B'/u + C_G'u + C_L' \cdot u$$

C_G' and C_L' account for the resistance to mass transfer in the gaseous and liquid phases, respectively. It may be noted that in this case a term involving column packing (A) is not required.

Although there is still great interest in the mathematical analysis of the operation of chromatography columns, in the majority of cases the best conditions for separating mixtures are usually determined empirically.

IV. PARTITION CHROMATOGRAPHY

A. COLUMN OPERATION

As indicated in a preceding discussion (Section III,B), partition chromatography is a countercurrent extraction process which depends upon the partition of solutes between a mobile phase and a water phase held in some way by a solid support.

The support material should be inert to the substances being separated but must adsorb and retain the stationary phase. For maximum efficiency it should be mechanically stable and expose a large surface area to the flowing phase. The materials most commonly used are silica gel, cellulose powder, starch, and diatomaceous earths (kieselguhr, Celite, etc.). These solids are capable of retaining between 0.5 and 1 ml of liquid phase per gram.

The columns and other apparatus used are much the same as for the adsorption technique. Because they minimize axial diffusion, long narrow columns (e.g., 25–30 cm long, 1 cm diameter) are preferable to short wide ones. The columns are packed by adding small portions of a slurry prepared by mixing the moist stationary phase with mobile phase, each addition being tamped down with a rammer (e.g., a glass rod with a flattened end). The excess solvent is either removed with a pipette or is allowed to run through the column. The final column should be packed uniformly throughout its length. The flow of solvent through such columns tends to be rather slow and is sometimes accelerated by the application of air pressure to the top.

In most applications the stationary phase is aqueous; the mobile phase is an organic liquid or mixture. Some typical separations are shown in Table 15.2.

In "reversed phase" chromatography, a hydrophobic support is used to retain an organic stationary phase, and development is achieved by a flow of an

TABLE 15.2

TYPICAL SEPARATIONS ON PARTITION COLUMNS

Separation	Mobile phase	Stationary phase	Support
C_1–C_4 alcohols	$CHCl_3$ or CCl_4	Water	Celite
Phenols	MeOH/n-BuOH/$CHCl_3$	Water	Cellulose
Fatty acids,	$CHCl_3$/n-BuOH	Water	Silica gel or
Acetylated amino acids			kieselguhr
DNP amino acids,			
DNP peptides			
Methoxy aromatic	$CHCl_3$/n-BuOH	$0.25\ M\ H_2SO_4$	Silica gel
acids			
Amino acids	n-PrOH or n-BuOH/HCl	Water	Starch
Purines, pyrimidines	n-PrOH/HCl	Water	Starch,
			Celite,
			kieselguhr
Lipids	Various	Water	Silica gel
Ammonia and amines	BuOH/H_2O	Water	Celite,
			kieselguhr
Sugars and derivatives	BuOH/H_2O	Water	Cellulose
Inorganic ions	Various, e.g., n-BuOH/	Water	Cellulose
	HCl; acetone/HCl		
Paraffins and cyclo			
paraffins	i-PrOH/benzene	Aniline	Silica gel
Lanthanides	HNO_3	Tri-n-butyl	Kieselguhr
		phosphate	
Corticosterones,			
17-hydroxy	Toluene	Propylene glycol	Cellulose

aqueous phase. This approach is useful when the substances to be separated are very soluble in organic solvents.

The samples are added at the top of the column and the separated substances are identified either by continuous monitoring of the eluate or by collecting suitable fractions for subsequent examination. The quantitative analysis of the isolated fractions is often facilitated by using larger samples and larger columns. However, this aspect has become less important in recent years with the introduction of moving wire gas chromatographs. With these instruments, a fine wire is drawn through the eluate leaving the column, and the attached film passes into the sample inlet of a gas chromatograph. In this way the recorder of the gas unit serves to monitor the composition of the liquid eluate.

Column partition chromatography is extremely useful for isolating pure components in preparative studies. By using a fairly large column (e.g., 50 cm long, 4 cm diameter) gram quantities of material can be separated satisfactorily.

For qualitative analysis of mixtures, speed, and convenience dictate that the thickness of packing should be minimal. With silica gel and alumina supports this ideal is achieved by means of the thin layer technique (cf. Chapter 14, Section II,C). For separations based on a cellulose support, packed columns are replaced by strips of filter paper. The modified technique is known as paper chromatography, and for many years this was the most widely used chromatographic technique.

B. Paper Chromatography

A separation by paper chromatography is achieved by placing a drop of the test solution (e.g., 0.02 ml) near one end of a strip of filter paper and a solvent is allowed to move over the strip by capillary flow. The mobile phase is usually saturated with water to prevent dehydration of the paper during the solvent flow and it may also contain additional reagents such as acids, complexing agents, etc.

As the mobile phase travels over the initial sample spot, the individual solutes become distributed between the moving solvent and the water film retained by the cellulose fibers. Under ideal conditions the components migrate, at differential rates, as zones just slightly larger than the original spot. The flow of solvent is stopped after travelling some predetermined distance and the position of the separated species is located by a visualization process. This normally involves spraying the paper with a reagent which causes the components to form a derivative which is either colored or fluoresces when exposed to ultraviolet light.

The isolated species are identified through their behavior in respect to the visualizing reagent and through the observed R_f values (cf. Chapter 14, Section II,C).

The three major variables in paper chromatography are:

1. Type of paper used.
2. Composition of the mobile phase.
3. Selectivity and sensitivity of the detection reagents.

Papers differ in respect to porosity and degree of hydration. These properties influence the rate of movement of the mobile phase and the magnitude of the phase volume ratio. The influence of these factors on the efficiency of separation processes has been discussed in preceding sections. In addition, the manufacturing process tends to cause paper to possess different flow rates in different directions; hence in comparative studies it is necessary to ensure that the grain is oriented in the same direction in each test. The cellulose fibers can absorb appreciable amounts of impurities during paper manufacture, and the presence of some acidic groups makes filter paper equivalent to a weak cation-exchanger of limited capacity. These two factors can cause some undesired effects, e.g., multiple spotting and tailing, and some workers have recommended that papers should be washed with various solvents to remove impurities before use. This is usually unnecessary if one uses the papers now prepared specifically for use in chromatography, and if these are stored in an atmosphere free of fumes.

The flow rate of solvent at a fixed temperature is controlled by the density and thickness of the paper. The flow rate also increases with decreasing viscosity of the solvent (and therefore with increases in temperature). To ascertain the conditions equivalent to the optimum flow rate indicated by the van Deemter equation, one should therefore investigate the effect of paper type, temperature, and solvent composition.

The composition of the mobile phase is dictated by the nature of the species to be separated. In most applications, this phase plays two roles. It acts as the mobile extracting agent and it serves to convert the individual components into chemical forms which differ in terms of distribution ratios.

Because of this dualism, the selection of an appropriate mobile phase for a new separation procedure has often to be determined by empirical experiments. However, selection of systems of potential value is facilitated by the tabulations presented in the "Handbook of Analytical Chemistry" and in monographs on paper chromatography.

These tabulations quote the observed R_f values of a wide range of inorganic and organic species using several different types of mobile phase. The lists are fairly comprehensive, and extensive. Besides quoting the composition of the mobile phase, the tables indicate the type of paper used, the developing technique used, and the reagent required to detect the isolated component.

For example, using descending chromatography on Whatman No. 1 paper at 21°C and a mobile phase composed of 2-butanone:dimethyl ketone:water: anhydrous formic acid (40:2:6:1), the R_f values of citric acid, tartaric acid, oxalic acid and crotonic acid are respectively equal to 0.48, 0.32, 0.66, and 0.93. Spraying the paper with dilute potassium permanganate serves to locate the zones due to crotonic, oxalic, and tartaric acids. Spraying with iron (III) chloride solution gives a yellow zone in the presence of tartaric and citric acids.

Thus, in this particular case, reference to a table indicates a procedure which should provide excellent separation of a mixture of these acids.

The information is organized in terms of types of chemical compounds; hence in studies where the only information available is a vague suspicion of the compound types (e.g., in natural product surveys), a useful starting point is to apply the chromatographic conditions recommended for the compound type suspected.

The R_f values and color reactions provide circumstantial evidence of the composition of isolated components. Confirmation may be obtained by subjecting a pure sample of the material tentatively identified to similar treatment. That is, a new chromatogram is run with alternate sample spots containing the mixture, and predicted components. Absolute confirmation sometimes requires isolation of the separated zone, extraction of the solute, and microanalysis (including an infrared absorption study) of the isolated chemical species.

In order to allow the paper to come to equilibrium with the vapors of the solvent, and to prevent complications arising from evaporation of the solvent during development, the paper chromatograms and solvent are usually totally enclosed in a sealed vessel. Some solvents, when mixed in the proportions recommended, give a two phase system. It is the custom in such cases to place the aqueous layer in a separate vessel at the bottom of the tank, the organic phase being retained in the solvent trough.

Since the solvent moves along the paper strip by capillary action, development does not require an overhead reservoir of solvent; thus there are three major development techniques.

In the ascending technique, the strip of paper is supported at its upper end, and the liquid is drawn up the paper from a reservoir of solvent located in the base of the chromatography tank.

In the descending technique the solvent flows from an elevated container down the paper. It possesses the advantage that the solvent can be allowed to flow off the paper if it is desired to eluate the separated components.

The technique of horizontal chromatography involves a slightly different principle. The sample spot is placed at or near the center of a piece of circular filter paper. The solvent is fed up to this central spot by a wick dipping in the solvent reservoir. The solvent spreads out radially, causing the sample components to spread into a series of concentric bands.

The principles of these three techniques are illustrated in Fig. 15.13.

With complex mixtures, greater separation of the components may be obtained by two-dimensional development. The mixture is first developed in one direction using solvent A. The paper is then dried, turned at right angles and developed in this new direction with another solvent B.

If the substances being chromatographed are only very sparingly soluble in water, they tend to move with the solvent front instead of being separated. Under these circumstances the reversed phase approach can be advantageous. Paper can be rendered water-repellent by treating it' with rubber latex or silicones. In this condition it absorbs the organic component of the solvent mixture, which then becomes the stationary phase. Reversed phase methods can be applied to all the various forms of paper chromatography.

The convenience of paper chromatography can be combined with the specificity of ion exchange by using ion-exchange papers. These papers are

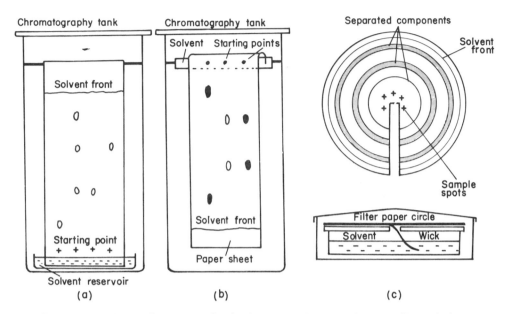

FIGURE 15.13   Paper chromatography development techniques. (a) Ascending technique. The paper is either supported at its upper end by a frame or is formed into a cylinder which is self-supporting; (b) descending technique with an elevated solvent reservoir; and (c) horizontal technique, showing a developed radial chromatogram. From "Fundamental Principles of Chemical Analysis," W. F. Pickering. Used with permission of Elsevier Publishing Company, Amsterdam, copyright © 1966.

prepared by modifying the cellulose through the incorporation of ion-exchange groups such as $-OPO_3H_2$, $-COOH$, $-C_2H_4NH_2$, etc., or by blending an ion-exchange resin with the cellulose used in making the sheets. The apparatus used is that of ordinary paper chromatography and the methods are in general the same. The one difference is that the eluting solution must contain some ion which can displace those to be separated.

The amount of substance which should be put on the paper for a chromatographic separation varies considerably; e.g., the separation of some metals and amino acids can be done on as little as 0.1 μg. However, most studies require 20–50 μg of material.

Because of the small amounts of material required, the simplicity of the technique, and the wide range of variable experimental conditions, paper chromatography has been used in almost every field of scientific study.

It is most widely employed in separations of a biochemical nature, e.g., for the separation of amino acids and peptides in the investigation of protein structures.

Other applications include the separation of alkaloids, steroids, and the various extracts obtained in natural product studies. The procedure is suitable for the separation of radioisotopically labeled compounds, and it has been proposed as a field method for the detection and estimation of metals in soils and geological specimens.

Paper chromatography is of great value for qualitative analysis but it is very limited in respect to quantitative studies, because in most cases the separated components have to be retrieved from the paper before analysis, and the precision and accuracy achieved is quite poor.

The mechanism of separations based on paper chromatography involves more processes than simple multiple partition between two liquid phases. Reaction with the cellulose and attached functional groups occurs regularly and solutes are actually distributed in a three-phase system during development of a chromatogram.

In the absence of a complete understanding of all the processes involved, the choice of solvent is rather arbitrary, and paper chromatography remains a technique in which the individual worker must find the best solution for his own problems.

V. GAS-LIQUID CHROMATOGRAPHY

A. COLUMN PACKINGS

Gas-liquid chromatography is the most elegant and useful of all the chromatographic methods.

The equipment used is similar to that described in connection with gas-solid chromatography (cf. Figure 14.5); the two differ only in the packing of the column.

In gas-liquid chromatography, a solid porous support (diatomaceous earth or crushed firebrick) is coated with a thin film of a liquid phase. The coatings are normally nonvolatile and chemically inert. The support, if not evenly coated, can introduce some undesirable effects such as tailing or catalysis. These effects can be minimized by treating the solids with the vapor of dimethyl-dichlorosilane prior to coating with the stationary phase.

As indicated by the van Deemter equation, particle size and packing are important. The particle sizes used normally fall within the range 30 to 120 mesh BSS and narrow fractions (80 to 100 or 100 to 120 mesh) are preferred because they lead to more even packing in the column.

A number of nonporous substances (which cannot hold more than 3% of their weight of liquid phase) have been found to give less tailing. Materials in this classification are glass beads or powder, sodium chloride crystals, metal helices, and granular polytetrafluorethylene.

A great variety of substances have been used as the coating liquid, and broadly speaking, the success of a given gas separation depends upon the choice of a satisfactory liquid.

The requirements for a good liquid stationary phase are:

1. It must possess a very low vapor pressure (<0.1 torr) at the temperature of operation, otherwise it will gradually bleed off the support;
2. It should be reasonably fluid over the desired range of operating temperatures;
3. It must be thermally stable;
4. It should be inert toward the solutes; or if it reacts, it should do so fast and reversibly;
5. It must possess some solvent power for the mixtures to be examined in the gas chromatograph.

The large number of liquids suitable for use as stationary phases provides a bewildering choice, but since many possess similar chemical and physical properties, it has been proposed that whenever possible, chemists should choose stationary phases from a restricted list such as that shown in Table 15.3.

For each material, there is an operating temperature above which loss of the stationary phase as a vapor becomes excessive. The choice of a particular liquid, therefore, depends on the column temperature to be used and on the chemical nature of the compounds to be separated.

There are some very selective stationary phases. For example, silver nitrate dissolved in polyethylene glycol forms loose adducts with olefins and is specific for such compounds. Tri-*o*-thymotide dissolved in tritolyl phosphate selectively retards straight chain organic compounds in the presence of branched chain compounds.

For high temperature studies, molten salt mixtures (e.g., the eutectic mixture of $NaNO_3$, KNO_3, and $LiNO_3$) have been introduced with apparent success.

Although the proportion of stationary phase does have an effect on the performance of a column, it is usually not critical.' The amount used is generally between 5 and 25% by weight, and is uniformly dispersed by making a slurry of the appropriate amounts of the solid and the stationary phase dissolved in a volatile solvent, e.g., ether. The solvent is then removed by heating and the solid carefully packed into the column to give a minimum of voids.

B. Capillary Columns

Excellent resolution of mixtures has been achieved using capillary columns whose inner wall is coated with a thin layer of stationary phase. The columns used vary in length from 20 to 300 feet or more with internal diameters of 0.01 to 0.05 inches. The materials commonly used are stainless steel, copper, nylon, and glass. There are several methods for applying the internal coating. In one of these, about one percent of the column length is filled with a 10% solution of the stationary phase in a volatile solvent such as ether. The column of liquid is then forced through the capillary, leaving a film on the walls, with the excess solvent being removed by a flow of gas through the column.

TABLE 15.3

STATIONARY PHASES FOR GAS-LIQUID CHROMATOGRAPHY

Stationary phase	Maximum operating temperature (°C)	Applications
Dimethyl formamide	20	Low boiling paraffins and olefins
n-Hexadecane	50	Lower hydrocarbons, fluorides
Silver nitrate in propylene glycol	50	Olefins, cyclic hydrocarbons
Polyethylene glycol	100	Aldehydes, ketones, epoxides, alcohols
Benzyldiphenyl	120	Aromatic compounds and halogen compounds in general
Dinonyl phthalate	130	Esters and ketones
Squalane (hexamethyl tetracosane)	150	Hydrocarbons, particularly branched chain and cyclic alkanes
Diglycerol	150	Alcohols, phenols, aliphatic amines
Polyethylene glycol adipate	150	Esters, fatty acids
Apiezon grease M	150	Hydrocarbons,
L	230	Fatty acids and esters, ethers, phenols, boranes
Carbowax 20M	250	Aromatics, alcohols, amines, ketones, essential oils
Silicone oils and greases	200–400 (depending on type)	General applications, steroids, alkaloids, pesticides, acids, methyl esters

Capillary columns are claimed to give greater resolution in less time using lower temperatures than packed columns. On the other hand, many refinements are required. For example, because of the small amount of stationary phase present on the walls, sample size is restricted to less than 1 μg, and the simplest way of ensuring the reproducible introduction of such a sample is to use a stream splitter. A small sample is first injected in the carrier gas stream by means of a microsyringe; the stream is then divided so that only a known small fraction of the gas plus sample finds its way to the column, the rest going to waste.

The very small samples used require that the detector units should have a small volume, high sensitivity and rapid response. In some cases the separation is so quick that the detector signals need to be fed to a cathode ray oscilloscope instead of a fast response recorder.

The care required in admitting the samples, and in detecting and recording, detracts from the popularity of this procedure. The main function of capillary column chromatography would seem to be the resolution of extremely complex mixtures and the speeding up of lengthy routine separations now performed on packed columns.

C.   SAMPLE INTRODUCTION AND DETECTION

While the composition of the column packing determines the ability of a chromatograph to effect separations, the factors which determine the analytical usefulness of the technique are sample introduction, and detection of the eluted components.

Consideration of the ideal chromatographic process indicates that the sample should be introduced into the first theoretical plate as a compact plug. The initial sample may be gaseous, liquid, or solid, but on introduction to the column there must be sufficient heat to convert it to the vapor form. The size of the sample is dictated by the capacity of the column and the sensitivity of the detector. For example, with a katharometer about 10 μl of liquid sample would be used; with an ionization detector this needs to be reduced to about 0.5 μl.

A sample of gas or liquid can be injected into the carrier stream through a serum cap at the top of the column by means of a hypodermic syringe. A swift, neat motion of the plunger is necessary to give the plug effect. Solid samples are more difficult to introduce, but these can be sealed into a thin walled glass vial which is inserted into a heated injector port and then crushed from the outside. An alternative method is to dissolve the solid sample (e.g., a grease) in a volatile solvent and then inject the solution with a syringe.

The purpose of detectors is to monitor the column effluent, measuring variations in its composition; many types have been proposed.

Differential detectors measure the instantaneous concentrations or the instantaneous rate of emergence of a component and give a zero signal when pure carrier gas is passing through them. Integral detectors, on the other hand, accumulate the sample components and the signal gives the total amount which has emerged up to a given instant. Signals from differential detectors are usually integrated electronically for quantitative analysis.

Most instruments are fitted with differential detectors and these may be divided into two major groups: high sensitivity, ionization detectors (limit 10^{-12} mole), and low sensitivity units (limit about 10^{-6} mole).

The principles associated with two low sensitivity units (thermal conductivity cells and flame temperature detector) have been discussed in Chapter 4, Section IV, and therefore need not be considered again.

The ionization detectors measure the increase in current produced when eluted substances passing through them are ionized.

In the flame ionization unit, combustion in a hydrogen flame is used to produce the ions responsible for the increase in current. The process is not completely understood but the ion current which flows between two oppositely charged electrodes is approximately proportional to the number of carbon atoms entering the flame. The burning jet usually forms the negative electrode of the cell, and the other electrode is a piece of brass or platinum wire mounted at some point near the tip of the flame.

In β-ray ionization detectors, the ionization of the sample molecule is achieved by subjecting the material to β radiation (e.g., from a ^{90}Sr source). The current flow is measured between electrodes maintained at a potential difference of 300 to 1200 V.

Direct sample ionization yields a low sensitivity (detection limit about 10^{-5} moles) but this can be increased manyfold by using argon as the carrier gas; β rays excite some of the argon atoms to a metastable energy level. The sample molecules are then ionized by collision with excited argon atoms. Metastable argon atoms have an excitation level of 11.7 eV; hence, substances with ionization potentials greater than this (e.g., H_2, N_2, O_2, CO_2, H_2O, fluorocarbons) are not detected.

Purified helium has an excitation level of 19.8 eV which means that gases not sensed in an argon detector can be identified by using helium as the carrier gas.

During ionization, electrons are produced, and in electron–capture detectors the signal depends on the capture of electrons by the various substances being sensed; this causes a reduction in the ion current. The detector resembles the argon and helium detector except that the cell geometry is different and that the potential applied across the electrodes is only about 20 V. This type of detector possesses a high degree of sensitivity and specificity for molecules which have high electron affinities such as oxygen, halogens, and compounds containing oxygen and the halogens.

An interesting and simple form of an integral detector is the Janak absorption method. The column is operated in the usual way but the carrier gas is carbon dioxide which is absorbed by a solution of potassium hydroxide as it leaves the column. Any nonabsorbed gases pass into a gas burette where their volume is measured at constant pressure. (The method is highly suitable for detecting gases with relatively low boiling points such as the low molecular weight hydrocarbons.) If the volume of gas in the burette is plotted against time, a stepped curve is obtained where the height of the steps is proportional to the amount of the particular constituent in the original mixture.

D. RETENTION VOLUMES

The detector units merely indicate that some component of the mixture has left the separation column. For positive identification of the components it is necessary to trap them as they emerge, and to identify each separately by techniques such as infrared and mass spectroscopy. Such a procedure is tedious, and for routine separations the components are usually identified by their chromatographic behavior. The volume of carrier gas required to transport a given component through the column is compared with the volume required for standard substances. Similarity of behavior, measured in terms of retention volumes, is taken as circumstantial evidence of the nature of the individual component.

The retention time, t_R, is defined as the time interval between the injection of the sample and the maximum of the elution peak (as shown in Fig. 15.14).

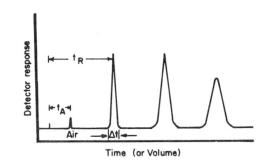

FIGURE 15.14   Typical isothermal gas chromatogram which shows the mode of measuring retention times (or volumes).

The retention volume, V_R, is proportional to the retention time and to the flow rate. The flow rate (F) is normally measured at the end of the column; hence the true flow rate (F_C) at the column temperature, T_C, is

$$F_C = F \times T_C/T_{\text{room}}$$

The retention volume at column temperature and atmospheric pressure is then:

$$V_R = t_R \cdot F_C$$

However, a pressure gradient exists down any column, and it is necessary to introduce an additional correction factor, j, to correct the gas volume for compressibility in the column.

$$j = 3(p_i^2/p_o^2 - 1)/2(p_i^3/p_o^3 - 1)$$

where p_i is the pressure of carrier gas at the column inlet, and p_o is the pressure at the column outlet.

The corrected absolute retention volume, V_R°, is thus:

$$V_R^\circ = t_R F_C j$$

It can be seen from Fig. 15.14 that at the point corresponding to V_R° (i.e., t_2) one half of the solute has been eluted and one half remains in the column. The one half remaining in the column must be distributed between the gas space in the column, V_G, and the volume of liquid phase, V_L, in accordance with the proportions required by a partition constant, k. If the concentrations of solute in the liquid and gas phase are C_L and C_G respectively, then:

$$k = \frac{C_L V_L}{C_G V_G} = K \cdot V_L / V_C$$

where $K = C_L / C_G$ is the partition coefficient. The magnitude of this coefficient is determined by the solute, solvent and temperature.

The ratio of the amount of solute remaining in the column, to the amount eluted from the column at the peak maximum, is given by the expression:

$$\frac{0.5}{0.5} = 1 = \frac{(C_G \times V_G) + (C_L \times V_L)}{(C_G \times V_R^\circ)} = \frac{V_G + KV_L}{V_R^\circ}$$

From this it may be deduced that

$$V_R^\circ = V_G + KV_L$$

Because V_G is a property of the column and the way it is packed, it is easily determined by noting the volume or time (t_A in Fig. 15.14) required for a trace of air injected along with the sample to reach the end of the column and make a small pip in the recorder signal.

The value for V_R° is specific for a single column having particular V_G and V_L values. For comparative purposes it is preferable to use a unit which can be applied by any chromatographer merely by inserting his own particular column parameters.

Such a quantity is the specific retention volume, V_g, which is the net retention volume of the substance per gram of stationary phase, at 0°C.

$$V_g = \frac{V_R^\circ - V_G}{W_L} \cdot \frac{273}{T_C} = \frac{j \cdot F_C \cdot (t_R - t_A)}{W_L} \cdot \frac{273}{T_C}$$

where W_L is the weight of liquid phase in the column.

If one recalls that $V_R^\circ = V_G + KV_L$, then

$$V_g = \frac{KV_L}{W_L} \cdot \frac{273}{T_C} = \frac{K}{\rho_L} \cdot \frac{273}{T_C}$$

since $W_L = V_L \cdot \rho_L$ where ρ_L is the density of the liquid phase.

The magnitude of the specific retention volume is thus directly related to the partition coefficient, a characteristic of the solute-solvent combination.

Relative retention values (α) are obtained by comparing the specific retention volumes (V_g) of the solute under consideration with some standard solute whose behavior on the particular column in use is precisely known.

$$\alpha = \frac{V_{g_1}}{V_{g_2}} = \frac{t_{R_1} - t_A}{t_{R_2} - t_A}$$

Whenever a relative retention is cited, it should be accompanied by information on the solute, liquid phase, temperature and standard solute.

The specific or relative retention volumes of a wide range of representative groups of compounds (on a series of stationary phases) are listed in the "Handbook of Analytical Chemistry" (Tables 10.58 to 10.78). Other extensive compilations of this useful data are to be found in the monographs and journals associated with gas chromatography. In fact, there are few common applications for which some appropriate information is not available, and if the composition of the mixture is known, suitable experimental conditions can usually be selected from a literature survey.

On the other hand, if the nature of the sample mixture is virtually unknown, optimum conditions may have to be found by trial and error.

The separation of components is a matter of relative volatility, and the separation factor (or relative retention value), α, can also be defined as

$$\alpha = V_{g_1}/V_{g_2} = a_2 p_2^\circ / a_1 p_1^\circ$$

where p_1° and p_2° represent the vapor pressures of solutes 1 and 2 respectively at a given temperature, and a_1 and a_2 represent primarily the activity coefficients of the sample components in the liquid phase.

For an ideal solution ($a = 1$), α is determined by the solute vapor pressures. On the other hand, separation of close-boiling mixtures depends on alteration of the a_2/a_1 ratio by solute-solvent interaction. The solute-solvent effects which influence the magnitude of a can involve forces between permanent dipoles, forces between a permanent and an induced dipole, and dispersion forces. Strong dipole–dipole interaction is observed wherever hydrogen bond formation is possible.

In other systems metal complexes are formed, e.g., solutions of silver nitrate in glycols selectively absorb olefins because of the weak organometallic complexes formed.

In general, nonpolar liquid phases are nonselective, i.e., in the absence of special forces between solute and solvent, the volatility of the solute is determined primarily by its vapor pressure. With this type of stationary phase, elution occurs in the order of the boiling points of the solutes.

With polar liquid phases, polar components of the mixture are preferentially retarded and the polarity of the liquid phase is important.

Interaction is also a function of chemical type, e.g., liquids containing aromatic rings tend to selectively detain aromatic compounds.

As an example of the effect of a change in solvent one may quote the separation of *n*-heptane and water. The separation factor for this pair varies by a factor of over one thousand on changing the stationary phase from a hydrocarbon to glycerol.

E.   THE EFFECT OF TEMPERATURE

The most suitable operating temperature for a gas–liquid column is that which yields sufficient vapor pressure of the solutes to ensure rapid transport through the column while retaining the ratio of vapor pressures, i.e., p_2/p_1, in the vapor phase at a value which permits separation.

The effect of temperature on the specific retention volume of an individual solute can be predicted from thermodynamic considerations.

It has been noted in a preceding section that the specific retention volume is given by

$$V_g = 273 \, K/T_C \rho_L$$

Taking logarithms of both sides

$$\log V_g = \log K + \log 273/\rho_L T_C$$

The partition coefficient, K, is an equilibrium constant and its temperature dependence can be derived from the Clausius-Clapeyron equation

$$\log K = - \Delta H/2.303 \, RT + C$$

in which H is the pertinent enthalpy change (ideally the heat of vaporization); R is the gas constant; T is the absolute temperature; and C is an integration constant.

Substitution in the specific retention volume equation yields

$$\log V_g = - \Delta H/2.303 \, RT_C + \log 273/\rho_L T_C + C$$

In a number of cases a plot of $\log V_g$ against the reciprocal of T gives a straight line. Better fits are obtained if the equation is written as a three-constant equation of the so-called Antoine type, i.e., $\log V_g = [-A/(t + B)] + C$.

The slope of the plot is $-A$; B is an adjustable quantity (value around 273) which is added to the centigrade temperature to provide the best straight line plot; and C is a constant representing the intercept of this plot on the $\log V_g$ axis.

The slope, A, varies with the nature of the solute and liquid phase, and is, therefore, a measure of the degree of interaction. The same solute on different liquid phases can yield different values of A, B, and C. Similarly, different solutes on the same liquid phase can be expected to give different values for these three terms.

For transport through a column at a reasonable rate, the vapor pressure of a sample component should exceed 10 mm Hg at the temperature of the column. Since, at any given temperature, the vapor pressure of high boiling point components of a mixture can be much less than this, excessively large retention volumes may be required for elution of all components.

To eliminate this problem, it is recommended that the operator temperature-program his chromatographic study whenever the components of a sample differ greatly in boiling point. This procedure involves eluting the more volatile components from a column operating at a temperature suitable for the separation of these species in a reasonable time. The column temperature is then increased and the next group of components eluted. This process is repeated at further higher temperatures to accelerate the movement of the highest boiling point components. In the simplest case the column temperature is increased linearly with time. Alternatively, the temperature rise may be interrupted at regular intervals to provide periods of isothermal development. The best program is that which separates all components as sharp peaks in a minimum of time.

There are two other parts of the chromatograph in which the temperature must be controlled; the first is the injector port. To provide a plug of sample, rapid vaporization of the sample is required and this is achieved by keeping the injector area at a relatively high temperature. The temperature used should be consistent with the thermal stability of the sample, although a modern trend with materials of low volatility is to cause pyrolysis in the injector zone and subsequently examine the products of thermal decomposition.

The other zone in which temperature control is necessary is the detector area, since the temperature of this unit must be high enough to prevent constituents condensing on it. On the other hand, the sensitivity of many detectors (e.g., katharometers) decreases with increasing temperature, so the optimum temperature lies just above the boiling point of the highest boiling sample component.

F.   Applications

Gas-liquid chromatography is an extremely popular analytical tool because of its speed, simplicity, and versatility, coupled with a moderate capital cost for the equipment.

The recorder unit provides three types of data—retention time, peak size, and peak shape. These can be combined to yield qualitative and quantitative analyses of mixtures, or basic thermodynamic data associated with solute-solvent interactions.

For qualitative analysis, there is no equal to the gas chromatograph for separating sample components, but unfortunately the identification of the isolated species is left to the ingenuity of the analyst.

Similarity of retention behavior with that of a known sample, using several different columns at more than one temperature, is almost certain proof of identity. Unfortunately, this method requires both patience and a very extensive stock of pure standard chemicals.

An alternative is to calculate the relative retention volumes of the separated materials and compare them with the values tabulated in the literature. If the general type of compound involved is known, identification can be facilitated by referring to family plots.

The equation relating the relative retention volume to vapor pressures and activity coefficients can be expressed in the logarithmic form

$$\log \alpha = \log (p_2^o/p_1^o) + \log (a_2/a_1)$$

For a homologous series the value of a_2/a_1 tends to remain fairly constant. At a given temperature, p_2° for the reference compound is also fixed. Hence,

$$\log \alpha = -\log p_1^{\circ} + \log a_2 p_2^{\circ}/a_1$$

From this equation, one may predict, that for members of a homologous series, there should be a linear relationship between the logarithm of the relative retention value and the logarithm of the pure solute vapor pressure.

The vapor pressure of a solute is related to its boiling point, and in a homologous series the boiling point tends to vary regularly with the carbon number.

Accordingly, it is not surprising that in practice a linear relationship has been observed between $\log \alpha$ for the various members of a series and

1. the number of carbon atoms in the molecule,
2. the boiling point of pure solute, and
3. the logarithm of the relative retention volume observed on a different stationary phase.

These linear plots are known as "family plots." The ultimate procedure for identifying the components separated on the chromatograph involves subjecting the isolated compounds to study by infrared spectroscopy or mass spectrometry.

For quantitative studies it is possible (in principle) to relate the area under an elution peak with the amount of component in the original sample. The area recorded on a chart can be determined by means of a planimeter, by cutting out the peak and weighing the paper, or by estimating the area geometrically by triangulation. Alternatively, the detector signals may be fed through an electronic integrator during the recording process.

The measured areas are related to concentration through preliminary calibration studies using samples of known size. The calibration curve is not necessarily linear and the figures can lack precision due to the difficulty in accurately introducing small volumes of standard solutions. The proportionality constant between area and concentration tends to vary from compound to compound; hence a calibration is required for each component.

Despite these limitations, the method is usually more accurate than alternative procedures, and fortunately, often all that is required is an indication of the relative proportions of the components in a mixture.

Gas-liquid chromatography has proved so versatile that it is difficult to keep abreast of the many fields of application, which currently include:

1. Analysis of petroleum and derived products. Typical examples are the analysis of mixtures of hydrocarbon gases; the determination of the ratio of branched to normal paraffins; the analysis of gasolines, waxes, and oxidation products; and studies of reformates, sulfur and nitrogen compounds.

2. Analysis of lacquer solvents (a blend of hydrocarbons, alcohol, ketones and other oxygenated compounds) and the volatile components of paints.
3. Biochemical and clinical applications including fatty acid and steroid analysis.
4. Studies of the nature of the compounds present in natural products.
5. Isolation and identification of the trace components in food, including those responsible for taste, color and odor.
6. The determination of the fragrant essences in cosmetics and the nature of other ingredients.
7. The analysis of herbicides, pesticides, fertilizers, and other materials of interest to agricultural chemists.
8. The analysis of miscellaneous manufactured products such as plastics, coal tar products, alcoholic beverages, rubber products, soaps and synthetic detergents.
9. Determination of thermodynamic data related to gas-liquid reactions such as activity coefficients, heats of mixing, etc.
10. Inorganic separations.

The number of inorganic separations so far accomplished are comparatively few because of the nonvolatile nature of most inorganic species. This limitation is being overcome by preparing volatile derivatives and/or by using molten salt mixtures as the stationary phase.

Some of the volatile derivatives which have been successfully separated are the tetramethyl derivatives of silicon, germanium, tin, and lead; the acetyl-acetonates of beryllium, aluminum, and chromium; boranes, phosphorus compounds; and metal chlorides such as $TiCl_4$ and $SbCl_3$.

Clearly gas chromatography can be used only for gases and volatile substances. The range is continually being extended through the use of higher column temperatures (e.g., $>400°C$), the formation of derivatives, and the classification of pyrolysis patterns. Nevertheless, there still remains a large number of compounds, such as inorganic salts, dyestuffs, and proteins which cannot be volatilized (or even heated) without decomposition. These are preferably separated by other forms of chromatography.

VI. ASSIGNMENTS AND QUESTIONS

Question 1
a. Quote the phase rule and define the symbols used in this equation.
b. Carefully distinguish between a partition coefficient and the distribution ratio.
c. Explain the process known as fractional crystallization and indicate possible applications of this process.
d. What do you understand by the term reversed phase chromatography?

e. How may the term separation factor be defined in (1) solvent extraction, and (2) gas-liquid chromatography?

f. In solvent extraction it is proposed that the separation factor, α, is preferably either >100 or <0.01. What practical implications are associated with this factor?

g. Explain why the number of separations of inorganic species by gas-liquid chromatography is so small.

h. Discuss the possible influence of kinetics (e.g., slow equilibration) on the efficiency of chromatographic separation processes.

Question 2

Outline the principles of the following techniques:

a. Zone refining.

b. Simple solvent extraction.

c. Continuous extraction.

d. The Craig countercurrent extraction procedure.

e. Horizontal paper chromatography.

f. Two-dimensional paper chromatography.

Question 3

Prepare an essay on one (or all) of the following techniques:

a. Partition chromatography on columns.

b. Inorganic paper chromatography.

c. Gas-liquid chromatography.

The essay(s) should concentrate on the practical aspects of the techniques, e.g., procedure for packing columns, methods of sample application, alternative means of component identification, etc. Theoretical principles need not be considered.

Question 4

a. Explain the statement: "In the solvent extraction of organic compounds one need only consider the pH of the aqueous phase and the nature of the organic solvent; for the extraction of inorganic species additional organic and/or inorganic chemicals have to be added to the system."

b. Tabulate the various types of solvent extraction systems used for the separation of metal ions.

c. Write brief notes on the importance of (1) solvent adducts; (2) liquid ion exchangers; and (3) ion pairing, in the solvent extraction of metal ions.

d. Use the "Handbook of Analytical Chemistry" or other reference books to prepare a summary of the experimental conditions required to separate, by solvent extraction, the first named element from the others listed.

i. Ag from Cu, other metals.

ii. Al from Ca, Cu, Fe, Sr, Y, Zn.

    iii. B from Si.
    iv. Ca from Ba, Sr.
     v. Co from elements in steel.
    vi. Hf from Zr.
   vii. Li from K, Na.
  viii. Sc from Al, Be, Cr, Ti.
    ix. U from Bi, Th.
     x. Zn from Co, Mn, Ni.

e. Explain in terms of chemical equilibria why the distribution coefficient of metal chelates is a complex function of pH.
f. What is an extraction curve, and how is it used?

Question 5
a. Write down the equations which relate the experimental parameters with the height of a theoretical plate in (1) a packed chromatographic column and (2) a Golay capillary column. Define each term, and indicate why there is an optimum flow rate for the mobile phase.
b. Explain how the HETP of a column is determined from experimental results.
c. What is temperature programming and what are the advantages gained by using this procedure?
d. Briefly describe the major types of differential detector units used in gas chromatographs and indicate why their behavior varies in some cases with the nature of the carrier gas.
e. Outline modifications of the standard procedure which permit high boiling point species to be separated by gas–liquid chromatography.

Question 6
    Using recent review articles and papers in journals, prepare a seminar entitled: "Modern Developments in Gas Chromatography" or debate the statement: "For quantitative inorganic analysis, solvent extraction has greater potentialities than all forms of chromatography."

Question 7
a. Partition chromatography on columns is recommended as a preparative procedure in organic chemistry.
    From available literature, outline three preparations where this approach has been adopted.
b. An important application of paper chromatography is the separation of amino acids. Outline the type of solvent systems which have proved useful for these separations and indicate how the separated components may be identified.
c. Paper chromatography is usually not recommended for quantitative analysis. Why?

d. Explain how a table of R_f values for a series of metal ions using a number of solvents might be used to develop a scheme of qualitative analysis based on paper chromatography.
e. Modern chemists are tending to use thin-layer techniques and gas-chromatography in preference to paper chromatography. Suggest reasons for this change in approach.

Question 8
a. The elution behavior of a solute on a gas-chromatogram can be described in terms of retention time, corrected retention volume, specific retention volume, and relative retention volume. Explain the relationship between these various terms.
b. Table 15.3 lists some applications of typical stationary phases in gas-liquid chromatography.

   Prepare a companion list arranged under the headings, Type of Sample, Recommended Stationary Phase, and Maximum Operating Temperature.

   For the sake of brevity the list may be restricted to no more than three phases per compound type and the sample types should include alcohols, aldehydes, ketones, esters; hydrocarbons, hydrocarbon isomers, fluorinated, and chlorinated hydrocarbons; aromatics and phenols; sulfur compounds; polar compounds; and high-boiling point compounds.
c. Using tables of Relative Retention Times of Representative Compounds (such as those in the "Handbook of Analytical Chemistry") prepare two groups of "family plots."

   In one plot, the logarithm of the relative retention times (on a particular solvent) for several homologous series (e.g., paraffins, olefins, alcohols, ketones, etc.) should be plotted against carbon number.

   In the second figure, the logarithms of the relative retention times of the various components of a homologous series on two different types of stationary phase should be plotted against each other.

   The systems selected should give near linear plots.
d. Explain how relative retention times and family plots may be used in the identification of the components of a mixture.

Question 9
a. Describe some of the uses of chromatography in geological studies. (Reference: A. S. Ritchie, "Chromatography in Geology," Elsevier, 1964.)
b. The lubricant used in a well-worn piece of machinery is considered to contain oxidation and polymerization products, together with finely divided fragments of metal from the moving parts and bearings. Indicate how a combination of solvent extraction, paper

chromatography and gas chromatography might be used to ascertain the degree of contamination of the lubricant.

Question 10

The systematic extraction, with different solvents, of the crushed roots of a local botanical specimen yields residues which are either mixed crystals or colored oils.

Outline the procedures which could be applied to determine the number and types of compound isolated in each extraction process, and suggest means of preparing pure samples of each component.

References

L. Alders, *Liquid-Liquid Extraction*, 2nd Ed., Elsevier, Amsterdam, 1959.

D. Ambrose and B. A. Ambrose, *Gas Chromatography*, Newnes, London, 1961.

E. Bayer, *Gas Chromatography*, Elsevier, Amsterdam, 1961.

E. W. Berg, *Physical and Chemical Methods of Separation*, McGraw-Hill, New York, 1963.

R. J. Block and G. Zweig, *A Practical Manual of Paper Chromatography and Electrophoresis*, Academic, New York, 1963.

H. P. Burchfield and E. E. Storss, *Biochemical Applications of Gas Chromatography*, Academic, New York, 1962.

S. Dal Nogare and R. S. Juvet, Jr., *Gas–Liquid Chromatography*, Wiley (Interscience), New York, 1962.

D. Dryrssen, J. O. Liljenzin and J. Rydberg, *Solvent Extraction Chemistry*, Wiley (Interscience), New York, 1967.

M. St. C. Flett, *Physical Aids to the Organic Chemist*, Elsevier, Amsterdam, 1962.

J. C. Giddings, *Dynamics of Chromatography; Principles and Theory*, Marcel Dekker, New York, 1965.

J. C. Giddings and R. A. Keller, (eds.), *Advances in Chromatography*, Volumes 1 to 7, Marcel Dekker, New York, 1966–1968.

E. Heftmann, (ed)., *Chromatography*, Reinhold, New York, 1961.

P. G. Jeffery and P. J. Kipping, *Gas Analysis by Gas Chromatography*, Macmillan, New York, 1964.

R. Kaiser, *Gas Chromatography*, Vols. I to III, Butterworth, London, 1963.

R. Kaiser and R. A. Keller, *Chromatography in the Gas Phase: Quantitative Techniques*, Marcel Dekker, New York, 1969.

A. I. M. Keulemanns, *Gas Chromatography*, Reinhold, New York, 1960.

C. E. H. Knapman, (ed.), *Gas Chromatography Abstracts*, Elsevier, Amsterdam, 1964–1967.

J. H. Knox, *Gas Chromatography*, Methuen, London, 1962.

E. Lederer and M. Lederer, *Chromatography*, 2nd Ed., Elsevier, Amsterdam, 1960.

M. Lederer, (ed.), *Chromatographic Reviews*, Vols. 1–8, Elsevier, Amsterdam, 1959–1966. Now published as a journal.

M. Macek and I. M. Hais, (eds.), *Stationary Phase in Paper and Thin-layer Chromatography*, Elsevier, Amsterdam, 1965.

C. J. O. R. Morris and P. Morris, *Separation Methods in Biochemistry*, Wiley (Interscience), New York, 1964.

G. H. Morrison and H. Freiser, *Solvent Extraction in Analytical Chemistry*, Wiley, New York, 1957.

R. Neher, *Steroid Chromatography*, Elsevier, Amsterdam, 1964.

R. L. Pecsok, (ed.), *Principles and Practice of Gas Chromatography*, Wiley, New York, 1959.

W. G. Pfann, *Zone Melting*, 2nd Ed., Wiley, New York, 1966.

F. H. Pollard and J. F. W. McOmie, *Chromatographic Methods of Inorganic Analysis*, Butterworth, London, 1953.

H. Purnell, *Gas Chromatography*, Wiley, New York, 1962.

A. S. Ritchie, *Chromatography in Geology*, Elsevier, Amsterdam, 1964.

L. R. Snyder, *Principles of Adsorption Chromatography*, Marcel Dekker, New York, 1968.

J. Stary, *The Solvent Extraction of Metal Chelates*, Macmillan (Pergamon) New York, 1964.

H. A. Szymanski, *Biomedical Applications of Gas Chromatography*, Plenum, New York, 1964.

R. E. Treybal, *Liquid Extraction*, 2nd Ed., McGraw-Hill, New York, 1962.

L. Zechmeister and L. Cholnoky, *Principles and Practice of Chromatography*, Chapman & Hall, London, 1941.

16

THE CHALLENGE OF AUTOMATION

I. Modern Instrumentation 587
   A. Current Trends 587
   B. The Role of Instruments 588
II. Automated Chemical Analysis 589
   A. Automated Measuring Procedures 589
   B. AutoAnalyzer Systems 591
   C. Automated Instruments for Specific Applications 593
III. Analytical Research 596
IV. Conclusions 599

I. MODERN INSTRUMENTATION

A. CURRENT TRENDS

In the preceding chapters emphasis has been placed on the *chemistry* of the processes associated with modern methods of chemical analysis. These methods make wide use of instrumental procedures and chemists must understand the general principles of the equipment they use.

However, it is not essential that all chemists should be capable of repairing the electronics or be able to design a modified version. On the other hand for the advancement of the discipline as a whole, it is desirable that some develop a marked interest in the physics and engineering associated with instruments. This aspect is probably best studied as an advanced subject entitled "Chemical Instrumentation."

The omission of instrument circuitry and detailed descriptions of particular equipment in this text is based on the conviction that the biggest advances in the next few decades will be in the field of instrumentation. The fundamental principles on which most analytical procedures are based are not likely to alter significantly in this period, although there may be a few new techniques added to the list and innumerable new applications proposed for all existing techniques.

Manufacturers are continually seeking to produce equipment which is superior in quality and response to any other product on the market. Instruments are becoming more compact through the use of solid-state circuitry, and by combining several units, apparatus capable of automatic or semiautomatic operation is now widely available. This trend is discussed in a little more detail in Section II.

Volume production and competition reduce costs, and it is often only a matter of time before it becomes economically feasible to use some current exotic technique for fairly mundane routine analysis. For example, although high resolution nuclear magnetic resonance spectrometers are extremely expensive and are used primarily for structure elucidation, low resolution instruments can now be constructed at a reasonable cost. As a result, at least one company is marketing an NMR quantity analyzer suitable for the determination of the water or liquid fat content of samples.

The problem of analyzing complex mixtures has been overcome by combining a separation unit with an analyzing unit. For example, the components of a solution may be separated on a liquid-solid chromatography column and the isolated zones detected by means of a recording ultraviolet spectrophotometer or a moving wire gas chromatograph. Another standard combination is a gas chromatograph-mass spectrometer unit. This combination has been extended by adding a high quality computer unit, and in the near future it is hoped that the computer will be programmed to print out the chemical structure of the individual materials isolated from the original sample. Smaller computers are being used to convert the signals from direct-reading emission spectrometers (UV-visible and X-ray types), mass spectrometers and gas chromatographs, directly into percentage compositions. New instruments for structure elucidation, such as far infrared interferometer units and recording optical rotatory dispersion instruments, include a small computer to convert the detector responses into graphs of standard form.

From these few examples, it is obvious that a computer must be considered as part of the analytical hardware now available and future chemists will require a sound background of computer programming and operation.

B.   The Role of Instruments

The introduction of instruments into chemical laboratories has been justified on many grounds, including:

1. Reduction in human tedium and human error.
2. Reduction in labor costs.
3. Increased speed of analysis and throughput.
4. Increased sensitivity or ability to perform multiple analyses on complex samples.
5. Ability to determine molecular as well as elemental composition.

One-pan balances or pH meters fitted with digital readouts are typical examples of instruments designed to reduce tedium and error. For repetitive titrimetric procedures there are automatic pipettes and burettes, the latter perhaps being connected to a unit which gives the volume required on a digital readout. By carefully selecting the strength of the titrant, the percentage composition may be read off the dial. In electroanalytical studies, automatic potentiostats relieve the operator from the boredom of manually adjusting rheostats, and electronic integrators measure the quantity of electricity consumed. Errors in reading instrument dials and the need for plotting results can both be minimized by feeding test signals on to chart recorders.

Many of the other factors are closely integrated. For example, the ability of certain instruments to perform multiple analyses increases throughput and speed of analysis and reduces labor costs.

Where technological advances demand greater sensitivity or a different type of analysis, new instrumental procedures are often the only answer. For each new application, a variety of instruments may be suitable; hence the analytical chemist must first define his instrument problem in terms of analysis demand, sensitivity limits, time requirements, and operator skills. The second step is to select the instrument which goes closest to meeting these specifications within the financial budget allowed.

With many instrumental techniques, the skill demanded of the operator is merely an ability to reproduce a set of conditions in a regular, repetitive manner. There is little intellectual challenge; hence the interest of the operator tends to waver. On the other hand, the logical sequence involved is usually amenable to automation, and for routine analysis the trend is to automated methods of analysis.

II.   AUTOMATED CHEMICAL ANALYSIS

A.   AUTOMATED MEASURING PROCEDURES

Any method of chemical analysis can usually be divided into four major segments.

1. Sample collection and preparation.
2. Separation of components.
3. Measurement of the response of the selected technique to the component
   . of interest.
4. Calculation of sample composition.

There are few situations where all stages can be simultaneously automated. In most systems, only one or two of the major steps proceed independently of the human operator.

One analytical procedure which might be classified as completely automated is neutron activation analysis in exploratory bore holes. A compact, high energy, isotope neutron source is mounted at the front of a testing plug. At the rear of the plug (well shielded from the source) is a gamma ray detector. The output of the detector is fed by a cable to a multichannel pulse height analyzer. As the plug is lowered slowly into the bore hole, the surrounding area is subjected to a short period of neutron activation. Any induced radioactivity is observed by the detector unit. The nature and intensity of the signals recorded by the pulse height analyzer provide semiquantitative indication of the nature of the underlying rock and mineral strata.

This procedure requires no specific sample collection or preparation, and modified versions could prove useful for monitoring outcrops in geological surveys or the composition of veins in mineral mines.

The absorption or emission of electromagnetic radiation by prepared samples has proved to be very amenable to automation. For example, there is now available a UV-visible spectrophotometer which automatically measures the absorbance of one to fifty samples (taken from test cells by an autosampler) at up to ten different wavelengths. The absorbance reading is presented on a built-in illuminated digital display and is simultaneously recorded by an electric typewriter. There is automatic calibration before every measurement which ensures high precision and a readout resolution of 0.001 Å.

While this represents a significant advance in automation of the measuring process, two rate and accuracy controlling steps remain. First, in order to produce the test solutions in a form suitable for spectrophotometric examination, some preliminary wet chemical manipulation has to be performed by associated operators. Second, a human operator is required to give the printed digital readings' some chemical significance. The testing of standard samples gives readings suitable for the preparation of calibration graphs, but in most cases, the relationship between test reading and chemical composition is ascertained by a nonautomated procedure.

On the other hand, the operator(s) need not be analytical chemists or skilled analysts; hence in the eyes of management this may be regarded as a worthwhile investment in terms of speed, throughput and recurring costs.

A computer can be added to the system to replace the "interpretative" operator but few techniques have successfully eliminated the need for manual sample preparation and/or manual sample introduction.

In many industries, solid samples are regularly delivered to the control laboratory through pneumatic tubes. After arrival, the samples may be prepared (e.g., by cutting, polishing, conversion of powders to pellets, etc.) by semiautomatic equipment. However, if the measuring unit is a direct reading UV-visible or X-ray emission spectrograph, the prepared sample has to be inserted manually into the instrument. Since multiple determinations are performed on each sample, this manual step does not significantly interfere with the speed and throughput of the equipment. However, it does leave a dull repetitive operation for the instrument operator, the tedium of which may only be broken when a change in the nature of the material being studied requires adjustment to other instrument dials.

B.   AutoAnalyzer Systems

The problem of automating the preliminary wet chemical procedures which precede measurements made by spectrophotometers, flame photometers, atomic absorption units, etc., has been overcome, and AutoAnalyzer systems (marketed by the Technicon Corporation) have been applied successfully to over two hundred different wet chemical determinations, involving both clinical and industrial applications.

The equipment consists of a number of units or modules and a typical system is shown schematically in Fig. 16.1.

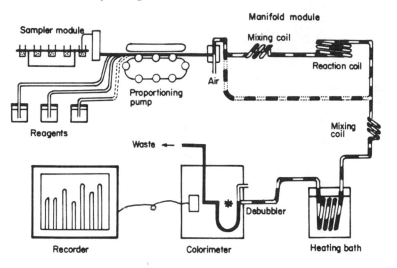

FIGURE 16.1   Flow sheet showing in schematic form the principal components of an Auto-Analyzer system.

The liquid samples (and standards) are loaded by an operator into the forty cups located on the turntable of a sampler module. At regular intervals (e.g., once a minute) a new cup is automatically presented to the system. Its contents are aspirated by a proportioning pump which continuously propels samples and reagents at precisely determined flow rates through a manifold positioned on top of the pump.

The proportioning pump consists of a series of chain driven rollers which depress up to twenty plastic tubes of various internal diameters at the same time, simulating peristaltic action in metering and transporting fluids through the system. The lumen of each tube is progressively occluded by the rollers, advancing gases or liquids in the exact proportions required. A roller leaves the pump platen every two seconds and each time, an air-bar rises and lets a measured quantity of air through. This adds air bubbles to the flowing stream in a precise timed sequence. The air bubbles act as barriers and divide each sample (and reagent stream) into a large number of discrete liquid segments. The air bubbles continually "scrub" the walls of the tubing and so reduce the possibility of contamination in succeeding segments of the same sample. Thus, if there is contamination from a preceding sample, it is restricted to the first few segments and the measurements made on the middle segments should be free of all interference problems.

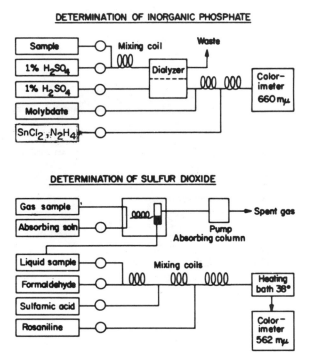

FIGURE 16.2   Typical flow systems for an AutoAnalyzer unit.

The manifold is a closed system of glass and plastic tubing, specific for a given determination, and interchangeable within minutes with other manifolds. In the manifold, samples and reagents are brought together under constant physical conditions.

Units for the manifolds are available which allow any standard operation to be performed. For example, the primary step in clinical analysis is usually dialysis; hence one of the units for this type of analysis is a miniature plastic dialyzer unit. Mixing and time delays are achieved by circulating the sample segments and added reagent through circular coils. Heating and/or incubation is achieved by means of short heating blocks or long path heating baths. Units are also available for the absorption of gases in liquids, the digestion of liquid samples, the separation of two liquid phases (for solvent extractions), and so on.

Reagents can be added in sequence, and in different volumes. The volume of each fluid aspirated into the system by the proportioning pump is determined by the inside diameter of its particular pump tube. The flow rate can be varied in nineteen steps from 0.015 ml min^{-1} to 3.9 ml min^{-1}.

Two typical flow systems are shown in Fig. 16.2.

The sample streams are debubbled just before the measuring unit in a special type of debubbler and flow cell combination. The detection device most widely used is a colorimeter, but if desired any type of sensor such as a UV spectrophotometer, fluorimeter, flame photometer, or scintillation counter may be substituted.

The output of the sensor unit is fed to a recorder and the resultant peaks are interpreted in quantitative analytical terms by the operator.

Until recently the AutoAnalyzer was used primarily for clinical studies, but the current applications extend into most fields of analysis and include air pollution studies, fertilizer analysis, food, and metal analysis. A recent development is a combined unit capable of simultaneously making twelve biochemical tests on a single, two milliliter sample of serum, with samples being processed at the rate of one a minute.

C. Automated Instruments for Specific Applications

While direct-reading spectrometers and AutoAnalyzers are often quoted as representing the trend in automation, equally useful units have evolved for more restricted classes of analysis. For example, several different designs have been proposed for the simultaneous determination of carbon, hydrogen, and nitrogen in organic compounds. In the dynamic systems, the products formed on combustion of the sample in oxygen are fed to a gas chromatograph. The separated components are subsequently detected by a katharometer and the amount of the constituent elements estimated from the relative areas of the recorder peaks. In another instrument, oxidation in the combustion tube is aided by chemicals and the products are flushed by a helium stream through a reduction tube where excess oxygen is removed and any oxides of nitrogen are reduced to molecular nitrogen. The remaining mixture (CO_2, H_2O, N_2, and He) is brought to thermal equilibrium under a pressure of about two atmospheres, before expanding through a sampling system into a series of thermal conductivity cells. Situated between the first pair of thermal conductivity cells is an adsorption trap containing a dehydrating reagent which removes the water vapor from the gas stream. The amount of hydrogen in the original sample is indicated by the difference in thermal conductivity caused by the removal of the water. A similar difference measurement is made with a second pair of conductivity cells mounted on either side of a trap which removes carbon dioxide. The nitrogen content of the helium-nitrogen mixture remaining is estimated by comparing the response of a conductivity cell with another through which pure helium is flowing. The sensor signals are all fed to a chart recorder and the peaks corrected to percentage composition by means of an appropriate calibration factor. After the insertion of the sample, the process proceeds automatically up to the interpretation of graph stage.

In many industries the determination of nitrogen is an essential routine analysis. In the automated analyzer the weighed sample is pyrolyzed in the presence of copper oxide at temperatures up to 1000°C. The molecular nitrogen so released is swept by a stream of high purity carbon dioxide into a glass nitrometer containing a specially prepared caustic solution. Here the carrier gas is absorbed and the scrubbed nitrogen collected for volumetric measurement. The actual volume of nitrogen is finally measured in a syringe, the plunger of which is driven by a micrometer screw linked to a precision digital counter.

The concept of an automated piece of equipment for selected single determinations has been adopted in many fields.

Thus, there is available an instrument for automatically determining the melting point of solids or boiling points of liquids. For the determination of melting points the operator inserts capillary tubes containing the solid into the instrument heating block which is capable of being heated at different rates. Mounted on one side of the capillary is a light; on the other is a photocell triggered to stop a digital temperature recorder at some fixed point of light intensity. As the substance under test changes from a solid to a liquid, its optical characteristics change from opaque to transparent. This change promotes the photocell response required to stop the temperature record.

For the determination of boiling points, the sample is placed in a special boiling tube which is inserted in the heating block. The tube is illuminated from beneath, and a photocell is mounted at right angles to the sample tube and light path. At the boiling point a continuous stream of bubbles is released throughout the sample. These bubbles reflect sufficient light into the photocell to cause it to trip the digital temperature readout.

With another piece of equipment it is possible to follow the distillation pattern of a sample, the percentage yield at each temperature being recorded on a graph.

In fuel analysis, one can also determine the flash point of an oil with the operator performing only two operations. He charges the test cup with sample and reads off the flash point from a digital display. The sample is heated at a slow rate and a test flame is automatically inserted through the shuttered lid above the sample at regular time intervals. When enough vapor has accumulated in the cup to cause a flash of flame, the flash is detected by a differential thermocouple. This signal stops the digital indicator which continually records the temperature of the sample during the test.

For the accurate determination of the calorific values of coal, coke, petroleum products, foodstuffs, etc, there is an automatic adiabatic bomb calorimeter. The calorimeter (which contains the sample) is surrounded by a water jacket through which water is circulated rapidly. The automatic control system

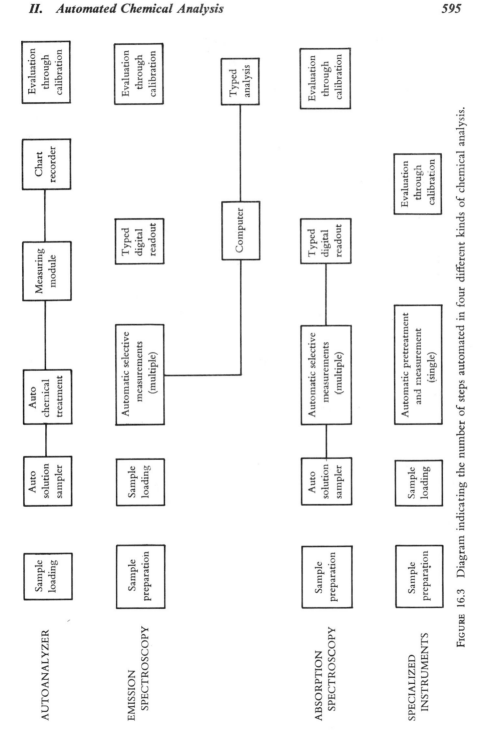

FIGURE 16.3   Diagram indicating the number of steps automated in four different kinds of chemical analysis.

prevents exchange of heat between the calorimeter and jacket by ensuring that the jacket is at the same temperature as the calorimeter vessel at every instant throughout the test. The temperature of the jacket water is measured accurately before and after combustion of the sample. The adiabatic operation eliminates the need to make cooling corrections, so that the calorific value can be calculated directly from the observed temperature change.

In this bid to provide automated equipment for particular processes, separation procedures have not been neglected. Thus for column chromatography there is a recycling unit which facilitates the isolation of components by automatically recycling the eluant through the column. In the "autoprep" type of gas chromatograph, individual eluates are directed to separate collection tubes or traps. Repetitive separation of the same initial mixture eventually provides a useful amount of each component. In some cases the separated material is used for subsequent analytical studies (e.g., infrared and NMR spectroscopy) while at other times it is retained as a pure chemical for use in synthesis.

Sampling has also received attention. Commercially available are samplers and sampling valves for use with moving streams of fine solids, liquids, and gases. The liquid and gaseous samples are sometimes fed directly to process control analyzers based on suitable sensor elements such as conductivity bridges, radiation absorption meters, etc. Solid samples normally require some form of pretreatment before being introduced to the analytical module, and in most cases this stage remains the duty of a human operator.

The semiautomatic nature of the current sophisticated systems is shown in diagrammatic form in Fig. 16.3.

The automation of equipment has changed the nature of the staff required for routine analysis, since the remaining manual operations can usually be performed by semiskilled labor.

However, it may be noted from the above examples that automated processes are based on simple analytical procedures. The chemistry associated with each type of determination has to be proved by qualified chemists in standard operations before automation can proceed.

The advent of automation should relieve analytical chemists from dull routine operations and free them to investigate the new systems and processes which may prove amenable to automation in the future.

Automation presents a challenge, since it must force more chemists into the fields of investigation and research. It should be the stimulant that provokes, not a palliative or excuse for decreased interest in analytical chemistry.

III.   ANALYTICAL RESEARCH

Analytical chemists are charged with the responsibility of providing methods of analysis suitable for the determination of the composition of any sample that may be encountered in scientific studies or industrial operations.

Since the number of problems evolved per day far exceeds the number of analytical chemists trained per year, chemists with all types of training and inclination are forced at irregular intervals to develop modified or new methods to meet a particular need. Their guide in this work must be the research papers submitted to the analytical journals by experienced industrial chemists, specialist research workers, and academics.

In many respects the boundaries of analytical research are not clearly defined. For example, an organic synthesis may yield a side product which proves to be a selective colorimetric reagent. Alternatively, a fundamental physical chemistry study may suggest a new instrumental method of analysis. In short, any type of chemical or physical research may yield data of potential analytical value, provided the scientist engaged in such work has a feeling for quantitative chemistry and is able to comprehend or predict some possible applications. For example, the technique of atomic absorption spectroscopy was conceived and developed by a physicist interested in atomic spectra.

For the purpose of discussion, research in analytical chemistry may be divided into a number of broad classifications.

1. Solution of a Specific Analytical Problem

The introduction of a new product by a manufacturing industry or the development of a new research project can pose problems of analysis which are not readily solved by reference to the literature.

The literature search may suggest several alternative procedures of potential value or the chemist may devise a new procedure based on his own experience in the field. In either case, the proposed procedure has to be tested on the sample of interest, and each difficulty systematically overcome.

The final result of this type of research is a detailed method of analysis which meets the specifications of the particular analytical problem which sponsored it.

Development of the method may require modification of equipment, modification of chemical procedures, and/or the introduction of new reagents or apparatus.

2. Evaluation of an Analytical Method

A method developed to meet a specific need may subsequently be investigated to evaluate the scope of the procedure. The aim in such research studies is to examine the effect of changes in concentration and matrix on the accuracy and precision of the method. Interest in interference effects generally predominates. Secondary developments from the work can be new direct applications of the original method, or modified procedures which overcome matrix or interference problems.

The evaluation of the practical applications of a new reagent can be considered to form a subgroup of this classification.

3.   Applications of a Technique

When a new analytical technique is introduced, its value is normally proved by the innovator through applications to a limited number of systems. This leaves a knowledge gap in respect to the number of areas in which the technique might prove useful. To bridge this gap it is considered good research to apply the technique to any and every conceivable field of chemical analysis. The number of possible applications is almost infinite and the full scope of a technique may never be ascertained.

In this classification, the term technique embraces instrumental and non-instrumental procedures. Since the aim of the work is to determine the range of applicability, new applications of the technique do not need to solve a specific problem or prove superior to alternative procedures. However, for maximum value to the journal reader, the advantages and disadvantages of the new application should be critically evaluated and if possible compared in quantitative terms with the results obtained using alternative procedures.

4.   Apparatus and Equipment Design

The availability of new materials (e.g., solid state electronic components), advances in related technological areas (e.g., radiation counting, computers) and improved workshop facilities prompt many chemists to improve on the equipment available to them. The changes introduced may be small (e.g., Teflon taps in burettes, octagonal bases on measuring cylinders, etc.) or large (e.g., evolution of the AutoAnalyzer or direct reading spectrograph).

5.   The Synthesis of New Reagents

Our knowledge of the mechanisms associated with indicator behavior, chelate formation, selectivity of functional groupings, etc. has now reached the stage where it is possible to predict the effect of substitution on the behavior of a particular indicator or reagent. To test these hypotheses, and, optimistically, introduce improved reactants, many new compounds have to be prepared. This type of research can prove of great interest to chemists who possess particular skill in chemical synthesis.

6.   Fundamental Data and Mechanism Studies

Many analytical procedures have been developed on an empirical basis and one might predict that improvements could follow better understanding of the reactions and phenomena involved.

Similarly, the time associated with many experiments can be saved if preliminary calculations demonstrate that particular reactions are not feasible. Unfortunately, much of the thermodynamic and kinetic data required for analytical systems are not available in the literature.

On these grounds, it is possible to classify many types of fundamental studies as being of interest to analytical chemists. Thus analytical research can include studies of the mechanism of homogeneous or heterogeneous reactions; the determination of the stability constants of complexes; the distribution coefficients of species in solvent extraction studies; the determination of overvoltage potentials; the effect of radiation on matter, and so on.

7. Innovation

While research of types 1 to 5 is responsible for the majority of publications and the steady advancement of analytical chemistry, the major advances in this discipline arise from the occasional new innovation.

The guiding theme for an innovator is the attitude:

"Here is an interesting reaction or physical phenomenon—how can I use it for quantitative analysis?"

The introduction of a new analytical technique is often the by-product of some general scientific program, and its evolution may be prompted by a range of stimuli, specialized experience, and environmental factors.

IV.   CONCLUSIONS

The aim of this chapter has been to indicate that analytical chemistry is a dynamic discipline in which research can take many different forms. The tedium and errors associated with standard routine analyses are gradually being eliminated through the introduction of new instruments and automation. At the same time, modern science and technology are creating new problems in quantitative analysis at a rate which will tax the skill and ingenuity of chemists for generations to come.

As indicated in the preface, the aim of this book was to provide an introduction to modern analytical chemistry. It is hoped that the reader has developed an appreciation of the concepts and principles used in the techniques associated with chemical analysis, and has been stimulated to supplement the minimum of basic details provided in this text with information gleaned from the monographs available on all topics.

The field of analytical chemistry is so broad that no single general text can do justice to the discipline, but if the reader has undertaken the various library exercises scattered throughout this book, he will have introduced himself to the source of all current knowledge.

Subject Index

A

Abbé refractometer, 235, 236, 237
Absolute retention volume, 574
Absorbance, in flames, 136
    of infrared, 157, 158, 165, 166
    of solutions, 47, 48, 127–131, 145, 147
    of X rays, 131, 133
    measurements, 150–153
Absorber velocity, 209
Absorptiometers, 122, 133
Absorption bands, (infrared), 160, 161, 164
Absorption edge, 104, 105, 132, 133, 134
Absorption, electromagnetic radiation, 7, 8, 46, 121–170, 241–243, 254, 590
    gases, 509, 592
    heat, 7, 11, 12, 55, 56, 57, 59, 66, 73, 82
    infrared radiation, 8, 121, 153–166
    microwaves, 15, 82, 84, 121, 169–171
    oscillations, 190, 191, 202
    radiation by atoms, 131–138
    recoilless emission, 209
    spectra, 121–127, 130, 140, 264, 493
    spectroscopy, 46, 121–166, 261
    UV light, 8, 106–109, 121, 122, 135–146, 165, 167, 242, 243, 264–268, 285, 548
    visible light, 8, 106–109, 121, 122, 135–139, 147–153, 167, 243, 264–268
    X rays, 8, 47, 104, 105, 131–134
Absorptivity, 108, 127, 128, 153
Accuracy, 37–43
    of absorbance readings, 130, 136, 138, 151, 160, 247
    of electrical methods, 440, 442, 451, 458, 459, 463, 482
    of emission techniques, 90, 106, 108
    of gravimetric methods, 406
    of kinetic methods, 420
    of paper chromatography, 569
    of refractive index readings, 235
    of titrimetry, 337
Acetic acid, as eluant, 530, 531
    as solvent, 328, 329, 330, 351, 352, 356
    dissociation of, 314, 339, 341, 342

Acetone, as solvent, 162, 262, 328, 357, 514
    mass spectrum, 279, 281, 282
    UV spectrum, 140, 141
Acetonitrile, as solvent, 357
Acetylene flames, 85, 95, 136, 137
Acid–base titrations, 6, 65, 109, 310–313, 336, 338–350, 440, 465
Acid, definition, 314, 338
    dissociation constant, 310–314, 329, 338, 339, 341–343, 345, 348, 350, 354, 355, 361, 363, 450
Acidic solvents, 351–354
Activated complex, 415, 433
Activated diffusion, 510, 511
Activation, analysis, 112–115
    cross section, 112, 113
    energy, 414, 415, 433, 434, 492
    overpotential, 477, 478
Active sites, 510, 511
Activity, coefficients, 300–303, 306, 316, 319, 400, 431, 436, 522 523, 558, 576, 578
    radioactive, 111, 112, 115, 116
    of species, 301, 305, 307, 323, 341, 350, 375, 376, 399, 400, 431, 434, 442, 446, 449, 450, 474, 475, 522, 523
Additive effects, structure, 260, 261
Adsorbents, 307, 492, 495–497, 499, 501, 502, 512, 538
Adsorption, 116, 167, 185, 187, 306–308, 319, 326, 327, 329, 371, 372, 404, 406, 430, 433, 446, 470, 479, 491–502, 507, 513, 514, 517, 530, 538
    analytical applications of, 492
    capacity, 497
    chromatography, 493–502, 558, 563
    equilibria, 491, 492
    gases and vapors, 491, 492, 510, 511, 538
    indicators, 326, 327, 369, 371, 372
    isotherms, 308, 327, 404, 491, 522
    maxima, polarography, 327, 473, 479
    sites, 326, 433, 491, 492
    solutes, 493
Aerosols, 84, 85, 95, 246
Alanine, 349

601

Alcohols, as solvents, 357, 405, 544
    determination of, 420, 492, 506
    mass spectra, 278, 281
Aldehydes, mass spectra, 280, 281
Alkaloids, 330, 356, 499, 568
Alpha dioximes, 396
Alpha particles, 110, 111, 112, 114, 115
Alumina, 497, 499, 501, 565
Aluminosilicates, 512
Amalgams, 392, 394, 442, 449, 461, 473, 475
Amino acids, 348–350, 356, 420, 505, 506, 517, 530, 568
Amino groups, 107, 349, 350
Ammonia, as base, 314, 321, 339, 341, 342
    as ligand, 315, 320, 358, 359, 360, 366
    as precipitant, 318, 320, 366, 403
    liquid (solvent), 351
    vapor, 160, 509, 512
Ammonium dihydrogen phosphate crystal, 103, 244
Amperometric titrations, 16, 365, 482
Amperostats, 465
Amphiprotic solvents, 351, 352
    species, 342, 506
Amplitude of oscillations, 154, 156, 226, 227, 229
Analyzed standards, 44
Analytical precipitates, 62, 302, 305, 316–322, 328, 344, 363, 367, 368, 399–407
Analytical research, 596–599
Analyzer crystals, 101, 102, 105, 133
Angle, of deviation, 123
    divergence, 214
    incidence, 228, 229, 230, 232, 233, 236
    reflection, 228, 229
    refraction, 228, 232
    rotation, 240, 241, 243
Angular frequency, 226
Angular momentum, 18, 77, 169, 182, 186, 189
    momentum quantum number, 76
Aniline point, 67
Anion exchangers, 515, 517–521, 525–530
    strongly basic, 519, 521, 525, 529
    weakly basic, 519, 520, 526
Anisotropic crystals, 231, 238, 250, 252
Anisotropic effects, 194, 202, 207, 231, 238, 274
Anode potential, 454
Anodic current, 433, 434, 470, 474
Anthracene, 107, 142, 208

Antibonding molecular orbitals, 138, 139, 140, 144
Antimony electrode, 354
Anti-Stokes lines, 167
Applications, of ESR, 207, 208
    gas chromatography, 578–580
    infrared, 163–166
    mass spectrometry, 219–221
    nonaqueous titrations, 356, 357
    organic reagents, 394, 395
    oxidation–reduction titrations, 384, 385
    polarography, 481
    precipitation titrations, 373, 374
Applied potential, electroanalytical methods, 437, 453, 454, 456–458, 461, 467, 469, 470, 472, 477, 482
    electromigration, 502, 504–508
    spectroscopy, 91, 100, 124, 214, 253
Aprotic solvents, 351, 352, 354
Arc, discharge, 90–93, 99
    mechanism, 91
Argon, 94, 95, 97, 501, 573
Arrhenius equation, 414, 511
Ascending chromatography, 567
Ascorbic acid, determination, 149, 150, 530
    reducing agent, 364, 366
Association constant, 303, 305
Association of ions, 4, 5, 128, 303–306
Asymmetry, 238, 239, 243, 270, 285, 287
    potential, 446
Atomic absorption, 95, 121, 123, 131–138
Atomic absorption spectrometer, 135, 136, 548, 591
Atomic absorption spectroscopy, 8, 90, 94, 135–138
Atomic fluorescence, 9, 106, 137
Atomic groupings, 122, 163, 395, 398
Atomic nuclei, 76, 110, 112, 114, 155, 181, 183, 189, 190, 192, 196, 200, 201, 204, 209, 210, 227, 272, 275, 277
Atomic number, 76, 77, 100, 101, 103, 104, 105, 131, 182
Atomic orbitals, 75, 138, 143
Atomic resonance radiation, 135
Atomic spectra, 76–81
Atomic susceptibility, 180, 182
Atomic vapor, 135, 137
Atomic volumes, 260
Atomic weight, 113, 131, 154, 180, 219
Atomizer-burners, 84, 85, 87, 90, 106, 136
Attenuated total reflectance, 162
AutoAnalyzer systems, 2, 25, 591–593

Automated chemical analysis, 589–596
Automated instruments, 593–596
Automated measuring procedures, 589–590
Automatic titrations, 451
Autoprotolysis, 352, 357
Autoprotolysis equilibrium constant, 351, 352
Auxochromes, 143, 265, 267
Avogadro's number, 113, 131, 186, 245, 301, 432
Azeotrope, 69

B

Back titration, 364, 373
Balances, 45, 60, 183, 588
Band, emissions, 82, 87
    width, 122, 123, 126, 130, 202
Base, definition, 314
    dissociation constant, 339–341, 348, 354
    line, 166
    peak, 218, 219, 279, 283
Basic solvents, 351, 353, 354
Bathochromic effect, 141
Beer–Lambert law, 7, 126–130, 147, 151–153, 165
Bending modes, 155, 156, 164, 269
Benzene, as solvent, 162, 194, 208, 262, 263, 329, 352, 357, 514, 518, 544, 546, 547
    spectral properties, 107, 142, 155, 267, 275, 284
Beta particles, 110, 111, 573
Beta-ray ionization detectors, 573
Beynon's Tables, 279, 280
Birefringence, 231
Bismuth(electrodeposition), 459
Bohr magneton, 179, 181, 182, 188, 201
    model of atom, 76
Boiling points of liquids, 59, 66–69, 86, 235, 260, 262, 263, 573, 576–579, 594
Boltzmann's constant, 78, 186, 202, 301, 431–433, 509
Bomb calorimeter, 65, 594
Bond, angle, 155, 170, 294
    axis, 155, 194
    strength, 154, 218, 220, 497
    structure, 187, 208, 210, 211
Bonding, electrons, 138–141, 188, 197
    modes, 491
    orbital, 139
Boron-11 nuclei, 190, 200

Bragg angle, 102–104, 133
Bragg's law (or equation), 102–104, 288, 289, 292
Breathalyzer, 11, 492
Bromine, as oxidant, 374, 393, 465
    determination of, 458
    diffusion, 511
    substituent compounds, 218, 219, 267
Bronsted-Lowry theory, 314, 338, 339
Buffer solution, 109, 129, 314, 316, 322, 338, 362, 364, 403, 450, 504–507, 529, 530
Bunn charts, 291
Burettes, 46, 440
Burner units, *see* Atomizer-burners
Butane, as fuel, 85
    mass spectra, 219, 280, 281

C

Calcium oxalate (precipitation), 317, 318
Calibration, in chemical microscopy, 252
    in light scattering, 245, 246
    in polarography, 481
    of absorption techniques, 128, 129, 136, 150–153, 159, 160, 165, 210, 590
    of emission techniques, 89, 90, 98, 99, 101, 105, 108
    of equipment, 38, 44, 45, 406, 449, 450
    of gas chromatograms, 579
    of kinetic methods, 417, 418, 423
    of refractometer, 235, 238
Calomel electrodes, 354, 443, 449, 471
Calorific value, 65, 594
Calorimetry, 65, 66
Capacitance, 437, 438, 480
Capacity, adsorbents, 497
    exchangers, 520, 524, 531
Capillaries, 510, 512
Capillary, columns, 563, 570–572
    film, 161
    flow, 499, 507, 510, 565, 567
    leak, 213
    tube, 468, 469, 471, 479, 563, 594
Carbohydrates, 286, 287
Carbon, determination, 185, 219, 492, 593
Carbon dioxide, 158, 403, 500
    absorption of, 348, 356, 492, 509, 512, 593, 594
    carrier gas, 501, 562, 573, 594
Carbon disulfide, 161, 168, 192
Carbonic acid, 342, 345

Carbonium ions, 218, 277
Carbon monoxide determination, 492, 500
Carbon tetrachloride, 161, 168, 169, 192, 274, 307, 352, 356, 514, 518, 543, 544, 546, 547
Carbonyl groups, 163, 193, 267, 269, 272, 273, 281–284, 287
Carboxyl groups, 349, 350, 361, 518, 519
Carrier gas, 47, 62, 221, 501, 538, 562, 571–574, 594
Catalysis, 415, 569
Catalysts, 109, 208, 221, 337, 364, 382, 385, 393, 407, 412, 415–419, 421, 514, 515
Catalytic currents, 478
Catalyzed reactions, 412, 415–419
Cathode potential, 454, 455, 457, 458, 461, 463
Cathode ray oscilloscope, 170, 212, 215, 480, 571
Cathode ray oscilloscope polarography, 480
Cathodic current, 433, 434, 473
Cation exchange, 327, 513, 525
Cation exchangers, 514, 517–519, 521, 523, 525, 526, 528–530, 566
    strongly acidic, 519, 521, 529
    weakly acidic, 59, 525, 566
Cell constant, 435
    electrolysis, 16, 466–468
    potential, 441, 442, 449
    thickness, 127, 134, 136, 137, 147, 151, 161, 166
Cells (sample), 122, 126, 127, 130, 136, 150, 152, 157, 159, 160, 170, 246
Cellulose, 499, 513, 517, 518, 524, 531, 563–566, 568, 569
Cellulose acetate, 513
Center of gravity, 154
Center of symmetry, 238
Centrifugal force, 215
Chain reactions, 412, 413
Change of state, 56
Characteristic frequencies (or lines), 76, 79, 80, 81, 83, 87, 88, 91, 96, 97, 100, 101, 106, 107, 122, 132, 137, 155, 163–167, 228, 268
Charcoal, 327, 497, 498, 501
Charged double layer, 470
Charge transfer peaks, 144
Chelate formation, 165, 360–362
Chelating agents, 86, 360, 544
    resins, 519, 527
Chelometric titrations, 360–362, 365

Chelons, 361, 362, 364, 365
Chemical bond, 138, 154, 180, 210
Chemical decomposition, 12, 69, 187
Chemical effects, in flame, 86
Chemical equilibrium, 128, 148, 308–327, 359, 535
Chemical exchange, 273, 275
Chemical microscopy, 13, 249
Chemical potential, 300, 523
Chemical separations, 90, 114
Chemical shifts, 194–196, 198, 200, 209–211, 271–273, 275
Chemiluminescence, 84, 109
Chemiluminescent indicators, 109
Chemisorption, 492
Chi-square values, 43
Chlorine, as oxidant, 374, 393, 465
    determination, 458
    substituent compounds, 218, 219, 267
Chloroform, 161, 168, 243, 262, 266, 352, 514, 516, 544, 547
Chromatic aberration, 234, 249
Chromatographic column, gas, 509, 560, 570, 573
    liquid, 493–496, 498, 556, 563, 565, 588
Chromatography, 20, 259, 493–502, 507, 509, 516, 528, 529, 560–580
Chromogenic reagent, 147, 148, 150
Chromophores, 141, 143, 146, 241, 242, 264, 265, 285, 287
    groups, 141, 142, 145, 267
Chronopotentiometry, 16, 466, 467
Circular dichroism, 241–243
Circularly polarized light, 229, 238, 241, 243, 244
Citric acid, 320, 530, 566
Classification of solvents, 350–353
Clausius–Clayperon equation, 577
Cleavage, 218
Coagulation, 247
Coal gas, 85, 136
Cobalt, colorimetric determination, 149, 150
Collimator, 101
Collisional broadening, 79
Colloidal precipitates, 369, 371, 372, 403, 404, 493
Colloids, 107, 254, 508
Color change interval, 312, 384
Color filter, 46, 107, 108, 150
Color-forming reagents, 147, 148, 150, 493, 503, 504

Colorimetry, 5, 7, 8, 124, 147–152, 246, 247, 336, 392, 395, 408, 509, 528
Column chromatography, 493–499
Column packing, 493, 501, 560, 569, 570
Column temperature, 501, 570
Combination tones, 157, 269
Combustion, 62, 65
Comparative costs, 21
Competing equilibria, 147, 308–327, 337, 344
Complex formation, 19, 20, 107, 303–306, 319, 320, 327, 348, 358, 377, 382, 383, 400, 469, 526, 527
Complexing agents, 20, 138, 147, 303, 314, 315, 316, 320, 325, 328, 358, 363, 402, 403, 460, 473, 475, 476, 481, 526, 529, 565
Complexometric titrations, 6, 65, 310, 314–316, 322, 336, 358–361, 440, 465, 482
Components of system, 536
Computers, 165, 217, 219, 291, 294, 309, 588, 590
Concentration gradient, 455, 467, 473, 495, 511, 521, 523
Condensation reactions, 514, 515
Conductance, 435, 436
   measurement, 437, 438, 465
Conductivity bridge, 437, 438, 596
Conductivity, of solutions, 435–440, 508
Conductometric titrations, 16, 439, 440, 482
Conformation, 243, 288
Conjugation, 141, 146, 165, 261, 265, 267, 268, 398
Conservation of momentum, 208
Constant current, 456, 458, 463, 464, 466, 467
Constant error, 42
Constant potential coulometry, 462
Constitutive correction factors, 180
Contamination, 45, 50, 97, 108, 213, 348, 395, 399, 404, 459, 496, 503
Continuous extraction, 549–553
Continuous extractors, 550
Continuum, 95, 100, 154
Controlled cathode potential electrogravimetry, 16, 394, 459–463
Controlled current coulometry, 463–465
Converging lens, 233, 234
Coordination (with ligands), 143, 268, 307, 358–360, 395, 396, 475, 476, 506, 544
Copper, electrodeposition, 459, 460
Coprecipitation, 116, 395, 406, 458, 528
Cotton effect, 241–244, 287

Coulomb's law, 178, 300, 350, 525
Coulometer, 462, 463
Coulometric analysis, 16, 462–467
   titrations, 462–467
Coulometry, at constant potential, 462, 463
   at constant current, 463–465
Counter-current extraction, 552–556, 560, 563
Coupling constant, 197, 198, 199, 203, 272, 275–277
Cracking pattern, 214
Craig apparatus, 554–556, 559
   tube, 554–556
Critical absorption wavelength, 132
Critical angle, 228, 235, 236
   rays, 235, 236
Critical solution temperature, 67
Crystal, analyzers, 102, 104, 105, 133
   axis, 202
   detector, 170, 206
   field effect, 144
   growth, 247, 254, 305, 318, 367, 369, 401–403
   lattice, 288, 403
   nuclei, 369, 401, 402
   planes, 288–291
   structure, 201, 252, 293
   systems, 290, 291
Crystal violet, 330, 356
Crystals, ionic, 307, 318
Crystallization, 539, 540
Cubic symmetry, 252, 289, 290, 291
Curie, 112
Curie–Weiss law, 181, 185, 186
Current density, 455, 456
Current–voltage relationships, 91, 467, 468, 472, 477, 479, 480
Curvature of field, 234

D

de Broglie equation, 253
Debye–Hückel equations, 301
Decay constant, 110, 112
   law, 110, 111
Decomposition potential, 452, 454, 457, 468
Degeneracy, 78, 157, 197, 202
Degree of cross-linking, 510, 512, 514, 515, 521, 525, 530
Degree of extraction, 542, 550
Degree of ionization (atoms), 79

Degree of swelling, 511, 513, 530
Degree of thermal dissociation, 86
Degrees of freedom, 39, 42, 43, 154, 536–539
Demasking, 321
Demineralized water, 528
Demountable cells, 160
Density, 131, 134, 184, 235, 237, 238, 260, 290, 401, 566
Depolarizers, 456, 460
Derivative polarography, 479, 480
Desalting, 512, 517
Descending chromatography, 566–568
Deshielding effects, 193, 194, 196, 210, 272, 274, 275
Detection limit, 88, 92, 105
Detection unit, 47, 103, 111, 122, 124, 130, 157, 213, 590
Detector response, 62, 152
Deuterated solvents, 192, 274
Deuterium labeling, 167, 284
Deuterium lamp, 123, 145
Deuterons, 112
Development of chromatogram, 493, 495, 560
Dextran, 513, 517, 518
dextro rotation, 239, 240, 286, 287
Dialysis, 512, 517, 592
Diamagnetic shielding, 192, 194
Diamagnetism, 181–188, 261
Diaphragms, 214, 234, 250
Diatomaceous earths, 563, 569
Diatomic molecules, 81, 169
Dichroism, 230, 241–243
Dielectric constant, 48, 207, 300–303, 306, 328, 350, 351, 353, 354, 400, 438, 497
Difference tones, 157
Differential absorption techniques, 133, 151, 152
Differential curve, 451
Differential gas detectors, 63, 493, 572, 578, 593
Differential migration, 493, 502
Differential reaction rates, 421–424
Differential thermal analysis, 12, 59–62
Differential thermogravimetric analysis, 62
Differential titrations, 355, 357
Differentiating solvents, 352–354
Diffraction, 102
Diffraction grating, 88, 96, 122, 123, 232
Diffraction pattern, 289–293
Diffractometer, 291

Diffusion, 508–517
    coefficient, 466, 468, 469, 473, 474, 511, 521, 561, 562
    control, 402, 430, 468, 473, 476, 511
    current, 16, 468, 469, 472, 477–479, 481, 482
    gases, 501, 509
    ions, 402, 430, 433, 453, 455, 466, 472, 473, 477, 479, 503–505, 521, 523
    layer, 467–469, 521
    species, 48, 491, 494, 499, 508–517, 560, 561
Digestion, 404
Dilatometry, 69
Dimethyl formamide, 162, 329, 351, 357
Dioxan, 162, 243, 266, 356
Dipolar ions, 349
Dipole–dipole interaction, 307, 491, 511, 576
Dipoles, 300; *see also* Dipole interactions
Dipping refractometer, 235, 237
Direct reading spectrometers, 25, 99, 593
Discs (pressed), 162, 166, 271
Dispersing power, 234
Dispersion, 96, 102, 123, 169, 225, 227, 232, 235, 242, 243
    unit, 46, 74, 96, 102, 159
Displacement development, 495, 496
Dissociation, constant, 350
    of complexes, 306, 322, 361
    effects, 128, 135, 525, 530
Distillation, 509, 594
Distortion aberration, 234
Distribution, between phases, 490, 492, 495, 516, 517, 542, 558, 563
    coefficients, 116, 306, 323, 325, 516, 517, 529, 537, 560
    isotherm, 495, 496, 522, 538, 558
    ratio, 306, 323–325, 542, 546, 547, 553–557, 559, 560, 566
Dithizone, 545, 547, 549
Diverging lenses, 233
Donnan equilibrium relationship, 524
    potential, 523
d Orbitals, 143, 144, 188, 210
Doping, 44
Doppler broadening, 79, 209
Double beam spectrometer, 89, 129, 137, 158, 159, 160
Double irradiation, 274
Double refraction, 231
Double resonance, 274

Dropping mercury electrode, 468-470, 473, 480, 481
Drop time, 469-471, 478–480
Drude equation, 242

E

EDTA, *see* Ethylenediaminetetraacetic acid
Effective Bohr magneton number, 186, 188
Effective (magnetic field), 192–194, 197
Effective magnetic moment, 186
Electrical arc, 90–92
Electrical conductivity, 16, 435–440, 508
Electrical dipole, 227
Electrical discharge, 90–95
Electrical double layer, 453
Electrical field, 162, 201, 202, 206, 207, 210, 214, 503, 504, 507, 523
Electrical potential, 16, 21, 63, 91, 100, 374
Electrical resistance, 63, 91, 124, 435, 437, 438, 454
Electrical spark, 91–93, 97
Electrical transformations, 429–482
Electric dipole moment, 156, 168, 169, 170
Electric vector of waves, 227, 229
Electroanalytical techniques, 16, 21, 47, 336, 408, 429–482, 589
Electrocapillary maximum, 470
Electrochemical cell, 376, 377, 379, 440–442, 448–450, 453, 454
Electrochromatography, 507, 508, 513
Electrodeposition, 5, 453–463
Electrode potential, 375–377, 431, 432, 434, 462, 466
Electrode types, 442–448
Electrography, 502–504
Electrogravimetry, 16, 456–458
Electrolysis, 16, 21, 437, 442, 452–458, 461–463, 465, 466, 468, 473, 502–504
    cell, 21, 466, 467, 468
    circuit, 19, 453, 459, 471, 473
Electromagnet, 190, 206, 254
Electromagnetic radiation, 74, 75, 121, 126, 141, 147, 156, 201, 226, 227, 229, 238, 590
Electromigration, 467, 502–508
Electron beam, 101, 212, 213, 217, 218, 253, 254
Electron capture detectors, 573
Electron clouds, 106, 139
Electron density, 193, 210
Electron density projection map, 294

Electron diffraction pattern, 254
Electronegativity, 165, 193, 194, 210, 218
Electron exchangers, 520
Electron gun, 253
Electronic charge, 179, 181, 201, 253, 301
Electronic energy levels, 139, 143
Electronic structure, 78, 106, 358
Electronic transitions, 75, 81, 91, 107, 122, 135, 138–144, 167, 203
Electron microscope, 253, 254
Electron microscopy, 13, 253, 254
Electron pairs, 181, 188, 192
Electron probe microscopy, 10, 100, 101, 254
Electron spin, 106, 186, 201
Electron spin resonance (ESR), 201–205
    spectrometer, 202, 205, 206
    spectroscopy, 15, 18, 177, 187, 205–208
    spectrum, 203, 206, 207
Electron transfer, 4, 5, 17, 19, 281, 374, 376, 377, 381–383, 393, 407, 430–434, 445, 463, 469, 477, 481
Electron volt, 110, 218, 573
Electroosmosis, 505
Electrophoresis, 350, 504–507
Electrophoretic bed, 505
Electrophoretogram, 505
Electroseparations, 458–460
Electrostatic attraction, 143, 215, 216, 299, 300, 304, 307, 401, 491, 511
Elliptically polarized light, 229, 230, 238, 241
Eluotropic series, 497
Elucidation of structure; *see also* Structure elucidation
    by UV–visible spectroscopy, 145, 264–268
    by infrared spectroscopy, 157, 163, 165, 166, 169, 268–271
    NMR, 192, 271–277
    mass spectrometry, 213, 218, 219, 220, 277–284
    optical activity, 243, 285–288
    X rays, 288–294
Eluting agent (eluant), 494, 495, 507, 508, 529, 568, 596
Eluting power, 497
Elution development, 493, 495, 496, 501, 516, 528, 529, 556
Elution peaks, 502, 517, 529, 558, 574, 579
Elution volume, 516

Elution zones, 493, 494, 498, 499, 508, 509, 556

Emission of UV–visible light, 10, 47, 73, 82, 84, 85, 86–100, 588, 590

Emission of radioactive particles, 110, 114

Emission of X rays, 9, 100, 101, 133

Emission spectra, 73–83, 90, 93, 98, 104, 138, 169

Emission spectroscopy, 7, 10, 90–100, 138, 169

Endothermic reactions, 56, 60, 64, 414

End point; *see* Equivalence point

Energy states, 74

Enthalpy, 55, 56
    changes, 12, 56, 64, 65, 415, 577

Entropy, 56

Environmental effects, 163–165, 190, 192, 194, 198, 202, 209, 269, 271

Enzymes, 416, 419, 420

Equilibrium, constants, 57, 129, 147, 304–306, 309, 312, 313, 325, 342, 359, 381, 382, 415, 431, 476, 522, 543, 577
    potential, 434, 477

Equivalence point, 309, 337
    detection of, in titrations, 310, 336, 353, 366
    acid–base, 311, 312, 329, 338, 339, 341, 343–345, 349
    amperometric, 482
    complexometric, 315, 322, 358–360, 363, 365
    conductometric, 440
    nonaqueous, 354
    potentiometric, 450–452
    precipitation, 326, 327, 367, 370–372
    redox, 374, 379, 382, 384, 385
    spectrophotometric, 153, 247

Equivalent conductance, 435

Equivalent ionic conductance, 435, 436, 439, 441

Eriochrome black T, 364–366

Errors, 38, 42, 50, 376, 377
    in absorption methods, 130, 131, 151, 152, 166
    in emission methods, 86, 99, 109
    in gravimetric analysis, 404
    in titrimetry, 341, 343, 344, 364, 365, 371, 465

Ethanol, as solvent, 262, 266, 313, 328, 405, 458, 512, 514
    determination, 492
    NMR spectrum, 193, 197, 273, 276

Ether (diethyl), 262, 307, 544, 546, 547, 549, 550, 570

Ethylene diamine, 329, 351, 357

Ethylenediaminetetraacetic acid (EDTA),
    acid dissociation constants, 343, 361, 362
    as masking agent, 86, 321, 322
    as titrant, 360–362, 365, 366

Eutectic point, 539, 540

Eutectic system, 539, 570

Exchange current, 477

Exchange equilibrium, 521, 522, 529, 549

Excitation, by accelerated particles, 10, 74, 84, 91, 100–101
    by flame, 84–90
    by heat, 10, 47, 74, 83–100
    by radiation, 74, 101–109
    by X rays, 9, 47, 74, 84, 100–105

Excited state, 77, 78, 82, 106, 121, 138, 139, 167, 201, 209, 210, 211, 226

Exothermic reaction, 56, 60, 64, 414

Extraction curves, 546–549

Extraction, of iron(III), 544
    of thorium, 550–553
    of uranium, 544, 549–553

Extraction systems, types of, 542–546

Extra ordinary ray, 231, 238

F

"Family" plots, 579

Faraday, laws, 16, 462
    (magneto-) method, 183, 184
    unit, 375, 431, 462, 466

Far infrared, 153, 169

Ferromagnetism, 180, 181, 183, 184, 185

Fick's law, 467, 511, 521

Field strength, electrical, 505
    magnetic, 179, 184, 189, 192–194, 197, 203, 277

Filaments, 63, 100

Filter, paper, 499, 504–506, 513, 516, 531, 550, 565–567
    photometer, 150

Filters, color, 46, 107, 150
    metal, 124, 133

Flame, characteristics, 84–87
    excitation, 84–90, 95, 135
    ionization gas detector, 573
    photometer, 87, 88, 106, 591, 593
    photometry, 10, 46, 86–90, 137, 138
    temperature, 85, 86, 95
    temperature detector, 62, 572

Flames, 46, 84–90, 95, 106, 135–138
Flash point of oil, 594
Flow rate, 494, 501, 517, 527, 559, 561, 562, 565, 574, 593
Fluorescein, 371. 372
Fluorescence, 7, 82, 83, 101, 106, 107, 137, 146, 168, 565
Fluorimetry, 9, 106–108, 593
Fluorine-19 nuclei, 190, 200
Fluorogenic reagent, 107
Flux (magnetic), 178, 179
Focal length, 233, 234, 249, 254
Focal plane, 74, 96
Focal point, 232, 233, 234
Force constant, 154
Formaldehyde, 141, 321, 367, 514, 515
Formal reduction potential, 376–378, 380–382, 384, 393, 434
Formation constants, 359, 360, 361
Formic acid, 351, 352, 530
Formvar, 104, 254
Fractional crystallization, 541
Fractionation, 68, 69, 491, 514, 541
Fragmentation pattern, 214, 218, 219, 220, 277, 280, 284, 285
Free electrophoresis, 504
Free energy change, 56, 338, 379–381, 392, 430, 431, 434
Freezing points, 59, 66, 262, 263, 539
Frequency, factor, 414, 415
    of radiation, 74, 76, 101, 121, 138, 140, 141, 154, 156, 157, 163, 166, 167, 168, 170, 205, 226, 227, 233, 268, 269
    shifts, 167
Freundlich isotherm, 308, 327, 404, 523
Frontal analysis, 495, 496
Fuel gas, 85, 88
Functional groups, 155, 243, 268, 269, 274, 285, 353, 357, 395, 396, 398, 424, 511, 513, 515, 518–521, 523, 525, 569
Fundamental vibrational frequency, 81
Fundamental vibrational mode, 155, 157
Fusion (with flux), 97, 105, 393

G

Gamma, counters, 111, 114, 211, 589
    magnetic term, 178
    photons, 208

Gamma rays, 75, 110, 111, 112, 114, 115, 208, 209
    spectrometry, 114, 211
Gamma-ray emitter, 209, 211
Gamma-ray nuclear resonance fluorescence, 208–212
Gas, chromatograph, 62, 63, 69, 500, 501, 588, 593, 596
    chromatography, 46, 47, 48, 69, 221, 493, 500–502, 510, 516, 538, 560, 569–580
    detectors, 494, 501, 562, 572, 573
    overvoltage, 457, 461
    sheathing, 92
Gaseous discharges, 93–95
Gaseous samples, mass spectra, 212, 213
    vibrational and rotational spectra, 154, 160, 166, 168, 169
Gas–liquid chromatography, 500, 502, 560, 569–580
    equilibria, 538, 559
Gas–solid chromatography, 493, 500–502, 510. 516, 538, 569
Gauss (unit), 178, 190, 203
Gaussian curve, 39, 40, 553, 560
Geiger counter, 43, 103, 111, 124
Gelatin, 247, 479
Gel filtration, 510, 516, 517, 530
Gels, 505, 507, 508, 512, 514, 516, 520, 531
Geometric optics, 231–234
Gibb's Phase Rule, 535–538
Glass, 123, 158, 184, 185, 232
Glass electrode, 354, 443–446, 449
Glueckauf equations, 560
Golay equation, 563
Goniometer, 101, 102
Gouy method, 183, 184, 261
Gradient elution, 495, 529
Graham's law, 509
Grating, 88, 96, 122, 123, 145
Gravimetric analysis, 4, 6, 152, 316, 336, 373, 395, 399, 405–407
Grazing incidence, 229
Grazing rays, 235–236
Gross sample, 49
Ground state, 78, 79, 81, 82, 106, 107, 109, 121, 135, 137, 138, 154, 156, 167, 201, 209, 211
Growth factor, 113
Gyromagnetic ratio, 189

H

Half-cell reactions, 377–379, 381, 382, 392, 430, 440–444, 454, 473
Half-life, 110–113, 115, 410
Half-neutralization potential, 356
Half-wave potential, 468, 472, 474–477, 480
Half-width (of spectral line), 79, 123, 135, 170
Harmonic motions, 154, 155, 157
Heat capacity, 55, 64, 65
Heat, of combustion, 57, 62
    fusion, 56, 59, 66
    reaction, 60
    solvation, 58, 400
    vaporization, 58, 67, 577
Heat transfer, 56, 59
HETP, 561–563
Helium, 105, 221, 501, 562, 573, 593
Helmholtz coils, 190, 192
Henry's law, 538
Heterogeneous equilibria, 305–308, 535–580
    reactions, 412, 430, 433
Hexagonal symmetry, 252, 290, 291
High frequency field, 93–95, 216
High spin complex, 188, 210
Hollow cathode tubes, 84, 94, 123, 135, 137
Homogeneous precipitation, 402, 403
Homologous pairs, 98, 99
    series, 235, 579
Hooke's law, 154
Horizontal chromatography, 567
Hot gas flames, 46, 47, 62, 84, 95
Hudson's isorotation rules, 286
Hudson's lactone rule, 287
Hydrated ionic radii, 328, 436, 525, 526, 529
Hydrocarbons, 85, 139, 146, 162, 201, 208, 218, 235, 238, 243, 260, 261, 272, 277–280, 500, 512, 573
Hydrochloric acid, 338, 339, 345–348, 352, 385, 394, 439, 529, 530, 544, 547, 548
Hydrogen, atoms, 260, 274, 278–285
    determination of, 492, 593
    electroproduction, 456, 457, 459–461
    gas, 85, 136, 221, 376, 501, 562, 573
Hydrogen bonds, 162, 165, 166, 196, 269–271, 274, 299, 576
Hydrogen electrode, 376, 379, 441, 444
Hydrogen ions, solvated, 304, 345, 351, 356, 361, 383, 527
Hydrogen lamp, 123, 145

Hydrogen overvoltage, 394, 455, 457, 461, 469
Hydrogen peroxide, 419, 478, 479, 481
Hydrogen sulfide, 318, 394, 403, 459
Hydrolysis, 241, 304, 318, 363, 402, 403, 450, 520, 543
Hydrous oxides, 344, 366, 402, 405, 518, 528
8-Hydroxyquinoline, 307, 324, 325, 397, 543, 545, 547
Hyperfine coupling constant, 203, 204
Hyperfine splitting (ESR), 203, 204
Hypochromic effect, 143

I

Ilkovic equation, 469, 470, 473, 478
Immiscible liquids, 67, 536, 537, 542, 553
Incident radiation, 102, 103, 105, 156, 157, 159, 166, 167, 170, 227–233, 242–244
Indicator, dissociation constant, 311, 312, 315, 316
    electrodes, 354, 356, 442, 444, 446, 449, 450, 464
    errors, 313
    selection, 309–316, 341, 345, 353, 354, 364, 395
Indicators, 41, 305, 336, 337, 341, 343, 344, 353, 356
    acid base, 310–313, 329, 345, 347, 348, 384
    complexometry, 41, 314–316, 360, 362–365
    precipitation, 326, 327, 367, 369–372
    redox, 383, 384
Inductance, 92, 94, 185
Induction, 94
Induction period, 418, 420
Infrared, absorption spectrometry, 8, 153–167, 220, 221, 265, 268–271, 596
    radiation, 75, 123, 124, 153–166
    spectra, 46, 61, 156, 159, 160, 162, 168, 268–271, 285, 499
    spectrometer, 69, 157, 159
Inhibition, 415, 419; *see also* Catalysis
Initial rates, 410, 417, 420
Inner complexes, 188, 396, 544
Instability (of complexes), 358; *see also* Stability of complexes
Integral heat of solution, 58

Intensity, of γ rays, 209
 of magnetization, 178, 179
 or microwaves, 202, 206, 207
 of radiation, 78–81, 86–89, 92, 94, 97–101, 103–105, 106–108, 122, 124, 127, 132, 133, 135, 137, 143, 144, 145, 148, 156, 157, 158, 160, 166, 168, 169, 170, 226, 229, 230, 244–246
 of rf signals, 189, 192, 197, 198
 of scattered light, 244, 245, 248
Interactions with magnetic fields, 177–221
Interchain bonding forces, 510, 511
Interconal zone, 85, 136
Interfaces, 229, 232, 370
Interfacial phenomena, 228, 229
Interference effects, 3, 23, 29, 64, 170, 309, 320, 353, 384, 407, 489, 527, 528
 in colorimetry, 149, 150, 392, 503, 528
 in electrochemical methods, 440, 448, 458, 459
 in flames, 85, 86, 90
 in kinetic studies, 407, 424
 in precipitation, 316, 373, 406
 in titrimetry, 305, 364, 366, 528
Interference filter, 88
Interference (of waves), 226, 227, 229, 237, 238, 247
Interference pattern, 161
Intermolecular bonding, 165, 166, 269, 270, 271, 274
Internal electrolysis, 465
Internal standardization, 44, 89, 98, 105, 169, 196
Internuclear distances, 170
Intramolecular bonding, 165, 269, 270, 271
Integral gas detectors, 572
Iodine, as titrant solution, 357, 374, 383, 385, 452, 492
 solvent extraction of, 307, 543
Ion acceleration, 214, 215
Ion association, 303, 544, 545
Ion association extraction systems, 544, 547, 548
Ion exchange, analytical applications, 526–531
 chromatography, 328, 528–530
 columns, 526, 527, 530, 531
 membranes, 520
 papers, 105, 567
 process, 305, 350, 445, 465, 491, 510, 511, 516–531, 535
 resins, 90, 514, 515, 518, 524, 526

Ion exchangers, 307, 513, 517–520, 522, 526
 selectivity of, 525, 526
Ion exclusion, 530
Ion formation (mass spectroscopy), 213, 218, 219, 221, 277, 281, 283
Ion fragments, 15, 18, 21, 213, 214, 215, 217, 218, 219, 220
Ion gun, 214
Ionic strength, 301, 302, 303, 317, 400, 436, 449, 529
Ionic susceptibility, 180, 186
Ionization, chamber, 213
 constant, 349, 350, 362
 detectors, 572, 573
 electrolytes, 338, 349, 350, 362, 506, 543
 potential, 79, 84, 86, 91, 217, 220, 221, 573
 vapors, 572, 573
 water, 310
Ionizing power, 111, 213
Ion pairs, 303, 307, 320, 326, 328, 350, 354, 476, 544
Ion selective electrodes, 446–450
Ions (in solution), 5, 6, 58, 299–306, 374
Ions (in spectral studies), 78, 85, 86, 93, 95
Iron(III) salts, 320–322, 344, 371, 383, 391, 460, 464, 478, 529, 566
Irradiation time, 113
Irreversible waves, 476–478
Isoelectric point, 350
Isomerism, 146, 165, 166, 167, 268
Isotope, abundance, 217, 219
 dilution, 116
Isotopes, 94, 110, 111, 112, 115, 116, 218, 219, 220, 278, 284, 589
Isotropic crystals, 252

J

Janak gas detector, 573

K

Karl Fischer reagent, 357, 452
Katharometer, 63, 572, 578, 593
Kerosene, 307, 518, 546
Kinetic current, 478
 energy, 132, 214

Kinetics, analytical applications of, 336, 406–424
 in chromatographic separations, 560, 561
 of complex formation, 358
 of electron transfer, 382, 392, 430, 432–434, 473, 477
 of precipitate formation, 399–403
 techniques used, 62, 407–409, 415
Klystron oscillator, 123, 170, 205

L

Lability of complexes, 358
Lactones, 287
Laminar-type flame, 85
Lande splitting factor, 186, 201, 202, 207
Langmuir isotherm, 308, 327, 491, 523
Laser beams, 168
Lattice patterns, 231
Law of Mass Action, 521, 524, 535
Law of Rational Indices, 290
Lead, electrodeposition, 457, 458
"Leading," 494, 496
Lenses, 232–234, 249, 250
Leveling effect, 352
levo Rotation, 239, 240, 287
Lewis acids, 357
Lewis theory, 314
Ligand field, 143, 144
 theory, 143, 188
Ligands, 129, 143, 144, 148, 307, 314, 315, 316, 327, 328, 363, 364, 395, 400, 403, 476, 506, 529, 544
Light scattering, by macromolecules, 247, 248
 effect of variables, 244, 245, 246
Light scattering photometer, 245
 photometry, 244–248
Limiting current, 464, 468, 469, 473, 478, 482
Linear absorption coefficient, 104, 131
Lines of force, 178, 179, 182, 194
Lines, spectral, 74, 78, 79, 87, 88, 90, 96, 97, 135, 154, 167, 168
Line width, 79, 170, 190, 206
Liquid anion exchangers, 518
Liquid cation exchangers, 518
Liquid ion exchangers, 307, 446–448, 518, 531, 544, 545
Liquid junction potential, 441, 446, 449, 450

Liquid–liquid extraction; *see* Solvent extraction
Liquid samples, 154, 160, 168, 184, 213, 234, 235, 237
Liquidus curve, 541
Liquid–vapor equilibria, 68, 536, 537
Logarithmic sector, 98
Lowry–Bronsted theory, 314, 329, 338
Low spin complex, 188, 210
Lyate ion, 352, 353
Lyonium ion, 352, 353

M

Macromolecules, 247, 248
Magnetic deflection, 215
Magnetic dipole, 11, 178, 179, 181
Magnetic field, 11, 15, 18, 94, 101, 109, 177–179, 181–185, 189, 190–194, 198, 200, 201, 202, 205, 206, 210, 211, 215, 216, 253
Magnetic flux, 178–179
Magnetic hyperfine splittings, 209, 210, 211
Magnetic induction, 179
Magnetic lenses, 101, 253
Magnetic moment, 18, 179, 181, 186, 189, 201
Magnetic permeability, 178
Magnetic pinch effects, 94
Magnetic pole, 178
Magnetic quantum number, 77
Magnetic resonance, 189
Magnetic separation unit, 215
Magnetic susceptibility, 11, 15, 177, 179–188
 methods of determinations, 183–185
 applications, 185–187
Magnetic titration, 187
Magnetic vector (of waves), 227, 229
Magnetism, 177–183
Magnifying power, 234, 249
Masking, 20, 148, 305, 316, 318, 320–322, 325, 366, 544
Masking agents, 148, 321, 322, 363, 364, 365, 395, 490, 546, 549
Masking ratio, 321
Mass absorption coefficients, 105, 131–134
Mass of electron, 179, 201
Mass spectra, 15, 21, 213, 215, 217, 218, 219, 277–284
Mass spectrometry, 15, 18, 165, 178, 212–221, 573, 579

Mass spectrometer, 45, 69, 213, 215, 216, 217, 219, 220, 221, 277, 279, 588

Mass susceptibility, 180, 181, 184, 185

Mass-to-charge ratio, 215, 217, 278, 281, 282

Mass transport, 467, 509, 510, 561

Matrix absorption, 104

Matrix effects, 90, 99, 101, 104, 105, 133, 134

Maxima (polarographic); *see* Adsorption maxima

Maxwell (unit), 178, 179

Mean ionic activity, 300, 522

Melting curves, 66, 541

Melting points, 12, 66, 192, 235, 251, 594

Membranes, 512, 520

Mercury, 394, 442, 458, 461, 468–471, 473, 478

Mercury cathode, 394, 461, 466, 468–471, 478.

Mercury cathode separations, 394, 461

Mercury vapor lamp, 107, 160, 167, 168

Metal chelates, 146, 307, 360, 361, 363, 364, 395, 397, 544, 545, 547, 548

Metal complexes, 129, 143, 147, 166, 304, 307, 315, 320, 325, 344, 358–360, 391, 420, 456, 476, 529, 531, 544, 576

Metal hydroxides, 319–321, 363

Metal oxides, 123, 393, 456, 459

Metal sulfides, 318, 319, 394

Metallochromic indicators, 41, 314, 363, 365

Method selection, 3, 22, 25–30

Methyl alcohol, 160, 243, 266, 357, 512, 514, 531

Methyl isobutyl ketone, 355, 357

Methyl orange, 345, 347, 348

Microanalysis, 284, 500

Microdiffusion analysis, 509

Microelectrode, 467, 468, 471

Microphotometer, 99, 216

Micropipette, 501

Microscope, 67, 234, 249–252

Microscope slide, 249, 251, 252

Microscopy, 249–254

Microsyringe, 571

Microwave signals, 15, 123, 169, 202, 205, 206

Microwave spectra, 170

Microwave spectrometers, 170

Microwave spectroscopy, 169, 170

Mid infrared, 153, 157

Mid-point potential, 355, 356

Migration current, 469

Migration of ions, 453, 463, 466, 467, 504, 523

Mirror optics, 158, 250

Mobile phase, 493–496, 502, 509, 512, 517, 529, 559–561, 565, 566

Modern instrumentation, 587–589

Modes of sample excitation, 83, 84

Moisture determination, 69, 357, 452

Molal boiling point constant, 67

Molality, 67

Molar absorption coefficient, 127, 130, 141, 145, 147, 151, 161, 242, 243, 265, 268, 395

Molar extinction coefficient; *see* Molar absorption coefficient

Molar refractivity, 237, 261

Molar susceptibility, 180, 182, 187, 261

Molar volume, 260, 261

Molecular absorption spectrum, 143, 144, 145, 150, 153

Molecular depression constant, 263

Molecular electronic levels, 106, 138–144

Molecular electronic transitions, 138–152

Molecular elevation, 262, 263

Molecular formulas, 213

Molecular ion, 212, 213, 215, 217, 218, 219, 278–280, 283, 284

Molecular magnetic anisotropy, 193

Molecular orbital, 139, 143, 144

Molecular rotation, 239, 285, 286

Molecular sieve, 501, 512, 513, 515, 516

Molecular size, 260, 516

Molecular spectra, 81–83, 87, 93, 95, 138–144

Molecular streaming, 510

Molecular structure, 21, 145, 163, 213

Molecular weight, 15, 145, 180, 213, 217, 218, 220, 235, 237, 239, 245, 247, 248, 260, 262, 263, 264, 278, 279, 284, 395, 401, 462, 508, 509, 512, 514, 516–518, 531, 562, 573

Molecular weight determination, 66, 67, 248, 262–264

Mole fraction, 59, 66, 170, 262, 423, 560

Moment of inertia, 82, 169, 170

Monochromatic radiation, 126, 130, 131, 133, 167, 168, 170, 231, 232, 235, 236, 239, 246, 289, 293

Monochromator, 74, 88, 107, 122, 123, 126, 130, 135, 137, 150, 157, 158, 243, 244

Monoclinic symmetry, 252, 291

Monoprotic acids, 312, 341

Morpholine, 329, 357
Mössbauer spectroscopy, 208–212
Mössbauer spectrum, 209, 211
Moving wire gas chromatographs, 564, 588
Mulls, 162, 271
Multicomponent analysis, 150, 153, 165, 169
Multiple distributions, 490, 492, 558–561
Multiple equilibria, 47; *see also* Competing
  equilibria
Multiple excitation, 104, 163
Multiple extractions, 549–553
Multiplicity, 77

N

Naphthalene, 107, 125, 142
Near infrared, 153
Nephelometry, 6, 245, 246, 247
Nernst diffusion layer, 467–469, 521
Nernst equation, 375, 379, 380, 384, 430–
  432, 434, 442, 443, 447, 449, 454, 466, 473,
  474
Neutralization, 338, 347
Neutron activation, 11, 112–115, 589
Neutron capture, 112
Neutrons, 110, 112, 114, 183
Nicol prisms, 229, 231, 239, 241, 244, 250,
  252
Nitric acid, 348, 352, 366, 371, 383, 385,
  393, 405, 458, 459, 544, 549–551
Nitrogen, as carrier gas, 160, 472, 501, 562
  determination, 593, 594
  organic compounds containing, 107,
    143, 205, 274, 275
Nitrous oxide, 95, 136, 137
NMR; *see* Nuclear magnetic resonance
Nonaqueous solvents, 137, 148, 328–330,
  348, 451, 476, 514, 543
Nonaqueous titrations, 5, 65, 329, 330, 336,
  348, 350–357, 440, 451
Nonbonded electrons, 107, 139, 143, 144,
  188, 281, 283
Nonlinear molecule, 154, 155
Nonpolar solvents, 143; *see also* Organic
  solvents
Normal incidence, 229, 230
Normal vibrational mode, 155, 156
Nuclear magnetic moment, 183, 189, 203,
  211
Nuclear magnetic resonance (NMR), 189–
  201
NMR spectrometer, 190, 191, 588

NMR spectroscopy, 15, 18, 165, 177, 183,
  189–201, 220, 221, 261, 265, 271–277, 596
NMR spectrum, 192, 193, 196–200, 271–
  274, 276, 277, 285
NMR with nuclei other than protons, 200,
  201
Nuclear particle flux, 112, 113
Nuclear radiation, 110
Nuclear radiation resonance, 209
Nuclear reactor, 113
Nuclei (atomic); *see* Atomic nuclei
Nuclei formation (precipitates), 247, 254,
  305, 318, 369, 401–403
Nuclides, 110–115
Number of distributions, 559

O

Objective lens, 249
Occluded impurities, 404
Octahedral symmetry, 143, 144, 188
Ocular lens, 249, 251
Ocular micrometer, 251
Oersted, 178, 183, 189, 191, 201, 202, 203,
  206, 215, 216
Ohm's law, 435
Optical aberrations, 234
Optical activity, 231, 239, 240, 241, 285–288
Optical anisotropy, 231, 238
Optical center, 233, 239
Optical density, 127
Optical microscopy, 249–253
Optical rotation, 238, 239, 240–243, 286
Optical rotatory dispersion (ORD), 14, 231,
  241–244, 287, 288, 588
  curve, 241–243, 287, 288
  peak, 241, 242, 243
  trough, 241, 242, 243
Optical rotatory power, 231, 238, 241, 242
Optical system, for infrared, 158
Optic axis, 231
Optics (geometric); *see* Geometric optics
Orbital electron motion, 138, 181, 182, 186
Orbital quantum number, 75, 76, 182
Order of reaction, 409, 410, 412, 422, 423
Ordinary ray, 231, 238
Origin of absorption spectra, 138–144
Organic precipitants, 328, 373, 392, 394–
  398, 406
Organic solvents, 67, 143, 323, 325, 395, 400,
  514, 530, 531, 543, 550, 553, 556
Organophosphorus compounds, 307, 546

Orthorhombic symmetry, 252, 291
Oscillating electric field, 156, 189
Oscilloscope; *see* Cathode ray oscilloscope
Osmotic pressure, 263
Outer complex, 188
Overtones, 157
Overvoltage, 454–457, 477
Oxidants (solid), 393
Oxidation and reduction; 187, 374–376, 433, 454, 462; *see also* Electron transfer
Oxidation–reduction titrations, 5, 65, 330, 336, 357, 374–385, 392, 440, 453, 465
Oxidation state, 185
Oxygen, compounds containing, 107, 143, 194, 274, 573
   determination of, 115, 185, 500, 573
   oxidation by, 85, 419, 492, 593
   production of, 457
   reduction of, 472, 481

P

Paper, 502, 504, 505; *see also* Filter paper
Paper chromatograms, 567
Paper chromatography, 565–569
Paper electrophoresis, 503, 507
Parachors, 260, 261, 262
Paramagnetism, 180, 181, 183–188, 202, 207, 208
Parent peak, 217–220, 279–281
Partial dispersion, 235, 237
Partial pressure, 86, 160, 170, 221
Particle counting, 251, 254
   mass, 91
   size, 49, 50, 104, 246, 326, 369, 372, 400, 401, 403, 433, 561, 562, 569
Partition chromatography, 327, 498, 517, 556, 557, 559, 563–569
   coefficient, 323, 543, 547, 560, 562, 575, 577
   isotherm, 306, 543
Pascal's additivity law, 180, 185, 187, 261
Path, length, 126, 127, 134, 136, 137, 147, 151, 152, 160, 244
   heights, 170, 218, 219
Peptization, 405
Percentage extraction, 543
Perchloric acid, 348, 352, 356, 392, 393
Permeability, constant, 511
   magnetic, 179
   of polymers, 508–512

Permitted transitions, 76, 77, 80, 82, 138, 155, 157
Persistent lines, 97
Phase angles, 226, 227, 229, 238
Phase boundary potential, 445, 446
Phase, definition, 490, 538
   diagram, 539
   difference, 226, 229, 238
   distributions; *see* Distribution between phases
   interfaces, 229, 232, 307, 370, 433, 434, 467, 468, 470, 473, 491, 523
   rule, 535–538
   transformations, 55, 56, 59, 201
Phase volume ratio, 323, 543, 557, 560, 562, 565
pH, calculation of, 338–343
   changes in, 341, 343, 344, 349, 457
   control, 129, 148, 314, 316, 318, 319, 339, 362, 366, 372, 395, 403, 459, 547
   determination, 311, 341, 441, 444, 446, 450
   effects, 109, 129, 146, 247, 304, 310, 312, 314, 316, 317, 322, 324, 327, 361, 362, 364, 366, 371, 391, 399, 400, 402, 458, 461, 506, 511, 518, 519, 529, 543, 547
Phenolphthalein, 345, 347, 348
Phosphorescence, 107, 109
Phosphoric acid, 343, 345–347, 457, 513, 528, 546
Phosphorus-31 nuclei, 190, 200
Photoelectric recording, 96, 99, 145, 169
Photoemission, 73, 132
Photographic recording, 74, 96, 98, 99, 111, 216, 293
Photomultiplier tube, 88, 99, 114, 124, 136, 137, 244, 246
Photons, 110, 167, 226
Photosensitive detectors, 88, 96, 107, 124, 136, 150, 152
Physical adsorption, 481, 492
Physical optics, 226–234
Pi, bonds, 139, 193, 194, 218
   electrons, 106, 107, 141, 218, 284
Pipette, 44
Planck's constant, 74, 179, 253, 433
Plane polarized light, 227, 229, 230, 231, 238, 239, 241, 245, 248, 250, 252
Plasmas, 84, 93–95
Plasma torch, 94, 95, 97
Plastic, 111, 146, 162, 254

Platinum electrodes, 376, 377, 437, 444, 452, 453, 457, 458, 461, 465, 482, 573
Poisson distribution, 43, 560
Polar groups, 143, 168, 398, 497
Polarimeter, 239, 240, 241, 243
Polarimetry, 14, 238–240, 243
Polarity, of adsorbents, 497, 510
Polarizability, 168, 227
Polarization, electrical, 437, 452, 466, 478
    of light, 227, 229–231
    potential, 437, 454, 455, 459
Polarized indicator electrodes, 452
Polarizing angle, 230
Polar molecules, 169
Polarograms, 470, 472, 480
Polarographic, characteristics of elements, 27
    circuit, 471, 472, 479
    techniques, 479–481
    wave equations, 473–476
    waves, 473–476, 480
Polarography, 16, 44, 48, 395, 467–482, 509
Polar solvents, 143, 243, 307, 308, 328, 497, 514, 520, 543
Polybasic acids, titration of, 345–348
Polychromatic beam, 130, 133, 232
Polycyclic compounds, 142
Polymer chain mobility, 510
Polymer chains, 510, 511
Polymerization, 306, 324, 446, 515, 543
Polymers, 166, 184, 187, 248, 254, 263, 505, 507, 510, 517
    types, 512–515
Polyprotic species, 312, 314, 342, 345
Polysaccharides, 513, 517
Polystyrene, 160, 515
Pores, 509, 510, 511, 516, 521, 530
    size, 513, 515
Porous cup, 97
Positrons, 110
Post-precipitation, 405
Potassium, use as spectroscopic buffer, 99
    determination of, 88, 90, 115
Potassium bromide, discs, 145, 166, 271
    optics, 124, 157, 162
Potassium chloride, in reference electrodes, 443–444
    salt bridge, 441, 471
    specific conductance of, 435
Potassium chromate, as indicator, 370
Potential changes, 381, 382

Potential difference, 253, 354, 377, 435, 523, 573
Potentiometer, 125, 377, 448, 449, 467
Potentiometric measurements, 48, 301, 365, 448–450, 465
    methods, 440–452
    titrations, 16, 353, 354, 363, 373, 379, 382, 450–452, 464, 519
Potentiostats, 460, 463, 489 ·
Powder camera, 291
    photographs, 291
Precipitate, formation; *see* Analytical precipitates
    properties of, 403–405
Precipitation titrations, 6, 65, 247, 326, 330, 336, 367–374, 440, 444, 465, 482
Precision, 38–43, 108, 130, 151, 152, 235, 245, 250, 369, 440, 482, 509, 569, 579, 590
Premixed flame, 85
Preoxidation, 385, 391–394
Preparation errors, 50
Pre-reduction, 385, 391–394
Pressure broadening, 166, 170
    gradient, 509, 510, 511, 574
Primary adsorbed layer, 370, 371, 372, 404
Primary combustion zone, 85
Primary standards, 43
Principal quantum numbers, 75, 76
Principal line powder, 98
Prism, 88, 96, 145, 157, 232, 233, 235, 236, 237
Probability, 40, 42, 43, 82, 99, 107, 139, 141, 197, 203, 248, 360, 400, 401, 510
Probes (NMR), 191
Problem definition, 23, 24
Prompt radiation, 114
Propane–air flames, 85, 136
Proportional counter, 103, 111, 124
Protogenic solvents, 351, 354
Proton magnetic resonance, 189–199; *see also* Nuclear magnetic resonance spectroscopy
Protons, accelerated particles, 112
    nuclear, 18, 110, 181, 183, 189, 190, 192, 194, 196, 203, 204, 271–277
    spin–spin coupling constants, 199
    solvated; *see* Hydrogen ions, solvated
    transfer, 4, 5, 129, 306, 314, 316, 323, 329, 341–343, 346, 347, 349, 356
Protophilic solvents, 351, 354
Pulfrich refractometer, 235, 237

Pulse height analyzer, 111, 114, 211, 590
Pulse polarograph, 480
Pyridine as solvent, 162, 329, 351, 356, 357
Pyrolysis, 69, 578, 594

Q

Quadrupole, moment, 210
  splittings, 209, 210, 211
Quanta of energy, 74, 81, 121, 155
Quantitative laws for radiation absorption, 126–131
Quantum numbers, atomic, 75–77
  molecular, 81, 82, 156, 188
Quantum yield, 108
Quartz, 103, 123, 124, 155, 158, 241
Quartz wedge, 241

R

Radiant power, 122, 124, 127, 131, 150, 158
Radiation, buffer, 86, 89, 99
  detectors, 43, 101, 209
  emission, 73–76
Radicals, 15, 85, 180, 187, 201, 207, 208, 210, 277, 278, 283, 284, 413
Radioactive decay, 110, 113
  isotopes, 115, 133, 569
  tracers, 115
Radioactivity, 11, 24, 110–112, 590
Radiochemical procedures, 110–116
Radiofrequency, detector, 190
  field, 189, 192, 196, 200, 206
  oscillations, 15, 93, 169
  oscillator, 190
Radiometric methods, 115
Raleigh scattering, 244
Raman, activity, 167
  spectroscopy, 167–169
  spectrum, 167–169, 271
Random sampling, 49, 50
Raoult's law, 262, 263
Rate constant, 400, 410, 414, 415, 417, 421, 433, 434, 477
Rate curves, 408, 410
Rate equation, 409–413, 415, 424, 433
Rate of, diffusion, 441, 478, 509, 512, 521
  dissolution, 399
  electrolysis, 454

  elution, 561, 565
  exchange, 521, 560
  extraction, 548
  growth, 401–403
  heat transfer, 63, 66
  migration, 504, 506, 510, 565
  mixing, 369
  nucleation, 401, 402
  permeation, 510
  precipitation, 367, 373, 399
  reaction, 129, 211, 336, 337, 358, 373, 382, 385, 392, 407–424, 432, 433, 476–478, 520, 521
  zone migration, 553, 554, 556
Rate theory, of chromatography, 560, 561
Reaction mechanisms, 409, 412, 413, 415, 419, 424, 442
Reaction rates, temperature effect, 414, 415
Real image, 233
Recoil energy, 209
Recoil velocity, 209
Redox ion exchangers, 519, 520
"Reduced" phase rule, 539, 540
Reference electrode, 354, 356, 376, 379, 443, 445, 449, 450, 459, 463, 464, 471, 473
Reflectance, 229
Reflected ray, 230
Reflection, of light, 11, 129, 158, 160, 163, 225, 227–230, 234, 237, 246
  of X rays, 288, 289, 293
Refrachor, 261
Refracted ray, 230, 232, 233, 234
Refraction, 225, 227–230, 232–237
Refractive index, 14, 123, 129, 162, 227–232, 234–238, 241, 244, 245, 247, 248, 251, 261, 504
Refractometer, 235, 237
Refractometry, 14, 234–237
Refractory oxides, 95, 137
Relative affinity, 522, 525–528
Relative free energies, 431
Relative retention volume, 575, 576, 578, 579
Releasing agents, 86'
Re-precipitation, 405
Reproducibility, 38–43, 91, 92, 93; *see also* Precision
Residual current, 453, 468, 470, 479
Resistance thermometer, 124
Resolution, 96, 167, 170, 217, 219, 254
Resolving power, 169, 217, 220

Resonance, 18, 189, 191, 192, 194, 202, 206, 207, 208
    frequency (NMR), 189, 192, 193, 194, 202
    line, 74, 83, 137
    line or peak (NMR), 193, 194, 197, 198, 273, 274, 275
    radiation, 94, 135
    shifts (NMR), 194
Resonant cavity, 205, 206, 207
Restoring force, 154
Retention time, 559, 574
Retention volume, 494, 502, 559, 573–577
Reversed phase chromatography, 564, 567
Reversible waves, 473–475
R_f values, 498–500, 565–567
Rhombohedral symmetry, 291
Ring, currents, 194
    discharge tube, 94
    rule, 279, 281, 285
    strain, 165
Rocking motions, 155, 156
Role of instruments, 588, 589
Rotating disc, 97
Rotating step sector, 98
Rotational energy changes, 81, 82, 122, 140, 154, 163, 167, 170
Rotational energy level, 82, 153, 154, 167, 169
Rotational quantum numbers, 82, 169
Rotational spectra, 154, 169
Rotation of molecules, 155
Rotation of polarized light, 11, 14, 225, 238–244

S

Saccharimeters, 241
Saha's equation, 79
Salt bridge, 377, 441, 443, 445, 449, 451
Salting-out chromatography, 531
Sample cell, 122, 126, 127, 136, 145, 150, 157, 160, 161, 170, 246
Sample characterization, 61, 67
Sample, dissolution, 4, 393
    introduction, 84, 501, 571
    preparation, 3, 50, 489
Sampling, 3, 29, 48–51, 489, 596
    procedure, 49
    valves, 501
Saturation(ESR), 202, 207

Scattering, of electrons, 254
    of light, 107, 127, 131, 162, 167, 169, 227, 229, 244–248, 264
    of X rays, 288
Schlieren effect, 504
Scintillation counter, 103, 111, 114, 124, 211, 593
Scissoring motion, 155, 156
Secondary combustion zone, 85
Secondary standards, 44
Secondary X rays, 47, 81, 100, 101, 104, 105, 133
Sedimentation, 264
Selection of methods, 25–30
Selection rules, 77, 78, 82, 211
Selective adsorption, 494, 530
Selective oxidation, 19, 391, 394
Selective precipitation, 19, 316–320, 322, 391, 395, 459
Selective reduction, 19, 391–394
Selective volatilization, 20, 79
Selectivity coefficient, 522
Selectivity, of exchangers, 525, 526
    of instruments, 46, 135
    of techniques, 329, 364, 395, 463, 489, 546
Selectivity ratio, 321, 322
Self-absorption, 47, 79, 80, 86, 92, 115
Self-reversal, 79, 80
Semiconductors, 124
Semipermeable membrane, 523
Sensitivity of instruments, 46, 63, 88, 98, 108, 129, 157, 158, 170
Sensitivity of techniques, 98, 108, 113, 128, 135, 136, 137, 138, 150, 160, 163, 200, 208, 245, 246, 395
Separation, based on electromigration, 502–508
    by vaporization, 68, 69
    factor, 556–560, 576
    procedures, 22, 29, 45, 489–491, 510, 516, 517
Separatory funnel, 490, 542, 548, 553, 556
Sephadex, 513, 516, 517
Sharpness of elution peaks, 559, 561
Shielding, effects, 192, 196, 200, 210, 274
    electrons, 192, 193, 194, 196
    parameter, 193
Sigma, bonds, 139, 218
    electrons, 106
Silica gel, 499, 501, 563, 565
Silicon carbide, 123

Silicon crystal detector, 170, 206
Silver chloride, optics, 124, 158, 163
   precipitates, 370, 371, 405, 449
Silver electrodes, 373, 444, 458
Silver–silver chloride electrode, 354, 379, 443, 445, 446
Sine curves, 227
Single crystal studies, 291, 293, 294
Singlet state, 106, 107, 109
Slits, 46, 88, 96, 98, 122, 145, 152, 158, 159
Sodium, determination of, 88, 90
   D line emission, 46, 74, 239, 243, 261
Sodium chloride optics, 123, 124, 157, 158, 162
Solid–liquid interfaces, 491, 502
Solid–liquid phase equilibria, 326, 327
Solid samples, 154, 160–162, 166, 168, 183, 184, 208, 237
Solid solution, 404, 536, 540, 541
Solid surfaces, 162, 167
Solubility, coefficient, polymers, 511
   product, 302, 305, 308, 316–320, 327, 367–370, 391, 400, 405, 450
Solubilization chromatography, 531
Solute–solvent interaction, 162, 165, 166, 185, 201, 576
Solvation, 328
Solvent, classification, 350–353
   effects, 196, 200, 247, 248, 266, 271, 273, 274, 328–330, 337, 357
   extraction, 19, 90, 138, 148, 150, 305–308, 323–326, 395, 500, 531, 537, 542–557, 559, 592
   front, 498, 500
   IR absorption bands, 161
   selection, 150, 152, 161, 168, 192, 353, 500
Source (of energy), 122, 123
Spark discharges, 92, 93, 97
Specific activity, 116
Specific conductance, 435
Specific dispersion, 235, 237
Specific rate constants, 400, 410, 414, 415, 417
Specific retention volume, 575, 577
Specific rotation, 239, 240
Specific susceptibility, 179, 180, 184, 185
Specific viscosity, 263
Spectral interference, 87, 88, 138
Spectrochemical series, 144
Spectrophotometers, 122, 150, 151, 160, 246, 591

Spectrophotometric methods, 128, 130
Spectrophotometric titrations, 153
Spectropolarimeters, 243
Spectroscopic buffer, 99
Spectroscopic notations, 77, 78
Spectroscopic splitting factor, 18
Speed of light, 75, 179, 209
Spherical aberration, 234
Spin angular momentum, 181, 186
Spin lattice relaxation, 190, 202
Spin lattice relaxation time, 190, 202
Spin magnetic moment, 181
Spin pairing, 181, 188
Spin quantum number, 18, 77, 182, 186, 189, 200, 201, 202, 210
Spin–spin, interaction, 185, 190, 196, 197, 198
   relaxation, 190
   splitting (or coupling), 196–198, 200, 271, 273–277
Splitting energy, 144
Splitting patterns, 143, 144, 197, 198, 207, 209, 210, 211, 271, 276
Square planar symmetry, 144, 358
Square wave polarograph, 480
Stability constants, 315, 316, 326, 358, 359, 361, 362, 363, 547
Stability of complexes, 148, 302, 320, 327, 358–360, 363, 364, 400, 476, 506, 527–529, 531, 547
Stabilizer, 247
Stage micrometer, 251
Stage refractometer, 251
Standard addition method, 89, 105
Standard chemicals, 337, 344, 498, 499, 502, 574, 575, 590
Standard deviation, 39, 40
Standard enthalpy change, 57
Standard free energy change, 57, 338, 377, 379–381, 392, 430, 431, 442
Standardization, 343, 374, 528
Standard reduction potentials, 375–378, 384, 392, 394, 431, 433, 442, 450, 454, 457, 466, 473, 474, 520
Standard samples, 30, 38, 43, 44, 62, 65, 88, 99, 101, 104, 108, 126, 128, 136, 152, 153, 166, 184, 185, 206
Standard spectra, 146, 163, 165, 169, 220, 221, 280
Standard states, 57, 375, 376, 379
Standinger's law, 263
Stark effect, 79

Stationary phase, 517, 560, 563, 564, 567, 570, 571, 576, 579
Stationary support, 505, 563, 569
Statistical weight, 78
Statistics, 39–43, 76, 344
Steady state, 416
Stepwise elution, 494
Steric effects, 143, 146, 268, 269, 398
Steroids, 146, 266, 267, 285, 287, 506, 568
Stoichiometry, 153
Stoke's lines, 167
Stretching vibrations, 154, 155, 156, 163, 164, 269, 270
Structure elucidation, 11, 13, 15, 146, 157, 163, 165, 166, 169, 171, 177, 187, 192, 200, 208, 211, 213, 218, 219, 220, 243, 259–294, 588
Structure of coordination compounds, 187, 188
Substituents, 145, 163, 239, 243, 265, 266, 267, 284, 286, 395, 398, 511, 519
Sugar degrees, 241
Sulfonamides, determination of, 330, 353, 356, 357
Sulfur, compounds containing, 143, 219, 274
    determination of, 134
Sulfur dioxide, 392, 393, 481, 512, 543
Sulfuric acid, 345, 348, 509, 515
Supersaturation, 369, 402, 403
Supporting electrolyte, 27, 467, 473, 476, 479, 481
Surface-active agent, 479
Surface area, 308, 369, 370, 372, 403, 404, 433, 435, 456, 466, 470, 491, 497, 563
Surface coatings, 502, 569
Surface energy, 401
Surface tension, 89, 260, 401, 405, 470
Swelling, of polymers, 511, 522
    pressure, 522
Symmetry, 156, 166, 168, 194, 198, 202, 203, 207, 210, 211, 227, 231, 276, 277, 293
Synergistic effect, 546
Syringes, 501

T

Tafel equation, 478
Tail flame (plasmas), 95
Tailing, 327, 495, 496, 502, 566, 569
Tau values, 195, 196
Telescope, 235

Temperature, effects, 48, 186, 414, 415, 576–578
    measurement, 62–67
    programming, 501, 577
Tesla coil, 94
Tetragonal symmetry, 144, 252, 290, 291
Tetrahedral symmetry, 144, 299, 512
Tetrahydrofuran, 357, 514
Tetramethylsilane, 194, 196
Theoretical plate, 556, 559–561, 572
Theoretical plate concept, 560
Thermal analysis, 12, 66
Thermal conductivity, 60, 63, 86, 593
Thermal conductivity detectors; *see* Katharometers
Thermal decomposition, 56, 61, 69, 185, 221, 536, 578
Thermal methods, miscellaneous, 68, 69
Thermal motion, of polymer chains, 510
Thermal neutrons, 112
Thermal transformations, 11, 12, 56, 59, 66; *see also* Thermal decomposition
Thermocouple, 60, 124, 594
Thermodynamic relationships, 55–59, 392
Thermogravimetric curves, 60, 406
Thermogravimetry, 12, 59–62, 491
Thermometric titration, 12, 64, 65
Thermostats, 501
Thin-layer chromatography, 493, 499, 500, 517, 565
Thorium, determination of, 115
    extraction of, 550–552
Thyraton, 93
Time-of-flight spectrometers, 215, 216, 221
Titration, curve, 311–315, 343, 345, 353–355, 360, 362, 368, 369, 379, 380, 381, 383, 439, 440, 451
    errors, 341, 343, 344, 348, 364, 370
Titrimetric procedures, 4, 152, 305, 309–316, 335–384, 395, 407, 408, 450, 509, 528, 589
Toluene, 514, 518, 546
"Trace analysis" differential technique, 152
Transfer coefficient, 434
Transfers, between phases, 553–557, 559, 561
Transition metal complexes, 143, 180, 186, 187, 208
Transition state, 414, 433
Transition time, 16, 466, 467
Transmission range, of materials to infra-red, 158

Transmittance, 127, 130, 131, 150–153, 159, 166
Transmitted beam, 228, 232
Transmitted intensity, 46, 126, 131, 230, 244
Transmitted ratio method, 151, 152
Transparent medium, 228, 232
Transport coefficient, 477
Triclinic symmetry, 252, 291
Triplet state, 106, 107, 109
Tube constant, 184
Tuned circuit, 94
Turbidimetric titrations, 247
Turbidimetry, 6, 245, 246, 247
Turbidity, 244, 246, 247, 360, 373
Turbulent flames, 85
Turning moment, 179
Twisting motion, 155, 156
Two-dimensional development, 567
Two-phase systems, 535–541
Tyndal effect, 244

U

Ultimate precision differential method, 152
Ultraviolet light, 8, 75, 124, 130, 144–146, 234
Ultraviolet spectrophotometer, recording, 588, 590, 593
Ultraviolet spectrophotometry, 8, 144–146, 165, 221, 264–268
Uncertainty principle, 190
Unit cell, 289, 290, 291, 293, 294
Unpaired electrons, 181, 186–188, 201, 202, 203, 206, 207, 208, 210
Unsaturated groups, 141
Uranium, determination of, 115
    extraction of, 546, 549–552

V

Vacuum cup, 97
Vacuum tube voltmeter, 449
Valence bond approach, 188
Valence electron, 77, 139
Van Deemter equation, 561, 562, 566, 569
van der Waals bonds, 307, 491, 525
Vapor, pressure, 20, 58, 169, 213, 262, 509, 537, 538, 570, 576–578
    density, 260
Vaporization, 68, 84, 86, 93, 137, 213

Vapors, atomic, 135, 137
    molecular, 160, 213, 492, 500, 502, 509–511, 538, 567, 570, 594
Variance, 42, 43, 49, 50, 537
Velocity of light, 75, 179, 209, 227, 238, 241
    mobile phase, 561
    particles, 91, 214, 215
    waves, 75, 226, 227, 231
Vibrational bands, 154, 163
Vibrational energy changes, 81, 82, 122, 140, 153–157, 167
Vibrational energy levels, 81, 154, 155, 163, 167
Vibrational frequencies, 81, 154, 168
Vibrational quantum number, 156, 167
Virtual image, 233
Viscosity, 89, 263, 469, 517, 566
Viscous flow, 510
Visible light, 8, 75, 130, 231, 234, 254
Visible spectrometry, 147–151, 166, 264–268
Visualization, 498, 499, 503, 565
Void space, 526, 570, 574
Volatility, 576
Volatilization, 86, 90, 91, 92, 98, 99, 213, 221, 405, 406
von Weimarn equation, 402

W

Wagging motion, 155, 156
Water, as solvent, 158, 160, 169, 194, 243, 262, 263, 266, 278, 300, 307, 323, 329, 350, 400, 499, 512, 530, 531, 556, 563
    determination of, 357, 452, 509, 588
    phase system, 536
    removal of, 406, 512, 593
    structure of, 299
Wave guide, 170, 205
Wave motion, 226, 227
Wave numbers, 75, 127, 157, 159
Wavelength of radiation, 75, 88, 96, 97, 98, 102, 104, 105, 107, 122, 125, 127, 128, 131–133, 140, 141, 143, 145, 150, 153, 165, 169, 205, 229, 239, 241–244, 246, 247, 267, 288, 292, 293
Weber, 178
Weight fraction, 104, 180, 185
Weiss constant, 181, 185, 188
Weston cell, 448, 449
Wheatstone bridge circuit, 437, 438

X

Xenon arc lamp, 107
X rays, 75, 76, 80, 100–105, 110, 124, 131–134, 254, 288–294
    absorption spectroscopy, 8, 131–134
    crystallography, 13, 288–294
    diffraction, 259, 264, 288–294
    diffraction patterns, 61, 289–293
    emission spectra, 80, 104, 105, 133
    emission spectroscopy, 9, 100–106
    generator or tube, 100, 101, 124, 133
    powder photographs, 291, 292, 293
    spectrometer, 100–102, 105

Z

Zeolites, 510, 512, 518
Zinc amalgam, 392, 394
Zinc, as reductant, 392–394
    determination of, 321, 365–367, 373, 458
Zirconium phosphate, 513, 518
Zone electrophoresis, 503–507, 517
Zone refining, 538–540
Zwitterions, 349, 350